TOPICS IN
M●DERN
PHYSICS
Theoretical Foundations

TOPICS IN MODERN PHYSICS
Theoretical Foundations

John Dirk Walecka
College of William and Mary, USA

World Scientific

NEW JERSEY · LONDON · SINGAPORE · BEIJING · SHANGHAI · HONG KONG · TAIPEI · CHENNAI

Published by

World Scientific Publishing Co. Pte. Ltd.

5 Toh Tuck Link, Singapore 596224

USA office: 27 Warren Street, Suite 401-402, Hackensack, NJ 07601

UK office: 57 Shelton Street, Covent Garden, London WC2H 9HE

British Library Cataloguing-in-Publication Data
A catalogue record for this book is available from the British Library.

TOPICS IN MODERN PHYSICS
Theoretical Foundations

ISBN 978-981-4436-88-5
ISBN 978-981-4436-89-2 (pbk)

Printed in Singapore by World Scientific Printers.

Dedicated to Christina, Connor, Ned, and Will

Preface

Our understanding of the physical world was revolutionized in the twentieth century—the era of "modern physics". A book on *Introduction to Modern Physics: Theoretical Foundations* was recently published by World Scientific Publishing Company [Walecka (2008)]. The goal of this book, aimed at the very best students, is to expose the reader to the foundations and frontiers of today's physics. Typically, students have to wade through several courses to see many of these topics, and I wanted them to have some idea of where they were going, and how things fit together, as they went along. Hopefully, they will then see more inter-relationships, and get more original insights, as they progress. The book assumes the reader has had a good one-year calculus-based freshman physics course along with a good one-year course in calculus. While it is assumed that mathematical skills will continue to develop, several appendices are included to bring the reader up to speed on any additional mathematics required at the outset. With very few exceptions, the reader should then find the text, together with the appendices and problems to be self-contained. The aim is to cover the chosen topics in sufficient depth that things "make sense" to students and that students achieve an elementary working knowledge of them.

After completing that book, it occurred to me that a second volume could be prepared that would significantly extend the coverage, while furthering the stated goals. The ground rules would be that anything covered in the text and appendices of the first volume would be fair game, while anything covered in the problems would first be re-summarized. The reader would now be assumed to be familiar with multi-variable calculus and linear algebra, with any additional required mathematics again covered in appendices. Those few results quoted without proof in the first volume would now be derived. The topics chosen would be those of wide applicability in all areas of physics. Again, an important goal would be to keep the entire coverage self-contained. This second book (Vol. II) has, in fact, now been completed and was recently published by WSPC as *Advanced Modern Physics: Theoretical Foundations* [Walecka (2010)]. As stated there, when finished, the reader should have an elementary working knowledge in the principal areas of theoretical physics of the twentieth century.

These two books provide an essentially linear progression, and many important modern physics topics were necessarily omitted in that presentation. The purpose of the present volume (Vol. III) is to branch out from the first two, and provide an analysis of several of those additional topics. After an introductory chapter, this book is divided into three parts:

- *Part 1—Quantum Mechanics:* Analytic solutions to the Schrödinger equation are first developed for some basic systems. The analysis is then formalized, concluding with a set of postulates that encompass the theory of quantum mechanics;
- *Part 2—Applications of Quantum Mechanics:* The following applications of the theory are then discussed: approximation methods for bound states, scattering theory, time-dependent perturbation theory, electromagnetic radiation and quantum electrodynamics;
- *Part 3—Relativistic Quantum Field Theory:* This part represents an extension of the material in Vol. II, and contains chapters on discrete symmetries, the Heisenberg picture, and the Feynman rules for QCD.

The present book is designed so that one can read through part 2 without recourse to Vol. II; some repetition of the material in chapter 2 of Vol. II was required in order to accomplish this. Over 130 accessible problems enhance and extend the coverage. All of the material in this book is taken from course lectures given over the years by the author at either Stanford University or the College of William and Mary.

I was again delighted when World Scientific Publishing Company, which had done an exceptional job with several of my previous books, showed enthusiasm for publishing this new one. I would like to thank Dr. K. K. Phua, Executive Chairman of World Scientific Publishing Company, and my editor Ms. Lakshmi Narayanan, for their help and support on this project.

My colleague Paolo Amore deserves special thanks for again agreeing to read the manuscript; his assistance on these three volumes has been invaluable. He and I are in the process of preparing a solution manual for the problems in Vol. I.

I believe the three volumes in this modern physics series, taken together, provide a base from which the very best students can learn and understand modern physics. I hope students and instructors alike can obtain as much pleasure from these books as I did in writing them.

Williamsburg, Virginia   *John Dirk Walecka*
October 1, 2012   *Governor's Distinguished CEBAF*
                                     Professor of Physics, emeritus
                                     College of William and Mary

Contents

Applications of Quantum Mechanics   125

Relativistic Quantum Field Theory   333

Problems and Appendices   409

Chapter 1

Introduction

As stated in the preface, our understanding of the physical world was revolutionized in the twentieth century—the era of "modern physics". A book on *Introduction to Modern Physics: Theoretical Foundations* was recently published by World Scientific Publishing Company [Walecka (2008)]. The goal of this book (Vol. I), aimed at the very best students, is to expose the reader to the foundations and frontiers of today's physics. Typically, students have to wade through several courses to see many of these topics, and I wanted them to have some idea of where they were going, and how things fit together, as they went along. A second book in this modern physics series (Vol. II) has also been published by WSPC as *Advanced Modern Physics: Theoretical Foundations* [Walecka (2010)]. Those few results quoted without proof in Vol. I are derived here. Again, an important goal is to keep the entire coverage self-contained. As stated in Vol. II, when finished, the reader should have an elementary working knowledge in the principal areas of theoretical physics of the twentieth century.

The goal of the present book (Vol. III) is to branch out and cover in some detail several modern physics topics omitted in the essentially linear progression in Vols. I and II.

Part 1 of this book is on *Quantum Mechanics*.[1] After a careful examination of the boundary conditions to be applied, the Schrödinger equation is solved analytically for several basic physical systems: the simple harmonic oscillator (actually solved in a variety of ways), the spherical box, the finite radial square well, hydrogen-like atoms where a negatively charged particle moves in the Coulomb potential of a point nucleus, and a periodic potential in one dimension, which provides an introduction to the properties of

---

[1] This book is designed so that parts 1 and 2 rely only on the material in Vol. I.

solids. A problem takes the reader through the important solution for the isotropic three-dimensional oscillator.

The theory is then formalized to linear hermitian operators acting in an abstract infinite-dimensional Hilbert space. Once one knows how to compute inner products, and use completeness, one knows how to navigate in this space. Operator methods are developed, and matrix mechanics, through which most practical calculations are carried out, is derived along the way. Ehrenfest's theorem, which provides the time derivative of the expectation value of an operator in the Schrödinger picture, is derived. A brief introduction is given to measurement theory, and part 1 concludes with a summary of the postulates of quantum mechanics.

Part 2 is on *Applications of Quantum Mechanics*. First, there is a discussion of approximation methods for bound states. The variational method is introduced, which provides an invaluable tool for obtaining insight and results for the properties of the low-lying states of *any* system, whether or not analytic solutions are available. An example is provided by the ground state in an exponential potential. A problem takes the reader through a variational calculation of the ground state of the H_2^+ molecular ion, which provides an understanding of the basic molecular bond.

Perturbation theory allows one to express the solution to a problem with hamiltonian $H_0 + H_1$ in terms of the solutions to the unperturbed problem with H_0 alone. We first discuss the situation where the unperturbed spectrum is non-degenerate. Two classes of degenerate perturbation theory are then discussed—case I, where the perturbation is diagonal in the appropriate degenerate subspaces, and case II, where the perturbation must first be diagonalized in those subspaces. A variety of applications are given, including motion in external magnetic and electric fields, the effect of finite nuclear size on hydrogen-like spectra, and the ground state of the two-electron He atom. In the latter problem, it is shown how first-order perturbation theory can also provide a variational calculation.

There are also continuum solutions to the Schrödinger equation, and a substantial portion of part 2 is concerned with scattering theory for central potentials in three dimensions, where the physics is extracted in terms of the ratio of probability fluxes and cross sections. The scattering problem here is formulated in terms of a Green's function and integral equation. This has two distinct advantages:

- The scattering boundary conditions of incident wave plus asymptotic outgoing scattered wave can be built into the Green's function;

- An integral equation provides an invaluable framework for obtaining an interated solution in terms of the strength of the potential.

Born's first approximation ("Born approximation") is derived, its validity examined, and several examples discussed. A diagrammatic expansion is presented for the entire Born series. A partial-wave analysis is developed by inserting a separated ansatz into the scattering integral equation, and showing how the resulting decoupled radial integral equation describes both an incoming and an asymptotically phase-shifted outgoing radial wave. The radial integral equation is converted to a differential equation with boundary conditions, and the scattering problem is analyzed both at low energy, where s-wave scattering dominates and the effective-range expansion is valid, and at high energy, where many partial waves contribute and the WKB approximation holds. The low-energy scattering problem is solved for various potentials. It is shown that asymptotically the WKB approximation reduces to a straight-line eikonal description of the scattering (the "Glauber approximation"), and examples are given. Although derived within a non-relativistic framework, the high-energy expressions for the cross sections can be recast in a form that involves only transverse quantities, and the Glauber approximation provides a framework for understanding scattering from hadronic targets at very high energy. The discussion of scattering theory concludes with a brief tour of the theory of nuclear reactions, based primarily on the material in [Blatt and Weisskopf (1952)].

We then move on to a discussion of time-dependent (or Dirac) perturbation theory. An example is given of inelastic scattering with a time-dependent perturbation $H_1(t)$. The problem of scattering through a time-independent H_1, where there is one particle in the continuum in the final state, is analyzed. The result is *Fermi's Golden Rule* for the transition rate, and the validity criteria here are examined in detail. It is shown how the results from time-independent scattering theory are reproduced with time-dependent perturbation theory, and the example of charged lepton scattering from a hydrogen-like atom with finite nucleus exhibits a broad range of physics. It is demonstrated how the first-order calculation of the transition amplitude can be extended to all orders in H_1.

The interaction of quantum systems with external electromagnetic fields is analyzed in detail within the context of time-dependent perturbation theory. An appendix justifies canonical quantization in this case by showing that Ehrenfest's theorem reproduces Newton's second law with the Lorentz force for particle motion, while reproducing Maxwell's equations in the Coulomb gauge for the external field in the source-free region.

The free radiation field is quantized in the Coulomb gauge, where there are only physical transverse photons, by putting the problem in normal modes and then quantizing the resulting simple harmonic oscillators. Some time is then spent formulating the full problem of quantum electrodynamics (QED), where the quantized electromagnetic field interacts with a charged particle whose motion in the fields is governed by non-relativistic quantum mechanics. The hamiltonian for the entire interacting system is developed in the Coulomb gauge, the Schrödinger equation written in abstract Hilbert space, and canonical quantization imposed. Canonical quantization is justified by showing that Ehrenfest's theorem now reproduces Newton's second law with the Lorentz force for particle motion, while reproducing the full set of Maxwell's equations with sources for the electromagnetic field. This guarantees the correct classical limit of the theory. After separating the zero-point energy of the field and the Coulomb self-energy of the particle, one has a hamiltonian that can be applied to a multitude of problems. Here we focus on five of them: photoemission and absorption, the lifetime of the $2p$-state in hydrogen, the dipole sum rule, the derivation of the Planck distribution using detailed balance, and the Wigner-Weisskopf theory of the line width. Several problems reinforce and expand the developments in the text.

Part 3 of this book focuses on some special topics in *Relativistic Quantum Field Theory*. Here the reader is assumed to be familiar with various parts of Vol. II, as listed in the text. First, the discussion of symmetries in relativistic quantum field theory is extended to the discrete symmetries of parity P, charge (or particle-antiparticle) conjugation C, and time reversal T. These symmetries are detailed for spin-0, spin-1/2 (Dirac), and massless spin-1 (photon) fields. Applications are given. After a brief discussion of Lorentz transformations, the CPT theorem is established for two representative classes of interactions. This theorem states that with a local Lorentz-invariant lagrangian density, the *combined* operation CPT, taken in any order, will always be a good symmetry. Two immediate consequences are established: particles and antiparticles must have identical masses, and they must have identical lifetimes.

The next topic is the Heisenberg picture. Here, as opposed to the Schrödinger picture where the time dependence is that of the state vector, the time dependence is put into the operators. Two theorems are proven that relate the Heisenberg picture to the interaction picture, where the free time dependence is placed in the operators. With the aid of Wick's theorem, these results allow one to analyze Heisenberg matrix elements in terms of Feynman diagrams. We discuss the role of the four-momentum, Lorentz transformations, discrete symmetries, and continuous symmetries in the Heisenberg picture, and then show how closed expressions can be obtained for propagators in this picture. This provides a framework for the derivation of general results such as the Lehmann representation.[2]

Finally, path integral methods are used to establish the Feynman rules for the Yang-Mills local gauge theory of quantum chromodynamics (QCD). It is first shown how the Faddeev-Popov identity can be employed to compute the generating functional for quantum electrodynamics (QED), from which the Feynman rules for the Green's functions follow. It is shown how the Faddeev-Popov determinant can be represented by "ghost" fields in the effective lagrangian, which decouple in the case of QED. After a review of QCD, the generating functional for that theory is similarly calculated, from which the Feynman rules again follow. Here the ghosts do couple to the gluons. With this introduction to the Feynman rules, together with the introduction to lattice gauge theory for QCD in [Walecka (2004)], the reader has the basic tools with which to work in this theory of the strong interactions binding quarks into the observed hadrons.

I believe the three volumes in this modern physics series taken together provide a clear, logical, self-contained, and comprehensive base from which the very best students can learn and understand the subject.[3] They will provide the reader with an elementary working knowledge in modern physics. With this overview and these tools in hand, development in depth and reach in this field can then be obtained from more detailed or advanced physics courses and texts.

There are indeed many good textbooks available with which readers can extend the depth and breadth of their understanding of quantum mechanics [Schiff (1968); Messiah (1999); Gasiorowicz (2003); Griffiths (2004); Sakurai and Napalitano (2010)]; at a more advanced level, there are [Landau and Lifshitz (1981); Shankar (1994); Merzbacher (1998); Gottfried and Yan (2004); Cohen-Tannoudji, Diu, and Laloe (2006)].

Familiarity with the material in the present volume provides a foundation for practical applications of quantum mechanics, the marvelously successful description we have of the microscopic world. It makes a fun and rewarding journey. Let us take it together.

---

[2] The Lehmann representation is derived elsewhere.
[3] Particularly, when supplemented with [Walecka (2007)].

PART 1
Quantum Mechanics

Chapter 2

Solutions to the Schrödinger Equation

We start by summarizing several properties of the Schrödinger equation as developed in Vol. 1.[1] We then investigate in some detail the boundary conditions to be applied to that equation, and proceed to solve the Schrödinger equation for several very basic systems.

2.1 The Schrödinger Equation and its Interpretation

We summarize three starting postulates of non-relativistic quantum mechanics:

2.1.1 *Postulates*

(1) There is a wave function $\Psi(\mathbf{x}, t)$ associated with a particle, which satisfies the Schrödinger equation

$$i\hbar\frac{\partial\Psi}{\partial t} = H\Psi \qquad ; \text{ Schrödinger eqn} \qquad (2.1)$$

Here $H(\mathbf{p}, \mathbf{x})$ is the hamiltonian, which for a non-relativistic particle moving in a static potential takes the form

$$H = \frac{\mathbf{p}^2}{2m} + V(\mathbf{x}) \qquad ; \text{ hamiltonian} \qquad (2.2)$$

(2) The canonical momenta and coordinates satisfy the canonical commutation relations

$$[p_i, x_j] = \frac{\hbar}{i}\delta_{ij} \qquad ; \text{ C. C. R.} \qquad (2.3)$$

---

[1]Volume 1 refers to [Walecka (2008)].

where the indices $(i, j) = (x, y, z)$ indicate the cartesian components of the vectors. A representation of these commutation relations in coordinate space is obtained by taking

$$\mathbf{p} = \frac{\hbar}{i} \boldsymbol{\nabla} \qquad \text{; coordinate representation}$$

$$(2.4)$$

(3) The probability of finding the particle in a volume d^3x at position $\mathbf{x}$ and time t is $\rho\, d^3x$, where $\rho = |\Psi(\mathbf{x}, t)|^2$ is the *probability density*

$$\rho = |\Psi(\mathbf{x}, t)|^2 \qquad \text{; probability density} \qquad (2.5)$$

It follows from the above that with a real potential, the continuity equation is satisfied

$$\frac{\partial \rho}{\partial t} + \boldsymbol{\nabla} \cdot \mathbf{S} = 0 \qquad \text{; continuity equation} \qquad (2.6)$$

where the *probability current* is given by[2]

$$\mathbf{S} = \frac{\hbar}{2im} \left[\Psi^* \boldsymbol{\nabla} \Psi - (\boldsymbol{\nabla} \Psi)^* \Psi \right] \qquad \text{; probability current} \qquad (2.7)$$

The last three relations provide the basis for interpreting the theory.[3]

2.1.2 *Boundary Conditions*

The Schrödinger equation in coordinate space takes the form of a partial differential equation

$$i\hbar \frac{\partial \Psi}{\partial t} = \left[-\frac{\hbar^2 \boldsymbol{\nabla}^2}{2m} + V(\mathbf{x}) \right] \Psi \qquad \text{; Schrödinger eqn} \qquad (2.8)$$

We proceed to investigate the *boundary conditions* that are to be associated with this differential equation. Since the physics follows from expressions that are *bilinear* in Ψ, some care must be taken in this task:

(1) We take the probability to be *normalized*, so that

$$\int |\Psi|^2 \, d^3x = 1 \qquad \text{; normalization} \qquad (2.9)$$

---

[2]See Probs. 2.1–2.2. As in Vol. I, we continue to refer to the current density $\mathbf{S}$ as the probability current.
[3]See also Prob. 2.3.

(2) If $V(\mathbf{x})$ is continuous, then the *physics* must also be continuous, so that

$$\rho(\mathbf{x}, t) = \Psi^\star \Psi \quad \text{must be continuous} \tag{2.10}$$

(3) There should be no *sources or sinks* of probability. Consider an arbitrary, very thin, volume $\mathcal{V}$ (Fig. 2.1).

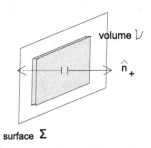

surface Σ

Fig. 2.1 No sources or sinks of probability. The thin volume will be denoted by $\mathcal{V}$, and the indicated surface by Σ.

Now compute

$$-\frac{d}{dt} \int_{\mathcal{V}} \rho\, d^3x = -\int_{\mathcal{V}} \left(\frac{\partial \rho}{\partial t}\right) d^3x = \int_{\mathcal{V}} (\boldsymbol{\nabla} \cdot \mathbf{S}) d^3x$$

$$= \int_{\Sigma} \mathbf{S} \cdot d\boldsymbol{\Sigma} \tag{2.11}$$

Here the continuity equation has been used, Gauss' law invoked, and the width of $\mathcal{V}$ shrunk to zero. The last integral then goes over both sides of the contained surface Σ in Fig. 2.1. The first integral in Eq. (2.11) must vanish if there are no sources or sinks, and therefore, if the contained surface is small,

$$S_\perp^{(+)} = S_\perp^{(-)} \tag{2.12}$$

where the projections of the current now point in the same direction. Hence the normal component of the current must be continuous across any surface. Thus with no sources or sinks, the probability current is *also* continuous

$$\mathbf{S} = \frac{\hbar}{2im} [\Psi^\star \boldsymbol{\nabla} \Psi - (\boldsymbol{\nabla} \Psi)^\star \Psi] \quad \text{must be continuous} \tag{2.13}$$

(4) It would appear from items (2) and (3) that the appropriate boundary conditions to go along with the Schrödinger equation are that with no

sinks or sources, the wave function and its gradient must everywhere be continuous

$$\Psi, \boldsymbol{\nabla}\Psi \quad \text{must be continuous} \tag{2.14}$$

Evidently

(a) These conditions are *sufficient* to make the appropriate bilinear expressions continuous;

(b) They are also *necessary*, provided we invoke the basic principle of *superposition*:

> *Basic Axiom of Superposition for Schrödinger Wave Motion: If* Ψ_1 *and* Ψ_2 *are acceptable solutions, then so is the linear combination* $\Psi = \alpha\Psi_1 + \beta\Psi_2$.

To now establish the necessity of Eqs. (2.14), consider two solutions Ψ_1 and Ψ_2 where Ψ_2 and $\boldsymbol{\nabla}\Psi_2$ are *any* continuous solution (one can always find such solutions). The principle of superposition implies that $\alpha\Psi_1 + \beta\Psi_2$ must again be an acceptable solution. Thus

- Since $(\alpha\Psi_1 + \beta\Psi_2)^\star(\alpha\Psi_1 + \beta\Psi_2)$ is continuous, Ψ_1 must also be continuous;
- Since $(\alpha\Psi_1 + \beta\Psi_2)^\star\boldsymbol{\nabla}(\alpha\Psi_1 + \beta\Psi_2) - [\boldsymbol{\nabla}(\alpha\Psi_1 + \beta\Psi_2)]^\star(\alpha\Psi_1 + \beta\Psi_2)$ is continuous, $\boldsymbol{\nabla}\Psi_1$ must then also be continuous.

These results now follow from the presence of the arbitrary cross terms. One cannot make use of any accidental cancellation of discontinuities in the bilinear expressions $(\rho, \mathbf{S})$ calculated with Ψ_1 alone.

2.1.3 *Stationary States*

Let us look for normal-mode solutions to the Schrödinger equation of the form

$$\Psi(\mathbf{x}, t) = \psi(\mathbf{x})e^{-iEt/\hbar} \qquad \text{; normal modes} \tag{2.15}$$

Substitution into Eq. (2.8) gives the time-independent Schrödinger equation

$$H\psi = \left[-\frac{\hbar^2\boldsymbol{\nabla}^2}{2m} + V(\mathbf{x})\right]\psi = E\psi \qquad \text{; time-independent S-eqn} \tag{2.16}$$

We make several comments, summarizing results from Vol. I:[4]

---

[4]See also Probs. 2.4–2.7.

- When appropriate boundary conditions are imposed, the time-independent Schrödinger equation becomes an *eigenvalue equation*, which has as its solution a set of eigenvalues and eigenfunctions

$$H\psi_n = E_n\psi_n \qquad ; n = 0, 1, 2, \cdots, \infty \qquad (2.17)$$

- Define the matrix element of the hamiltonian by

$$\langle \psi_m | H | \psi_n \rangle \equiv \int d^3x\, \psi_m^\star(\mathbf{x}) H \psi_n(\mathbf{x}) \qquad (2.18)$$

The hamiltonian is *hermitian* if

$$\langle \psi_m | H | \psi_n \rangle = \langle \psi_n | H | \psi_m \rangle^\star \qquad ; \text{ hermitian} \qquad (2.19)$$

If the hamiltonian is hermitian, the eigenvalues E_n are *real*.

- The eigenfunctions belonging to different eigenvalues are now orthogonal, and the degenerate eigenfunctions can always be orthogonalized. We are free to choose the normalization, and the eigenfunctions can be taken to be *orthonormal*

$$\langle \psi_m | \psi_n \rangle \equiv \int d^3x\, \psi_m^\star(\mathbf{x}) \psi_n(\mathbf{x}) = \delta_{mn} \qquad ; \text{ orthonormal} \quad (2.20)$$

- It is another postulate of quantum mechanics that the eigenfunctions of an hermitian hamiltonian form a complete set[5]

$$\sum_{n=0}^{\infty} \psi_n(\mathbf{x}) \psi_n^\star(\mathbf{y}) = \delta^{(3)}(\mathbf{x} - \mathbf{y}) \qquad ; \text{ complete} \qquad (2.21)$$

This sum will, in general, receive contributions from both the bound and scattering states in the potential V.

Substitution of Eq. (2.15) into Eqs. (2.5) and (2.7), with a real energy E, reduces the probability density and probability current to the forms

$$\rho = \psi^\star \psi \qquad\qquad\qquad\quad ; \text{ stationary states}$$

$$\mathbf{S} = \frac{\hbar}{2im} \left[\psi^\star \boldsymbol{\nabla} \psi - (\boldsymbol{\nabla}\psi)^\star \psi \right] \qquad (2.22)$$

These are *independent of time*, and hence the reference to the normal modes as *stationary states*. For these states, the boundary conditions in Eqs. (2.14)

---

[5]Here $\delta^{(3)}(\mathbf{x} - \mathbf{y})$ is a *Dirac delta-function* (see Vol. I). The relation in Eq. (2.21) can be demonstrated for operators of the Sturm-Liouville type (see, for example, [Fetter and Walecka (2003a)]); see also Prob. 2.7.

take the form

$$\psi, \boldsymbol{\nabla}\psi \ \text{ must be continuous} \tag{2.23}$$

2.1.3.1 *Example: One-Dimension*

As an example, consider a particle incident on a potential barrier in one dimension. In this case the time-independent Schrödinger equation reads

$$-\frac{\hbar^2}{2m}\frac{d^2\psi}{dx^2} = (E - V)\psi \tag{2.24}$$

Define

$$\frac{2m}{\hbar^2}(E - V) \equiv \varepsilon - v \equiv k^2 \qquad ; v < \varepsilon$$
$$\equiv -\kappa^2 \qquad ; v > \varepsilon \tag{2.25}$$

Consider the two cases:

(1) If $v(x) < \varepsilon$, the Schrödinger equation reduces to

$$\frac{d^2\psi}{dx^2} = -k^2(x)\psi \qquad ; v(x) < \varepsilon \tag{2.26}$$

With a constant v, the solution to this equation takes the form

$$\psi = a\,e^{ikx} + b\,e^{-ikx} \qquad ; \text{constant } v < \varepsilon$$
$$= A\sin kx + B\cos kx \qquad k = \sqrt{\varepsilon - v} \tag{2.27}$$

(2) If $v(x) > \varepsilon$, the Schrödinger equation becomes

$$\frac{d^2\psi}{dx^2} = \kappa^2(x)\psi \qquad ; v(x) > \varepsilon \tag{2.28}$$

If v is constant, the solution to this equation takes the form

$$\psi = c\,e^{\kappa x} + d\,e^{-\kappa x} \qquad ; \text{constant } v > \varepsilon$$
$$\kappa = \sqrt{v - \varepsilon} \tag{2.29}$$

Now suppose the potential is, in fact, a function of position in case (1), as illustrated in Fig. 2.2. Any real physical potential is *continuous*, as we have assumed. As a mathematical idealization, we can let $v(x)$ get sharper and sharper. The wave function ψ remains continuous during this process.

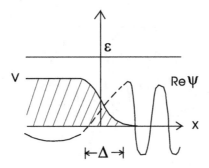

Fig. 2.2 Behavior of the wave function as it passes over a continuous potential barrier with $v(x) < \varepsilon$. We now let the barrier sharpen until it becomes a step.

What about the derivative? Integrate the second derivative over the region Δ in Fig. 2.2. Then

$$\int_{-\Delta/2}^{\Delta/2} \frac{d^2\psi}{dx^2} dx = \int_{-\Delta/2}^{\Delta/2} \frac{d}{dx}\left(\frac{d\psi}{dx}\right) dx = \frac{d\psi}{dx}\bigg|_{\Delta/2} - \frac{d\psi}{dx}\bigg|_{-\Delta/2}$$

$$= -\int_{-\Delta/2}^{\Delta/2} k^2(x)\psi(x)\, dx \tag{2.30}$$

The final integrand in Eqs. (2.30) remains bounded as v gets sharper and sharper, and the integral then goes to zero as $\Delta \to 0$. Hence

$$\frac{d\psi}{dx}\bigg|_{\Delta/2} - \frac{d\psi}{dx}\bigg|_{-\Delta/2} = 0 \qquad ; \Delta \to 0 \tag{2.31}$$

We conclude that the slope is *also* continuous if ψ is continuous and v is finite.

> $\psi, \boldsymbol{\nabla}\psi$ *are continuous even if* $V(x)$ *is discontinuous, as long as* $V(x)$ *is finite.*

What happens now at an infinite barrier (a "wall"), obtained as $v \to \infty$ in Fig. 2.3? Consider the local solutions to the Schrödinger equation as given above:

(1) Region I with $k = \sqrt{\varepsilon}$:

$$\psi_I = \sin kx + A\cos kx \qquad ; x > 0 \tag{2.32}$$

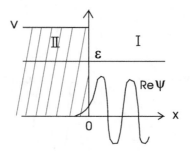

Fig. 2.3   As above with a finite step potential, but now below the barrier with $v > \varepsilon$.

An overall norm N appropriate to the problem at hand can be determined later.

(2) Region II with $\kappa = \sqrt{v - \varepsilon}$;

$$\psi_{II} = Be^{\kappa x} + Ce^{-\kappa x} \qquad\qquad ; x < 0 \qquad\qquad (2.33)$$

The solution $e^{-\kappa x}$ blows up as $x \to -\infty$, and one must take $C = 0$ to produce a normalizable wave function.

We proceed to match the wave functions and their derivatives at $x = 0$, as indicated by the above analysis,

$$\sin kx + A \cos kx = Be^{\kappa x}$$
$$k\left(\cos kx - A \sin kx\right) = \kappa Be^{\kappa x} \qquad ; x = 0 \qquad\qquad (2.34)$$

The solutions to these equations with $x = 0$ are

$$A = B = \frac{k}{\kappa}$$
$$\psi_I = N\left(\sin kx + \frac{k}{\kappa}\cos kx\right)$$
$$\psi_{II} = N\left(\frac{k}{\kappa}e^{\kappa x}\right) \qquad\qquad\qquad (2.35)$$

Now let the barrier become infinitely high so that $\kappa \to \infty$. We conclude the following:

$$\psi_{II}(x \le 0) \to 0 \qquad ; \kappa \to \infty$$
$$\psi_I(0) \to 0$$
$$\psi_I'(0) \to Nk \qquad\qquad\qquad (2.36)$$

Hence we arrive at the additional boundary condition that the wave function must vanish at a wall, while the slope is discontinuous there.

At an infinite barrier ("wall") $\psi = 0$, and $\mathbf{n} \cdot \boldsymbol{\nabla}\psi$ can take any value.

Note that the probability density $\rho = |\psi|^2$ vanishes at the wall (see Fig. 2.4).

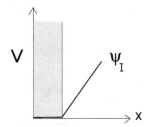

Fig. 2.4   Behavior of the wave function at a wall. Note that $\rho = |\psi|^2$ vanishes at the wall.

The problems of a particle in a three-dimensional box, and reflection and transmission of a particle at a potential barrier in one-dimension, were discussed in detail in Vol. I. We proceed to explicitly solve the Schrödinger equation in several other important cases.

2.2   Simple Harmonic Oscillator

The simple harmonic oscillator presents one of the most important problems in quantum mechanics:

- Any mechanical system performing small oscillations about equilibrium can be reduced to a system of uncoupled simple harmonic oscillators (molecules, crystals, any normal-mode problem);
- The normal modes of the electromagnetic field are uncoupled oscillators;
- Oscillators provide a basis for the expansion of *any* quantum field.

The potential for the simple harmonic oscillator is (see Fig. 2.5)

$$V(x) = \frac{1}{2}kx^2 \qquad ; \text{ simple harmonic oscillator} \qquad (2.37)$$

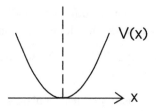

Fig. 2.5   Simple harmonic oscillator potential $V(x) = kx^2/2$.

The time-independent Schrödinger equation then takes the form

$$-\frac{\hbar^2}{2m}\frac{d^2\psi}{dx^2} + \frac{1}{2}kx^2\psi = E\psi \qquad (2.38)$$

Since the walls of this potential go up to infinity in both directions, all the acceptable solutions to this equation represent localized, bound states. Define the classical oscillator angular frequency and dimensionless energy by

$$\omega_0^2 \equiv \frac{k}{m} \qquad ; \varepsilon \equiv \frac{E}{\hbar\omega_0/2} \qquad (2.39)$$

Then

$$-\frac{\hbar}{m\omega_0}\frac{d^2\psi}{dx^2} + \frac{m\omega_0}{\hbar}x^2\psi = \varepsilon\psi \qquad (2.40)$$

Now introduce the dimensionless length[6]

$$\xi \equiv \left(\frac{m\omega_0}{\hbar}\right)^{1/2}x = \left(\frac{mk}{\hbar^2}\right)^{1/4}x \qquad ; \text{ dimensionless} \qquad (2.41)$$

The simple harmonic oscillator equation then takes the form

$$\frac{d^2\psi}{d\xi^2} + (\varepsilon - \xi^2)\psi = 0 \qquad ; \text{ dimensionless} \qquad (2.42)$$

The goal is to find the normalizable solutions to this dimensionless second-order differential equation.[7]

---

[6]Note the one-fourth power in the second relation, and in the first of Eqs. (2.92).
[7]See, for example, [Schiff (1968)].

2.2.1 *Eigenfunctions and Eigenvalues*

We start by extracting the asymptotic form of the solutions for large ξ^2, where the differential equation takes the form

$$\frac{d^2\psi}{d\xi^2} \sim \xi^2\psi \qquad ; \xi^2 \to \infty \qquad (2.43)$$

The solutions to this equation are

$$\psi \sim e^{\pm\xi^2/2}$$
$$\psi' \sim \pm\xi e^{\pm\xi^2/2}$$
$$\psi'' \sim \xi^2 e^{\pm\xi^2/2} \qquad ; \xi^2 \to \infty \qquad (2.44)$$

Now $e^{+\xi^2/2}$ is *not* normalizable, and we therefore want the other solution that dies off as $\xi^2 \to \infty$. Let us then take out this behavior and look for solutions of the form

$$\psi \equiv H(\xi)e^{-\xi^2/2} \qquad (2.45)$$

This defines the new function $H(\xi)$, and so far, there is no loss of generality.

Now substitute Eq. (2.45) into Eq. (2.42). One has

$$\psi' = H'e^{-\xi^2/2} - H\xi e^{-\xi^2/2}$$
$$\psi'' = \left(H'' - H'\xi - H'\xi - H + H\xi^2\right)e^{-\xi^2/2}$$
$$\left(\varepsilon - \xi^2\right)\psi = \left(\varepsilon - \xi^2\right)He^{-\xi^2/2} \qquad (2.46)$$

The terms in $H\xi^2$ cancel in Eq. (2.42), as does the factor $e^{-\xi^2/2}$, and the new function $H(\xi)$ thus satisfies the differential equation

$$\frac{d^2H}{d\xi^2} - 2\xi\frac{dH}{d\xi} + (\varepsilon - 1)H = 0 \qquad (2.47)$$

We observe that $\xi = 0$ is a regular point of this differential equation, and one can therefore look for a power-series solution of the form

$$H(\xi) = \xi^\sigma \sum_{p=0}^{\infty} a_p\xi^p \qquad ; a_0 \neq 0 \qquad (2.48)$$

Substitution into Eq. (2.47), and division by ξ^σ, leads to

$$\sum_{p=0}^{\infty}\left\{(\sigma + p)(\sigma + p - 1)a_p\xi^{p-2} - [2(\sigma + p) + 1 - \varepsilon]a_p\xi^p\right\} = 0 \qquad (2.49)$$

This is a power series which vanishes for all ξ, and hence the coefficient of each power must vanish

$$\sigma(\sigma - 1)a_0 = 0 \qquad ; \xi^{-2}$$
$$(\sigma + 1)\sigma \, a_1 = 0 \qquad ; \xi^{-1}$$
$$(\sigma + p + 2)(\sigma + p + 1)a_{p+2} - [2(\sigma + p) + 1 - \varepsilon]a_p = 0 \qquad ; \xi^p \qquad ; p \geq 0$$
$$(2.50)$$

The first relation is known as the "indicial equation", and since $a_0 \neq 0$, it determines the possible values of σ, which are $\sigma = (0, 1)$. The last of Eqs. (2.50) is a two-term recursion relation

$$\frac{a_{p+2}}{a_p} = \frac{2(\sigma + p) + 1 - \varepsilon}{(\sigma + p + 2)(\sigma + p + 1)} \qquad ; p \geq 0 \qquad (2.51)$$

Given a_p, this relation determines a_{p+2}. With $\sigma + p \equiv \nu$, it becomes

$$\frac{a_{\nu+2-\sigma}}{a_{\nu-\sigma}} = \frac{2\nu + 1 - \varepsilon}{(\nu + 2)(\nu + 1)} \qquad ; \sigma + p \equiv \nu \qquad (2.52)$$

Let us now discuss the two possible cases:

(1) $\underline{\sigma = 0}$: The first solution to the indicial equation, the first of Eqs. (2.50), is $\sigma = 0$. In this case, the second of Eqs. (2.50) implies that a_1 is *arbitrary*. Thus the power series solution in this case takes the form

$$H = a_0 \left(1 + \frac{a_2}{a_0}\xi^2 + \frac{a_4}{a_2}\frac{a_2}{a_0}\xi^4 + \frac{a_6}{a_4}\frac{a_4}{a_2}\frac{a_2}{a_0}\xi^6 + \cdots \right) +$$
$$a_1\xi \left(1 + \frac{a_3}{a_1}\xi^2 + \frac{a_5}{a_3}\frac{a_3}{a_1}\xi^4 + \frac{a_7}{a_5}\frac{a_5}{a_3}\frac{a_3}{a_1}\xi^6 + \cdots \right) \qquad (2.53)$$

In both factors

$$\frac{a_{\nu+2}}{a_\nu} = \frac{2\nu + 1 - \varepsilon}{(\nu + 2)(\nu + 1)} \qquad (2.54)$$

This holds for both the even ν in the first series and the odd ν in the second.

(2) $\underline{\sigma = 1}$: The second solution to the indicial equation is $\sigma = 1$. In this case, the second of Eqs. (2.50) implies that a_1 *vanishes*, and hence there are no even powers of ξ in this solution, which now takes the form

$$H = a_0\xi \left(1 + \frac{a_2}{a_0}\xi^2 + \frac{a_4}{a_2}\frac{a_2}{a_0}\xi^4 + \frac{a_6}{a_4}\frac{a_4}{a_2}\frac{a_2}{a_0}\xi^6 + \cdots \right) \qquad (2.55)$$

The required ratios are here given by

$$\frac{a_{\nu+1}}{a_{\nu-1}} = \frac{2\nu + 1 - \varepsilon}{(\nu+2)(\nu+1)} \tag{2.56}$$

Since ν is now odd, the generated series is *identical to the second series in Eq. (2.53)*.[8]

Let us discuss these results:

(1) We have generated two distinct power series solutions to the differential Eq. (2.47):
   - The first series in Eqs. (2.53), which is *even* under $\xi \to -\xi$;
   - The second series in Eqs. (2.53), which is *odd* under $\xi \to -\xi$.

(2) Consider the behavior of these series for very large ν, where from Eq. (2.54)

$$\frac{a_{\nu+2}}{a_\nu} \sim \frac{2}{\nu} \qquad ; \nu \to \infty \tag{2.57}$$

We cannot live with this, as the resulting infinite series is then not normalizable. To see this, call $\nu \equiv 2p$, where p runs over the integers, so that

$$\frac{a_{2(p+1)}}{a_{2p}} \sim \frac{1}{p} \qquad ; p \to \infty \tag{2.58}$$

The remainder in the first series in Eq. (2.53), for example, for all those terms with $p \geq m$ then takes the form

$$R_m \sim A_{2m}\xi^{2m}\left[1 + \frac{1}{m}\xi^2 + \frac{1}{m(m+1)}\xi^4 + \frac{1}{m(m+1)(m+2)}\xi^6 + \cdots\right]$$

$$= A_{2m}(m-1)!\,\xi^2 \sum_{l=0}^{\infty} \frac{\xi^{2(l+m-1)}}{(l+m-1)!} \tag{2.59}$$

But now

$$\sum_{l=0}^{\infty} \frac{\xi^{2(l+m-1)}}{(l+m-1)!} = \sum_{\nu=0}^{\infty} \frac{\xi^{2\nu}}{\nu!} + \text{polynomial in } \xi^2 \tag{2.60}$$

and

$$\sum_{\nu=0}^{\infty} \frac{\xi^{2\nu}}{\nu!} = e^{\xi^2} \tag{2.61}$$

---

[8] Just write out a few terms.

Since $e^{\xi^2} e^{-\xi^2/2} = e^{+\xi^2/2}$, the resulting infinite series solution *cannot be normalized, and therefore the series must terminate to yield a finite polynomial*. From Eq. (2.54), the condition on ε that one or the other of the series terminate is

$$\varepsilon - 1 = 2n \qquad ; n = 0, 1, 2, \cdots, \infty \qquad (2.62)$$

If this condition is satisfied, then from the recursion relation, all the subsequent coefficients $\{a_{n+2}, a_{n+4}, a_{n+6}, \cdots\}$ in that series will vanish identically.

We note that the second asymptotic solution $e^{+\xi^2/2}$ *must* be contained in our power series solutions, since there is no lack of generality in what we have done so far.

(3) For these polynomial solutions, labeled now by the integer n,

$$\frac{d^2 H_n}{d\xi^2} - 2\xi \frac{dH_n}{d\xi} + 2n H_n = 0 \qquad (2.63)$$

- If n is *even*, we must use the finite *even* series in Eq. (2.53), with $H_n(\xi) = H_n(-\xi)$;
- If n is *odd*, we must use the finite *odd* series in Eq. (2.53), with $H_n(\xi) = -H_n(-\xi)$.

Thus

$$H_n(-\xi) = (-1)^n H_n(\xi) \qquad (2.64)$$

Equations (2.63) and (2.64) characterize the finite *Hermite polynomials*, up to a normalization constant.

(4) From Eqs. (2.39) and (2.62), the energies of these normalizable eigenfunctions are

$$E_n = \hbar\omega_0 \left(n + \frac{1}{2}\right) \qquad ; n = 0, 1, 2, \ldots, \infty \qquad (2.65)$$

2.2.2　*Generating Function*

Many properties of the eigenfunctions can be derived with the aid of a *generating function*, similar to that used for the Legendre polynomials in electrostatics

$$\frac{1}{|\mathbf{r} - \mathbf{r}'|} = \frac{1}{(r^2 + r'^2 - 2rr' \cos\theta)^{1/2}}$$

$$= \sum_{l=0}^{\infty} \frac{r_<^l}{r_>^{l+1}} P_l(\cos\theta) \qquad ; \text{Legendre polynomials} \quad (2.66)$$

Consider the following generating function for the Hermite polynomials

$$S(\xi, s) \equiv e^{-s^2 + 2s\xi} = e^{-(s-\xi)^2 + \xi^2} \qquad ; \text{generating function} \quad (2.67)$$

Take the following partial derivatives

$$\frac{\partial S}{\partial \xi} = 2sS$$

$$\frac{\partial^2 S}{\partial \xi^2} = 4s^2 S$$

$$\frac{\partial^2 S}{\partial \xi^2} - 2\xi \frac{\partial S}{\partial \xi} = 4s(s - \xi)S$$

$$2s \frac{\partial S}{\partial s} = -4s(s - \xi)S \qquad (2.68)$$

The function $S(\xi, s)$ therefore satisfies the following differential equation

$$\frac{\partial^2 S}{\partial \xi^2} - 2\xi \frac{\partial S}{\partial \xi} + 2s \frac{\partial S}{\partial s} = 0 \qquad (2.69)$$

Now $S(\xi, s)$ is analytic in s, and it therefore has a power-series expansion

$$S(\xi, s) = \sum_{n=0}^{\infty} \frac{h_n(\xi)}{n!} s^n \qquad (2.70)$$

Substitute this expansion into Eq. (2.69)

$$\sum_{n=0}^{\infty} \left(\frac{d^2 h_n}{d\xi^2} - 2\xi \frac{dh_n}{d\xi} + 2nh_n \right) \frac{s^n}{n!} = 0 \qquad (2.71)$$

If a power series is equal to zero, then each coefficient must vanish. Thus

$$\frac{d^2 h_n}{d\xi^2} - 2\xi \frac{dh_n}{d\xi} + 2nh_n = 0 \qquad (2.72)$$

Now it follows from Eq. (2.70) that

$$h_n(\xi) = \left. \frac{\partial^n S}{\partial s^n} \right|_{s=0} \qquad (2.73)$$

One observes from Eq. (2.67) that

- This expression is even in ξ, if n is even;
- This expression is odd in ξ, if n is odd.

Hence

$$h_n(-\xi) = (-1)^n h_n(\xi) \tag{2.74}$$

We previously argued that Eqs. (2.72) and (2.74) serve to define the Hermite polynomials up to a normalization constant, and therefore

$$H_n(\xi) = C_n h_n(\xi) \tag{2.75}$$

where $H_n(\xi)$ are the oscillator wave functions and $h_n(\xi)$ are the Hermite polynomials, as defined here. We discuss the normalization constant C_n below, but first we observe that many useful properties of the Hermite polynomials $h_n(\xi)$ now follow directly from the generating function.

(1) *Recursion Relations*: Substitution of the expansion in Eq. (2.70) into the first of Eqs. (2.68) leads to

$$\frac{\partial S}{\partial \xi} = 2sS$$

$$\sum_{n=0}^{\infty} \frac{h_n'(\xi)}{n!} s^n = \sum_{n=0}^{\infty} \frac{2h_n(\xi)}{n!} s^{n+1} \tag{2.76}$$

The coefficients of $s^n/n!$ can now be equated to give the first recursion relation

$$\frac{dh_n(\xi)}{d\xi} = 2n h_{n-1}(\xi) \tag{2.77}$$

A similar approach employing the fourth of Eqs. (2.68) gives

$$\frac{\partial S}{\partial s} = -2(s - \xi)S$$

$$\sum_{n=0}^{\infty} \frac{n h_n(\xi)}{n!} s^{n-1} = \sum_{n=0}^{\infty} \frac{-2(s - \xi) h_n(\xi)}{n!} s^n \tag{2.78}$$

The coefficients of $s^n/n!$ then provide the second recursion relation

$$h_{n+1}(\xi) = 2\xi h_n(\xi) - 2n h_{n-1}(\xi)$$
$$= 2\xi h_n(\xi) - \frac{dh_n(\xi)}{d\xi} \tag{2.79}$$

The second line follows from Eq. (2.77).

(2) *Hermite Polynomials*: It follows from Eqs. (2.70) and (2.67) that the Hermite polynomials are given by[9]

$$h_n(\xi) = \frac{\partial^n S}{\partial s^n}\Big|_{s=0} = e^{\xi^2}\frac{\partial^n}{\partial s^n}e^{-(s-\xi)^2}\Big|_{s=0}$$

$$= (-1)^n e^{\xi^2}\frac{\partial^n}{\partial \xi^n}e^{-(s-\xi)^2}\Big|_{s=0} \qquad (2.80)$$

One can proceed to set $s = 0$ in this expression to give

$$h_n(\xi) = (-1)^n e^{\xi^2}\frac{d^n}{d\xi^n}e^{-\xi^2} \qquad ;\text{ Hermite polynomials} \qquad (2.81)$$

For example

$$h_0(\xi) = 1$$
$$h_1(\xi) = 2\xi$$
$$h_2(\xi) = e^{\xi^2}\frac{d}{d\xi}\left(-2\xi e^{-\xi^2}\right) = 4\xi^2 - 2 \qquad ;\text{ etc.} \qquad (2.82)$$

(3) *Normalization*: The harmonic oscillator eigenfunctions and eigenvalues are now given by

$$\psi_n(x) = C_n h_n(\xi)e^{-\xi^2/2} \qquad ;\text{ eigenfunctions}$$

$$E_n = \hbar\omega_0\left(n + \frac{1}{2}\right) \qquad ;\text{ eigenvalues}$$

$$n = 0, 1, 2, \cdots, \infty \qquad (2.83)$$

For each value of n there is a distinct eigenvalue E_n and a distinct polynomial $h_n(\xi)$. Thus the spectrum is *non-degenerate*, and the wave functions satisfy the orthonormality relation in Eq. (2.20)[10]

$$\int \psi_l^\star(x)\psi_n(x)\,dx = \delta_{ln} \qquad (2.84)$$

Set $l = n$, and insert the above to give

$$|C_n|^2 \int_{-\infty}^{\infty} [h_n(\xi)]^2 e^{-\xi^2}\,dx = |C_n|^2\left(\frac{\hbar}{m\omega_0}\right)^{1/2}\int_{-\infty}^{\infty}[h_n(\xi)]^2 e^{-\xi^2}\,d\xi = 1 \qquad (2.85)$$

---

[9]Notice the nice trick played with the partial derivatives in the second line.
[10]Recall this is a *Kronecker delta*, with $\delta_{ln} = 1$ if $l = n$, and $\delta_{ln} = 0$ if $l \neq n$.

Now use Eq. (2.81) to write the integral as

$$\int_{-\infty}^{\infty} [h_n(\xi)]^2 e^{-\xi^2} \, d\xi = \int_{-\infty}^{\infty} e^{-\xi^2} (-1)^{2n} \left[e^{\xi^2} \frac{d^n}{d\xi^n} e^{-\xi^2} \right] \left[e^{\xi^2} \frac{d^n}{d\xi^n} e^{-\xi^2} \right] d\xi$$

$$= \int_{-\infty}^{\infty} \left[\frac{d^n}{d\xi^n} e^{-\xi^2} \right] \left[e^{\xi^2} \frac{d^n}{d\xi^n} e^{-\xi^2} \right] d\xi \qquad (2.86)$$

The eigenfunctions in the oscillator are all bound states, and the wave functions fall off exponentially in both directions as $\xi \to \pm\infty$. Consequently, one can carry out n partial integrations on this expression with vanishing boundary contributions to give

$$\int_{-\infty}^{\infty} [h_n(\xi)]^2 e^{-\xi^2} \, d\xi = (-1)^n \int_{-\infty}^{\infty} e^{-\xi^2} \frac{d^n}{d\xi^n} \left[e^{\xi^2} \frac{d^n}{d\xi^n} e^{-\xi^2} \right] d\xi \quad (2.87)$$

The quantity in square brackets is a polynomial of the form

$$e^{\xi^2} \frac{d^n}{d\xi^n} e^{-\xi^2} = a_n \xi^n + a_{n-1} \xi^{n-1} + \cdots + a_0 \qquad (2.88)$$

We need to take n derivatives of this term in Eq. (2.87). The only non-vanishing contribution then comes from the leading term $a_n \xi^n$, which arises from repeatedly differentiating the exponential in Eq. (2.88)

$$a_n \xi^n = (-2\xi)^n \qquad (2.89)$$

Hence

$$\int_{-\infty}^{\infty} [h_n(\xi)]^2 e^{-\xi^2} \, d\xi = (-1)^n (-2)^n n! \int_{-\infty}^{\infty} e^{-\xi^2} \, d\xi$$

$$= 2^n n! \sqrt{\pi} \qquad (2.90)$$

The normalization condition in Eq. (2.85) then reads

$$|C_n|^2 \left(\frac{\hbar}{m\omega_0} \right)^{1/2} 2^n n! \sqrt{\pi} = 1 \qquad (2.91)$$

Now the phase of the eigenfunctions is a matter of *convention*, and we are free to take the C_n to be real and positive. Thus, finally,

$$C_n = \left(\frac{m\omega_0}{\hbar} \right)^{1/4} \left(\frac{1}{2^n n! \sqrt{\pi}} \right)^{1/2}$$

$$\psi_n(x) = C_n h_n(\xi) e^{-\xi^2/2} \qquad (2.92)$$

(4) *Transition Dipole Moment*: In many applications, for example in radiative transitions, one requires the transition dipole moment

$$\langle l|x|n \rangle \equiv \int \psi_l^*(x) x \psi_n(x) dx \qquad (2.93)$$

We observe that

- If $l = n$, the integrand is odd, and the matrix element vanishes;[11]
- The wave functions are now real, and this matrix element is the same as $\langle n|x|l \rangle$. It is therefore no loss of generality to assume $l > n$ in evaluating this expression.

Insertion of the explicit form of the wave functions as above leads to

$$\langle l|x|n \rangle = C_l C_n \left(\frac{\hbar}{m\omega_0} \right) (-1)^{n+l} \times$$
$$\int_{-\infty}^{\infty} e^{-\xi^2} \left[e^{\xi^2} \frac{d^l}{d\xi^l} e^{-\xi^2} \right] \xi \left[e^{\xi^2} \frac{d^n}{d\xi^n} e^{-\xi^2} \right] d\xi \qquad (2.94)$$

The integral can be partially integrated l times to give

$$\langle l|x|n \rangle = C_l C_n \left(\frac{\hbar}{m\omega_0} \right) (-1)^{n+2l} \int_{-\infty}^{\infty} e^{-\xi^2} \frac{d^l}{d\xi^l} \left\{ \xi \left[e^{\xi^2} \frac{d^n}{d\xi^n} e^{-\xi^2} \right] \right\} d\xi \qquad (2.95)$$

The use of Eq. (2.88) reduces this to

$$\langle l|x|n \rangle = C_l C_n \left(\frac{\hbar}{m\omega_0} \right) (-1)^{n+2l} \int_{-\infty}^{\infty} e^{-\xi^2} \frac{d^l}{d\xi^l} \left(a_n \xi^{n+1} + \cdots + a_0 \xi \right) d\xi \qquad (2.96)$$

If we recall that $l > n$, then the only non-vanishing result is for $l = n+1$. With the use of Eq. (2.89) one finds

$$\langle n+1|x|n \rangle = C_{n+1} C_n \left(\frac{\hbar}{m\omega_0} \right) 2^n (n+1)! \sqrt{\pi} \qquad (2.97)$$

Thus we arrive at the result

$$\langle n+1|x|n \rangle = \left(\frac{\hbar}{2m\omega_0} \right)^{1/2} \sqrt{n+1} \qquad (2.98)$$

This is the only non-vanishing dipole matrix element with $l \geq n$ for the harmonic oscillator.

---

[11]This is a *parity* selection rule.

Since the wave functions are real, they can be interchanged, and one can then then let $n \to n-1$, to obtain the only non-zero matrix element with $l \leq n$

$$\langle n-1|x|n \rangle = \left(\frac{\hbar}{2m\omega_0} \right)^{1/2} \sqrt{n} \qquad (2.99)$$

We proceed to discuss some aspects of the quantum mechanics of the harmonic oscillator that have almost unlimited applicability.

2.2.3 *Raising and Lowering Operators*

It follows from Eqs. (2.92) that the normalization constants C_n and C_{n-1} are related by

$$C_{n-1} = \sqrt{2n}\, C_n \qquad (2.100)$$

The corresponding Hermite polynomials satisfy the recursion relation in Eq. (2.77). The eigenfunctions thus satisfy the relation

$$\begin{aligned}
\psi_{n-1} &= C_{n-1} h_{n-1} e^{-\xi^2/2} \\
&= \sqrt{2n}\, C_n \frac{1}{2n} \left(\frac{dh_n}{d\xi} \right) e^{-\xi^2/2} \\
&= \frac{C_n}{\sqrt{2n}} \left(\frac{d}{d\xi} + \xi \right) \left(h_n e^{-\xi^2/2} \right) \qquad (2.101)
\end{aligned}$$

This allows us to define a *lowering operator* by

$$\begin{aligned}
a &\equiv \frac{1}{\sqrt{2}} \left(\frac{d}{d\xi} + \xi \right) \qquad ; \text{ lowering operator} \\
a\psi_n &= \sqrt{n}\, \psi_{n-1} \qquad (2.102)
\end{aligned}$$

Note, in particular, that this operator annihilates the ground state

$$a\psi_0 = 0 \qquad (2.103)$$

Given an operator A in quantum mechanics, the *adjoint* operator $A^\dagger$ is defined through the relation

$$\int (A\psi_l)^\star \psi_n\, dx \equiv \int \psi_l^\star A^\dagger \psi_n dx \qquad ; \text{ defines } A^\dagger \qquad (2.104)$$

for any (ψ_l, ψ_n). We note the following:

- An hermitian operator is *self-adjoint*

$$A^\dagger = A \qquad ; \text{ hermitian} \qquad (2.105)$$

Examples of hermitian operators are $\{x, p = (\hbar/i)\partial/\partial x, H = p^2/2m + V(x), \text{ etc.}\}$;

- The operator a is *not* self-adjoint. Since the solutions are well-behaved at infinity, partial integrations can be carried out with impunity. Thus

$$\frac{1}{\sqrt{2}} \int \left[\left(\frac{d}{d\xi} + \xi \right) \psi_l \right]^\star \psi_n \, dx = \frac{1}{\sqrt{2}} \int \psi_l^\star \left(-\frac{d}{d\xi} + \xi \right) \psi_n \, dx$$

$$(2.106)$$

The adjoint operator $a^\dagger$ is therefore identified as

$$a^\dagger = \frac{1}{\sqrt{2}} \left(-\frac{d}{d\xi} + \xi \right) \qquad (2.107)$$

What does this operator do to the eigenfunctions? Use

$$C_n = \sqrt{2(n+1)} \, C_{n+1} \qquad (2.108)$$

Then compute

$$\begin{aligned}
a^\dagger \psi_n &= \frac{C_n}{\sqrt{2}} \left(-\frac{d}{d\xi} + \xi \right) \left(h_n e^{-\xi^2/2} \right) \\
&= \sqrt{n+1} \, C_{n+1} \left(-\frac{dh_n}{d\xi} + 2\xi h_n \right) e^{-\xi^2/2} \\
&= \sqrt{n+1} \, C_{n+1} h_{n+1} e^{-\xi^2/2} \qquad (2.109)
\end{aligned}$$

The last equality follows from the recursion relation in Eq. (2.79). Hence $a^\dagger$ is the *raising operator*, satisfying

$$a^\dagger = \frac{1}{\sqrt{2}} \left(-\frac{d}{d\xi} + \xi \right) \qquad ; \text{ raising operator}$$

$$a^\dagger \psi_n = \sqrt{n+1} \, \psi_{n+1} \qquad (2.110)$$

We proceed to investigate *properties* of the raising and lowering operators in Eqs. (2.102) and Eqs. (2.110):

(1) *Matrix Elements*: The matrix elements of the raising and lowering operators follow from Eqs. (2.102) and (2.110) and the orthonormality of

the eigenfunctions

$$\int \psi_l^* a \psi_n = \sqrt{n} \int \psi_l^* \psi_{n-1} \, dx = \sqrt{n} \, \delta_{l,n-1}$$

$$\int \psi_l^* a^\dagger \psi_n = \sqrt{n+1} \int \psi_l^* \psi_{n+1} \, dx = \sqrt{n+1} \, \delta_{l,n+1} \quad (2.111)$$

Thus

$$\langle n-1|a|n \rangle = \sqrt{n} \qquad ; \text{ only non-zero M.E.}$$
$$\langle n+1|a^\dagger|n \rangle = \sqrt{n+1} \qquad (2.112)$$

These are the only non-zero matrix elements.

(2) *Create Normalized States*: Normalized states can be created by re-peated application of Eq. (2.110), starting from ψ_0

$$\psi_n = \frac{1}{\sqrt{n}} a^\dagger \cdots \frac{1}{\sqrt{3}} a^\dagger \frac{1}{\sqrt{2}} a^\dagger \frac{1}{\sqrt{1}} a^\dagger \psi_0 \qquad (2.113)$$

Thus

$$\psi_n = \frac{1}{\sqrt{n!}} \left(a^\dagger \right)^n \psi_0 \qquad (2.114)$$

(3) *Commutation Relations*: Suppose one lets a and then $a^\dagger$ act on an arbitrary wave function Ψ

$$a^\dagger a \Psi = \frac{1}{2} \left(-\frac{d}{d\xi} + \xi \right) \left(\frac{d}{d\xi} + \xi \right) \Psi$$

$$= \frac{1}{2} \left(-\frac{d^2}{d\xi^2} + \xi^2 - 1 \right) \Psi \qquad (2.115)$$

Now let them operate in the opposite order

$$a a^\dagger \Psi = \frac{1}{2} \left(\frac{d}{d\xi} + \xi \right) \left(-\frac{d}{d\xi} + \xi \right) \Psi$$

$$= \frac{1}{2} \left(-\frac{d^2}{d\xi^2} + \xi^2 + 1 \right) \Psi \qquad (2.116)$$

Take the difference of these two results

$$\left(a a^\dagger - a^\dagger a \right) \Psi = \Psi \qquad (2.117)$$

The l.h.s. is just the *commutator* of a and $a^\dagger$, and since Ψ is arbitrary

$$[a, a^\dagger] = 1 \qquad ; \text{ commutation relations} \qquad (2.118)$$

These *commutation relations* serve to characterize the raising and lowering operators.

(4) *Number Operator:* The effect of $a^\dagger a$ acting on the eigenfunction ψ_n can be explicitly calculated through the use of Eqs. (2.102) and (2.110)

$$a^\dagger a \psi_n = a^\dagger \left(\sqrt{n} \psi_{n-1} \right) = \sqrt{n}\sqrt{n}\, \psi_n \qquad (2.119)$$

Thus

$$a^\dagger a \psi_n = n \psi_n \qquad ; \text{ number operator}$$
$$n = 0, 1, 2, \cdots, \infty \qquad (2.120)$$

We refer to $a^\dagger a$ as the *number operator*, and ψ_n is an eigenstate of the number operator with eigenvalue n.

(5) *Some Commutators:* With the aid of Eq. (2.118), one can immediately calculate some useful additional commutation relations

$$[a^\dagger a, a] \equiv a^\dagger a a - a a^\dagger a = -[a, a^\dagger]a = -a$$
$$[a^\dagger a, a^\dagger] \equiv a^\dagger a a^\dagger - a^\dagger a^\dagger a = a^\dagger [a, a^\dagger] = a^\dagger \qquad (2.121)$$

(6) *Oscillator Hamiltonian:* A re-writing of the oscillator hamiltonian in terms of the raising and lowering operators plays a central role in quantum mechanics. Use Eq. (2.41) to write

$$H = -\frac{\hbar^2}{2m}\frac{d^2}{dx^2} + \frac{1}{2}m\omega_0^2 x^2 = \frac{\hbar\omega_0}{2}\left(-\frac{d^2}{d\xi^2} + \xi^2 \right) \qquad (2.122)$$

It follows from Eqs. (2.115)–(2.116) that

$$H = \hbar\omega_0 \frac{1}{2}\left(a^\dagger a + a a^\dagger \right) \qquad ; \text{ harmonic oscillator}$$
$$= \hbar\omega_0 \left(a^\dagger a + \frac{1}{2} \right) \qquad (2.123)$$

In the last form, the hamiltonian for the harmonic oscillator has been *explicitly diagonalized*.

2.2.4 Factorization Method

We start over and solve the harmonic oscillator with the *factorization method*. This is a very elegant method that uses only the general operator properties developed above.

We wish to solve the time-independent Schrödinger equation

$$H\psi = E\psi \tag{2.124}$$

Use the hamiltonian in the form of Eq. (2.123)

$$\left(2a^\dagger a + 1\right)\psi = \varepsilon\psi \qquad ; \text{S-eqn in factored form} \tag{2.125}$$

Now proceed through the following series of steps:

(1) Multiply this equation on the left by a, and use

$$
\begin{aligned}
a\left(a^\dagger a\right) &= \left(a^\dagger a\right)a - [a^\dagger a, a] \\
&= \left(a^\dagger a + 1\right)a
\end{aligned}
\tag{2.126}
$$

where the last equality follows from the first of Eq. (2.121). Equation (2.125) then becomes

$$\left[\left(2a^\dagger a + 1\right) + 2\right](a\psi) = \varepsilon(a\psi)$$

$$\text{or;} \qquad \left(2a^\dagger a + 1\right)(a\psi) = (\varepsilon - 2)(a\psi) \tag{2.127}$$

One concludes that if ψ is an eigenstate with eigenvalue ε, then $a\psi$ is again an eigenstate with eigenvalue $\varepsilon - 2$.

(2) The eigenvalue ε_n follows from the form of H in Eq. (2.122) and the orthonormality of the eigenfunctions

$$\left(-\frac{d^2}{d\xi^2} + \xi^2\right)\psi_n = \varepsilon_n\psi_n$$

$$\varepsilon_n = \int \psi_n^\star \left(-\frac{d^2}{d\xi^2} + \xi^2\right)\psi_n\, dx \tag{2.128}$$

Since the eigenfunctions are localized, a partial integration can be carried out to give

$$\varepsilon_n = \int \left(\left|\frac{d\psi_n}{d\xi}\right|^2 + \xi^2|\psi_n|^2\right) dx \geq 0 \tag{2.129}$$

The eigenvalues are *positive definite*.[12]

The lowering process in Eq. (2.127) must therefore eventually *terminate*, otherwise one could produce eigenstates with negative eigenvalues. Thus, the lowering process must eventually reach an eigenstate ψ_0

---

[12]The condition $\varepsilon_n \geq 0$ follows directly from the factored form in Eq. (2.125) and the existence of the adjoint (see Prob. 2.12).

satisfying

$$a\psi_0 = 0 \qquad ; \text{ground state} \qquad (2.130)$$

For this *ground state*, Eq. (2.125) becomes

$$\left(2a^\dagger a + 1\right)\psi_0 = \psi_0 = \varepsilon_0 \psi_0 \qquad (2.131)$$

Hence the lowest eigenvalue is

$$\varepsilon_0 = 1 \qquad ; \text{ground state} \qquad (2.132)$$

(3) Multiply Eq. (2.125) on the left by $a^\dagger$ and carry out a similar calculation

$$a^\dagger \left(a^\dagger a\right) = \left(a^\dagger a\right) a^\dagger - [a^\dagger a, a^\dagger]$$
$$= \left(a^\dagger a - 1\right) a^\dagger$$
$$\left(2a^\dagger a + 1\right)\left(a^\dagger \psi\right) = (\varepsilon + 2)(a^\dagger \psi) \qquad (2.133)$$

One concludes that if ψ is an eigenstate with eigenvalue ε, then $a^\dagger \psi$ is again an eigenstate with eigenvalue $\varepsilon + 2$. The eigenstates of H are then constructed from ψ_0 as[13]

$$\psi_n \sim \left(a^\dagger\right)^n \psi_0 \qquad ; \text{eigenfunctions}$$
$$n = 0, 1, 2, \cdots, \infty \quad (2.134)$$

These satisify

$$\left(2a^\dagger a + 1\right)\psi_n = (1 + 2n)\psi_n$$
$$E_n = \hbar\omega_0 \left(n + \frac{1}{2}\right) \quad ; \text{eigenvalues} \qquad (2.135)$$

(4) The *ground state* satisfies the following first-order differential equation

$$a\psi_0 = \frac{1}{\sqrt{2}}\left(\frac{d}{d\xi} + \xi\right)\psi_0 = 0 \qquad (2.136)$$

The unique solution for the ground state in coordinate space is

$$\psi_0 = C_0 e^{-\xi^2/2} \quad ; \text{ground state} \qquad (2.137)$$

These operator methods serve to provide an essential abstraction for quantum mechanics.

---

[13] The normalization of Eq. (2.114) can again be obtained with operator methods (see Prob. 2.13).

2.3   Central-Force Problem

The time-independent Schrödinger equation in three dimensions with a central potential $V(r)$ is[14]

$$\left[-\frac{\hbar^2}{2m_0}\nabla^2 + V(r) \right]\psi = E\psi \qquad ; \text{ central force} \qquad (2.138)$$

The analysis is now just classical continuum mechanics,[15] *except for the boundary conditions.*

The spherical coordinates (r, θ, ϕ), along with the associated orthogonal unit vectors $(\mathbf{e}_r, \mathbf{e}_\theta, \mathbf{e}_\phi)$, are shown in Fig. 2.6.

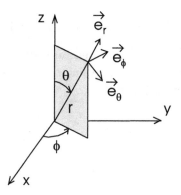

Fig. 2.6 Spherical coordinates (r, θ, ϕ) and associated orthogonal unit vectors $(\mathbf{e}_r, \mathbf{e}_\theta, \mathbf{e}_\phi)$.

The gradient and laplacian in spherical coordinates are given by

$$\boldsymbol{\nabla} = \mathbf{e}_r\frac{\partial}{\partial r} + \mathbf{e}_\theta\frac{1}{r}\frac{\partial}{\partial \theta} + \mathbf{e}_\phi\frac{1}{r\sin\theta}\frac{\partial}{\partial \phi}$$

$$\nabla^2 = \frac{1}{r^2}\frac{\partial}{\partial r}r^2\frac{\partial}{\partial r} + \frac{1}{r^2\sin\theta}\frac{\partial}{\partial \theta}\sin\theta\frac{\partial}{\partial \theta} + \frac{1}{r^2\sin^2\theta}\frac{\partial^2}{\partial \phi^2} \qquad (2.139)$$

---

[14]The generic rest mass in this central-force discussion will be denoted m_0; for electrons m_0 is m_e, for protons, m_p. This notation avoids any later confusion with the z-component of the angular momentum.

[15]We assume the reader is familiar with this (see [Fetter and Walecka (2003a)]; see also Vol. I).

2.3.1 *Separation of Variables*

We use the method of *separation of variables* and look for a solution to Eq. (2.138) of the form[16]

$$\psi(r, \theta, \phi) = R(r)P(\theta)\Phi(\phi) \qquad ; \text{ separation of variables} \quad (2.140)$$

Substitute this expression in Eq. (2.138), divide by ψ, and multiply by r^2

$$r^2 \left[\frac{1}{R} \frac{1}{r^2} \frac{d}{dr} r^2 \frac{dR}{dr} + \frac{2m_0}{\hbar^2}[E - V(r)] \right] +$$

$$\left[\frac{1}{P} \frac{1}{\sin\theta} \frac{d}{d\theta} \sin\theta \frac{dP}{d\theta} + \frac{1}{\sin^2\theta} \frac{1}{\Phi} \frac{d^2\Phi}{d\phi^2} \right] = 0$$

$$(2.141)$$

The first term on the l.h.s. is only a function of r, while the second is only a function of (θ, ϕ). This relation can only hold for all (r, θ, ϕ) if both terms are constant. Define this constant to be $\alpha(\alpha + 1)$. Then

$$r^2 \left[\frac{1}{R} \frac{1}{r^2} \frac{d}{dr} r^2 \frac{dR}{dr} + \frac{2m_0}{\hbar^2}[E - V(r)] \right] = \alpha(\alpha + 1)$$

$$\frac{1}{P} \frac{1}{\sin\theta} \frac{d}{d\theta} \sin\theta \frac{dP}{d\theta} + \frac{1}{\sin^2\theta} \frac{1}{\Phi} \frac{d^2\Phi}{d\phi^2} = -\alpha(\alpha + 1) \qquad (2.142)$$

Similarly, multiply the second of Eqs. (2.142) by $\sin^2\theta$. The second term on the l.h.s. is then only a function of ϕ, while the rest is only a function of θ. Both must again be constant. Define this constant as m^2. Thus

$$\sin^2\theta \left[\frac{1}{P} \frac{1}{\sin\theta} \frac{d}{d\theta} \sin\theta \frac{dP}{d\theta} + \alpha(\alpha + 1) \right] = m^2$$

$$\frac{1}{\Phi} \frac{d^2\Phi}{d\phi^2} = -m^2 \qquad (2.143)$$

We examine each of these relations in turn.

2.3.2 *Azimuthal-Angle Equation*

Start with the azimuthal-angle equation

$$\frac{d^2\Phi}{d\phi^2} = -m^2\Phi \qquad (2.144)$$

---

[16]The general solution is then constructed through superposition.

The solutions to this equation are $e^{\pm im\phi}$. What are the boundary conditions to be applied? Suppose we demand that as the azimuthal angle is increased by 2π, the wave function comes back to where it started

$$\Phi(\phi + 2\pi) = \Phi(\phi) \qquad ; \text{ p.b.c.} \qquad (2.145)$$

This are just *periodic boundary conditions in* ϕ. The corresponding condition on m is $m = 0, \pm 1, \pm 2, \cdots$. The normalized solutions obeying periodic boundary conditions are

$$\Phi_m(\phi) = \frac{1}{\sqrt{2\pi}} e^{im\phi} \qquad ; \ m = 0, \pm 1, \pm 2, \cdots$$
$$\qquad ; \text{ p.b.c.} \qquad (2.146)$$

These solutions are orthonormal, and we know the eigenfunctions are complete on the interval $[0, 2\pi]$ since one now simply has a complex Fourier series.

Is the above really justified? All we know is that the *probability density and current in Eqs. (2.22) must be single-valued under* $\phi \to \phi + 2\pi$. Let us go back to the superposition principle, which is what we previously used to go from the bilinear expressions $(\rho, \mathbf{S})$ to the wave function. Take the linear combination $\Phi = ae^{im_1\phi} + be^{im_2\phi}$. Then

(1) $|ae^{im_1\phi} + be^{im_2\phi}|^2$ must be single-valued under $\phi \to \phi + 2\pi$, which implies

$$\Delta m = \text{integer} \qquad ; \text{ all } m \qquad (2.147)$$

(2) $|ae^{im\phi} + be^{-im\phi}|^2$ must be single-valued under $\phi \to \phi + 2\pi$, which implies

$$2m = \text{integer} \qquad ; \text{ all } m \qquad (2.148)$$

There are *two* possibilities here

$$m = 0, \pm 1, \pm 2, \pm 3, \cdots$$
$$\text{or;} \quad m = \pm 1/2, \pm 3/2, \pm 5/2, \cdots \qquad (2.149)$$

While with the second set the wave function $\Phi_m(\phi)$ is *double-valued* under $\phi \to \phi + 2\pi$, the physical quantities $(\rho, \mathbf{S})$ remain single-valued and well-behaved.

(3) If we ask that $\Phi \sim 1$, a constant independent of ϕ, be an acceptable solution, then $|a + be^{im\phi}|^2$ must be single valued. This implies

$$m = \text{integer} \qquad ; \text{ all } m \qquad (2.150)$$

This is *all* we can demand at this stage of the game.

2.3.3 *Polar-Angle Equation*

Consider next the polar-angle equation

$$\frac{1}{\sin\theta}\frac{d}{d\theta}\sin\theta\frac{dP}{d\theta} + \left[\alpha(\alpha+1) - \frac{m^2}{\sin^2\theta}\right]P = 0 \qquad (2.151)$$

Change variables to

$$x \equiv \cos\theta \qquad\qquad\qquad ; \ 1 - x^2 = \sin^2\theta$$
$$\frac{d}{d\theta} = \frac{dx}{d\theta}\frac{d}{dx} = -\sin\theta\frac{d}{dx} \qquad ; \ \frac{d}{dx} = -\frac{1}{\sin\theta}\frac{d}{d\theta} \qquad (2.152)$$

Equation (2.151) then takes the form

$$\frac{d}{dx}(1-x^2)\frac{dP}{dx} + \left[\alpha(\alpha+1) - \frac{m^2}{1-x^2}\right]P = 0 \qquad (2.153)$$
$$; \text{ associated Legendre eqn}$$

This is the *associated* Legendre equation, where *Legendre's equation* is given by

$$\frac{d}{dx}(1-x^2)\frac{dP}{dx} + \alpha(\alpha+1)P = 0 \quad ; \text{ Legendre's eqn} \qquad (2.154)$$

From Eq. (2.150), m is an integer. It is then simply a matter of algebra to show that if one has a solution P_α to Legendre's equation, a corresponding solution P_α^m to the associated Legendre equation is obtained as

$$P_\alpha^m \equiv (1-x^2)^{|m|/2}\frac{d^{|m|}}{dx^{|m|}}P_\alpha \quad ; \text{ integer } m \qquad (2.155)$$

Consider then, solutions to Legendre's equation

$$(1-x^2)\frac{d^2P}{dx^2} - 2x\frac{dP}{dx} + \alpha(\alpha+1)P = 0 \qquad (2.156)$$

The points $x = \pm 1$ are singular points of this differential equation (Fig. 2.7).

Fig. 2.7   Singular points of Legendre's equation, where $x = \cos\theta$.

A fundamental system of solutions to this equation, obtained either with an integral representation or power series, is given by[17]

$$Q_\alpha(x) \;\; ; \text{ which has a term } \sim \ln\frac{1+x}{1-x} \text{ for all } \alpha$$

$$P_\alpha(x) \;\; ; \text{ which has a term } \sim \ln\frac{1+x}{1-x} \text{ for all } \alpha \text{ \underline{except} if}$$

$$\alpha = 0, 1, 2, 3, \cdots, \infty \tag{2.157}$$

Both members (P_α, Q_α) of the fundamental system contain a term in $\ln(1+x)/(1-x)$ for all α, except if α is a positive integer or zero. relabel the index by $\alpha \equiv l$. In this case P_l reduces to the finite *Legendre polynomial*

$$\alpha \equiv l = 0, 1, 2, 3, \cdots, \infty$$

$$P_l(x) = \frac{1}{2^l\, l!}\frac{d^l}{dx^l}(x^2 - 1)^l \qquad ; \text{ Legendre polynomial} \tag{2.158}$$

Unless $\alpha = 0, 1, 2, 3, \cdots, \infty$, the solutions have a logarithmic singularity at $x = \pm 1$. Why can't we tolerate a logarithmic singularity, say $\ln(1-x)$?

- The wave function ψ becomes infinite there, but *there is no requirement in quantum mechanics that ψ must be finite!* The only requirement is that

$$\int |\psi|^2 d^3x < \infty \tag{2.159}$$

  and a logarithmic singularity is integrable;
- We can use our *superposition principle* to show that the singularity acts as a *probability source*, and it can be ruled out on these grounds. The argument goes as follows:

---

[17]See appendix D in [Fetter and Walecka (2003a)].

(1) Take two solutions, where the first is assumed to be acceptable

$$\Psi_1 = f_1(r) \cdot 1 \cdot e^{-iE_1 t/\hbar}$$
$$\Psi_2 = f_2(r) \cdot \ln(1-x) \cdot e^{-iE_2 t/\hbar} \qquad (2.160)$$

Superposition states that the linear combination

$$\Psi = A\Psi_1 + B\Psi_2 \qquad (2.161)$$

must again be an acceptable solution. Since we are here focused on the polar-angle dependence, write this as

$$\Psi = a + b \ln(1-x) \qquad (2.162)$$

(2) Make a gaussian pillbox along the positive z-axis as indicated in Fig. 2.8.

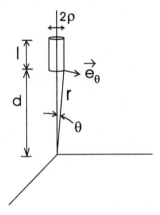

Fig. 2.8 Gaussian pillbox along the positive z-axis, where we let $\rho \to 0$. It is assumed here that $l \ll d$.

As $\rho \to 0$, with $l \ll d$, the angle to a point on the cylindrical surface is given by

$$x = \cos\theta \approx 1 - \frac{\rho^2}{2d^2} \qquad ; \ln(1-x) \approx \ln\frac{\rho^2}{2d^2} \qquad (2.163)$$

Now compute the total outgoing flux from the pillbox

$$\int_{\text{Box}} \mathbf{S} \cdot d\mathbf{A} \qquad ; \text{outgoing flux} \qquad (2.164)$$

(3) There is a negligible contribution from the ends of the box since

$$(\pi\rho^2)\ln\rho^2 \to 0 \qquad\qquad ; \rho \to 0 \qquad\qquad (2.165)$$

(4) On the cylindrical surface, the component of the gradient in the $\mathbf{e}_\theta$ direction is $\nabla_\theta = (1/r)\partial/\partial\theta$, and

$$\frac{\partial}{\partial\theta}\ln(1-x) = \frac{1}{1-x}\sin\theta = \frac{\sqrt{1-x^2}}{1-x} = \sqrt{\frac{1+x}{1-x}} \qquad (2.166)$$

Thus the component of the flux in the $\mathbf{e}_\theta$-direction is

$$S_\theta \approx \frac{\hbar}{2im_0d}\left\{[a^\star + b^\star\ln(1-x)]\left(b\sqrt{\frac{1+x}{1-x}}\right) - \right.$$
$$\left.\left(b^\star\sqrt{\frac{1+x}{1-x}}\right)[a + b\ln(1-x)]\right\}$$
$$= \frac{\hbar}{m_0d}\sqrt{\frac{1+x}{1-x}}\,\mathrm{Im}(a^\star b) \qquad (2.167)$$

With the use of Eqs. (2.163), this becomes

$$S_\theta \approx \frac{\hbar}{m_0d}\sqrt{\frac{2}{\rho^2/2d^2}}\,\mathrm{Im}(a^\star b) = \frac{2\hbar}{m_0\rho}\mathrm{Im}(a^\star b) \qquad (2.168)$$

As $\rho \to 0$, the integrated outgoing flux from the pillbox is therefore

$$\int_{\text{Box}} \mathbf{S}\cdot d\mathbf{A} = (2\pi\rho l)S_\theta \qquad\qquad ; \rho \to 0$$
$$\frac{1}{l}\int_{\text{Box}} \mathbf{S}\cdot d\mathbf{A} = \frac{4\pi\hbar}{m_0}\mathrm{Im}(a^\star b) \qquad\qquad (2.169)$$

The outgoing flux per unit length is a finite result independent of ρ in the limit $\rho \to 0$. The z-axis thus acts as a line source of probability in this case.[18]

- We conclude that we we must discard the logarithmic singularities, and choose only $\alpha = l = 0, 1, 2, 3, \cdots, \infty$.

---

[18]Notice that it is essential to use the superposed solution, with both the a and b terms present, to arrive at this conclusion.

In *summary*, the acceptable azimuthal- and polar-angle eigenfunctions and eigenvalues for the central-force problem in spherical coordinates are

$$\Phi_m(\phi) = \frac{1}{\sqrt{2\pi}} e^{im\phi} \qquad\qquad ; m = 0, \pm 1, \pm 2, \cdots$$

$$P_l^m(x) \equiv (1 - x^2)^{|m|/2} \frac{d^{|m|}}{dx^{|m|}} P_l(x)$$

$$P_l(x) = \frac{1}{2^l \, l!} \frac{d^l}{dx^l} (x^2 - 1)^l \qquad ; l = 0, 1, 2, \cdots \qquad (2.170)$$

where $x = \cos\theta$. Two comments:

- Since the Legendre polynomial $P_l(x)$ is of degree l, the associated Legendre polynomial $P_l^m(x)$ vanishes unless

$$|m| \le l \qquad (2.171)$$

- The associated Legendre polynomials are complete in the interval $-1 \le x \le 1$ for *each m*.[19]

It is much more convenient to introduce the *spherical harmonics* defined by

$$Y_{lm}(\theta, \phi) \equiv (-1)^m \left[\frac{2l + 1}{4\pi} \frac{(l - m)!}{(l + m)!} \right]^{1/2} P_l^m(\cos\theta) e^{im\phi} \quad ; m \ge 0$$

$$Y_{lm}^\star \equiv (-1)^m Y_{l,-m} \qquad\qquad ; \text{ defines others}$$

$$(2.172)$$

They have the following useful properties:

(1) The spherical harmonics are *orthonormal*

$$\int_0^{2\pi} d\phi \int_0^\pi \sin\theta \, d\theta \, Y_{l'm'}^\star(\theta, \phi) Y_{lm}(\theta, \phi) = \delta_{ll'} \delta_{mm'} \qquad (2.173)$$

(2) The spherical harmonics are *complete* on the unit sphere

$$F(\theta, \phi) = \sum_{l=0}^\infty \sum_{m=-l}^l a_{lm} Y_{lm}(\theta, \phi) \qquad (2.174)$$

---

[19]They are solutions to a Sturm-Liouville problem.

2.3.4 *Angular Momentum*

The angular momentum is given by

$$\mathcal{L} = \mathbf{r} \times \mathbf{p} \tag{2.175}$$

In quantum mechanics, in the coordinate representation where $\mathbf{p} = (\hbar/i)\boldsymbol{\nabla}$, this takes the form

$$\mathcal{L} = \frac{\hbar}{i}\mathbf{r} \times \boldsymbol{\nabla} \equiv \hbar\mathbf{L} \tag{2.176}$$

With the $\hbar$ extracted, the new angular momentum $\mathbf{L}$ is now dimensionless[20]

$$\mathbf{L} = \frac{1}{i}\mathbf{r} \times \boldsymbol{\nabla} \tag{2.177}$$

It follows that

$$L_x = \frac{1}{i}\left[y\frac{\partial}{\partial z} - z\frac{\partial}{\partial y}\right]$$
$$L_y = \frac{1}{i}\left[z\frac{\partial}{\partial x} - x\frac{\partial}{\partial z}\right]$$
$$L_z = \frac{1}{i}\left[x\frac{\partial}{\partial y} - y\frac{\partial}{\partial x}\right] \tag{2.178}$$

The operators can again be characterized through their *commutation relations*. For example[21]

$$[L_x, L_y] = -\left[y\frac{\partial}{\partial z} - z\frac{\partial}{\partial y}, \; z\frac{\partial}{\partial x} - x\frac{\partial}{\partial z}\right]$$
$$= -y\frac{\partial}{\partial x} + x\frac{\partial}{\partial y} = iL_z \tag{2.179}$$

Thus

$$[L_x, L_y] = iL_z \qquad \text{; and cyclic permutations} \tag{2.180}$$

It is readily verified that this relation holds for all cyclic permutations of the indices (x, y, z), It is re-written in terms of the completely antisymmetric tensor ϵ_{ijk} as

$$[L_i, L_j] = i\epsilon_{ijk}L_k \qquad ; \; (i, j, k) = 1, 2, 3 \tag{2.181}$$

---

[20] Angular momenta in units of $\hbar$ will be denoted with roman type.

[21] As before, one computes the commutator by letting this expression act on an arbitrary wave function.

where we here adopt the convention that repeated Latin indices are summed from one to three.

The *square* of the angular momentum is given by

$$\mathbf{L}^2 = L_x^2 + L_y^2 + L_z^2 \tag{2.182}$$

One can then compute, for example,

$$
\begin{aligned}
[\mathbf{L}^2, L_x] &= [L_x^2 + L_y^2 + L_z^2, L_x] \\
&\equiv L_y[L_y, L_x] + [L_y, L_x]L_y + L_z[L_z, L_x] + [L_z, L_x]L_z \\
&= -iL_yL_z - iL_zL_y + iL_zL_y + iL_yL_z = 0
\end{aligned} \tag{2.183}
$$

This vanishes identically, and $\mathbf{L}^2$ commutes with each component of the angular momentum

$$[\mathbf{L}^2, L_i] = 0 \quad ; \; i = x, y, z \tag{2.184}$$

The angular momentum operator in spherical coordinates follows from the first of Eqs. (2.139) and Fig. 2.6

$$\mathbf{L} = \frac{1}{i} \mathbf{r} \times \boldsymbol{\nabla} = \frac{1}{i}\left[\mathbf{e}_\phi \frac{\partial}{\partial \theta} - \mathbf{e}_\theta \frac{1}{\sin\theta} \frac{\partial}{\partial \phi} \right] \tag{2.185}$$

This only involves angles. It also follows from Fig. 2.6 that the relations between the unit vectors $(\mathbf{e}_x, \mathbf{e}_y, \mathbf{e}_z)$ and $(\mathbf{e}_r, \mathbf{e}_\theta, \mathbf{e}_\phi)$ are

$$
\begin{aligned}
\mathbf{e}_z \cdot \mathbf{e}_\phi &= 0 & &; \; \mathbf{e}_z \cdot \mathbf{e}_\theta = -\sin\theta \\
\mathbf{e}_x \cdot \mathbf{e}_\phi &= -\sin\phi & &; \; \mathbf{e}_x \cdot \mathbf{e}_\theta = \cos\theta \cos\phi \\
\mathbf{e}_y \cdot \mathbf{e}_\phi &= \cos\phi & &; \; \mathbf{e}_y \cdot \mathbf{e}_\theta = \cos\theta \sin\phi
\end{aligned} \tag{2.186}
$$

Thus, with $L_z = \mathbf{e}_z \cdot \mathbf{L}$ and so on, one finds

$$
\begin{aligned}
L_z &= \frac{1}{i} \frac{\partial}{\partial \phi} \\
L_x &= \frac{1}{i}\left[-\sin\phi \frac{\partial}{\partial \theta} - \cos\phi \cot\theta \frac{\partial}{\partial \phi} \right] \\
L_y &= \frac{1}{i}\left[\cos\phi \frac{\partial}{\partial \theta} - \sin\phi \cot\theta \frac{\partial}{\partial \phi} \right]
\end{aligned} \tag{2.187}
$$

We can also work out

$$
\begin{aligned}
\mathbf{L}^2 &= -(\mathbf{r} \times \boldsymbol{\nabla}) \cdot (\mathbf{r} \times \boldsymbol{\nabla}) \\
&= -\left(\mathbf{e}_\phi \frac{\partial}{\partial \theta} - \mathbf{e}_\theta \frac{1}{\sin\theta} \frac{\partial}{\partial \phi} \right) \cdot \left(\mathbf{e}_\phi \frac{\partial}{\partial \theta} - \mathbf{e}_\theta \frac{1}{\sin\theta} \frac{\partial}{\partial \phi} \right)
\end{aligned} \tag{2.188}
$$

For this, we need to differentiate the unit vectors, which change as the angles change. A contemplation of Fig. 2.6 convinces one that[22]

$$\frac{\partial \mathbf{e}_\phi}{\partial \theta} = 0 \qquad ; \quad \frac{\partial \mathbf{e}_\phi}{\partial \phi} = -\mathbf{e}_\theta \cos\theta - \mathbf{e}_r \sin\theta$$

$$\frac{\partial \mathbf{e}_\theta}{\partial \theta} = -\mathbf{e}_r \qquad ; \quad \frac{\partial \mathbf{e}_\theta}{\partial \phi} = \mathbf{e}_\phi \cos\theta \qquad\qquad (2.189)$$

Hence[23]

$$\mathbf{L}^2 = -\left[\frac{1}{\sin\theta} \frac{\partial}{\partial \theta} \sin\theta \frac{\partial}{\partial \theta} + \frac{1}{\sin^2\theta} \frac{\partial^2}{\partial \phi^2} \right] \qquad (2.190)$$

2.3.4.1 *Eigenfunctions and Eigenvalues*

The differential operator on the r.h.s. of Eq. (2.190) is just the one we have been working with in the separated Schrödinger equation, and it allows the laplacian in Eq. (2.139) to be re-written as

$$\nabla^2 = \frac{1}{r^2} \frac{\partial}{\partial r} r^2 \frac{\partial}{\partial r} - \frac{\mathbf{L}^2}{r^2} \qquad (2.191)$$

It is clear from Eqs. (2.151), (2.158), and the first of (2.187) that the spherical harmonics in Eq. (2.172) satisfy the following differential equations

$$L_z Y_{lm} = \frac{1}{i} \frac{\partial}{\partial \phi} Y_{lm} = m Y_{lm} \qquad\qquad ; \quad |m| \le l$$

$$\mathbf{L}^2 Y_{lm} = -\left[\frac{1}{\sin\theta} \frac{\partial}{\partial \theta} \sin\theta \frac{\partial}{\partial \theta} - \frac{m^2}{\sin^2\theta} \right] Y_{lm}$$

$$= l(l+1) Y_{lm} \qquad\qquad ; \quad l = 0, 1, 2, \cdots, \infty \quad (2.192)$$

Hence we conclude that the spherical harmonics are the *eigenfunctions of the angular momentum operators* $(L_z, \mathbf{L}^2)$

$$L_z Y_{lm} = m Y_{lm} \qquad ; \quad m = 0, \pm 1, \pm 2, \cdots, \pm l$$

$$\mathbf{L}^2 Y_{lm} = l(l+1) Y_{lm} \qquad ; \quad l = 0, 1, 2, \cdots, \infty \qquad (2.193)$$

The corresponding *eigenvalues* are just those we have derived previously.

---

[22]Note $d(\mathbf{e}_\theta \cdot \mathbf{e}_\phi) = d\mathbf{e}_\theta \cdot \mathbf{e}_\phi + \mathbf{e}_\theta \cdot d\mathbf{e}_\phi = 0$.

[23]Equations (2.187) and (2.190) are also derived algebraically through the transformation from cartesian to spherical coordinates (see Prob. 2.15).

2.3.4.2 *Hermiticity*

As observables, we want $L_i, \mathbf{L}^2, H = -\hbar^2 \nabla^2 / 2m_0 + V(r)$ to all be *hermitian* operators. What are the implications for partial integrations over the angular coordinates? Consider $\mathbf{L}^2$, which also appears in ∇^2,

$$\int d\Omega \psi_b^* \mathbf{L}^2 \psi_a = -\int_0^\pi \sin\theta \, d\theta \int_0^{2\pi} d\phi \psi_b^* \left(\frac{1}{\sin\theta} \frac{\partial}{\partial\theta} \sin\theta \frac{\partial}{\partial\theta} + \frac{1}{\sin^2\theta} \frac{\partial^2}{\partial\phi^2} \right) \psi_a \tag{2.194}$$

- Periodic boundary conditions on ϕ allow the second term on the r.h.s. to be partially integrated twice on ϕ for fixed θ, with vanishing boundary contributions;
- One would like to be able to similarly partially integrate the first term on the r.h.s. twice on θ at fixed ϕ.

Suppose we explicitly include the contributions at the boundaries at each step. Then from the two partial integrations one has

$$\int_0^\pi d\theta \psi_b^* \left(\frac{\partial}{\partial\theta} \sin\theta \frac{\partial}{\partial\theta} \right) \psi_a = \left[\psi_b^* \sin\theta \left(\frac{\partial\psi_a}{\partial\theta} \right) \right]_0^\pi - \left[\left(\frac{\partial\psi_b}{\partial\theta} \right)^* \sin\theta \psi_a \right]_0^\pi$$
$$+ \int_0^\pi \sin\theta \, d\theta \left[\frac{1}{\sin\theta} \frac{\partial}{\partial\theta} \sin\theta \left(\frac{\partial\psi_b}{\partial\theta} \right) \right]^* \psi_a \tag{2.195}$$

The operator $\mathbf{L}^2$ will be hermitian if the boundary terms go away. If the wave functions are finite polynomials in $x = \cos\theta$, then the boundary terms *do* disappear because of the presence of the factor $\sin\theta$ in each of them.[24] In contrast, if we were to allow the singular solutions for (P_α, Q_α) in Eqs. (2.157) in the wave functions, then there would be *finite remaining contributions from the boundaries*. Consider, for example, a contribution at $\theta = 0$

$$\sin\theta \frac{\partial}{\partial\theta} \ln(1-x) = \frac{\sin^2\theta}{1-x} = \frac{1-x^2}{1-x} = 1+x$$
$$\to 2 \qquad\qquad ; x \to 1 \tag{2.196}$$

We conclude that if the operator $\mathbf{L}^2$ is to be hermitian, then the wave functions must be restricted to polynomials and the values $\alpha \equiv l =$

---

[24]The condition $\sin\theta \, (\partial\psi/\partial\theta) = 0$ at a boundary is known as the *natural boundary condition* in the Sturm-Liouville problem.

$0, 1, 2, \cdots, \infty$. In *summary*[25]

$$L_i, \mathbf{L}^2, H \text{ hermitian} \implies \alpha \equiv l = 0, 1, 2, \cdots, \infty \qquad (2.197)$$

This is essentially the same argument as the previous one on the z-axis becoming a line source of probability; however, in the author's opinion, this argument is less physical.

2.3.5 *Radial Equation*

The separated solutions in the central-force problem now take the form

$$\psi(r, \theta, \phi) = R(r) Y_{lm}(\theta, \phi) \qquad ; \text{ central-force} \quad (2.198)$$

From Eqs. (2.142) and (2.158), the radial function $R(r)$ satisfies the differential equation

$$\frac{1}{r^2} \frac{d}{dr} r^2 \frac{dR}{dr} + \left[\frac{2m_0}{\hbar^2} [E - V(r)] - \frac{l(l+1)}{r^2} \right] R(r) = 0 \qquad (2.199)$$

Three comments:

- The *separation constant* here is $l(l + 1)$;
- It enters in the combination $V(r) + \hbar^2 l(l+1)/2m_0 r^2$, and the additional term acts as an *angular momentum barrier* in the potential;
- This is where the exact nature of the central potential $V(r)$ first plays a role.

The following relations can be used to re-write the radial equation

$$\frac{1}{r^2} \frac{d}{dr} r^2 \frac{dR}{dr} = \frac{d^2 R}{dr^2} + \frac{2}{r} \frac{dR}{dr}$$
$$= \frac{1}{r} \frac{d^2}{dr^2} (rR) \qquad (2.200)$$

Define $R(r) \equiv u(r)/r$, then from Eq. (2.199) and the second of Eqs. (2.200), $u(r)$ satisfies the following relation

$$R(r) \equiv \frac{u(r)}{r}$$

$$\frac{d^2 u}{dr^2} + \left[\frac{2m_0}{\hbar^2} [E - V(r)] - \frac{l(l+1)}{r^2} \right] u = 0 \qquad ; \text{ radial eqn} \quad (2.201)$$

---

[25]For L_i, see Prob. 2.16.

An alternate way to proceed is to define $R(r) \equiv J(r)/r^{1/2}$ and use

$$R(r) \equiv \frac{J(r)}{r^{1/2}}$$

$$\frac{dR}{dr} = \frac{1}{r^{1/2}}J' - \frac{1}{2r^{3/2}}J$$

$$\frac{d^2R}{dr^2} = \frac{1}{r^{1/2}}J'' - \frac{2}{2r^{3/2}}J' + \frac{3}{4r^{5/2}}J \qquad (2.202)$$

Then from Eq. (2.199) and the first of Eqs. (2.200), one has

$$\left[\frac{1}{r^{1/2}}J'' - \frac{1}{r^{3/2}}J' + \frac{3}{4r^{5/2}}J\right] + \frac{2}{r}\left[\frac{1}{r^{1/2}}J' - \frac{1}{2r^{3/2}}J\right]$$

$$+ \left[\frac{2m_0}{\hbar^2}[E - V(r)] - \frac{l(l+1)}{r^2}\right]\frac{J}{r^{1/2}} = 0 \qquad (2.203)$$

With a combination of terms, and multiplication by $r^{5/2}$, this simplifies to

$$R(r) \equiv \frac{J(r)}{r^{1/2}}$$

$$r^2\frac{d^2J}{dr^2} + r\frac{dJ}{dr} + \left[\frac{2m_0 r^2}{\hbar^2}[E - V(r)] - \left(l + \frac{1}{2}\right)^2\right]J = 0 \qquad ; \text{ radial eqn}$$

$$(2.204)$$

Equations (2.201) and (2.204) are fully equivalent ways of re-writing the radial Eq. (2.199) in the separated form of the Schrödinger equation with a central potential $V(r)$, where the wave function is that given in Eq. (2.198).

2.3.5.1 *Free Particle with $V = 0$*

Consider a free particle with $V(r) = 0$. Define

$$\frac{2m_0 E}{\hbar^2} \equiv k^2 \qquad (2.205)$$

With $kr \equiv z$ and $(l + 1/2) \equiv \nu$, the radial Eq. (2.204) then takes the form

$$z^2\frac{d^2J}{dz^2} + z\frac{dJ}{dz} + (z^2 - \nu^2)J = 0 \qquad ; z \equiv kr \qquad ; \nu \equiv l + \frac{1}{2}$$

$$; \text{ Bessel's eqn} \qquad (2.206)$$

This is *Bessel's equation*. It has the power-series solution

$$J_\nu(z) = \left(\frac{z}{2}\right)^\nu \sum_{p=0}^{\infty} \frac{(-1)^p}{p!\,\Gamma(\nu + p + 1)}\left(\frac{z}{2}\right)^{2p} \qquad (2.207)$$

If ν is non-integer, then a fundamental system of solutions to Bessel's equation is obtained as

$$\{J_\nu(z),\, J_{-\nu}(z)\} \text{ fundamental system if } \nu \neq \text{integer} \qquad (2.208)$$

The *spherical Bessel* and *spherical Neumann* functions are defined as

$$j_l(z) \equiv \sqrt{\frac{\pi}{2z}}\, J_{l+\frac{1}{2}}(z) \qquad\qquad ;\text{ spherical Bessel function}$$

$$n_l(z) \equiv (-1)^{l+1}\sqrt{\frac{\pi}{2z}}\, J_{-l-\frac{1}{2}}(z) \qquad ;\text{ spherical Neumann function}$$

$$(2.209)$$

They have the following properties:[26]

$$j_0(z) = \frac{\sin z}{z} \qquad\qquad ; n_0(z) = -\frac{\cos z}{z} \qquad\qquad (2.210)$$

$$j_1(z) = \frac{\sin z}{z^2} - \frac{\cos z}{z} \qquad ; n_1(z) = -\frac{\cos z}{z^2} - \frac{\sin z}{z}$$

$$j_l(z) = (-z)^l\left(\frac{1}{z}\frac{d}{dz}\right)^l \frac{\sin z}{z} \quad ; n_l(z) = -(-z)^l\left(\frac{1}{z}\frac{d}{dz}\right)^l \frac{\cos z}{z}$$

$$j_l(z) \to \frac{z^l}{(2l+1)!!} \qquad\qquad ; n_l(z) \to -\frac{(2l-1)!!}{z^{l+1}} \qquad\qquad ; z \to 0$$

Note that $n_l(z)$ is singular as $z \to 0$ for all l. Can we tolerate this singularity? Once again, we can use superposition to show that in its presence the origin becomes a *probability source*, and the singular solution can be ruled out on these grounds. The argument goes as follows:[27]

(1) Surround the origin with a small spherical pillbox (see Fig. 2.9).

(2) Consider the following superposed wave function, where the first term is assumed acceptable

$$\Psi = [aj_l(kr) + bn_l(kr)]Y_{lm}(\theta, \phi) \qquad (2.211)$$

(3) Compute the radial flux through the surface of the pillbox

$$\int_{\text{Box}} \mathbf{S} \cdot d\mathbf{A} = \int S_r\, r^2 d\Omega \equiv r^2\sigma_r \qquad (2.212)$$

---

[26]See, for example, [Schiff (1968); Fetter and Walecka (2003a)]. Here $(2l+1)!! = 1 \cdot 3 \cdot 5 \cdots (2l+1)$, and $(-1)!! \equiv 1$.

[27]Note that all but $n_0(z)$ are also *non-normalizable*.

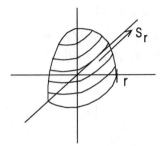

Fig. 2.9 Spherical gaussian pillbox around the origin, where we let $r \to 0$. S_r is the radial flux through the surface.

Here

$$
\sigma_r = \frac{\hbar}{2im_0} \left[(aj_l + bn_l)^* \left(a\frac{\partial}{\partial r}j_l + b\frac{\partial}{\partial r}n_l \right) - \right.
$$
$$
\left. \left(a\frac{\partial}{\partial r}j_l + b\frac{\partial}{\partial r}n_l \right)^* (aj_l + bn_l) \right]
$$
$$
= \frac{\hbar k}{m_0}\text{Im}\,(a^\star b) \left[j_l(z)\frac{\partial n_l(z)}{\partial z} - n_l(z)\frac{\partial j_l(z)}{\partial z} \right] \qquad (2.213)
$$

The quantity in square brackets in the last line is just the *wronskian*[28] of the two solutions in the fundamental system

$$
j_l(z)\frac{\partial n_l(z)}{\partial z} - n_l(z)\frac{\partial j_l(z)}{\partial z} = \frac{1}{z^2} \qquad ; \text{wronskian} \qquad (2.214)
$$

Hence

$$
\sigma_r = \frac{1}{r^2}\frac{\hbar}{m_0 k}\text{Im}\,(a^\star b)
$$
$$
\int_{\text{Box}} \mathbf{S}\cdot d\mathbf{A} = \frac{\hbar}{m_0 k}\text{Im}\,(a^\star b) \qquad (2.215)
$$

This expression is finite in the limit $r \to 0$

The origin thus acts as a point source of probability, and for a free particle, we can rule out the presence of the solution that is singular at the origin on these grounds.

---

[28] For the role of the wronskian, see Prob. 2.19.

2.3.5.2 *Spherical Box with $V = -V_0$*

Consider a particle in a spherical box of radius a, with infinite walls, and a potential inside it of $V = -V_0$ (Fig. 2.10).

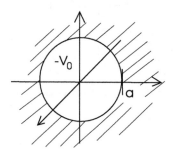

Fig. 2.10   Spherical box of radius a, and a potential inside it of $V = -V_0$.

Define

$$\frac{2m_0}{\hbar^2}(E + V_0) \equiv \kappa^2 \tag{2.216}$$

Then from Eqs. (2.204) and (2.209), the solution to the radial equation inside the box for a given l is

$$R_l(r) = Aj_l(\kappa r) + Bn_l(\kappa r) \tag{2.217}$$

Now apply the boundary conditions:

(1) The previous argument on the origin becoming a point source if the singular solution is retained is still valid, and hence $B = 0$;
(2) The wave function must vanish at a wall, which implies

$$j_l(\kappa a) = 0 \qquad \text{; eigenvalue equation} \tag{2.218}$$

This is an *eigenvalue equation* for κ. Define

$$j_l(X_{nl}) = 0 \qquad ; n = 1, 2, \cdots, \infty$$
$$; l = 0, 1, 2, \cdots, \infty \tag{2.219}$$

where X_{nl} is the nth zero of the lth spherical Bessel function, excluding the origin. Then

$$\kappa_{nl} a = X_{nl} \qquad \text{; eigenvalues} \tag{2.220}$$

The energy eigenvalues and eigenfunctions for a particle in a spherical box are thus given by

$$E_{nl} = -V_0 + \frac{\hbar^2 \kappa_{nl}^2}{2m_0} = -V_0 + \frac{\hbar^2}{2m_0 a^2} X_{nl}^2 \qquad ; n = 1, 2, \cdots, \infty$$

$$; l = 0, 1, 2, \cdots, \infty$$

$$\psi_{nlm} = N_{nl}\, j_l \left(X_{nl} \frac{r}{a} \right) Y_{lm}(\theta, \phi) \qquad\qquad (2.221)$$

Some comments:

- The eigenvalues E_{nl} and radial wave functions $R_{nl}(r)$ depend on (nl). They are independent of m;
- The zeros of the spherical Bessel functions are tabulated.[29] The eigenvalue spectrum for the first few states is shown in Fig. 2.11. Here we use the spectroscopic notation $l = s, p, d, f, \cdots$ for $l = 0, 1, 2, 3, \cdots$.

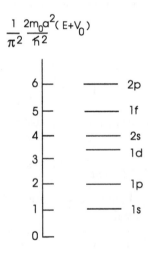

Fig. 2.11 First few energy levels in a spherical box of radius a, with a potential $V = -V_0$ inside it.

- A plot of the first few eigenfunctions is shown in 2.12;[30]

- The normalization constant N_{nl} can be calculated from formulas given

---

[29] See, for example, [Morse and Feshbach (1953)] p.1576. They can also be calculated with any good math program, such as Mathcad11.

[30] See also Prob. 2.18.

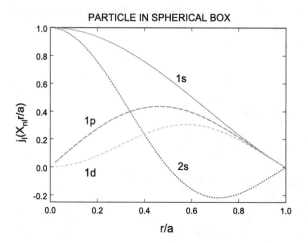

Fig. 2.12   Radial wave functions $j_l\left(X_{nl}r/a\right)$ as a function of r/a for the first few levels of a particle in a spherical box.

in Schiff[31]

$$N_{nl} = \left[\frac{2}{a^3 j_{l+1}^2 (X_{nl})}\right]^{1/2} \tag{2.222}$$

- This picture provides a decent first approximation to the *single-nucleon structure of nuclei*.[32]

2.3.5.3   *Finite Square-Well Potential*

Consider the *s*-wave ($l = 0$) bound-state solutions with $E = -|E|$ in a finite attractive square-well potential

$$V(r) = -V_0 \qquad ; r \leq a \tag{2.223}$$

There are two regions (see Fig. 2.13):

(1) *In Region I:* Define

$$\frac{2m_0}{\hbar^2}[E - V(r)] = \frac{2m_0}{\hbar^2}(V_0 - |E|) \equiv \kappa^2 \geq 0 \tag{2.224}$$

---

[31] See [Schiff (1968)] pp. 85–87, and Prob. 2.17.
[32] The notation (nl) for the levels is the *nuclear physics notation*.

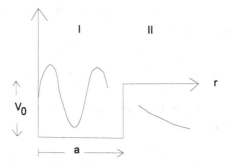

Fig. 2.13   Finite square-well potential $V(r) = -V_0$ for $r \leq a$, and s-wave solution $u_0(r)$ in regions I and II for a bound-state with $E = -|E|$.

From Eq. (2.201), the s-wave solution $u_0(r)$ in region I satisfies

$$\frac{d^2 u_0}{dr^2} + \kappa^2 u_0 = 0$$

$$u_0 = A \sin \kappa r + B \cos \kappa r \qquad (2.225)$$

The second term gives a radial wave function $R_0 = u_0/r$ that is too singular at the origin by our previous argument, and thus $B = 0$.

(2) *Region II*: Call

$$\frac{2m_0}{\hbar^2} E = \frac{2m_0}{\hbar^2}(-|E|) \equiv -\alpha^2 \qquad (2.226)$$

From Eq. (2.201), the s-wave function $u_0(r)$ in region II satisfies

$$\frac{d^2 u_0}{dr^2} - \alpha^2 u_0 = 0$$

$$u_0 = C e^{-\alpha r} + D e^{\alpha r} \qquad (2.227)$$

The second term is not normalizable, and so $D = 0$.

The boundary conditions in this case are that the wave function and its derivative must be continuous at $r = a$. Therefore

$$A \sin \kappa a = C e^{-\alpha a}$$

$$A \kappa \cos \kappa a = -\alpha C e^{-\alpha a} \qquad (2.228)$$

The ratio of these two equations again gives an eigenvalue equation for κ

$$\kappa a \cot \kappa a = -\alpha a \qquad ; \text{ eigenvalue equation} \qquad (2.229)$$

The just-bound 1s-state gives a reasonable first-approximation to the ground-state of the deuteron ${}^{2}_{1}\text{H}$.[33]

2.4 Hydrogen Atom

The Coulomb potential for the one-electron atom with a point nuclear charge Z is (Fig. 2.14)

$$V(r) = -\frac{Ze^2}{r} \qquad \text{; point Coulomb potential} \qquad (2.230)$$

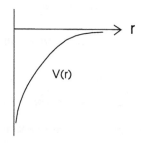

Fig. 2.14 Coulomb potential $V(r) = -Ze^2/r$ for the one-electron atom with a point nuclear charge Z.

It introduces two new features into the analysis:

- The potential is singular as $r \to 0$;
- It falls off only slowly as $r \to \infty$.

This is a central-force problem, and we again look for solutions of the form

$$\psi(r, \theta, \phi) = \frac{u(r)}{r} Y_{lm}(\theta, \phi) \qquad (2.231)$$

2.4.1 Radial Equation

From Eq. (2.201), the radial function $u(r)$ satisfies the following differential equation

$$\frac{d^2u}{dr^2} + \left[\frac{2m_0}{\hbar^2}[E - V(r)] - \frac{l(l+1)}{r^2} \right] u = 0 \qquad ; R = \frac{u}{r} \quad (2.232)$$

---

[33] Compare Prob. 2.20.

We go to dimensionless variables. The fundamental constants appropriate to this problem are[34]

$$\frac{e^2}{\hbar c} = \alpha = \frac{1}{137.0} \qquad\qquad \text{; fine-structure constant}$$

$$m_e c^2 = 0.511 \text{ MeV} \qquad\qquad \text{; electron rest mass}$$

$$a_0 = \frac{\hbar^2}{m_e e^2} = \frac{1}{\alpha} \frac{\hbar}{m_e c} = 0.529 \text{ Å} \qquad\qquad \text{; Bohr radius}$$

$$R = \frac{1}{2} \frac{e^2}{a_0} = \frac{1}{2} \alpha^2 m_e c^2 = 13.6 \text{ eV} \qquad\qquad \text{; Rydberg} \qquad (2.233)$$

The following units of length and energy, which exhibit the Z-dependence, will then be used

$$r_0 \equiv \frac{a_0}{Z} \qquad\qquad \text{; unit of length}$$

$$E_0 \equiv Z^2 R \qquad\qquad \text{; unit of energy} \qquad (2.234)$$

The dimensionless variables are taken to be

$$\sigma \equiv \frac{r}{r_0} \qquad ; \epsilon = -\frac{E}{E_0} = \frac{|E|}{E_0} \qquad\qquad (2.235)$$

where we here focus on the bound states with $E = -|E|$. After multiplication by r_0^2, Eq. (2.232) then takes the form

$$\frac{d^2 u}{d\sigma^2} + \left[r_0^2 \frac{2 m_e}{\hbar^2} \frac{Z e^2}{2 r_0} \left(-\epsilon + \frac{2}{\sigma} \right) - \frac{l(l+1)}{\sigma^2} \right] u = 0 \qquad\qquad (2.236)$$

This reduces to

$$\frac{d^2 u}{d\sigma^2} + \left[-\epsilon + \frac{2}{\sigma} - \frac{l(l+1)}{\sigma^2} \right] u = 0 \qquad ; \text{ radial S-eqn} \qquad (2.237)$$

The first step in seeking a solution to Eq. (2.237) is to again take out the asymptotic behavior as $\sigma \to \infty$

$$u'' \sim \epsilon u \qquad\qquad ; \sigma \to \infty \qquad\qquad (2.238)$$

The solutions to this equation are

$$u = e^{\pm\sqrt{\epsilon}\sigma} \qquad\qquad (2.239)$$

The normalizable solution dictates the minus sign, and we therefore define

$$u \equiv f(\sigma) e^{-\sqrt{\epsilon}\sigma} \qquad\qquad (2.240)$$

_____

[34]In this book, we work in c.g.s. units (see appendix K of Vol. I).

Once again, there is no loss of generality here. One now has

$$u' = \left(f' - \sqrt{\epsilon} f \right) e^{-\sqrt{\epsilon}\sigma}$$
$$u'' = \left(f'' - 2\sqrt{\epsilon} f' + \epsilon f \right) e^{-\sqrt{\epsilon}\sigma} \tag{2.241}$$

Substitution into Eq. (2.237), and cancellation of the factor $e^{-\sqrt{\epsilon}\sigma}$, gives

$$\frac{d^2 f}{d\sigma^2} - 2\sqrt{\epsilon}\frac{df}{d\sigma} + \left[\frac{2}{\sigma} - \frac{l(l+1)}{\sigma^2} \right] f = 0 \tag{2.242}$$

The behavior in the other limit $\sigma \to 0$ is dictated by the angular momentum barrier, and this can similarly be extracted through the re-definition

$$f \equiv \sigma^{l+1} F \tag{2.243}$$

It follows that

$$f' = (l+1)\sigma^l F + \sigma^{l+1} F'$$
$$f'' = l(l+1)\sigma^{l-1} F + 2(l+1)\sigma^l F' + \sigma^{l+1} F'' \tag{2.244}$$

Thus, after a combination of terms and division by σ^l,

$$\sigma\frac{d^2 F}{d\sigma^2} + \left[2(l+1) - 2\sqrt{\epsilon}\sigma \right]\frac{dF}{d\sigma} + \left[2 - 2\sqrt{\epsilon}(l+1) \right] F = 0 \tag{2.245}$$

Now introduce a new variable

$$\rho \equiv 2\sqrt{\epsilon}\sigma \tag{2.246}$$

A collection of these results leads to the following expression for the radial wave function in the hydrogen atom

$$R(r) = N\rho^l e^{-\rho/2} F(\rho)$$
$$\rho\frac{d^2 F}{d\rho^2} + [2l + 2 - \rho]\frac{dF}{d\rho} + \left[\frac{1}{\sqrt{\epsilon}} - (l+1) \right] F = 0 \tag{2.247}$$

Any remaining constants have been absorbed into the normalization constant N. Here ρ is given by Eq. (2.246) and (ϵ, σ) by Eqs. (2.234)–(2.235).

2.4.1.1 Confluent Hypergeometric Function

The radial equation for the hydrogen atom has now been reduced to the standard form of the *confluent hypergeometric equation*

$$\rho \frac{d^2 F}{d\rho^2} + (b - \rho)\frac{dF}{d\rho} - aF = 0 \qquad ; \text{ confluent hypergeometric eqn}$$

(2.248)

where (a, b) are constants.

The point $\rho = 0$ is a singular point of this differential equation. One seeks a series solution of the form

$$F = \rho^s \sum_{q=0}^{\infty} c_q \rho^q \qquad ; \ c_0 \equiv 1 \qquad (2.249)$$

Substitution of this expression in Eq. (2.248), and division by ρ^s, gives

$$\sum_q \left[(s+q)(s+q-1)c_q\rho^{q-1} + b(s+q)c_q\rho^{q-1} - (s+q)c_q\rho^q - ac_q\rho^q \right] = 0$$

(2.250)

This is a vanishing power series, and each term in the power series must again be equal to zero. The coefficient of ρ^{-1} leads to the *indicial equation*

$$[s(s-1) + bs]\, c_0 = 0 \qquad ; \ \rho^{-1}$$
$$s(s+b-1) = 0 \qquad \text{indicial eqn} \qquad (2.251)$$

The second line follows since $c_0 \equiv 1$. Consider the two cases:

(1) *Regular Solution* $(s = 0)$: The case $s = 0$ gives the *regular solution*, which is called F. For this solution, the behavior at the origin is

$$F \to 1 \qquad ; \ \rho \to 0 \qquad (2.252)$$

The coefficient of the power ρ^q then yields a two-term recursion relation[35]

$$[(q+1)q + b(q+1)]\, c_{q+1} = (a+q)c_q \qquad ; \ \rho^q \qquad (2.253)$$

which is re-written as

$$\frac{c_{q+1}}{c_q} = \frac{a+q}{b+q}\frac{1}{q+1} \qquad ; \ q \geq 0 \qquad (2.254)$$

---

[35]The motivation for explicitly extracting the asymptotic behavior in this and the preceding problem is to reduce the remaining analysis to a two-term recursion relation.

With $c_0 = 1$, this generates the *confluent hypergeometric series*

$$F(a|b|\rho) = 1 + \frac{a}{b}\frac{\rho}{1} + \frac{a\,(a+1)}{b\,(b+1)}\frac{\rho^2}{2!} + \frac{a\,(a+1)\,(a+2)}{b\,(b+1)\,(b+2)}\frac{\rho^3}{3!} + \cdots$$

$$; \text{ confluent hypergeometric series} \quad (2.255)$$

(2) *Irregular Solution* ($s = 1 - b$): The case $s = 1 - b$ gives the *irregular solution*, called G. The behavior of this solution at the origin is

$$G \to \rho^{1-b} \qquad ; \rho \to 0 \qquad (2.256)$$

2.4.1.2   *Boundary Conditions*

A comparison of Eqs. (2.247) and (2.248) indicates that the radial equation for the bound states in the hydrogen atom reduces to the confluent hypergeometric equation where the parameters (a, b) have the values

$$a = l + 1 - \frac{1}{\sqrt{\epsilon}}$$

$$b = 2l + 2 \qquad (2.257)$$

We must now impose the appropriate boundary conditions on the problem:

(1) Consider the behavior at the origin where $\rho \to 0$. From Eq. (2.256), the irregular solution goes as

$$G \sim \rho^{-2l-1} \qquad ; \rho \to 0$$

$$\rho^l G \sim \frac{1}{\rho^{l+1}} \qquad (2.258)$$

Our previous analysis on the origin becoming a point source of probability indicates that this behavior of the radial wave function is *too singular for any l*, and hence the irregular solution must be *discarded*;

(2) Consider the behavior of the regular solution in Eq. (2.255) for $\rho \to \infty$. From Eq. (2.254), one has for large q

$$\frac{c_{q+1}}{c_q} \sim \frac{1}{q} \qquad ; q \to \infty \qquad (2.259)$$

Hence, far enough out in the series, the remainder behaves as

$$R_m \sim A_m \rho^m \left[1 + \frac{1}{m}\rho + \frac{1}{m(m+1)}\rho^2 + \cdots \right]$$

$$= A_m \rho^m \sum_{p=0}^{\infty} \frac{(m-1)!}{(p+m-1)!}\rho^p \qquad (2.260)$$

The series can be re-written as

$$\sum_{p=0}^{\infty} \frac{(m-1)!}{(p+m-1)!} \rho^p = \rho^{-m+1}(m-1)! \sum_{s=m-1}^{\infty} \frac{\rho^s}{s!} \qquad (2.261)$$

The remaining series is

$$\sum_{s=m-1}^{\infty} \frac{\rho^s}{s!} = e^{\rho} + \text{polynomial in } \rho \qquad (2.262)$$

The behavior $e^{-\rho/2}e^{\rho} = e^{\rho/2}$ leads to a solution that is *not normalizable*, and hence the series in Eq. (2.255) must *terminate* to yield a finite polynomial.

Once again, the other asymptotic solution $e^{\rho/2}$ *must* be present in our series solution, since so far there has been no loss of generality.

The condition that the confluent hypergeometric series terminate is that the parameter a be a negative integer or zero. Hence the demand that the regular solution be normalizable leads to the relation

$$l + 1 - \frac{1}{\sqrt{\epsilon}} = -\nu \qquad ; \nu = 0, 1, 2, \cdots, \infty \qquad (2.263)$$

Note that F now contains powers of ρ up through ρ^{ν}.

The condition in Eq. (2.263) is re-written as

$$\frac{1}{\sqrt{\epsilon}} = \nu + l + 1 \equiv n \qquad ; n = 1, 2, 3, \cdots \infty \qquad (2.264)$$

principal quantum number

Here $n \equiv \nu + l + 1 = 1, 2, \cdots, \infty$ is known as the *principal quantum number*. Thus

$$\epsilon_n = \frac{1}{n^2} \qquad ; n = 1, 2, 3, \cdots \infty \qquad (2.265)$$

From Eqs. (2.233)–(2.235), this implies the bound-state energy eigenvalues are given by

$$E_n = -\frac{1}{2} \left(Z^2 \alpha^2 \right) m_e c^2 \frac{1}{n^2} \qquad ; n = 1, 2, 3, \cdots \infty \qquad (2.266)$$

The corresponding wave functions for the bound-states of the hydrogen atom are given by

$$\psi_{nlm}(r, \theta, \phi) = N_{nl} \, \rho^l e^{-\rho/2} F(l + 1 - n | 2l + 2 | \rho) \, Y_{lm}(\theta, \phi) \qquad (2.267)$$

where F is the confluent hypergeometric series in Eq. (2.255). Here

$$\rho = 2\sqrt{\epsilon}\sigma = \frac{2}{n}\sigma = \frac{2}{n}\frac{r}{r_0} = \frac{2Z}{n}\frac{r}{a_0} \tag{2.268}$$

where a_0 is the Bohr radius. The quantum numbers take the values

$$n = 1, 2, 3, \cdots, \infty$$
$$l = 0, 1, 2, \cdots, n-1$$
$$m = 0, \pm1, \pm2, \cdots, \pm l \tag{2.269}$$

Several comments:

- Equation (2.266) is just the *Bohr spectrum for the hydrogen atom!*
- The energy only depends on the principal quantum number $n = \nu+l+1$; all the other levels with given n, and $l = 0, 1, 2, \cdots, n-1$ and $m = 0, \pm1, \pm2, \cdots, \pm l$, are *degenerate*;
- The spectrum is illustrated in Fig. (2.15). Here the *atomic physics notation* (n, l), where n is the principal quantum number, is used for the levels;

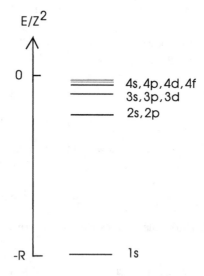

Fig. 2.15   First few levels in the spectrum of the one–electron atom with a point nuclear charge Z; there are an infinite number of bound states. This is now the atomic physics notation (n, l) where n is the principal quantum number. Here R is the Rydberg of Eq. (2.233).

- Because the potential falls off so slowly as $r \to \infty$, there are an infinite number of bound states in this one-electron problem with the point Coulomb potential of Eq. (2.230);[36]
- In the wave function in Eq. (2.267), $F(l+1-n|2l+2|\rho)$ is a finite polynomial of degree $\nu = n - l - 1$.

2.4.1.3 Laguerre Polynomials

The radial wave functions for the hydrogen atom can be related to the *Laguerre polynomials*, which are defined as follows

$$L_\kappa(\rho) \equiv e^\rho \frac{d^\kappa}{d\rho^\kappa} \left(\rho^\kappa e^{-\rho} \right) \qquad ; \text{ Laguerre polynomials} \qquad (2.270)$$

This is a polynomial of degree κ. One can explicitly construct it with the aid of the *Leibnitz formula* for multiple derivatives of a product

$$\frac{d^\kappa}{d\rho^\kappa}(AB) = \sum_{m=0}^{\kappa} \binom{\kappa}{m} \left(\frac{d^m}{dx^m} A \right) \left(\frac{d^{\kappa-m}}{dx^{\kappa-m}} B \right) \qquad ; \text{ Leibnitz}$$

$$\binom{\kappa}{m} \equiv \frac{\kappa!}{m!(\kappa-m)!} \qquad ; \text{ binomial coefficient} \qquad (2.271)$$

Thus

$$L_\kappa(\rho) = \kappa! + \binom{\kappa}{1} \left[\kappa(\kappa-1) \cdots 2\rho \right] (-1) + \binom{\kappa}{2} \left[\kappa(\kappa-1) \cdots 3\rho^2 \right] (-1)^2$$

$$+ \binom{\kappa}{3} \left[\kappa(\kappa-1) \cdots 4\rho^3 \right] (-1)^3 + \cdots \qquad (2.272)$$

Here the first term places all the derivatives on the ρ^κ, the second places all but one on the ρ^κ, and so on. This is written out as

$$L_\kappa(\rho) = \kappa! \times \qquad (2.273)$$
$$\left[1 + (-\kappa)\rho + \frac{(-\kappa)(-\kappa+1)}{2!} \frac{\rho^2}{2!} + \frac{(-\kappa)(-\kappa+1)(-\kappa+2)}{3!} \frac{\rho^3}{3!} + \cdots \right]$$

A comparison with the confluent hypergeometric series in Eq. (2.255) permits the identification

$$L_\kappa(\rho) = \kappa! F(-\kappa|1|\rho) \qquad (2.274)$$

---

[36]These bound states are of ever-increasing size. For the effect of the singular nature of the potential at the origin, see Prob. 2.23.

The *associated Laguerre polynomials* are defined as follows

$$L_\kappa^q(\rho) \equiv \frac{d^q}{d\rho^q} L_\kappa(\rho) \qquad ; \text{ associated Laguerre polynomials} \quad (2.275)$$

This is a polynomial of degree $\kappa - q$. Consider the first one

$$L_\kappa^1(\rho) = \frac{d}{d\rho} \left[e^\rho \frac{d^\kappa}{d\rho^\kappa} \left(\rho^\kappa e^{-\rho} \right) \right]$$

$$= e^\rho \frac{d^\kappa}{d\rho^\kappa} \left(\rho^\kappa e^{-\rho} \right) + e^\rho \frac{d^\kappa}{d\rho^\kappa} \left(\kappa \rho^{\kappa-1} e^{-\rho} - \rho^\kappa e^{-\rho} \right)$$

$$= e^\rho \frac{d^\kappa}{d\rho^\kappa} \left(\kappa \rho^{\kappa-1} e^{-\rho} \right) \tag{2.276}$$

Now repeat this process q times to obtain an alternate form of the associated Laguerre polynomials

$$L_\kappa^q(\rho) = e^\rho \frac{d^\kappa}{d\rho^\kappa} \left[\frac{\kappa!}{(\kappa - q)!} \rho^{\kappa-q} e^{-\rho} \right] \qquad ; \text{ alternate form} \quad (2.277)$$

Take one derivative of the confluent hypergeometric series in Eq. (2.255) to obtain

$$\frac{d}{dx} F(a|b|x) = \frac{a}{b} \left[1 + \frac{(a+1)}{(b+1)} x + \frac{(a+1)(a+2)}{(b+1)(b+2)} \frac{x^2}{2!} + \cdots \right] \tag{2.278}$$

Hence

$$\frac{d}{dx} F(a|b|x) = \frac{a}{b} F(a+1|b+1|x) \tag{2.279}$$

It then follows from Eqs. (2.274)–(2.275) that

$$L_\kappa^q(\rho) = \kappa! \left[\frac{d^q}{d\rho^q} F(-\kappa|1|\rho) \right] \tag{2.280}$$

$$= \kappa! \left[\left(\frac{-\kappa}{1} \right) \left(\frac{-\kappa+1}{2} \right) \cdots \left(\frac{-\kappa+q-1}{q} \right) F(-\kappa+q|1+q|\rho) \right]$$

Therefore the associated Laguerre polynomial is written in terms of the confluent hypergeometric series as

$$L_\kappa^q(\rho) = (-1)^q \frac{(\kappa!)^2}{(\kappa - q)! q!} F(-\kappa+q|1+q|\rho) \tag{2.281}$$

We are now in a position to identify with the hydrogen wave functions in Eq. (2.267)

$$q \equiv 2l + 1 \qquad\qquad ; \ \kappa \equiv n + l$$
$$\implies \quad -\kappa + q = -n + l + 1 \qquad ; \ 1 + q = 2l + 2 \qquad (2.282)$$

Thus the hydrogen wave functions are written in terms of associated Laguerre polynomials as

$$\psi_{nlm}(\rho, \theta, \phi) = \bar{N}_{nl} \, \rho^l e^{-\rho/2} L_{n+l}^{2l+1}(\rho) \, Y_{lm}(\theta, \phi)$$
$$\bar{N}_{nl} = N_{nl}(-1)^{2l+1} \frac{(n-l-1)!(2l+1)!}{[(n+l)!]^2} \qquad (2.283)$$

Here the constants are lumped into a new normalization constant $\bar{N}_{nl}$, and we henceforth adopt the phase convention that the N_{nl} are real and positive.[37]

2.4.1.4 *Normalization*

We know the eigenfunctions are orthonormal

$$\int d^3r \, \psi^\star_{n'l'm'}(\mathbf{r}) \psi_{nlm}(\mathbf{r}) = \delta_{nn'} \delta_{ll'} \delta_{mm'} \qquad (2.284)$$

The $\delta_{ll'} \delta_{mm'}$ follows from the orthonormality of the spherical harmonics, and the $\delta_{nn'}$ then follows from the orthogonality of eigenstates of different energy. It follows that the normalization condition is

$$\int |\psi_{nlm}|^2 r^2 dr \, d\Omega = |\bar{N}_{nl}|^2 \left(\frac{na_0}{2Z}\right)^3 \int_0^\infty \rho^{2l} e^{-\rho} \left[L_{n+l}^{2l+1}(\rho)\right]^2 \rho^2 d\rho = 1 \qquad (2.285)$$

We claim that

$$\int_0^\infty \rho^{q+1} e^{-\rho} \left[L_\kappa^q(\rho)\right]^2 d\rho = \frac{(\kappa!)^3}{(\kappa - q)!}(2\kappa - q + 1) \qquad (2.286)$$

This is demonstrated as follows. From Eq. (2.277), the integral is

$$\int_0^\infty \rho^{q+1} e^{-\rho} \left[L_\kappa^q(\rho)\right]^2 d\rho = \left[\frac{\kappa!}{(\kappa - q)!}\right]^2 \mathcal{I} \qquad (2.287)$$
$$\mathcal{I} \equiv \int_0^\infty d\rho \, \rho^{q+1} e^\rho \frac{d^\kappa}{d\rho^\kappa}\left(\rho^{\kappa-q} e^{-\rho}\right) \frac{d^\kappa}{d\rho^\kappa}\left(\rho^{\kappa-q} e^{-\rho}\right)$$

---

[37] See [Schiff (1968)].

The remaining integral is

$$\mathcal{I} = \int_0^\infty d\rho \, \frac{d^\kappa}{d\rho^\kappa} \left(\rho^{\kappa-q} e^{-\rho} \right) \left[\rho^{q+1} e^\rho \frac{d^\kappa}{d\rho^\kappa} \left(\rho^{\kappa-q} e^{-\rho} \right) \right] \qquad (2.288)$$

This can be partially integrated κ times to give[38]

$$\mathcal{I} = (-1)^\kappa \int_0^\infty d\rho \, \rho^{\kappa-q} e^{-\rho} \frac{d^\kappa}{d\rho^\kappa} \left[\rho^{q+1} e^\rho \frac{d^\kappa}{d\rho^\kappa} \left(\rho^{\kappa-q} e^{-\rho} \right) \right] \qquad (2.289)$$

The quantity in square brackets is a polynomial in ρ of degree $\kappa + 1$

$$\rho^{q+1} e^\rho \frac{d^\kappa}{d\rho^\kappa} \left(\rho^{\kappa-q} e^{-\rho} \right) = a_{\kappa+1} \rho^{\kappa+1} + a_\kappa \rho^\kappa + \cdots \qquad (2.290)$$

The first term comes from differentiating the exponential κ times, and the second from differentiating it $\kappa - 1$ times, with the correct factor given by Leibnitz' formula in Eq. (2.271). Thus

$$I = \int_0^\infty d\rho \, \rho^{\kappa-q} e^{-\rho} \left[(\kappa + 1)! \rho - \binom{\kappa}{1} (\kappa - q) \kappa! \right] \qquad (2.291)$$

Now use

$$\int_0^\infty d\rho \, \rho^m e^{-\rho} = m! \qquad (2.292)$$

Thus

$$\begin{aligned} \mathcal{I} &= (\kappa + 1)!(\kappa - q + 1)! - \kappa(\kappa - q)\kappa!(\kappa - q)! \\ &= \kappa!(\kappa - q)! \left[(\kappa + 1)(\kappa - q + 1) - \kappa(\kappa - q) \right] \end{aligned} \qquad (2.293)$$

This reduces to

$$\mathcal{I} = \kappa!(\kappa - q)!(2\kappa - q + 1) \qquad (2.294)$$

A combination of Eqs. (2.294) and (2.287) establishes the result in Eq. (2.286).

With the use of the result in Eq. (2.286), the normalization constants in Eqs. (2.285) and (2.283) are given by

$$\begin{aligned} |\bar{N}_{nl}|^2 &= \left(\frac{2Z}{na_0} \right)^3 \frac{(n-l-1)!}{[(n+l)!]^3} \frac{1}{2n} \\ |N_{nl}|^2 &= \left(\frac{2Z}{na_0} \right)^3 \frac{(n+l)!}{(n-l-1)![(2l+1)!]^2} \frac{1}{2n} \end{aligned} \qquad (2.295)$$

---

[38] There is enough convergence at infinity, and powers of ρ at the origin, that the boundary terms again disappear.

2.4.1.5 *Probability Density at the Origin*

As an application, we use Eq. (2.267) to calculate the probability density at the origin for a negatively charged lepton of mass m_0 in an atomic orbit about the nucleus. This has relevance for, among other things, the following contact interactions:

- The hyperfine interaction between atomic electrons and the nucleons in the nucleus;
- The capture of negative muons by the nucleons in the nucleus through the weak interaction.

Only $l = 0$ states yield a non-vanishing result for the wave function at the origin, and it follows from Eq. (2.255) that

$$F(a|b|0) = 1 \qquad (2.296)$$

From Eq. (2.295) and Prob. 2.14, one has[39]

$$|N_{n0}|^2 = \left(\frac{2Z}{na_0}\right)^3 \frac{1}{2} = \frac{4Z^3}{n^3} \alpha^3 \left(\frac{m_0 c}{\hbar}\right)^3$$

$$|Y_{00}|^2 = \frac{1}{4\pi} \qquad (2.297)$$

Thus the answer is

$$|\psi_{n00}(0)|^2 = \frac{1}{\pi} \frac{Z^3 \alpha^3}{n^3} \left(\frac{m_0 c}{\hbar}\right)^3 \qquad (2.298)$$

2.5 Periodic Potentials

We next examine the quantum dynamics of a particle in a periodic potential in one dimension

$$V(x + a) = V(x) \qquad ; \text{ periodic potential} \qquad (2.299)$$

where a is the period of the potential (see Fig. 2.16). This problem serves as a prototype for the behavior of electrons in solids.

We start by observing that *in the stationary states, the physics must similarly be invariant under translation by a.* This leads immediately to Bloch's theorem [Bloch (1928)].

---

[39] Here $a_0 = \hbar^2/m_0 e^2$.

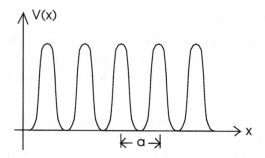

Fig. 2.16   Periodic potential in one dimension, with period a.

2.5.1   Bloch's Theorem

The stationary-state probability density and current must now satisfy

$$\rho(x+a) = \rho(x)$$
$$S_x(x+a) = S_x(x) \tag{2.300}$$

This implies

$$|\psi(x+a)|^2 = |\psi(x)|^2$$
$$\frac{\hbar}{2im}\left[\psi^\star\frac{\partial\psi}{\partial x} - \left(\frac{\partial\psi}{\partial x}\right)^\star\psi\right]_{x+a} = \frac{\hbar}{2im}\left[\psi^\star\frac{\partial\psi}{\partial x} - \left(\frac{\partial\psi}{\partial x}\right)^\star\psi\right]_{x} \tag{2.301}$$

The first of Eqs. (2.301) implies that

$$\psi(x+a) = e^{i\phi}\psi(x) \qquad ; \ \phi \text{ real} \tag{2.302}$$

where the phase ϕ is real. Now substitute this expression into the second of Eqs. (2.301). The left-hand-side becomes

$$\text{l.h.s.} = \frac{\hbar}{2im}\left[\left(e^{i\phi}\psi\right)^\star\left(i\frac{\partial\phi}{\partial x}e^{i\phi}\psi + e^{i\phi}\frac{\partial\psi}{\partial x}\right) - \left(i\frac{\partial\phi}{\partial x}e^{i\phi}\psi + e^{i\phi}\frac{\partial\psi}{\partial x}\right)^\star\left(e^{i\phi}\psi\right)\right] \tag{2.303}$$

This reduces to

$$\text{l.h.s.} = \frac{\hbar}{2im}\left[\psi^\star\frac{\partial\psi}{\partial x} - \left(\frac{\partial\psi}{\partial x}\right)^\star\psi\right]_x + \frac{\hbar}{m}|\psi|^2\frac{\partial\phi}{\partial x} \tag{2.304}$$

With the observation that $|\psi|^2 \neq 0$, the second of Eqs. (2.301) then gives

$$\frac{\partial \phi}{\partial x} = 0 \tag{2.305}$$

This implies that ϕ is a constant independent of position, Define this constant to be qa. Then

$$\phi \equiv qa \qquad ; \text{ constant}$$
$$\psi(x + a) = e^{iqa}\psi(x) \tag{2.306}$$

For the stationary states, any attempt to satisfy the boundary conditions between cells (see below), indicates that this relation must be the same from cell to cell.

Suppose one now writes the global stationary-state solution to the Schrödinger equation in the periodic potential as

$$\psi_q(x) \equiv e^{iqx}u_q(x) \qquad ; \text{ global solution} \tag{2.307}$$

where the constant q serves to label the eigenfunctions.[40] Substitution into the second of Eqs. (2.306) then gives

$$\psi_q(x + a) = e^{iq(x+a)}u_q(x + a)$$
$$= e^{iqa}\psi_q(x) = e^{iqa}e^{iqx}u_q(x) \tag{2.308}$$

Hence

$$u_q(x + a) = u_q(x)$$
$$\psi_q(x) = e^{iqx}u_q(x) \qquad ; \text{ Bloch's theorem} \tag{2.309}$$

This is Bloch's theorem: *The general form of the global stationary-state solution to the Schrödinger equation in a periodic potential is e^{iqx} times a function $u_q(x)$ that is strictly periodic in x.*

2.5.2 *Periodic Boundary Conditions*

The problem is simplified if the two ends of the potential are joined together and *periodic boundary conditions* imposed

$$\psi(x + L) = \psi(x + Na) = \psi(x) \quad ; \text{ p.b.c.}$$
$$; \ L = Na \tag{2.310}$$

Here $L = Na$ is the total length of the chain (see Fig. 2.17).

---

[40]There is clearly no loss of generality here.

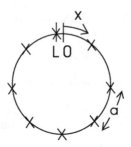

Fig. 2.17   Periodic boundary conditions with a periodic potential in one dimension. The coordinate x denotes the position along the chain. The total length of the chain is $L = Na$ where a is the period of the potential (here $N = 8$).

Then from the second of Eqs. (2.306), employed N times,

$$\psi(x + Na) = e^{iNqa}\psi(x) = \psi(x)$$
$$e^{iNqa} = 1 \qquad (2.311)$$

This relation implies that

$$qa = \frac{2\pi p}{N} \qquad ; p = 0, \pm 1, \pm 2, \cdots \qquad (2.312)$$

Two comments:

- The index q labels the states. There is one state for each p, or each q;
- Suppose one translates by

$$p \to p + N$$
$$qa \to qa + 2\pi \qquad (2.313)$$

The defining relation in Eq. (2.306) remains unchanged, and one does not generate a new, independent solution. Hence

We can take $-\pi \leq qa \leq \pi$, or translate by 2π in either direction; this leads to the same states, the same wave functions, and the same energies.[41]

Thus Eq. (2.312) should read

$$qa = \frac{2\pi p}{N} \qquad ; p = 0, \pm 1, \pm 2, \cdots, \pm N/2 \qquad (2.314)$$

The situation is illustrated in Fig. 2.18.

----

[41] See Eqs. (2.329) and (2.334). One or the other of the values $\pm N/2$ should be retained, but not both; it is assumed here that N is even.

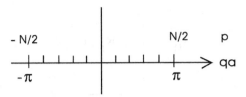

Fig. 2.18 Values of qa, with which the states are labeled. The corresponding values of the integer p in Eqs. (2.314) are indicated above the axis.

2.5.3 *Kronig-Penney Model*

A simple version of the Kronig-Penney model employs a series of delta-functions for the periodic potential

$$V(x) = \frac{\hbar^2}{2m} \frac{\lambda}{a} \sum_{n=1}^{N} \delta(x - na) \qquad ; \text{ Kronig-Penney model} \quad (2.315)$$

This allows one to obtain a relatively straightforward analytic solution for the problem.[42]
Define

$$E \equiv \frac{\hbar^2 k^2}{2m} \tag{2.316}$$

The time-independent, stationary-state Schrödinger equation then becomes

$$\left[-\frac{d^2}{dx^2} + \frac{\lambda}{a} \sum_{n=1}^{N} \delta(x - na) \right] \psi = k^2 \psi \tag{2.317}$$

Between the potentials, the solutions to this equation are $\{\sin kx, \cos kx\}$. Consider the region R_n in Fig. 2.19. Let us choose to write the linear combination of solutions in this region in the convenient fashion

$$\psi_n(x) = A_n \sin k(x - na) + B_n \cos k(x - na) \qquad ; x \text{ in } R_n \quad (2.318)$$

This defines the coefficients $\{A_n, B_n\}$. In the neighboring region R_{n+1}, the solution then takes the form

$$\psi_{n+1}(x) = A_{n+1} \sin k[x - (n+1)a] + B_{n+1} \cos k[x - (n+1)a]$$
$$; x \text{ in } R_{n+1} \quad (2.319)$$

---

[42]We assume here that $\lambda > 0$.

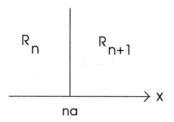

Fig. 2.19 Neighboring regions in the Kronig-Penney model. Within the regions, the solutions are written in the form of Eqs. (2.318) and (2.319). There is a potential $V = (\hbar^2\lambda/2ma)\delta(x-na)$ between the two regions at $x = na$, and the boundary conditions must be matched there.

The boundary conditions must now be satisfied between the two regions at $x = na$:

(1) The wave function must be continuous at the boundary[43]

$$\psi_+ = \psi_- \tag{2.320}$$

(2) The Schrödinger equation in the vicinity of the boundary is

$$\frac{d^2\psi}{dx^2} + k^2\psi = \frac{\lambda}{a}\delta(x - na)\psi \tag{2.321}$$

Integrate this expression across the potential

$$\int_{na-\epsilon}^{na+\epsilon} dx \left[\frac{d}{dx}\frac{d\psi}{dx} + k^2\psi\right] = \frac{\lambda}{a}\psi \tag{2.322}$$

The second term on the l.h.s. goes away as $\epsilon \to 0$, and thus there is a discontinuity of the slope at $x = na$ of

$$\left.\frac{d\psi}{dx}\right|_+ - \left.\frac{d\psi}{dx}\right|_- = \frac{\lambda}{a}\psi \tag{2.323}$$

Here, on the r.h.s., $\psi = \psi_+ = \psi_-$.

Now substitute the expressions for the wave functions on either side in Eqs. (2.318) and (2.319) into the boundary condtions at $x = na$ in Eqs. (2.320) and (2.323)

$$-A_{n+1}\sin ka + B_{n+1}\cos ka = B_n$$

$$k\left[A_{n+1}\cos ka + B_{n+1}\sin ka\right] - kA_n = \frac{\lambda}{a}B_n \tag{2.324}$$

---

[43] See Probs. 2.32–2.33.

These equations are readily solved for $\{A_{n+1}, B_{n+1}\}$:

- Multiply the first by $-\sin ka$, the second by $(1/k)\cos ka$ and add;
- Multiply the first by $\cos ka$, the second by $(1/k)\sin ka$ and add.

$$A_{n+1} = B_n \left(-\sin ka + \frac{\lambda}{ka}\cos ka\right) + A_n \cos ka$$

$$B_{n+1} = B_n \left(\cos ka + \frac{\lambda}{ka}\sin ka\right) + A_n \sin ka \qquad (2.325)$$

Now for x in R_n, one has from Eq. (2.318)

$$\psi_n(x) = A_n \sin k(x - na) + B_n \cos k(x - na) \qquad (2.326)$$

and for $x + a$ in R_{n+1}, one has from Eq. (2.319)

$$\psi_{n+1}(x + a) = A_{n+1} \sin k[(x + a) - (n + 1)a] + B_{n+1} \cos k[(x + a) - (n + 1)a] \qquad (2.327)$$

Thus for $x + a$ in R_{n+1}

$$\psi_{n+1}(x + a) = A_{n+1} \sin k(x - na) + B_{n+1} \cos k(x - na) \qquad (2.328)$$

The defining Eq. (2.306) can now be employed to relate the coefficients $\{A_{n+1}, B_{n+1}\}$ in Eq. (2.328) to $\{A_n, B_n\}$ in Eq. (2.326)

$$\psi(x + a) = e^{iqa}\psi(x)$$
$$\implies \quad A_{n+1} = e^{iqa} A_n$$
$$B_{n+1} = e^{iqa} B_n \qquad (2.329)$$

Substitution into Eqs. (2.325) then gives

$$\left(e^{iqa} - \cos ka\right) A_n + \left(\sin ka - \frac{\lambda}{ka}\cos ka\right) B_n = 0$$

$$\sin ka A_n + \left(\cos ka + \frac{\lambda}{ka}\sin ka - e^{iqa}\right) B_n = 0 \qquad (2.330)$$

The condition that these homogeneous algebraic equations for $\{A_n, B_n\}$ have a non-trivial solution is that the determinant of their coefficients must vanish

$$\left(e^{iqa} - \cos ka\right)\left(\cos ka + \frac{\lambda}{ka}\sin ka - e^{iqa}\right) - \sin ka\left(\sin ka - \frac{\lambda}{ka}\cos ka\right)$$
$$= 0 \qquad (2.331)$$

This simplifies to

$$e^{2iqa} - e^{iqa}\left(2\cos ka + \frac{\lambda}{ka}\sin ka\right) + 1 = 0 \qquad (2.332)$$

Multiplication by e^{-iqa} gives

$$2\cos qa = 2\cos ka + \frac{\lambda}{ka}\sin ka \qquad (2.333)$$

This is now an *eigenvalue equation* for k

$$\cos qa = \cos ka + \frac{\lambda}{2}\frac{\sin ka}{ka} \qquad \text{; eigenvalue equation} \qquad (2.334)$$

The corresponding energy is given by Eq. (2.316), and the appropriate values of qa are those in Eq. (2.314).

2.5.3.1 Eigenvalues

Introduce the variable $\xi \equiv ka$, so that

$$\xi \equiv ka \qquad ; E = \frac{\hbar^2}{2ma^2}\xi^2 \qquad (2.335)$$

The eigenvalue Eq. (2.334) then takes the form

$$\cos qa = F(\xi)$$
$$F(\xi) = \cos\xi + \frac{\lambda}{2}\frac{\sin\xi}{\xi} \qquad (2.336)$$

The possible values of qa are given in Eq. (2.314).

Equations (2.336) are solved graphically in Fig. 2.20. Several comments:

- To aid in the plotting, we can use the values in Table 2.1 calculated from $F(\xi)$ in Eq. (2.336) and

$$F'(\xi) = -\sin\xi + \frac{\lambda}{2}\left[\frac{\cos\xi}{\xi} - \frac{\sin\xi}{\xi^2}\right] \qquad (2.337)$$

- Pick a value of qa with $-\pi \leq qa \leq \pi$ and determine $\cos qa$. The intersection with the curve $F(\xi)$ then gives a series of values for ξ;

Table 2.1 Values of $F(\xi)$ and corresponding $F'(\xi)$.

ξ	$F(\xi)$	$F'(\xi)$
π	-1	$-\lambda/2\pi$
2π	$+1$	$+\lambda/4\pi$
3π	-1	$-\lambda/6\pi$

- Since $-1 \le \cos qa \le 1$, the solution must lie within the shaded boxes in Fig. 2.20. There are regions of ξ for which no solution is possible;

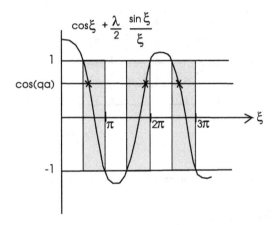

Fig. 2.20 Graphical solution to the eigenvalue Eq. (2.336) for the Kronig-Penney model.

- This leads to both *energy bands* and *energy gaps* in the eigenvalue spectrum, as illustrated in Fig. 2.21;
- For large N, the bands are densely populated;
- Since the analysis is unaffected by a translation of qa by 2π in either direction, one can redraw Fig. 2.21 by translating that portion of the second band in the interval $[-\pi, 0]$ by $qa \to qa + 2\pi$, and similarly translating that part in the interval $[0, \pi]$ by $qa \to qa - 2\pi$. A repetition of this translation process for the higher bands gives rise to the *extended band structure* shown in Fig. 2.22;
- The band structure for electron motion in solids plays a central role in solid-state physics.

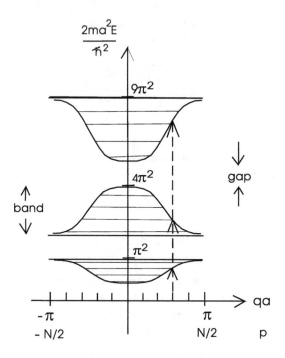

Fig. 2.21   Energy bands and energy gaps in the Kronig-Penney model.

2.5.3.2   *Limiting Cases*

Two limiting cases of these results are of particular interest:

(1) *Tight Binding,* $\lambda \to \infty$. In this case, independent of qa, the cross-over points in Fig. 2.20 become sharp at

$$\xi = ka = n\pi \qquad ; \; n = 1, 2, 3, \cdots \qquad (2.338)$$

The bands in Fig. 2.22 become horizontal lines located at

$$\frac{2ma^2 E}{\hbar^2} = \xi^2 = n^2 \pi^2 \qquad (2.339)$$

These are just the energy levels of a particle in a one-dimensional box of size a! A particle inserted into a region between the potentials is trapped in that region.[44]

---

[44]See Prob. 2.34.

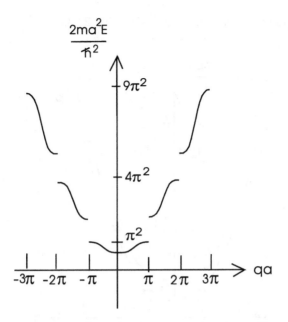

Fig. 2.22   Extended band structure in the Kronig-Penney model.

(2) *Free Particle*, $\lambda \to 0$. The solution to the eigenvalue Eq. (2.336) becomes

$$\xi = ka = qa \qquad (2.340)$$

The eigenfunctions in Fig. 2.17 reduce to the free-particle solutions

$$\psi_q(x) = \frac{1}{\sqrt{L}} e^{iqx} \qquad (2.341)$$

and the energies become the free-particle energies

$$E = \frac{\hbar^2 q^2}{2m} \qquad (2.342)$$

To satisfy periodic boundary conditions with $L = Na$, one has

$$q = \frac{2\pi p}{L} = \frac{2\pi p}{N}\frac{1}{a} \qquad ; p = 0, \pm 1, \cdots, \pm\infty \qquad (2.343)$$

This gives a parabola

$$\frac{2ma^2 E}{\hbar^2} = (qa)^2 = \left(\frac{2\pi p}{N}\right)^2 \qquad ; p = 0, \pm 1, \cdots, \pm\infty \qquad (2.344)$$

The two limiting cases are sketched in Fig. 2.23.

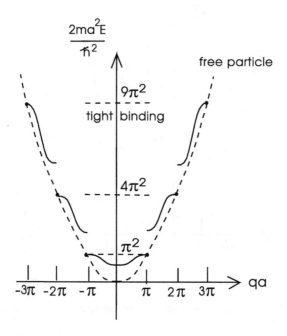

Fig. 2.23   Limiting cases of the band structure in the Kronig-Penney model: (1) Tight-binding ($\lambda \to \infty$) in Eq. (2.339); (2) Free particle ($\lambda \to 0$) in Eq. (2.344).

The task of making accurate numerical plots in Figs. 2.21 and 2.22 for various values of λ is assigned as Prob. 2.31.

Chapter 3

Formal Developments

We proceed to formalize the presentation of quantum mechanics, which will allow us to focus on the general structure of the theory. We start with a review of linear vector spaces.[1]

3.1 Linear Vector Spaces

Consider the ordinary three-dimensional linear vector space in which we live. Introduce an orthonormal set of basis vectors $\mathbf{e}_i$ with $i = 1, 2, 3$ satisfying

$$\mathbf{e}_i \cdot \mathbf{e}_j = \delta_{ij} \qquad ; (i, j) = 1, 2, 3 \qquad (3.1)$$

An arbitrary vector $\mathbf{v}$ is a physical quantity that has a direction and length in this space. It can be expanded in the basis $\mathbf{e}_i$ according to (see Fig. 3.1)

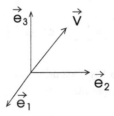

Fig. 3.1   An arbitrary vector $\mathbf{v}$ in a three-dimensional linear vector space with an orthonormal set of basis vectors $(\mathbf{e}_1, \mathbf{e}_2, \mathbf{e}_3)$. It can be written as $\mathbf{v} = \sum_{i=1}^{3} v_i \mathbf{e}_i$.

---

[1]In order to achieve continuity in the current presentation, some repetition of the material in chapter 2 of Vol. II is required here.

$$\mathbf{v} = \sum_{i=1}^{3} v_i \mathbf{e}_i \qquad\qquad ; v_i = \mathbf{e}_i \cdot \mathbf{v} \qquad (3.2)$$

The vector $\mathbf{v}$ can now be characterized by its components (v_1, v_2, v_3) in this basis.[2]

Vectors have the following properties:

(1) Addition of vectors, and multiplication of a vector by a constant, are expressed in terms of the components by

$$\mathbf{a} + \mathbf{b} : \ (a_1 + b_1, a_2 + b_2, a_3 + b_3)$$
$$\gamma \mathbf{a} : \ (\gamma a_1, \gamma a_2, \gamma a_3) \qquad\qquad ; \text{linear space} \qquad (3.3)$$

These properties characterize a *linear space*;

(2) The dot product, or inner product, of two vectors is defined by

$$\mathbf{a} \cdot \mathbf{b} \equiv a_1 b_1 + a_2 b_2 + a_3 b_3 \qquad\qquad ; \text{dot product} \qquad (3.4)$$

The *length* of the vector is then determined by

$$|\mathbf{v}| = \sqrt{\mathbf{v}^2} = \sqrt{\mathbf{v} \cdot \mathbf{v}} = (v_1^2 + v_2^2 + v_3^2)^{1/2} \qquad ; \text{length} \qquad (3.5)$$

One says that there is an *inner-product norm* in the space;

(3) Suppose one goes to a new orthonormal basis $\boldsymbol{\alpha}_i$ where the vector $\mathbf{v}$ has the components $\mathbf{v} : (\bar{v}_1, \bar{v}_2, \bar{v}_3)$. Then the components are evidently related by

$$\mathbf{v} = \sum_{i=1}^{3} v_i \mathbf{e}_i = \sum_{i=1}^{3} \bar{v}_i \boldsymbol{\alpha}_i$$
$$\implies \quad v_i = \sum_{j=1}^{3} \bar{v}_j (\mathbf{e}_i \cdot \boldsymbol{\alpha}_j) = \sum_{j=1}^{3} \bar{v}_j [\boldsymbol{\alpha}_j]_i \qquad (3.6)$$

These arguments are readily extended to *n-dimensions* by simply increasing the number of components

$$\mathbf{v} : \ (v_1, v_2, v_3, \cdots, v_n) \qquad\qquad ; \text{n-dimensions} \qquad (3.7)$$

The extension to *complex vectors* is accomplished through the use of the linear multiplication property with a complex γ. The positive-definite

---

[2]This will be denoted by $\mathbf{v} : (v_1, v_2, v_3)$.

norm is then correspondingly defined through $|\mathbf{v}|^2 \equiv \mathbf{v}^\star \cdot \mathbf{v}$,

$$\gamma\mathbf{v}: \ (\gamma v_1, \gamma v_2, \gamma v_3, \cdots, \gamma v_n) \qquad ; \ \text{complex vectors}$$

$$|\mathbf{v}|^2 \equiv \mathbf{v}^\star \cdot \mathbf{v} = |v_1|^2 + |v_2|^2 + \cdots + |v_n|^2 \qquad (3.8)$$

3.2 Hilbert Space

The notion of a *Hilbert space* involves the generalization of these concepts to a space with an infinite number of dimensions.

Let us start with an example. Recall the set of plane waves in one spatial dimension in an interval of length L satisfying periodic boundary conditions (Fig. 3.2)

$$\phi_n(x) = \frac{1}{\sqrt{L}}e^{ik_n x} \qquad ; \ k_n = \frac{2\pi n}{L} \qquad ; \ n = 0, \pm 1, \pm 2, \cdots$$

$$; \ \text{basis vectors} \qquad (3.9)$$

These will be referred to as the *basis vectors*.

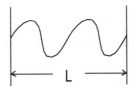

Fig. 3.2   Periodic boundary conditions in one dimension.

The basis vectors are orthonormal and satisfy

$$\int_0^L dx\, \phi_m^\star(x)\phi_n(x) = \delta_{mn} \qquad ; \ \text{orthonormal}$$

$$\equiv \langle \phi_m | \phi_n \rangle \qquad ; \ \text{inner product} \qquad (3.10)$$

This relation allows us to define the *inner product* of two basis vectors, denoted in the second line by $\langle \phi_m | \phi_n \rangle$.[3] The positive-definite *inner-product norm* of the basis vectors is then given by

$$|\phi_n|^2 = \langle \phi_n | \phi_n \rangle = \int_0^L dx\, |\phi_n(x)|^2 \qquad ; \ (\text{"length"})^2 \qquad (3.11)$$

---

[3]The notation, and most of the analysis in this chapter, is due to Dirac [Dirac (1947)].

An arbitrary function $\psi(x)$ can be expanded in this basis according to

$$\psi(x) = \sum_{n=-\infty}^{\infty} c_n \phi_n(x) \qquad ; \text{ expansion in complete set} \quad (3.12)$$

This is, after all, just a complex Fourier series. The orthonormality of the basis vectors allows one to solve for the coefficients c_n

$$c_n = \langle \phi_n | \psi \rangle = \int_0^L dx \, \phi_n^\star(x) \psi(x) \tag{3.13}$$

Any piecewise continuous function can actually be expanded in this set, and the basis functions are *complete* in the sense that[4]

$$\text{Lim}_{N \to \infty} \int_0^L dx \, \left| \psi(x) - \sum_{n=-N}^{N} c_n \phi_n(x) \right|^2 = 0 \qquad ; \text{ completeness} \quad (3.14)$$

Just as with an ordinary vector, the function $\psi(x)$ can now be characterized by the expansion coefficients c_n

$$\psi(x) \; : (c_{-\infty}, \cdots, c_{-1}, c_0, c_1, \cdots, c_\infty)$$
$$\text{or;} \qquad \psi \; : \{c_n\} \tag{3.15}$$

Addition of functions and multiplication by constants are defined in terms of the coefficients by

$$\psi^{(1)} + \psi^{(2)} \; : \{c_n^{(1)} + c_n^{(2)}\}$$
$$\gamma \psi \; : \{\gamma c_n\} \qquad ; \text{ linear space} \quad (3.16)$$

This function space is again a *linear space*.

The norm of ψ is given by

$$|\psi|^2 = \langle \psi | \psi \rangle = \int_0^L dx \, |\psi(x)|^2 = \sum_{n=-\infty}^{\infty} |c_n|^2 \qquad ; (\text{norm})^2 \quad (3.17)$$

which, in the case of Fourier series, is just Parseval's theorem.

The function $\psi(x)$ in Eq. (3.17) is said to be square-integrable. The set of all square-integrable functions ($\mathcal{L}^2$) forms a *Hilbert space*. Mathematicians define a Hilbert space as follows:

(1) It is a linear space;
(2) There is an inner-product norm;

---

[4] This is all the completeness we will need for the physics in this volume.

(3) The space is complete in the sense that every Cauchy sequence converges to an element in the space.

The above analysis demonstrates, through the expansion coefficients c_n, the isomorphism between the space of all square-integrable functions $(\mathcal{L}^2)$ and the ordinary infinite-dimensional complex linear vector space (l^2) discussed at the beginning of this section.

3.2.1 *Direct Analogy*

A more direct analogy to the infinite-dimensional complex linear vector space (l^2) is obtained through the following identification

$$v_i \to \psi_x \qquad ; \text{ coordinate space} \qquad (3.18)$$

We now use the coordinate x as a subscript, and we note that it is here a *continuous* index. The square of the norm then becomes

$$\sum_i v_i^\star v_i \to \sum_x \psi_x^\star \psi_x \equiv \int dx\, \psi^\star(x)\psi(x) \qquad (3.19)$$

The sum over the continuous index has here been appropriately defined through a familiar integral. With this notation, the starting expansion in Eq. (3.12) takes the form

$$\psi_x = \sum_{n=-\infty}^{\infty} c_n [\phi_n]_x \qquad (3.20)$$

3.2.2 *Abstract State Vector*

Equation (3.20) can be interpreted in the following manner:

This is just one component of the abstract vector relation

$$|\psi\rangle = \sum_n c_n |\phi_n\rangle \qquad ; \text{ abstract vector relation} \qquad (3.21)$$

The quantity $|\psi\rangle$ is now interpreted as a vector in an infinite-dimensional, abstract Hilbert space. It can be given a concrete representation through the component form in Eqs. (3.20) and (3.12), using the particular set of basis vectors in Eqs. (3.9).

As before, one solves for the expansion coefficients c_n by simply using the orthonormality of the basis vectors in Eq. (3.10)

$$c_n = \langle \phi_n | \psi \rangle = \sum_x [\phi_n]_x^\star \psi_x = \int dx\, \phi_n^\star(x)\psi(x) \qquad (3.22)$$

3.3   Linear Hermitian Operators

Consider an operator L in Hilbert space. Given $\psi(x)$, then $L\psi(x)$ is some new state in the space. L is a *linear operator* if it satisfies the condition

$$L(\alpha\phi_1 + \beta\phi_2) = \alpha(L\phi_1) + \beta(L\phi_2) \qquad ;\ \text{linear operator} \quad (3.23)$$

for any (ϕ_1, ϕ_2) in the space. L is *hermitian* if it satisfies the relation

$$\int dx\, \phi_1^\star(x)L\phi_2(x) = \int dx\, [L\phi_1(x)]^\star \phi_2(x) = \left[\int dx\, \phi_2^\star(x)L\phi_1(x) \right]^\star$$

$$;\ \text{hermitian} \qquad (3.24)$$

A shorthand for these relations is as follows

$$\langle \phi_1 | L | \phi_2 \rangle = \langle L\phi_1 | \phi_2 \rangle = \langle \phi_2 | L | \phi_1 \rangle^\star \qquad ;\ \text{shorthand} \qquad (3.25)$$

We now make the important observation that *if one knows the matrix elements of* L

$$L_{mn} \equiv \int dx\, \phi_m^\star(x)L\phi_n(x) \equiv \langle m | L | n \rangle \qquad ;\ \text{matrix elements} \quad (3.26)$$

in any complete basis, then one knows the operator L. Let us prove this assertion. Let $\psi(x)$ be an arbitrary state in the space. If one knows the corresponding $L\psi(x)$, then L is determined. Expand $\psi(x)$ in the complete basis

$$\psi(x) = \sum_n c_n \phi_n(x) \qquad ;\ \text{complete basis} \quad (3.27)$$

As above, the coefficients c_n follow from the orthonormality of the eigenfunctions ϕ_n

$$c_n = \int dx\, \phi_n^\star(x)\psi(x) \qquad ;\ \text{known} \qquad (3.28)$$

These coefficients are thus determined for any given ψ. Now compute[5]

$$L\psi(x) = \sum_n c_n[L\phi_n(x)] \tag{3.29}$$

The expansion in a complete basis can again be invoked to write the state $L\phi_n(x)$ as

$$L\phi_n(x) = \sum_m \beta_{mn}\phi_m(x) \tag{3.30}$$

and the orthonormality of the eigenfunctions allows one to identify

$$\beta_{mn} = \int dx\, \phi_m^\star(x)L\phi_n(x) = L_{mn} \tag{3.31}$$

Hence

$$L\phi_n(x) = \sum_m \phi_m(x)L_{mn}$$
$$\implies \quad L\psi(x) = \sum_n \sum_m \phi_m(x)L_{mn}c_n \tag{3.32}$$

This is now a *known quantity*, and thus we have established the equivalence

$$L \longleftrightarrow L_{mn} \qquad ; \text{ equivalent} \tag{3.33}$$

This equivalence is the basis of *matrix mechanics*, as discussed below.

3.3.1   *Eigenstates*

The *eigenfunctions* of a linear operator are defined by the relation

$$L\phi_\lambda(x) = \lambda\phi_\lambda(x) \qquad ; \text{ eigenfunctions}$$
$$\lambda \text{ is eigenvalue} \tag{3.34}$$

Here the operator simply reproduces the function and multiplies it by a constant, the *eigenvalue*. If L is an *hermitian* operator, then the following results hold:[6]

- The eigenvalues λ are real;
- The eigenfunctions corresponding to different eigenvalues are orthogonal.

---

[5]It is assumed here that there is enough convergence that one can operate on this series term by term.

[6]See Probs. 2.4–2.5

We give two examples from Vol. I:

(1) *Momentum.* The momentum operator in one dimension in coordinate space is

$$p = \frac{\hbar}{i} \frac{\partial}{\partial x} \qquad ; \text{ momentum} \qquad (3.35)$$

With periodic boundary conditions, the eigenfunctions are just those of Eq. (3.9), and

$$p\phi_k(x) = \hbar k \, \phi_k(x) \qquad ; k = \frac{2\pi n}{L},$$
$$n = 0, \pm 1, \pm 2, \cdots \qquad (3.36)$$

p is hermitian with these boundary conditions, and as we have seen, these eigenfunctions are both orthonormal and complete.

(2) *Hamiltonian.* In one dimension in coordinate space the hamiltonian is given by

$$H = \frac{-\hbar^2}{2m} \frac{\partial^2}{\partial x^2} + V(x) \qquad ; \text{ hamiltonian} \qquad (3.37)$$

We assume that $V(x)$ is real. The eigenstates are

$$H u_{E_n}(x) = E_n \, u_{E_n}(x) \quad ; \text{ eigenstates,}$$
$$n = 1, 2, \cdots, \infty \qquad (3.38)$$

In general, there will be both bound-state and continuum solutions to this equation. With the choice of periodic boundary conditions in the continuum, the hamiltonian is hermitian, and the energy eigenvalues E_n are real. The eigenstates of this hermitian operator *also* form a complete set, so that one can similarly expand an arbitrary $\psi(x)$ as

$$\psi(x) = \sum_n a_n \, u_{E_n}(x) \quad ; \text{ complete set} \qquad (3.39)$$

For the present purposes, one can simply add two additional quantum mechanics postulates to the three stated at the beginning of chapter 2:

(4) Observables are represented with linear hermitian operators;
(5) The eigenfunctions of a linear hermitian operator form a complete set.[7]

---

[7] A proof of completeness for any operator of the Sturm-Liouville type is contained in [Fetter and Walecka (2003a)]. The use of ordinary riemannian integration in the definition of the inner product in Eq. (3.19), and the notion of completeness expressed in Eq. (3.14), represent the extent of the mathematical rigor in the present discussion.

3.4 Matrix Mechanics

Before pursuing the abstraction of these results, we pause to present some concrete applications of the equivalence relation in Eq. (3.33). It was from this equivalence relation that Heisenberg (1925) developed his version of quantum mechanics. The method of *matrix mechanics* presented here provides an alternative to solving the differential Schrödinger equation, and today most practical calculations in atomic, molecular, nuclear, and solid-state physics are actually carried out in this fashion.

(1) We wish to solve the stationary-state Schrödinger equation

$$H = -\frac{\hbar^2}{2m}\frac{d^2}{dx^2} + V(x)$$

$$Hu(x) = Eu(x) \qquad\qquad \text{; stationary states} \qquad (3.40)$$

(2) Expand the solution in some *complete, orthonormal set* of solutions $\phi_n(x)$ obeying the appropriate boundary conditions

$$u(x) = \sum_{n=1}^{N} c_n \phi_n(x) \qquad (3.41)$$

- Completeness implies that as $N \to \infty$, the full solution can be reproduced;
- In addition, one can get arbitrarily close to the full solution with a *large, finite N*.

(3) Substitute the expansion in Eq. (3.41) into the Schrödinger Eq. (3.40)

$$\sum_n c_n [H\phi_n(x)] = E\left[\sum_n c_n \phi_n(x)\right] \qquad (3.42)$$

Now carry out an integration $\int \phi_m^\star(x)\, dx$.[8] This leads to

$$\sum_n \langle m|H|n\rangle c_n = E\sum_n c_n \delta_{mn} \qquad (3.43)$$

---

[8] It is this integration that provides a practical smoothing operation for the differential operator of the Schrödinger equation, useful in numerical work. In this respect, matrix mechanics as presented here converts the numerical solution of a differential equation to the numerical solution of a set of linear homogeneous algebraic equations. Although presented here in one dimension, the analysis holds in any number of them.

where the matrix elements of the hamiltonian are defined by

$$\int \phi_m^\star(x) H \phi_n(x)\, dx \equiv \langle m|H|n \rangle$$

$$\equiv H_{mn} \qquad ; \text{ hamiltonian matrix} \quad (3.44)$$

Since the hamiltonian H is an hermitian operator, H_{mn} will be an hermitian matrix[9]

$$H_{mn}^\star \equiv H_{nm}^\dagger = H_{nm} \qquad ; \text{ hermitian matrix} \quad (3.45)$$

Equations (3.43) present a set of N *linear, homogeneous, algebraic equations* for the expansion coefficients c_n[10]

$$\sum_{n=1}^{N} \left(\langle m|H|n \rangle - E\delta_{mn} \right) c_n = 0 \quad ; m = 1, 2, \cdots, N \quad (3.46)$$

Let us write these equations out in detail

$$(H_{11} - E)c_1 + H_{12}c_2 + \cdots + H_{1N}c_N = 0$$
$$H_{21}c_1 + (H_{22} - E)c_2 + \cdots + H_{2N}c_N = 0$$
$$\vdots \qquad\qquad\qquad \vdots$$
$$H_{N1}c_1 + H_{N2}c_2 + \cdots + (H_{NN} - E)c_N = 0 \quad (3.47)$$

(4) For a non-trivial solution to these equations, the determinant of the coefficients of the c_n must vanish

$$\det\left(H_{mn} - E\delta_{mn}\right) \equiv \det\left(\underline{H} - E\underline{I}\right) = 0 \quad (3.48)$$

where we subsequently denote a matrix by underlining the symbol. Several comments:

- The determinant is an Nth-order polynomial in E, and there are N roots to Eq. (3.48). Label them as $E^{(s)}$, with $s = 1, 2, \cdots, N$;
- When $E = E^{(s)}$, one of Eqs. (3.47) will be linearly dependent, say the last one. Discard it, and define the $N - 1$ ratios

$$X_n^{(s)} \equiv \frac{c_n^{(s)}}{c_N^{(s)}} \qquad ; n = 1, 2, \cdots, N - 1 \quad (3.49)$$

[9]Note the adjoint matrix is defined by $H_{nm}^\dagger = \left(H_{nm}^T\right)^\star = H_{mn}^\star$.

[10]The discussion now closely parallels that of normal modes in [Fetter and Walecka (2003a)]. The reader is assumed to be familiar with the basic elements of linear algebra required here.

Now divide the remaining $N - 1$ of Eqs. (3.47) by $c_N^{(s)}$

$$(H_{11} - E^{(s)})X_1^{(s)} + H_{12}X_2^{(s)} + \cdots + H_{1,N-1}X_{N-1}^{(s)} = -H_{1N}$$
$$H_{21}X_1^{(s)} + (H_{22} - E^{(s)})X_2^{(s)} + \cdots + H_{2,N-1}X_{N-1}^{(s)} = -H_{2N}$$
$$\vdots \qquad\qquad \vdots$$
$$H_{N-1,1}X_1^{(s)} + \cdots + (H_{N-1,N-1} - E^{(s)})X_{N-1}^{(s)} = -H_{N-1,N}$$

$$(3.50)$$

This is a set of $N - 1$ linear, *inhomogeneous* algebraic equations for the $N - 1$ ratios $X_n^{(s)}$, with a non-trivial solution.[11]

• We have now obtained the ratios $X_n^{(s)} = c_n^{(s)}/c_N^{(s)}$, but it is the coefficients $c_n^{(s)}$ themselves that are required for the new wave functions in Eq. (3.41). The overall *scale* of the new eigenfunctions, that is the magnitude of $c_N^{(s)}$, is set by the normalization condition

$$\int |u_s(x)|^2 dx = \sum_{n=1}^{N} |c_n^{(s)}|^2 = 1 \qquad (3.51)$$

The overall *phase* of the new eigenfunctions, that is the phase of $c_N^{(s)}$, is again a matter of convention.

• The new eigenfunctions actually satisfy the *orthonormality* relation

$$\int u_s^\star(x)u_t(x)\,dx = \sum_{n=1}^{N} c_n^{(s)\star}c_n^{(t)} = \delta_{st} \qquad (3.52)$$

The orthogonality follows since the matrix H_{mn} is hermitian. To see this, first write out the linear equations with the eigenvalue $E^{(s)}$, and their complex conjugate,

$$\sum_n H_{mn}c_n^{(s)} = E^{(s)}c_m^{(s)}$$
$$\sum_n H_{mn}^\star c_n^{(s)\star} = E^{(s)\star}c_m^{(s)\star} \qquad (3.53)$$

Now carry out $\sum_m c_m^{(s)\star}$ on the first equation and $\sum_m c_m^{(s)}$ on the second, interchange indices $m \leftrightharpoons n$ in the second pair of sums, and

---

[11]It is assumed here that the determinant of the coefficients of the $X_n^{(s)}$ on the l.h.s. of Eqs. (3.50) is non-zero. If it does vanish, this process must be repeated.

subtract

$$\sum_m \sum_n c_m^{(s)\star} \left(H_{mn} - H_{nm}^{\star} \right) c_n^{(s)} = \left(E^{(s)} - E^{(s)\star} \right) \sum_m |c_m^{(s)}|^2$$
$$= 0 \tag{3.54}$$

The final equality follows from the hermiticity of H_{mn}. It is a consequence of these equations that the eigenvalues are *real*

$$E^{(s)} = E^{(s)\star} \qquad \text{; real} \tag{3.55}$$

Similarly, one can write

$$\sum_n H_{mn} c_n^{(s)} = E^{(s)} c_m^{(s)}$$
$$\sum_n H_{mn}^{\star} c_n^{(t)\star} = E^{(t)\star} c_m^{(t)\star} \tag{3.56}$$

Now carry out $\sum_m c_m^{(t)\star}$ on the first equation and $\sum_m c_m^{(s)}$ on the second, interchange indices $m \leftrightarrows n$ in the second sums, and subtract

$$\sum_m \sum_n c_m^{(t)\star} \left(H_{mn} - H_{nm}^{\star} \right) c_n^{(s)} = \left(E^{(s)} - E^{(t)\star} \right) \sum_m c_m^{(t)\star} c_m^{(s)}$$
$$= 0 \tag{3.57}$$

As a result, the eigenvectors $(c_1^{(s)}, c_2^{(s)}, \cdots, c_N^{(s)})$ corresponding to distinct eigenvalues are *orthogonal*[12]

$$\sum_m c_m^{(t)\star} c_m^{(s)} = 0 \qquad ; E^{(s)} \neq E^{(t)} \tag{3.58}$$

The degenerate eigenvectors can again be orthogonalized with the Schmidt procedure.[13] Since one is always free to normalize the solutions to homogeneous equations, one can take the eigenvectors to be *orthonormal*, which is just the second of Eqs. (3.52)

$$\sum_{m=1}^{N} c_m^{(t)\star} c_m^{(s)} = \delta_{st} \tag{3.59}$$

(5) The *modal matrix* is obtained by placing the eigenvectors down its columns

---

[12] Note that now $E^{(t)\star} = E^{(t)}$.

[13] Compare Prob. 3.1.

$$\underline{M} \equiv \begin{bmatrix} \underline{c}^{(1)} & \underline{c}^{(2)} & \cdots & \underline{c}^{(N)} \\ \downarrow & \downarrow & \cdots & \downarrow \\ & & & \end{bmatrix} \qquad ; \text{ modal matrix} \qquad (3.60)$$

In terms of components, it is explicitly given by

$$M_{mn} \equiv c_m^{(n)} \qquad ; \text{ modal matrix} \qquad (3.61)$$

Since the eigenvectors satisfy the orthonormality relation in Eq. (3.59), the modal matrix satisfies

$$\sum_m M_{mn}^\star M_{mp} = \sum_m c_m^{(n)\star} c_m^{(p)} = \delta_{np} \qquad (3.62)$$

This result can be expressed in *matrix notation* as

$$\underline{M}^\dagger \underline{M} = \underline{1} \qquad (3.63)$$

Multiplication by the inverse matrix $\underline{M}^{-1}$ on the right then implies that the modal matrix is *unitary*[14]

$$\underline{M}^\dagger = \underline{M}^{-1} \qquad ; \text{ unitary} \qquad (3.64)$$

(6) It is readily established that the modal matrix diagonalizes the hamiltonian matrix

$$\left(\underline{M}^{-1} \underline{H} \underline{M} \right)_{pq} = \left(\underline{M}^\dagger \underline{H} \underline{M} \right)_{pq}$$
$$= \sum_m \sum_n M_{np}^\star H_{nm} M_{mq}$$
$$= \sum_m \sum_n c_n^{(p)\star} H_{nm} c_m^{(q)} = E^{(q)} \delta_{pq} \qquad (3.65)$$

Here the first of Eqs. (3.56) and Eq. (3.59) have been used to obtain the last equality.[15] In matrix notation this result reads

$$\underline{M}^{-1} \underline{H} \underline{M} = \underline{H}_D = \begin{bmatrix} E^{(1)} & & & \\ & E^{(2)} & & \\ & & \ddots & \\ & & & E^{(N)} \end{bmatrix} \qquad (3.66)$$

Several comments:

---

[14]It is assumed here that $\underline{M}$ is a non-singular matrix.

[15]In detail, $\sum_n c_n^{(p)\star} \left[\sum_m H_{nm} c_m^{(q)} \right] = \sum_n c_n^{(p)\star} \left[E^{(q)} c_n^{(q)} \right] = E^{(q)} \delta_{pq}$.

- The modal matrix is a *unitary matrix that diagonalizes the hamiltonian matrix*;[16]
- The eigenvalues of the diagonalized matrix are the stationary-state energies $E^{(s)}$, with $s = 1, \cdots, N$;
- The modal matrix gives the corresponding eigenfunctions in Eq. (3.41)

$$u_s(x) = \sum_{n=1}^{N} c_n^{(s)} \phi_n(x) = \sum_{n=1}^{N} M_{ns} \phi_n(x) \tag{3.67}$$

The coefficients $c_n^{(s)}$ are just the overlap of the eigenfunctions $u_s(x)$ with the basis functions $\phi_n(x)$

$$c_n^{(s)} = M_{ns} = \int \phi_n^\star(x) u_s(x)\, dx \tag{3.68}$$

- Suppose one now computes the following matrix element of H

$$\int u_t^\star(x) H u_s(x)\, dx = \int \left[\sum_n c_n^{(t)\star} \phi_n^\star(x)\right] H \left[\sum_m c_m^{(s)} \phi_m(x)\right] dx$$

$$= \sum_m \sum_n c_n^{(t)\star} H_{nm} c_m^{(s)}$$

$$= E^{(s)} \delta_{ts} \tag{3.69}$$

where the first of Eqs. (3.56) and Eq. (3.59) have again been used to obtain the final equality.

Furthermore

$$\int u_t^\star(x) u_s(x)\, dx = \sum_m \sum_n c_n^{(t)\star} \delta_{nm} c_m^{(s)} = \delta_{ts} \tag{3.70}$$

We have diagonalized the hamiltonian H, and orthogonalized the solutions $u_E(x)$, as best we can in this finite basis;

- Completeness implies that in the limit $N \to \infty$, the exact wave functions can be obtained through this procedure

$$\mathrm{Lim}_{N\to\infty} \sum_{n=1}^{N} c_n \phi_n(x) = u(x) \qquad ; \text{completeness} \tag{3.71}$$

After this interlude, let us return to our efforts to formalize the Schrödinger equation and quantum mechanics. The familiar eigenstates of *momentum* $\phi_k(x)$ were presented and discussed in Eqs. (3.35)–(3.36). To

---

[16]I always liked to refer to the modal matrix as the "magic matrix".

proceed, we will need the corresponding eigenstates of *position*, which may not be as familiar.

3.5 Eigenstates of Position

The position operator x in one dimension is an hermitian operator. Consider the eigenstates of x with eigenvalues ξ so that

$$x\psi_\xi(x) = \xi\,\psi_\xi(x) \qquad ; \text{ position operator} \qquad (3.72)$$

The solution to this equation, in coordinate space, is just a Dirac delta-function

$$\psi_\xi(x) = \delta(x - \xi) \qquad ; \text{ eigenstates of position} \quad (3.73)$$

It is readily verified that

$$x\psi_\xi(x) = x\delta(x - \xi) = \xi\delta(x - \xi) = \xi\,\psi_\xi(x) \qquad (3.74)$$

On the interval $[0, L]$, with periodic boundary conditions, the eigenvalues ξ run continuously over this interval. As to the orthonormality of these eigenfunctions, one can just compute

$$\int dx\,\psi_{\xi'}^\star(x)\psi_\xi(x) = \int dx\,\delta(x - \xi')\delta(x - \xi) = \delta(\xi - \xi') \qquad (3.75)$$

Hence

$$\int dx\,\psi_{\xi'}^\star(x)\psi_\xi(x) = \delta(\xi - \xi') \quad ; \text{ orthonormality} \qquad (3.76)$$

We make some comments on this result:

- One cannot avoid a continuum normalization here, since the position eigenvalue ξ is truly continuous;
- In contrast, in one dimension with periodic boundary conditions on this interval, the eigenfunctions of momentum in Eq. (3.36) have a *denumerably infinite set of discrete eigenvalues*. This proved to be an essential calculational tool in Vol. I;

- To make the analogy between coordinate space and momentum space closer, one can take L to infinity.[17] Define

$$\psi_k(x) = \left(\frac{L}{2\pi}\right)^{1/2} \phi_k(x) = \frac{1}{\sqrt{2\pi}} e^{ikx} \tag{3.77}$$

Then

$$\int dx\, \psi_{k'}^\star(x)\psi_k(x) = \frac{1}{2\pi} \int dx\, e^{i(k-k')x}$$

$$\to \delta(k - k') \qquad\qquad ; L \to \infty \tag{3.78}$$

In this limit *both* the momentum and position eigenfunctions have a continuum norm.

3.6   Abstract Hilbert Space

Recall the previous Eqs. (3.12) and (3.20), which represent an expansion in a complete set,

$$\psi(x) = \sum_n c_n \phi_n(x)$$

$$\text{or ;} \qquad \psi_x = \sum_n c_n [\phi_n]_x \tag{3.79}$$

This can be viewed as the component form of the abstract vector relation

$$|\psi\rangle = \sum_n c_n |\phi_n\rangle \qquad ; \text{ abstract vector relation} \tag{3.80}$$

Just as an ordinary three-dimensional vector $\mathbf{v}$ has a meaning independent of the basis vectors in which it is being decomposed, one can think of this as a vector pointing in some direction in the abstract, infinite-dimensional Hilbert space. Equations (3.79) then provide a component form of this abstract vector relation.

3.6.1   *Inner Product*

The inner product in this abstract space is provided by Eq. (3.19)

$$\langle \psi_a | \psi_b \rangle = \sum_x [\psi_a]_x^\star [\psi_b]_x \equiv \int dx\, \psi_a^\star(x)\psi_b(x) \qquad ; \text{ inner product} \tag{3.81}$$

---

[17] As shown in Vol. I, Fourier series are converted to Fourier integrals in this limit; one first uses the periodic boundary conditions to convert the interval to $[-L/2, L/2]$.

Thus, from Eqs. (3.79)

$$c_n = \langle \phi_n | \psi \rangle = \sum_x [\phi_n]_x^\star \psi_x \equiv \int dx\, \phi_n^\star(x) \psi(x) \qquad (3.82)$$

We note the following important inner products:

$$\langle \xi' | \xi \rangle = \int dx\, \psi_{\xi'}^\star(x) \psi_\xi(x) = \delta(\xi' - \xi)$$

$$\langle k' | k \rangle = \int dx\, \phi_{k'}^\star(x) \phi_k(x) = \delta_{kk'} \qquad ; \text{ with p.b.c.}$$

$$\langle \xi | k \rangle = \int dx\, \psi_\xi^\star(x) \phi_k(x) = \frac{1}{\sqrt{L}} e^{ik\xi} \qquad (3.83)$$

The last relation follows directly from the wave functions in Eqs. (3.36) and (3.73).[18]

3.6.2 *Completeness*

The statement of completeness with the set of coordinate space eigenfunctions $\phi_p(x)$, where p denotes the eigenvalues of a linear hermitian operator, is[19]

$$\sum_p \phi_p(x) \phi_p^\star(y) = \delta(x - y) \qquad ; \text{ completeness} \qquad (3.84)$$

Insert this relation in the definition of the inner product in Eq. (3.81)

$$\begin{aligned}
\langle \psi_a | \psi_b \rangle &= \int dx\, \psi_a^\star(x) \psi_b(x) \equiv \int dx\,dy\, \psi_a^\star(x) \delta(x - y) \psi_b(y) \\
&= \sum_p \int dx\, \psi_a^\star(x) \phi_p(x) \int dy\, \phi_p^\star(y) \psi_b(y) \\
&= \sum_p \langle \psi_a | \phi_p \rangle \langle \phi_p | \psi_b \rangle \qquad\qquad (3.85)
\end{aligned}$$

Here Eq. (3.84) has been used in the second line, and the definition of the inner product used in the third. This relation can be summarized by writing the abstract vector relation

$$\sum_p | \phi_p \rangle \langle \phi_p | = 1_{\text{op}} \qquad ; \text{ completeness} \qquad (3.86)$$

---

[18]See also Eq. (3.9); note that the subscript n on $k_n = 2\pi n/L$ is suppressed here.
[19]See Eq. (2.21) and Prob. 2.7.

This unit operator 1_{op} can be inserted into any inner product, leaving that inner product unchanged. This relation follows from the completeness of the wave functions $\phi_p(x)$ providing the coordinate space components of the abstract state vectors $|\phi_p\rangle$.

3.6.3 *Linear Hermitian Operators*

A linear hermitian operator L_{op} takes one abstract vector $|\psi\rangle$ into another $L_{op}|\psi\rangle$. The eigenstates of L_{op}, as before, are defined by

$$L_{op}|\phi_\lambda\rangle = \lambda|\phi_\lambda\rangle \qquad\qquad ; \text{ eigenstates} \quad (3.87)$$

For example:

$$
\begin{aligned}
p_{op}|k\rangle &= \hbar k|k\rangle &&; \text{ momentum} \\
x_{op}|\xi\rangle &= \xi|\xi\rangle &&; \text{ position} \\
(L_z)_{op}|m\rangle &= m|m\rangle &&; \text{ z-component of angular momentum} \\
H_{op}|\psi\rangle &= E|\psi\rangle &&; \text{ hamiltonian} \qquad ; \textit{ etc.} \qquad (3.88)
\end{aligned}
$$

In coordinate space, the *adjoint operator* $L^\dagger$ is defined by

$$\int d\xi\, \psi_a^\star(\xi)L^\dagger\psi_b(\xi) \equiv \int d\xi\, [L\psi_a(\xi)]^\star\psi_b(\xi) = \left[\int d\xi\, \psi_b^\star(\xi)L\psi_a(\xi)\right]^\star \quad (3.89)$$

The adjoint operator in the abstract Hilbert space is defined in exactly the same manner

$$\langle\psi_a|L_{op}^\dagger|\psi_b\rangle \equiv \langle L_{op}\psi_a|\psi_b\rangle = \langle\psi_b|L_{op}|\psi_a\rangle^\star \qquad ; \text{ adjoint} \quad (3.90)$$

Note that it follows from this definition that if γ is some complex number, then

$$[\gamma L_{op}]^\dagger = \gamma^\star L_{op}^\dagger \qquad\qquad (3.91)$$

An operator is *hermitian* if it is equal to its adjoint

$$L_{op}^\dagger = L_{op} \qquad\qquad ; \text{ hermitian}$$

$$\implies \quad \langle\psi_a|L_{op}|\psi_b\rangle = \langle L_{op}\psi_a|\psi_b\rangle = \langle\psi_b|L_{op}|\psi_a\rangle^\star \qquad (3.92)$$

With an hermitian operator, one can just let it act on the state on the left when calculating matrix elements.

3.6.4 *Schrödinger Equation*

We have previously formulated quantum mechanics in coordinate space, where the momentum p is given by $p = (\hbar/i)\partial/\partial x$. It was observed in Prob. 4.8 of Vol. I that one could equally well work in momentum space, where the position x is given by $x = i\hbar\partial/\partial p$. It was also observed there that the commutation relation $[p, x] = \hbar/i$ is independent of the particular representation. Our goal in this section is to *abstract the Schrödinger equation* and similarly free it from any particular component representation.

We claim the following:

> To get the time-independent Schrödinger equation in the coordinate representation, one projects the abstract operator relation $H_{op}|\psi\rangle = E|\psi\rangle$ onto the basis of eigenstates of position $|\xi\rangle$.

We show this through the following set of steps:

(1) First project $|\psi\rangle$ onto an eigenstate of position $|\xi\rangle$

$$\langle\xi|\psi\rangle = \sum_x [\psi_\xi]_x^\star [\psi]_x = \int dx\, \psi_\xi^*(x)\psi(x) = \int dx\, \delta(\xi - x)\psi(x)$$

$$\langle\xi|\psi\rangle = \psi(\xi) \qquad\qquad ; \text{wave function} \qquad\qquad (3.93)$$

This is simply the familiar coordinate space wave function $\psi(\xi)$;

(2) Compute the matrix element of the potential $V_{op} = V(x_{op})$ between eigenstates of position

$$\langle\xi|V_{op}|\xi'\rangle = \langle\xi|V(x_{op})|\xi'\rangle = V(\xi')\langle\xi|\xi'\rangle = V(\xi)\delta(\xi - \xi') \quad (3.94)$$

(3) Now compute the matrix elements of the kinetic energy T_{op}. This is readily accomplished by invoking the completeness relation for the eigenstates of momentum [see Eq. (3.86)]

$$\sum_k |k\rangle\langle k| = 1_{op} \qquad ; \text{completeness} \qquad\qquad (3.95)$$

With the insertion of this relation (twice), one finds

$$\langle\xi|T_{op}|\xi'\rangle = \frac{1}{2m}\langle\xi|p_{op}^2|\xi'\rangle = \frac{1}{2m}\sum_k\sum_{k'}\langle\xi|k\rangle\langle k|p_{op}^2|k'\rangle\langle k'|\xi'\rangle$$

$$= \frac{\hbar^2}{2m}\sum_k\sum_{k'}\langle\xi|k\rangle k^2\delta_{kk'}\langle k'|\xi'\rangle = \frac{\hbar^2}{2m}\sum_k\frac{k^2}{L}e^{ik(\xi-\xi')}$$

$$= -\frac{\hbar^2}{2m}\frac{\partial^2}{\partial\xi^2}\sum_k\frac{1}{L}e^{ik(\xi-\xi')} = -\frac{\hbar^2}{2m}\frac{\partial^2}{\partial\xi^2}\delta(\xi-\xi') \quad (3.96)$$

The final relation follows from the completeness of the momentum wave functions.

(4) Make use of the statement of completeness of the abstract eigenstates of position, which is

$$\int d\xi\,|\xi\rangle\langle\xi| = 1_{op} \qquad ; \text{completeness} \quad (3.97)$$

Note that the sum here is actually an integral because the position eigenvalues are continuous.[20]

(5) The operator form of the time-independent Schrödinger equation is

$$H_{op}|\psi\rangle = (T_{op} + V_{op})|\psi\rangle = E|\psi\rangle \qquad ; \text{S-equation} \quad (3.98)$$

A projection of this equation on the eigenstates of position gives

$$\langle\xi|H_{op}|\psi\rangle = E\langle\xi|\psi\rangle = E\psi(\xi) \qquad (3.99)$$

Now insert Eq. (3.97) in the expression on the l.h.s., and use the results from Eqs. (3.94) and (3.96)

$$\langle\xi|H_{op}|\psi\rangle = \int d\xi'\,\langle\xi|H_{op}|\xi'\rangle\langle\xi'|\psi\rangle$$

$$= \int d\xi'\left[-\frac{\hbar^2}{2m}\frac{\partial^2}{\partial\xi^2} + V(\xi)\right]\delta(\xi-\xi')\psi(\xi')$$

$$= \left[-\frac{\hbar^2}{2m}\frac{\partial^2}{\partial\xi^2} + V(\xi)\right]\int d\xi'\,\delta(\xi-\xi')\psi(\xi')$$

$$= \left[-\frac{\hbar^2}{2m}\frac{\partial^2}{\partial\xi^2} + V(\xi)\right]\psi(\xi) \qquad (3.100)$$

---

[20]See Prob. 3.2.

Thus, in summary,

$$\left[-\frac{\hbar^2}{2m}\frac{\partial^2}{\partial \xi^2} + V(\xi)\right]\psi(\xi) = E\,\psi(\xi) \qquad ; \text{S-equation} \qquad (3.101)$$

This is just the time-independent Schrödinger equation in the coordinate representation. It is the component form of the operator relation of Eq. (3.98) in a basis of eigenstates of position.

As a function of time, the state vector $|\Psi(t)\rangle$ varies in the abstract Hilbert space, and the *time-dependent* Schrödinger equation is imposed. Thus we arrive at

$$|\Psi(t)\rangle \qquad\qquad ; \text{time-dependent state vector}$$

$$i\hbar\frac{\partial}{\partial t}|\Psi(t)\rangle = \hat{H}|\Psi(t)\rangle \qquad ; \text{Schrödinger equation}$$

$$[\hat{p}_i, \hat{x}_j] = \frac{\hbar}{i}\delta_{ij} \qquad\qquad ; \text{C.C.R.} \qquad (3.102)$$

Several comments:

- From now on we will use a caret over a symbol to denote an operator in the abstract Hilbert space, for example $H_{\text{op}} \equiv \hat{H}$, etc.;
- The last of Eqs. (3.102) is just the abstract form of the canonical commutation relations of Eqs. (2.3);
- The hamiltonian here is

$$\hat{H} = \frac{\hat{\mathbf{p}}^2}{2m} + \hat{V}(\hat{\mathbf{x}}) \qquad ; \text{hamiltonian} \qquad (3.103)$$

If the potential $\hat{V}(\hat{\mathbf{x}})$ has no explicit time dependence, then, as before, one can look for *stationary-state* solutions to the time-dependent Schrödinger equation of the form

$$|\Psi(t)\rangle = e^{-iEt/\hbar}|\psi\rangle \qquad ; \text{stationary-states} \qquad (3.104)$$

Substitution into the second of Eqs. (3.102) then gives

$$e^{-iEt/\hbar}E|\psi\rangle = \hat{H}e^{-iEt/\hbar}|\psi\rangle = e^{-iEt/\hbar}\hat{H}|\psi\rangle \qquad (3.105)$$

where the last equality holds since E, the energy, is simply some real number. Hence

$$\hat{H}|\psi\rangle = E|\psi\rangle \qquad ; \text{stationary-states} \qquad (3.106)$$

and we recover Eq. (3.98);

- Equations (3.102) are now *independent of any particular representation.*

We proceed to show that the Schrödinger equation in various *representations* is obtained by projecting the abstract form in Eqs. (3.102) and (3.106) onto the appropriate basis.

3.7 Representations

It is first useful to *summarize* the previous results on the eigenstates of momentum and position in the current notation. In the abstract Hilbert space, with three spatial dimensions and periodic boundary conditions,

- *Eigenstates of Momentum*

$$\hat{\mathbf{p}}|\mathbf{k}\rangle = \hbar\mathbf{k}|\mathbf{k}\rangle \qquad ; \; k_i = \frac{2\pi n_i}{L} \qquad ; \; i = 1, 2, 3$$
$$; \; n_i = 0, \pm 1, \pm 2, \cdots$$

$$\langle \mathbf{k}'|\mathbf{k}\rangle = \delta_{\mathbf{k}',\mathbf{k}} \qquad ; \text{ Kronecker delta}$$

$$\sum_{\mathbf{k}} |\mathbf{k}\rangle\langle\mathbf{k}| = \hat{1} \qquad ; \text{ unit operator} \qquad (3.107)$$

- *Eigenstates of Position*

$$\hat{\mathbf{x}}|\boldsymbol{\xi}\rangle = \boldsymbol{\xi}|\boldsymbol{\xi}\rangle \qquad ; \; \xi_i \text{ continuous on interval } [0, L]$$

$$\langle \boldsymbol{\xi}'|\boldsymbol{\xi}\rangle = \delta^{(3)}(\boldsymbol{\xi}' - \boldsymbol{\xi}) \qquad ; \text{ Dirac delta-function}$$

$$\int d^3\xi \, |\boldsymbol{\xi}\rangle\langle\boldsymbol{\xi}| = \hat{1} \qquad ; \text{ unit operator} \qquad (3.108)$$

- *Inner Product*

$$\langle \boldsymbol{\xi}|\mathbf{k}\rangle = \langle\mathbf{k}|\boldsymbol{\xi}\rangle^\star = \frac{1}{\sqrt{L^3}}\exp\left(i\mathbf{k}\cdot\boldsymbol{\xi}\right) \qquad (3.109)$$

Here we focus on the stationary states. The extension to the time-dependent behavior is then immediate.

3.7.1 *Coordinate Representation*

The matrix element of the hamiltonian between eigenstates of position was calculated in Eqs. (3.94) and (3.96). In three dimensions this reads

$$\langle \boldsymbol{\xi}|\hat{H}|\boldsymbol{\xi}'\rangle = \left[-\frac{\hbar^2}{2m}\nabla_\xi^2 + V(\boldsymbol{\xi})\right]\delta^{(3)}(\boldsymbol{\xi} - \boldsymbol{\xi}') \qquad (3.110)$$

Now take the projection of Eq. (3.106) on the eigenstates of position $|\boldsymbol{\xi}\rangle$, and use completeness,

$$\langle\boldsymbol{\xi}|\hat{H}|\psi\rangle = \int d^3\xi' \langle\boldsymbol{\xi}|\hat{H}|\boldsymbol{\xi}'\rangle\langle\boldsymbol{\xi}'|\psi\rangle = E\langle\boldsymbol{\xi}|\psi\rangle \qquad (3.111)$$

As before, $\langle\boldsymbol{\xi}|\psi\rangle$ is the coordinate-space wave function

$$\langle\boldsymbol{\xi}|\psi\rangle = \psi(\boldsymbol{\xi}) \qquad ; \text{ coordinate-space wave function} \quad (3.112)$$

Thus, we again obtain the familiar Schrödinger equation in the coordinate representation

$$\left[-\frac{\hbar^2}{2m}\nabla_\xi^2 + V(\boldsymbol{\xi})\right]\psi(\boldsymbol{\xi}) = E\psi(\boldsymbol{\xi}) \qquad ; \text{ S-eqn in coordinate rep} \quad (3.113)$$

3.7.2 Momentum Representation

Suppose, instead, we take the projection of Eq. (3.106) on the eigenstates of momentum $|\mathbf{k}\rangle$. The momentum-space wave function $\langle\mathbf{k}|\psi\rangle$ is defined as

$$\langle\mathbf{k}|\psi\rangle \equiv A(\mathbf{k}) \qquad ; \text{ momentum-space wave function} \quad (3.114)$$

The use of completeness expresses this as

$$A(\mathbf{k}) = \int d^3\xi\, \langle\mathbf{k}|\boldsymbol{\xi}\rangle\langle\boldsymbol{\xi}|\psi\rangle = \frac{1}{L^{3/2}}\int d^3\xi\, e^{-i\mathbf{k}\cdot\boldsymbol{\xi}}\,\psi(\boldsymbol{\xi}) \qquad (3.115)$$

This is just the *Fourier transform* of the coordinate-space wave function.

To proceed, we require the matrix elements of the kinetic and potential energies between eigenstates of momentum.

(1) The matrix element of $\hat{T}$ is obtained immediately as

$$\langle\mathbf{k}|\hat{T}|\mathbf{k}'\rangle = \frac{1}{2m}\langle\mathbf{k}|\hat{\mathbf{p}}^2|\mathbf{k}'\rangle = \frac{\hbar^2\mathbf{k}^2}{2m}\delta_{\mathbf{k},\mathbf{k}'} \qquad (3.116)$$

(2) The matrix element of $\hat{V}$ is obtained by using completeness (twice)

$$\begin{aligned}
\langle\mathbf{k}|\hat{V}|\mathbf{k}'\rangle &= \int d^3\xi \int d^3\xi'\, \langle\mathbf{k}|\boldsymbol{\xi}\rangle\langle\boldsymbol{\xi}|\hat{V}|\boldsymbol{\xi}'\rangle\langle\boldsymbol{\xi}'|\mathbf{k}'\rangle \\
&= \frac{1}{L^3}\int d^3\xi\, e^{-i(\mathbf{k}-\mathbf{k}')\cdot\boldsymbol{\xi}}\,V(\boldsymbol{\xi}) \\
&\equiv \frac{1}{L^3}\tilde{V}(\mathbf{k}-\mathbf{k}') \qquad (3.117)
\end{aligned}$$

Here $\tilde{V}(\mathbf{k}-\mathbf{k}')$ is the Fourier transform of the potential.

A combination of these two results gives the appropriate matrix element of the hamiltonian

$$\langle \mathbf{k}|\hat{H}|\mathbf{k}'\rangle = \frac{\hbar^2 \mathbf{k}^2}{2m}\delta_{\mathbf{k},\mathbf{k}'} + \frac{1}{L^3}\tilde{V}(\mathbf{k}-\mathbf{k}') \qquad (3.118)$$

Note that the kinetic energy is diagonal in the momentum representation, while the potential couples states of different momentum.

The Schrödinger equation in the momentum representation is now obtained by taking the projection of Eq. (3.106) on the eigenstates of momentum $|\mathbf{k}\rangle$, and again using completeness,

$$\langle \mathbf{k}|\hat{H}|\psi\rangle = \sum_{\mathbf{k}'}\langle \mathbf{k}|\hat{H}|\mathbf{k}'\rangle\langle \mathbf{k}'|\psi\rangle = E\langle \mathbf{k}|\psi\rangle \qquad (3.119)$$

Thus

$$\sum_{\mathbf{k}'}\left[\frac{\hbar^2 \mathbf{k}^2}{2m}\delta_{\mathbf{k},\mathbf{k}'} + \frac{1}{L^3}\tilde{V}(\mathbf{k}-\mathbf{k}')\right]A(\mathbf{k}') = EA(\mathbf{k}) \qquad (3.120)$$

This is re-written as

$$\left(E - \frac{\hbar^2 \mathbf{k}^2}{2m}\right)A(\mathbf{k}) = \frac{1}{L^3}\sum_{\mathbf{k}'}\tilde{V}(\mathbf{k}-\mathbf{k}')A(\mathbf{k}') \quad ; \text{ S-eqn in momentum rep}$$

$$\rightarrow \frac{1}{(2\pi)^3}\int d^3k'\,\tilde{V}(\mathbf{k}-\mathbf{k}')A(\mathbf{k}') \qquad ; L \rightarrow \infty \quad (3.121)$$

The second line converts the sum to an integral in the large-volume limit.[21]

Some comments:

- This is the Schrödinger equation in the *momentum representation*;
- It is an *integral equation* for $A(\mathbf{k})$;
- Although the Schrödinger equation now appears in a completely different guise, it describes the *same physics*;
- The Schrödinger equation in the momentum representation is just the Fourier transform of the Schrödinger equation in the coordinate representation.[22]

---

[21] In a big box with p.b.c., $\sum_{\mathbf{k}} \rightarrow [L^3/(2\pi)^3]\int d^3k$ as $L \rightarrow \infty$ (see Vol. I).
[22] See Prob. 3.3.

3.8 Operator Methods

The following is one of the most useful operator identities in quantum mechanics

$$e^{i\hat{A}}\hat{B}e^{-i\hat{A}} \equiv \hat{B} + i[\hat{A},\hat{B}] + \frac{i^2}{2!}[\hat{A},[\hat{A},\hat{B}]] + \frac{i^3}{3!}[\hat{A},[\hat{A},[\hat{A},\hat{B}]]] + \cdots$$

; operator identity   (3.122)

This relates the expression on the l.h.s. to *repeated commutators*, which can be evaluated in a manner independent of representation through the use of the canonical commutation relations. We will be content here to establish this identity by evaluating the first few terms, just to see how it goes.[23] The l.h.s. is

$$\left(1 + i\hat{A} + \frac{i^2}{2!}\hat{A}^2 + \cdots\right)\hat{B}\left(1 - i\hat{A} + \frac{(-i)^2}{2!}\hat{A}^2 + \cdots\right)$$

$$\equiv \hat{B} + i\left(\hat{A}\hat{B} - \hat{B}\hat{A}\right) + \frac{i^2}{2!}\left[\hat{A}\left(\hat{A}\hat{B} - \hat{B}\hat{A}\right) - \left(\hat{A}\hat{B} - \hat{B}\hat{A}\right)\hat{A}\right] + \cdots$$

$$= \hat{B} + i[\hat{A},\hat{B}] + \frac{i^2}{2!}[\hat{A},[\hat{A},\hat{B}]] + \cdots \qquad (3.123)$$

3.8.1 *Translation Operator*

As an example, consider the *translation operator* in one dimension

$$\hat{U}(a) = \exp\left(-\frac{i}{\hbar}\hat{p}a\right) \qquad ; \text{ translation operator}$$

a is real number   (3.124)

Here a is some real number. Since $\hat{p}$ is hermitian, this is a *unitary* operator

$$\hat{U}(a)^\dagger = \hat{U}(a)^{-1} \qquad ; \text{ unitary} \qquad (3.125)$$

It follows from Eq. (3.122) that

$$e^{i\hat{p}a/\hbar}\,\hat{x}\,e^{-i\hat{p}a/\hbar} = \hat{x} + \left(\frac{ia}{\hbar}\right)[\hat{p},\hat{x}] + \frac{1}{2!}\left(\frac{ia}{\hbar}\right)^2[\hat{p},[\hat{p},\hat{x}]] + \cdots$$

$$= \hat{x} + a \qquad (3.126)$$

The term linear in a follows from the canonical commutation relation

$$[\hat{p},\hat{x}] = \frac{\hbar}{i} \qquad (3.127)$$

---

[23]For the general proof, see Prob. 3.6.

and since this expression is a c-number, all the remaining commutators vanish.[24] Hence

$$\hat{U}(a)^{-1}\,\hat{x}\,\hat{U}(a) = \hat{x} + a \qquad ; \text{ translation operator} \qquad (3.128)$$

Now let $|\xi\rangle$ be an eigenstate of position

$$\hat{x}|\xi\rangle = \xi|\xi\rangle \qquad ; \text{ eigenstate of position} \qquad (3.129)$$

We claim that the operator $\hat{U}(a)$ translates the eigenvalue by a distance a

$$\hat{U}(a)|\xi\rangle = |\xi + a\rangle \qquad ; \text{ translation operator} \qquad (3.130)$$

To see this, re-write Eq. (3.128) as

$$\hat{x}e^{-i\hat{p}a/\hbar} = e^{-i\hat{p}a/\hbar}\,(\hat{x} + a) \qquad (3.131)$$

Apply this expression to the eigenstate $|\xi\rangle$

$$\begin{aligned}
\hat{x}\left[e^{-i\hat{p}a/\hbar}|\xi\rangle\right] &= e^{-i\hat{p}a/\hbar}\left[(\hat{x} + a)|\xi\rangle\right] \\
&= (\xi + a)\left[e^{-i\hat{p}a/\hbar}|\xi\rangle\right] \qquad (3.132)
\end{aligned}$$

This establishes the result in Eq. (3.130).[25]

3.8.2   *Inner Product $\langle\xi|k\rangle$*

The following inner product can be established using operator properties alone

$$\langle\boldsymbol{\xi}|\mathbf{k}\rangle = \frac{1}{\sqrt{L^3}}\,\exp\,(i\mathbf{k}\cdot\boldsymbol{\xi}) \qquad (3.133)$$

To see this, consider the following matrix element

$$\begin{aligned}
\langle\mathbf{k}|e^{-i\hat{\mathbf{p}}\cdot\mathbf{a}/\hbar}|\boldsymbol{\xi}\rangle &= \langle\mathbf{k}|\boldsymbol{\xi} + \mathbf{a}\rangle \\
&= e^{-i\mathbf{k}\cdot\mathbf{a}}\langle\mathbf{k}|\boldsymbol{\xi}\rangle \qquad (3.134)
\end{aligned}$$

The first equality follows from the three-dimensional form of Eq. (3.130), and the second by letting the hermitian operator $\hat{\mathbf{p}}$ act on the eigenstate on the left. Now let $\boldsymbol{\xi} \to 0$, and the last equality in Eq. (3.134) becomes

$$\langle\mathbf{k}|\mathbf{a}\rangle = e^{-i\mathbf{k}\cdot\mathbf{a}}\langle\mathbf{k}|0\rangle \qquad (3.135)$$

This expression can be analyzed as follows:

---

[24]A "c-number" is a *classical number* (as opposed to a quantum operator).

[25]Note the norm is preserved since $\langle\xi + a|\xi + a\rangle = \langle U\xi|\hat{U}|\xi\rangle = \langle\xi|\hat{U}^\dagger\hat{U}|\xi\rangle = \langle\xi|\xi\rangle$.

- Observe that this relation is true for any **a**;
- Re-label **a** → **ξ**;
- Recall $\langle \boldsymbol{\xi} | \mathbf{k} \rangle = \langle \mathbf{k} | \boldsymbol{\xi} \rangle^\star$;
- Check the choice of normalization $\langle 0 | \mathbf{k} \rangle \equiv L^{-3/2}$ in Eq. (3.133) using completeness

$$\langle \mathbf{k}' | \mathbf{k} \rangle = \int d^3 \xi \, \langle \mathbf{k}' | \boldsymbol{\xi} \rangle \langle \boldsymbol{\xi} | \mathbf{k} \rangle$$

$$= \frac{1}{L^3} \int_{\text{Box}} e^{i(\mathbf{k}-\mathbf{k}')\cdot \boldsymbol{\xi}} = \delta_{\mathbf{k},\mathbf{k}'} \tag{3.136}$$

This establishes the result in Eq. (3.133).

3.8.3   *Simple Harmonic Oscillator*

In operator form, the hamiltonian and canonical commutation relation for the one-dimensional simple harmonic oscillator are

$$\hat{H} = \frac{1}{2m}\hat{p}^2 + \frac{1}{2}k\hat{x}^2 \qquad ; \text{hamiltonian}$$

$$[\hat{p}, \hat{x}] = \frac{\hbar}{i} \qquad ; \text{C.C.R.} \tag{3.137}$$

Define the *creation and destruction operators* by[26]

$$\hat{a} \equiv \left(\frac{m\omega_0}{2\hbar}\right)^{1/2} \hat{x} + \frac{i}{(2m\omega_0\hbar)^{1/2}}\,\hat{p} \qquad ; \text{destruction operator}$$

$$\hat{a}^\dagger = \left(\frac{m\omega_0}{2\hbar}\right)^{1/2} \hat{x} - \frac{i}{(2m\omega_0\hbar)^{1/2}}\,\hat{p} \qquad ; \text{creation operator} \tag{3.138}$$

It follows from Eqs. (3.137) that

$$[\hat{a}, \hat{a}^\dagger] = 1$$

$$\hat{H} = \hbar\omega_0 \left(\hat{a}^\dagger \hat{a} + \frac{1}{2}\right) \tag{3.139}$$

The *number operator* is defined by

$$\hat{N} \equiv \hat{a}^\dagger \hat{a} \qquad ; \text{number operator} \tag{3.140}$$

It is *hermitian*[27]

$$\hat{N}^\dagger = \hat{N} \qquad ; \text{hermitian} \tag{3.141}$$

---

[26] Recall Prob. 2.10; remember that both $(\hat{p}, \hat{x})$ are hermitian operators, and $k \equiv m\omega_0^2$.
[27] Recall Prob. 2.11.

This has the following immediate consequences:

- The eigenvalues are *real*

$$\hat{N}|n\rangle = n|n\rangle \qquad ; \text{ eigenstates}$$
$$n \text{ real} \qquad\qquad (3.142)$$

- The eigenstates corresponding to different eigenvalues are *orthogonal*, and since the states are to be normalized

$$\langle n'|n\rangle = \delta_{n,n'} \qquad ; \text{ orthonormal} \qquad (3.143)$$

From our general principles of quantum mechanics, the eigenstates also form a *complete set*

$$\sum_n |n\rangle\langle n| = \hat{1} \qquad ; \text{ complete} \qquad (3.144)$$

The *spectrum* of the number operator is then obtained through the following series of arguments:

(1) Consider the following matrix element of $\hat{N}$, and use completeness,

$$\langle n|\hat{N}|n\rangle = \langle n|\hat{a}^\dagger\hat{a}|n\rangle$$
$$= \sum_m \langle n|\hat{a}^\dagger|m\rangle\langle m|\hat{a}|n\rangle = \sum_m \langle m|\hat{a}|n\rangle^*\langle m|\hat{a}|n\rangle \quad (3.145)$$

The last equality follows from the definition of the adjoint. Thus[28]

$$n = \langle n|\hat{N}|n\rangle = \sum_m |\langle m|\hat{a}|n\rangle|^2 \geq 0 \qquad (3.146)$$

(2) The following commutators follow directly from the first of Eqs. (3.139)

$$[\hat{a}^\dagger\hat{a}, \hat{a}] = \hat{a}^\dagger\hat{a}\hat{a} - \hat{a}\hat{a}^\dagger\hat{a} = -[\hat{a}, \hat{a}^\dagger]\hat{a} = -\hat{a}$$
$$[\hat{a}^\dagger\hat{a}, \hat{a}^\dagger] = \hat{a}^\dagger\hat{a}\hat{a}^\dagger - \hat{a}^\dagger\hat{a}^\dagger\hat{a} = \hat{a}^\dagger[\hat{a}, \hat{a}^\dagger] = \hat{a}^\dagger \qquad (3.147)$$

Thus

$$[\hat{N}, \hat{a}] = -\hat{a}$$
$$[\hat{N}, \hat{a}^\dagger] = \hat{a}^\dagger \qquad (3.148)$$

---

[28] Again, we are assuming that the operators are *hermitian* and that the states are *normalizable* (see our previous discussion of solutions to the Schrödinger equation).

These are re-written as

$$\hat{N}\hat{a} = \hat{a}(\hat{N} - 1)$$
$$\hat{N}\hat{a}^\dagger = \hat{a}^\dagger(\hat{N} + 1) \tag{3.149}$$

(3) Now let $\hat{N}$ act on the state $\hat{a}|n\rangle$, use the first of Eqs. (3.149), and replace the operator by its eigenvalue

$$\hat{N}\hat{a}|n\rangle = \hat{a}(\hat{N} - 1)|n\rangle = (n - 1)\hat{a}|n\rangle \tag{3.150}$$

Hence $\hat{a}|n\rangle$ is again an eigenstate of $\hat{N}$ with eigenvalue $n - 1$. Thus

- $\hat{a}$ *lowers* the eigenvalue by 1;
- Eventually, one must arrive at

$$\hat{a}|0\rangle = 0 \qquad ; \text{ ground state} \tag{3.151}$$

If this condition is *not* satisfied, then the lowering operation will produce a state that violates the positivity condition $n \geq 0$ in Eq. (3.146);

- The lowest eigenvalue is thus $n = 0$, since

$$\hat{N}|0\rangle = \hat{a}^\dagger\hat{a}|0\rangle = 0 \tag{3.152}$$

(4) Let $\hat{N}$ act on the state $\hat{a}^\dagger|n\rangle$, use the second of Eqs. (3.149), and replace the operator by it eigenvalue

$$\hat{N}\hat{a}^\dagger|n\rangle = \hat{a}^\dagger(\hat{N} + 1)|n\rangle = (n + 1)\hat{a}^\dagger|n\rangle \tag{3.153}$$

Hence $\hat{a}^\dagger|n\rangle$ is again an eigenstate of $\hat{N}$ with eigenvalue $n + 1$.

- Thus $\hat{a}^\dagger$ *raises* the eigenvalue by 1;
- The spectrum of the number operator $\hat{N}$ is thus obtained as the positive integers and zero

$$\hat{N}|n\rangle = n|n\rangle \qquad ; n = 0, 1, 2, 3, \cdots \tag{3.154}$$

(5) Based on the above arguments, Eq. (3.146) actually takes the form

$$n = |\langle n - 1|\hat{a}|n\rangle|^2 \tag{3.155}$$

With an appropriate choice of the *relative phases of the states*, the square root of this relation gives

$$\langle n - 1|\hat{a}|n\rangle = \sqrt{n} \tag{3.156}$$

Now take the complex conjugate of this result, and let $n \to n+1$. This gives[29]

$$\langle n+1|\hat{a}^{\dagger}|n\rangle = \sqrt{n+1} \qquad (3.157)$$

(6) The oscillator hamiltonian is expressed in terms of the number operator according to Eq. (3.139)

$$\hat{H} = \hbar\omega_0\left(\hat{N} + \frac{1}{2}\right) \qquad (3.158)$$

The energy eigenvalue spectrum follows immediately

$$\hat{H}|n\rangle = E_n|n\rangle$$
$$E_n = \hbar\omega_0\left(n + \frac{1}{2}\right) \qquad ; n = 0,1,2,\cdots,\infty \quad (3.159)$$

(7) To go to the coordinate representation, use[30]

$$\langle x|\hat{H}|x'\rangle = \left(-\frac{\hbar^2}{2m}\frac{d^2}{dx^2} + \frac{1}{2}m\omega_0^2 x^2\right)\delta(x-x') \qquad (3.160)$$

$$\langle x|n\rangle \equiv \psi_n(x) = C_n h_n(\xi)e^{-\xi^2/2} \qquad ; \xi \equiv \left(\frac{m\omega_0}{\hbar}\right)^{1/2}x$$

(8) Equations (3.138) can be inverted to give

$$\hat{x} = \left(\frac{\hbar}{2m\omega_0}\right)^{1/2}(\hat{a} + \hat{a}^{\dagger})$$

$$\hat{p} = -im\omega_0\left(\frac{\hbar}{2m\omega_0}\right)^{1/2}(\hat{a} - \hat{a}^{\dagger}) \qquad (3.161)$$

It follows from Eqs. (3.158) and (3.148) that

$$\frac{\hat{p}}{m} = \frac{i}{\hbar}[\hat{H}, \hat{x}] \qquad (3.162)$$

The *time dependence* of the observables described by these operators arises from their expectation value in the state $|\Psi(t)\rangle$. The Schrödinger

---

[29]In detail, $\langle n-1|\hat{a}|n\rangle^{\star} = \langle n|\hat{a}^{\dagger}|n-1\rangle = \sqrt{n}$.

[30]Here the eigenvalues are labeled by (x, x'). One can use the differential equation in coordinate space to verify that no eigenstates have been missed by the formal arguments.

Eq. (3.102), and the hermiticity of $\hat{H}$, then give

$$\frac{d}{dt}\langle\Psi(t)|\hat{x}|\Psi(t)\rangle = \langle\Psi(t)|\frac{i}{\hbar}[\hat{H},\hat{x}]|\Psi(t)\rangle$$

$$= \langle\Psi(t)|\frac{\hat{p}}{m}|\Psi(t)\rangle \qquad (3.163)$$

This is an example of *Ehrenfest's theorem*, as discussed below, which helps establish the classical limit of quantum mechanics (see Prob. 3.5).

3.9 Pictures

Quantum mechanics can be formulated by placing all the time dependence in the state vector, as we have done so far, placing all of it in the operators, or some of both. These are referred to as different *pictures*. We start the by reviewing the familiar *Schrödinger picture*.

3.9.1 *Schrödinger Picture*

In the Schrödinger picture, there is a time-dependent state vector $|\Psi(t)\rangle$ in the abstract Hilbert space that satisfies the Schrödinger equation

$$i\hbar\frac{\partial}{\partial t}|\Psi(t)\rangle = \hat{H}|\Psi(t)\rangle \qquad ;\text{ Schrödinger equation} \qquad (3.164)$$

The basic structure of the theory is then as follows:

- Assume the hamiltonian $\hat{H}$ is independent of time. A formal solution to the Schrödinger equation can then be written as

$$|\Psi(t)\rangle = e^{-i\hat{H}t/\hbar}|\psi\rangle \qquad ;\hat{H}\text{ time-independent}$$
$$|\Psi(0)\rangle = |\psi\rangle \qquad (3.165)$$

- If $|\psi\rangle$ is a stationary-state solution to the time-independent Schrödinger equation, then $|\Psi(t)\rangle$ has a well-defined time dependence

$$\hat{H}|\psi\rangle = E|\psi\rangle \qquad ;\text{ stationary-state}$$
$$|\Psi(t)\rangle = e^{-iEt/\hbar}|\psi\rangle \qquad (3.166)$$

- The expectation value of any operator $\hat{O}(\hat{p},\hat{x};t)$ constructed from $(\hat{p},\hat{x})$, where $\hat{O}$ may also contain an additional *explicit* time dependence, is given by[31]

$$\langle \hat{O} \rangle \equiv \langle \Psi(t)|\hat{O}|\Psi(t)\rangle \qquad ; \text{ expectation value} \quad (3.167)$$

The essential time dependence of this expectation value arises through that of the state vector $|\Psi(t)\rangle$.

In the coordinate representation, this expression takes the familiar form

$$\langle \hat{O} \rangle = \int d^3\xi \, \Psi(\boldsymbol{\xi}, t)^\star O\left(\frac{\hbar}{i}\boldsymbol{\nabla}_\xi, \, \boldsymbol{\xi}\, ; t\right)\Psi(\boldsymbol{\xi}, t) \; ; \text{ S-picture} \qquad (3.168)$$

- The *adjoint* of the Schrödinger Eq. (3.164) can be written in the following shorthand

$$-i\hbar\frac{\partial}{\partial t}\langle \Psi(t)| = \langle \Psi(t)|\hat{H}^\dagger \qquad ; \text{ adjoint equation}$$

$$= \langle \Psi(t)|\hat{H} \qquad\qquad\qquad (3.169)$$

This expression has meaning in terms of the definition of inner products, on which it will be used. The second line follows since $\hat{H}$ is hermitian. The time derivative of the expectation value in Eq. (3.167) then arises from differentiating all the time dependence, and using Eqs. (3.164) and (3.169)

$$\frac{d}{dt}\langle \hat{O} \rangle = \langle \frac{\partial \hat{O}}{\partial t} \rangle + \frac{1}{i\hbar}\langle \Psi(t)|\hat{O}\hat{H} - \hat{H}\hat{O}|\Psi(t)\rangle$$

$$= \langle \Psi(t)|\left(\frac{\partial \hat{O}}{\partial t} + \frac{i}{\hbar}[\hat{H}, \hat{O}]\right)|\Psi(t)\rangle \qquad (3.170)$$

This is *Ehrenfest's theorem*. It expresses the time derivative of an expectation value in the Schrödinger picture in terms of a commutator of the hamiltonian $\hat{H}$ with $\hat{O}$, and any additional explicit time dependence in $\hat{O}$. It can be used to *define* the time derivative $d\hat{O}/dt$ in the Schrödinger picture, where the underlying operators $(\hat{x}, \hat{p})$ corresponding to the canonical degrees of freedom themselves have *no explicit time dependence*

$$\frac{d}{dt}\langle \hat{O} \rangle \equiv \langle \Psi(t)|\frac{d\hat{O}}{dt}|\Psi(t)\rangle \qquad (3.171)$$

$$\frac{d\hat{O}}{dt} = \frac{\partial \hat{O}}{\partial t} + \frac{i}{\hbar}[\hat{H}, \hat{O}] \quad ; \text{ Ehrenfest's theorem}$$

---

[31]The operator $\hat{O}$ may, for example, describe an interaction with an external field with a prescribed dependence on time. It is assumed here that $|\Psi(t)\rangle$ is normalized.

- The transition matrix element between stationary states has the time dependence

$$\langle \Psi_b(t)|\hat{O}|\Psi_a(t)\rangle = e^{i(E_b - E_a)t/\hbar}\langle \psi_b|\hat{O}|\psi_a\rangle \quad ; \text{stationary states} \quad (3.172)$$

In *summary*, in the Schrödinger picture:[32]

(1) The state vector $|\Psi(t)\rangle$ is *time-dependent* and satisfies the Schrödinger equation;
(2) The operator $\hat{O}(\hat{p}, \hat{x})$ is *time-independent*.[33]

3.9.2 Unitary Transformations

Although formulated in the abstract Hilbert space, all of the *physics* in quantum mechanics is in the inner products, or *in the matrix elements*. Let $\hat{U}$ be a unitary operator in the space, so that

$$\hat{U}^\dagger = \hat{U}^{-1} \quad ; \text{unitary} \quad (3.173)$$

Suppose $\hat{O}$ is an arbitrary operator. Define a *unitary transformation* by

$$\hat{O}_u \equiv \hat{U}\hat{O}\hat{U}^{-1} \quad ; \text{unitary transformation}$$
$$|\Psi(t)_u\rangle \equiv \hat{U}|\Psi(t)\rangle \quad (3.174)$$

Matrix elements of $\hat{O}$ are then left *invariant* under this transformation

$$\langle \Psi(t)_u|\hat{O}_u|\Psi(t)_u\rangle = \langle \Psi(t)|\hat{U}^\dagger \hat{U}\hat{O}\hat{U}^{-1}\hat{U}|\Psi(t)\rangle$$
$$= \langle \Psi(t)|\hat{O}|\Psi(t)\rangle \quad ; \text{invariant} \quad (3.175)$$

One has the freedom of making arbitrary unitary transformations in quantum mechanics, and the *physics is left unchanged!*

3.9.3 Heisenberg Picture

We assume $\hat{H}$ is hermitian and independent of time. The *Heisenberg picture* is then obtained with the time-dependent unitary transformation

$$\hat{U}(t) \equiv \exp\left(\frac{i}{\hbar}\hat{H}t\right) \quad ; \hat{U}^\dagger = \hat{U}^{-1} \quad (3.176)$$

---

[32]This is quantum mechanics as we have developed it so far.
[33]There may be an additional explicit time dependence in $\hat{O}(\hat{p}, \hat{x}; t)$.

Thus, with the aid of Eqs. (3.165), one has

$$\hat{O}_H(t) = e^{i\hat{H}t/\hbar}\,\hat{O}\,e^{-i\hat{H}t/\hbar} \qquad ; \text{ Heisenberg picture}$$

$$|\Psi_H\rangle = e^{i\hat{H}t/\hbar}\,|\Psi(t)\rangle = |\Psi(0)\rangle = |\psi\rangle \qquad (3.177)$$

In *summary*, in the Heisenberg picture,[34]

(1) All of the time-dependence is placed in the *operators* $\hat{O}_H(t)$;
(2) The state vectors $|\Psi_H\rangle = |\psi\rangle$ are *time-independent*.

3.9.4 *Constants of the Motion*

Ehrenfest's theorem implies the following:

*If the operator $\hat{O}$ has no explicit time dependence, and if it com-
mutes with the hamiltonian so that $[\hat{H}, \hat{O}] = 0$, then $d\hat{O}/dt = 0$ and
$\hat{O}$ represents a constant of the motion.*

We give an example. In the central-force problem, in spherical coordinates, the angular momentum operators are given by Eqs. (2.187) and
(2.190)

$$L_z = \frac{1}{i}\frac{\partial}{\partial\phi}$$

$$\mathbf{L}^2 = -\left[\frac{1}{\sin\theta}\frac{\partial}{\partial\theta}\sin\theta\frac{\partial}{\partial\theta} + \frac{1}{\sin^2\theta}\frac{\partial^2}{\partial\phi^2}\right] \qquad (3.178)$$

The corresponding hamiltonian is

$$H = -\frac{\hbar^2}{2m}\left[\frac{1}{r^2}\frac{d}{dr}r^2\frac{d}{dr} - \frac{\mathbf{L}^2}{r^2}\right] + V(r) \qquad (3.179)$$

It is evident that

$$[H, \mathbf{L}^2] = [H, L_z] = 0$$

$$\implies \quad \mathbf{L}^2, L_z \text{ are constants of the motion} \qquad (3.180)$$

3.10 Relation to Classical Mechanics

The most direct connection between quantum and classical mechanics is
achieved through the *Poisson brackets*, and we briefly review that topic[35]

---

[34] For an example of the Heisenberg picture, see Probs. 3.7–3.8.
[35] See [Fetter and Walecka (2003a)].

3.10.1 *Poisson Brackets*

Given a function $F(q_1, \cdots, q_n, p_1, \cdots, p_n; t)$ of canonical coordinates and momenta in classical mechanics, and similarly for G, the Poisson bracket of the pair is defined by

$$\{F, G\}_{\text{P.B.}} = \sum_{\sigma=1}^{n} \left[\frac{\partial F}{\partial q_\sigma} \frac{\partial G}{\partial p_\sigma} - \frac{\partial F}{\partial p_\sigma} \frac{\partial G}{\partial q_\sigma} \right] \qquad ; \text{ Poisson bracket} \qquad (3.181)$$

The partial derivative here implies that all the other variables in F and G are to be held fixed. The Poisson bracket evidently has the following properties:

(1) It is odd under interchange of its arguments

$$\{F, G\}_{\text{P.B.}} = -\{G, F\}_{\text{P.B.}} \qquad (3.182)$$

(2) Consider the Poisson bracket with the hamiltonian

$$\{H, F\}_{\text{P.B.}} = \sum_{\sigma=1}^{n} \left[\frac{\partial H}{\partial q_\sigma} \frac{\partial F}{\partial p_\sigma} - \frac{\partial H}{\partial p_\sigma} \frac{\partial F}{\partial q_\sigma} \right] \qquad (3.183)$$

Now substitute *Hamilton's equations*

$$\frac{\partial H}{\partial q_\sigma} = -\frac{dp_\sigma}{dt} \qquad ; \text{ Hamilton's equations}$$

$$\frac{\partial H}{\partial p_\sigma} = \frac{dq_\sigma}{dt} \qquad \sigma = 1, 2, \cdots, n \qquad (3.184)$$

This gives

$$\{H, F\}_{\text{P.B.}} = -\sum_{\sigma=1}^{n} \left[\frac{\partial F}{\partial p_\sigma} \frac{dp_\sigma}{dt} + \frac{\partial F}{\partial q_\sigma} \frac{dq_\sigma}{dt} \right] \qquad (3.185)$$

The total differential of the function F is given by

$$dF = \sum_{\sigma=1}^{n} \left[\frac{\partial F}{\partial p_\sigma} dp_\sigma + \frac{\partial F}{\partial q_\sigma} dq_\sigma \right] + \frac{\partial F}{\partial t} dt \qquad ; \text{ total differential} \qquad (3.186)$$

Divide this result by dt. Equation (3.185) then states that

$$\{H, F\}_{\text{P.B.}} = -\left(\frac{dF}{dt} - \frac{\partial F}{\partial t} \right) \qquad (3.187)$$

This expresses the total time derivative of a function F in terms of its Poisson bracket with the hamiltonian

$$\frac{dF}{dt} = \frac{\partial F}{\partial t} - \{H, F\}_{\text{P.B.}} \qquad (3.188)$$

The reader should compare this result with Ehrenfest's theorem in Eq. (3.171).

(3) Equations (3.188) and (3.183) can now be turned around to *derive* the full set of Hamiltons's equations, since they imply[36]

$$\frac{dq_\sigma}{dt} = -\{H, q_\sigma\}_{\text{P.B.}} = \frac{\partial H}{\partial p_\sigma} \qquad ; \text{Hamilton's equations}$$

$$\frac{dp_\sigma}{dt} = -\{H, p_\sigma\}_{\text{P.B.}} = -\frac{\partial H}{\partial q_\sigma} \qquad \sigma = 1, 2, \cdots, n$$

$$\frac{dH}{dt} = \frac{\partial H}{\partial t} \qquad (3.189)$$

Thus Hamilton's equations, and the Poisson bracket framework in Eqs. (3.183) and (3.188), provide *equivalent descriptions of classical mechanics.*

(4) Consider the basic Poisson bracket

$$\{p_\alpha, q_\beta\}_{\text{P.B.}} = \sum_{\sigma=1}^{n} \left[\frac{\partial p_\alpha}{\partial q_\sigma} \frac{\partial q_\beta}{\partial p_\sigma} - \frac{\partial p_\alpha}{\partial p_\sigma} \frac{\partial q_\beta}{\partial q_\sigma} \right] = -\sum_{\sigma=1}^{n} \delta_{\alpha\sigma} \delta_{\sigma\beta} \qquad (3.190)$$

Hence

$$\{p_\alpha, q_\beta\}_{\text{P.B.}} = -\delta_{\alpha\beta} \qquad ; \text{canonical values}$$

$$\{p_\alpha, p_\beta\}_{\text{P.B.}} = \{q_\alpha, q_\beta\}_{\text{P.B.}} = 0 \qquad (3.191)$$

These expressions should be compared with the canonical commutation relations in Eqs. (3.102).

(5) It follows that the Poisson-bracket formulation of classical mechanics has the same formal structure as quantum mechanics, and one can make the *transition from classical to quantum mechanics through the replacement*[37]

$$\{ \ , \ \}_{\text{P.B.}} \to \frac{1}{i\hbar} [\ , \ ] \qquad ; \text{classical} \to \text{quantum mechanics} \qquad (3.192)$$

The formal structure of classical mechanics is preserved; however, *dynamical variables now become linear hermitian operators in the abstract*

---

[36] Note the quantities (p_σ, q_σ) have no explicit time dependence.
[37] See [Dirac (1947)].

Hilbert space satisfying canonical commutation relations. In particular

$$[\hat{p}_\alpha, \hat{q}_\beta] = \frac{\hbar}{i}\delta_{\alpha\beta} \qquad ; \text{ canonical commutation relations} \qquad (3.193)$$

3.11 Measurements

We must establish the relation between these formal developments and physical measurements. Measurement theory is a deep and extensive topic, and we certainly shall not do justice to it here. No attempt is made to consider implications for very complex objects with a myriad of degrees of freedom.[38] Rather, the discussion here focuses on simple systems where measurement theory is really quite intuitive.

3.11.1 *Coordinate Space*

We start in coordinate space and abstract later. An observable F is represented by a linear hermitian operator $(H, p, x, L_z, \text{etc.})$ with an (assumed) complete set of eigenstates[39]

$$Fu_{f_n}(x) = f_n\, u_{f_n}(x) \quad ; \text{ eigenstates, } n = 1, 2, \cdots, \infty$$
$$\text{eigenvalues, } f_1, f_2, \cdots, f_\infty \qquad (3.194)$$

Let $\Psi(x, t)$ be an arbitrary wave function. At a given time t, its spatial dependence can be expanded in the complete set of wave functions $u_{f_n}(x)$

$$\Psi(x, t) = \sum_n a_{f_n}(t)\, u_{f_n}(x) \qquad (3.195)$$

The wave function is assumed to be normalized so that

$$\int dx\, |\Psi(x, t)|^2 = \sum_n |a_{f_n}(t)|^2 = 1 \qquad (3.196)$$

We can *measure* the expectation value of F given by

$$\langle F \rangle = \int dx\, \Psi^\star(x, t)F\Psi(x, t) = \sum_n \sum_{n'} a_{f_n}^\star(t)a_{f_{n'}}(t) \int dx\, u_{f_n}^\star(x)Fu_{f_{n'}}(x)$$
$$= \sum_n \sum_{n'} a_{f_n}^\star(t)a_{f_{n'}}(t)f_n\delta_{nn'} \qquad (3.197)$$

---

[38]Schrödinger's cat, for example (see [Wikipedia (2011)]).
[39]For clarity, we present the following arguments in one dimension.

Hence

$$\langle F \rangle = \sum_n |a_{f_n}(t)|^2 f_n \qquad ; \text{ expectation value} \qquad (3.198)$$

If one is in a stationary state so that

$$\Psi(x, t) = \psi(x) e^{-iEt/\hbar} \qquad ; \text{ stationary state} \qquad (3.199)$$

then the wave function $\psi(x)$ can be expanded in the $u_{f_n}(x)$ with time-independent coefficients a_{f_n}

$$\psi(x) = \sum_n a_{f_n} u_{f_n}(x) \qquad ; \text{ completeness} \qquad (3.200)$$

It follows, as above, that in this case

$$1 = \sum_n |a_{f_n}|^2$$

$$\langle F \rangle = \sum_n |a_{f_n}|^2 f_n \qquad ; \text{ stationary state} \qquad (3.201)$$

If one is also in an eigenstate $u_{f_n}(x)$ of F, then

$$\langle F \rangle = f_n \qquad ; \text{ in eigenstate} \qquad (3.202)$$

Equations (3.198) and (3.196) suggest that one should interpret the quantity $|a_{f_n}(t)|^2$ as the *probability of measuring the value f_n at the time t if a system is in the state $\Psi(x, t)$*. Based on this argument, we make the following *measurement postulates*:

(1) If one makes a precise measurement of F, then one *must observe one of the eigenvalues f_n*;
(2) If one is in an arbitrary state $\Psi(x, t)$, then $|a_{f_n}(t)|^2$ is the probability that one will observe the value f_n for F at the time t, where[40]

$$a_{f_n}(t) = \int dx \, u_{f_n}^\star(x) \Psi(x, t) \qquad (3.203)$$

As an example, consider the free-particle wave packet

$$\Psi(x, t) = \frac{1}{\sqrt{2\pi}} \int dk \, A(k) e^{i(kx - \omega_k t)} \qquad ; \text{ free particle} \qquad (3.204)$$

---

[40]Alternatively, if one has a large number of identical systems with wave function $\Psi(x, t)$, then the fraction of measurements yielding f_n will be $|a_{f_n}(t)|^2$.

The probability density in coordinate space is $|\Psi(x,t)|^2$. The Fourier transform of Eq. (3.204) gives

$$\frac{1}{\sqrt{2\pi}} \int dx\, e^{-ikx} \Psi(x,t) = A(k)e^{-i\omega_k t} \qquad (3.205)$$

For localized wave packets, one can take $u_p(x) = e^{ikx}/\sqrt{2\pi}$ as the eigenstates of momentum. Then, consistent with our interpretation in Vol. I,

$$|A(k)|^2 = \left| \int dx\, u_p^*(x) \Psi(x,t) \right|^2 \qquad ; p = \hbar k \qquad (3.206)$$

is the *probability density in momentum space*.[41]

3.11.2 *Abstract Form*

One can now proceed to *abstract* these results:

(1) The quantity F is represented with a linear hermitian operator $\hat{F}$ with eigenstates

$$\hat{F}|f_n\rangle = f_n|f_n\rangle \qquad ; n = 1, 2, \cdots, \infty \qquad (3.207)$$

If one makes a precise measurement of F, then one will observe one of the eigenvalues f_n.

(2) An arbitrary state $|\Psi(t)\rangle$ can be expanded in the (assumed) complete set of eigenstates of $\hat{F}$ according to

$$|\Psi(t)\rangle = \sum_n a_{f_n}(t)|f_n\rangle \qquad (3.208)$$

Then with a normalized state, the probability that a measurement will yield the value f_n is

$$|a_{f_n}(t)|^2 = |\langle f_n|\Psi(t)\rangle|^2 \qquad (3.209)$$

In particular, $|\langle \xi|\Psi(t)\rangle|^2 = |\Psi(\xi,t)|^2$ is the *probability density that one will observe the value ξ if one makes a measurement of the position x.* This is how we have used the wave function $\Psi(\xi,t)$. Now everything stands on the same footing, and the above contains all our previous assumptions concerning the physical interpretation of the theory.

---

[41] See Prob. 4.8 in Vol. I; note $p = \hbar k$.

3.11.3 *Reduction of the Wave Packet*

If a particle moves in a classical orbit, its position can be measured and one finds a value q. If the measurement is repeated a short time Δt later, such that $|\Delta q| \ll |q|$, one must again find the value q. *Measurements must be reproducible.* How does this show up in quantum mechanics?

If one measures the quantity F at the time t and finds a value f_n, then if F is measured again right away, one must again find the value f_n. *This is an assumption of the reproducibility of measurements.*

Suppose the system is in the state

$$\Psi(x,t) = \sum_n a_{f_n}(t) u_{f_n}(x) \tag{3.210}$$

If one measures F at the time t_0 and finds a value f_n, then right after this measurement, the wave function must be such as *to again give the value f_n*, and it must again be normalized. Thus, with no degeneracy, the effect of this measurement is to *reduce* the wave function to the form[42]

$$\Psi(x,t_0)' = \frac{a_{f_n}(t_0)}{|a_{f_n}(t_0)|} \, u_{f_n}(x) \tag{3.211}$$

This result can be abstracted and extended to lead to an additional measurement postulate:

(3) If, at the time t_0, one observes a value f for the quantity F which lies in the inteval $f' \leq f \leq f''$, then the state vector is reduced to

$$|\Psi(t_0)\rangle' = \frac{\sum'_n a_{f_n}(t_0)|f_n\rangle}{\left(\sum'_n |a_{f_n}(t_0)|^2\right)^{1/2}} \qquad ; \text{ where } f' \leq f_n \leq f'' \tag{3.212}$$

Here $\sum'_n$ implies $f' \leq f_n \leq f''$.

Although this postulate may at first seem very mysterious, a little reflection will convince the reader that a measurement does indeed provide a great deal of information about a system, in particular, this type of information. We briefly discuss, as an example, the classic Stern-Gerlach experiment.

---

[42] We assume "pure pass measurements" that do not modify the coefficients $a_{f_n}(t)$.

3.11.4 *Stern-Gerlach Experiment*

The first moral here is that in applying measurement theory, one must always discuss the specific measurement in detail.[43] Consider, for illustration, a spinless, positively-charged particle in a metastable p-state in a neutral atom, where there is no Lorentz force on the atom. There are three possible values of L_z, the angular momentum in the z-direction, $m = 0, \pm 1$. This atom has a magnetic moment, and if placed in a magnetic field which determines the z-direction, and which also *varies* in the z-direction, it will feel a force in the z-direction of

$$F_z = \mu_z \frac{dB_z}{dz} \tag{3.213}$$

This force acts differently on the different m components, and can be used to separate them. Suppose a beam of these atoms, produced, say, in an oven, is passed through an appropriate inhomogeneous magnet as sketched in Fig. 3.3.

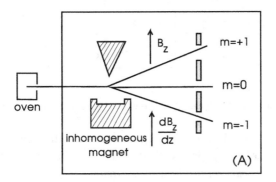

Fig. 3.3 Sketch of the Stern-Gerlach experiment. We will refer to the entire boxed unit as detector (A).

We then note the following:

- The beam will subsequently *split into three separate components with* $m = 0, \pm 1$. Each beam can be caused to pass through a separate slit as shown in Fig. 3.3. *This illustrates that one observes the eigenvalues of* L_z.

---

[43] See chapter IV of [Gottfried (1966)] for a thorough discussion of the measurement process.

- Initially, the internal wave function of an atom can be written

$$\psi_{\text{int}}(\mathbf{x}, t) = R_{np}(r) \sum_{m=0,\pm 1} c_m(t) Y_{1m}(\theta, \phi) \qquad (3.214)$$

If the center-of-mass of the atom goes through the top slit (this will happen with probability $|c_{+1}(t_0)|^2$ where t_0 is the time it goes through the magnet), then the *internal* wave function of the atom must be[44]

$$\psi_{\text{int}}(\mathbf{x}, t) = \frac{c_{+1}(t_0)}{|c_{+1}(t_0)|} R_{np}(r) Y_{11}(\theta, \phi) e^{-iE_{np}(t-t_0)/\hbar} \qquad (3.215)$$

If a second detector identical to (A) in Fig. 3.3 is placed after the top slit, the beam will be observed to pass through and emerge from *its* top slit with unit probability (see Fig. 3.4). *This illustrates the reproducibility of the measurement.*

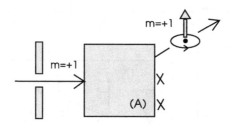

Fig. 3.4   Detector (A) placed after the upper beam with $m = +1$ in Fig. 3.3.

- If one looks for a beam emerging from the middle and bottom slits of the second detector, there will be none. *This illustrates the reduction of the wave packet by the first measurement.*[45]
- The top beam emerging in Fig. 3.4 has $L_z = +1$.

  - Now suppose we proceed to measure L_x by sending that top beam though a detector (A) oriented in the x-direction, as illustrated in Fig. 3.5. All three components of L_x will be observed;
  - If one of the emerging beams with a given L_x is sent through a detector (A) oriented in the z-direction to again measure L_z, all three components of L_z will be seen!

One concludes that an intermediate measurement of L_x destroys the knowledge of L_z.

---

[44]Again, we assume a "pure pass measurement" here.

[45]The first measurement essentially *prepares* the system in a given state.

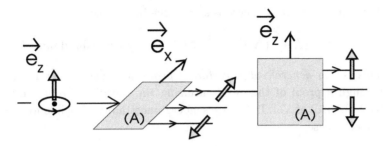

Fig. 3.5 In this figure, $(\mathbf{e}_z, \mathbf{e}_x)$ are unit vectors indicating the directions. The beam emerging in Fig. 3.4, with $L_z = +1$, is sent into a detector (A) now oriented in the x-direction to measure L_x. All three components of L_x will be observed. One emerging beam, with a given L_x, is then re-analyzed with detector (A) oriented in the z-direction to measure L_z. All three values of L_z will be seen!

Whenever you run into apparent paradoxes in discussing the measurement process, you should always return to this simple and fundamental example of the analysis.

We must now face the issue of just what observables *can* be simultaneously specified in quantum mechanics.

3.11.5 *Simultaneous Measurements and Compatible Observables*

In the previous example, one had

$$[\hat{L}_z, \hat{L}_x] = i\hat{L}_y \neq 0 \tag{3.216}$$

Suppose $(\hat{A}, \hat{B}, \hat{C})$ are hermitian operators and

$$[\hat{A}, \hat{B}] = i\hat{C} \tag{3.217}$$

It then follows that:

(1) The expectation values of all three operators are real

$$\langle \hat{A} \rangle = \langle \Psi | \hat{A} | \Psi \rangle = \langle \hat{A} \rangle^\star \qquad ; \; etc. \tag{3.218}$$

(2) The mean-square-deviation from the expectation value is given by

$$(\Delta A)^2 \equiv \langle \Psi | \left(\hat{A} - \langle \hat{A} \rangle \right)^2 | \Psi \rangle$$
$$= \langle \Psi | \hat{A}^2 | \Psi \rangle - \langle \Psi | \hat{A} | \Psi \rangle^2 \qquad ; \; etc. \tag{3.219}$$

These mean-square-deviations satisfy the inequality

$$(\Delta A)^2 (\Delta B)^2 \geq \frac{1}{4} (\Delta C)^2 \qquad ; \text{ generalized u.p.} \quad (3.220)$$

This is the *generalized uncertainty principle*. The reader is guided through a proof of this result utilizing the coordinate representation in Prob. 4.10 of Vol. I. The Heisenberg uncertainty principle follows as a special case

$$[\hat{p}, \hat{x}] = \frac{\hbar}{i}$$

$$\implies \quad (\Delta p)^2 (\Delta x)^2 \geq \frac{\hbar^2}{4} \qquad ; \text{ Heisenberg u.p.} \quad (3.221)$$

It is evident from Eq. (3.220) that a *necessary* requirement for observable compatibility is the following

> If $\hat{A}$ and $\hat{B}$ are to be simultaneously specified, they must commute.

(3) This is also a *sufficient* condition for the compatibility of observables

> If $\hat{A}$ and $\hat{B}$ commute, they are compatible observables.

If the operators $\hat{A}$ and $\hat{B}$ commute, there are several consequences:

(a) One can find *simultaneous eigenstates* of the operators;
(b) In matrix language, $\underline{A}$ and $\underline{B}$ can be *simultaneously diagonalized*;[46]
(c) One can simultaneously measure and know both of them;[47]
(d) A measurement of one of them will not, necessarily, destroy knowledge of the other.

We will prove the results in (3a)–(3b) in two steps:

I. No Degeneracy: Assume first that there is no degeneracy in the problem, and that

$$[\hat{A}, \hat{B}] = 0 \tag{3.222}$$

The proof then goes as follows:

- Let this commutator act on an eigenstate $|b_n\rangle$ of $\hat{B}$ satisfying

$$\hat{B}|b_n\rangle = b_n|b_n\rangle \qquad ; n = 1, 2, \cdots, \infty$$
$$\langle b_m|b_n\rangle = \delta_{mn} \tag{3.223}$$

---

[46] Recall the equivalence in Eq. (3.33).

[47] Use a basis of common eigenstates, and then reduce the wave function by measurement as discussed in the previous section.

This gives

$$\hat{B}\hat{A}|b_n\rangle = \hat{A}\hat{B}|b_n\rangle = b_n\hat{A}|b_n\rangle \tag{3.224}$$

Hence $\hat{A}|b_n\rangle$ is again an eigenstate of $\hat{B}$ with eigenvalue b_n;
- If there is no degeneracy, then $\hat{A}|b_n\rangle \propto |b_n\rangle$, or

$$\hat{A}|b_n\rangle = a_n|b_n\rangle \qquad ; a_n \text{ real} \tag{3.225}$$

Here the quantity a_n is real if $\hat{A}$ is hermitian. Thus, given an eigenstate of $\hat{B}$, it is also an eigenstate of $\hat{A}$;
- Therefore, we have $|a_n\,b_n\rangle$ as simultaneous eigenstates of $\hat{A}$ and $\hat{B}$.

II. Degeneracy. Now go over to matrix language, and consider the matrix element

$$\langle b_m|\hat{A}\hat{B} - \hat{B}\hat{A}|b_n\rangle = 0 \tag{3.226}$$

Insert a complete set of eigenstates of $\hat{B}$. The l.h.s. becomes

$$\text{l.h.s.} = \sum_p \left[\langle b_m|\hat{A}|b_p\rangle\langle b_p|\hat{B}|b_n\rangle - \langle b_m|\hat{B}|b_p\rangle\langle b_p|\hat{A}|b_n\rangle \right]$$

$$= \sum_p \left[\langle b_m|\hat{A}|b_p\rangle b_n\delta_{pn} - b_p\delta_{mp}\langle b_p|\hat{A}|b_n\rangle \right]$$

$$= (b_n - b_m)\langle b_m|\hat{A}|b_n\rangle \tag{3.227}$$

Thus Eq. (3.226) becomes

$$(b_n - b_m)\langle b_m|\hat{A}|b_n\rangle = 0 \tag{3.228}$$

Therefore

(1) With no degeneracy, if $n \neq m$ then $b_n \neq b_m$ and $A_{mn} = 0$. *Hence $\underline{A}$ is a diagonal matrix in the basis that diagonalizes $\underline{B}$*

$$\underline{A} = \begin{bmatrix} A_{11} & & & \\ & A_{22} & & \\ & & A_{33} & \\ & & & \ddots \end{bmatrix} \tag{3.229}$$

This reproduces the result obtained above.

(2) Suppose there is a degeneracy, and
- The first eigenvalue b_1 occurs v_1 times;
- The second eigenvalue b_2 occurs v_2 times, *etc.*

Now re-label the eigenstates of $\hat{B}$ in the subspace of given b_n

$$\hat{B}|b_{n,\nu}\rangle = b_n|b_{n,\nu}\rangle \qquad ; \nu = 1, 2, \cdots, v_n$$
$$\langle b_{n,\nu}|b_{n,\nu'}\rangle = \delta_{\nu\nu'} \tag{3.230}$$

Then by Eq. (3.228), the matrix element $\langle b_{m,\nu'}|\hat{A}|b_{n,\nu}\rangle$ vanishes if $m \neq n$, and *the matrix $\underline{A}$ is block diagonal*

$$\underline{A} = \begin{bmatrix} \underline{A}^{(1)} & & & \\ & \underline{A}^{(2)} & & \\ & & \underline{A}^{(3)} & \\ & & & \ddots \end{bmatrix} \tag{3.231}$$

Here $\underline{A}^{(n)}$ is an hermitian $v_n \times v_n$ matrix.

(3) The problem has now been reduced to diagonalizing the hermitian submatrices

$$A^{(n)}_{\nu'\nu} \equiv \langle b_{n,\nu'}|\hat{A}|b_{n,\nu}\rangle \tag{3.232}$$

in the degenerate subspaces. But this is exactly the problem we solved in our discussion of matrix mechanics! Construct the modal matrix for $\hat{A}$ in the nth subspace

$$\underline{M}^{(n)} \equiv \begin{bmatrix} \underline{c}_1^{(n)} & \underline{c}_2^{(n)} & \cdots & \underline{c}_{v_n}^{(n)} \\ \downarrow & \downarrow & \cdots & \downarrow \end{bmatrix} \qquad ; \text{ modal matrix} \tag{3.233}$$

Then

$$[\underline{M}^{(n)}]^{-1}\underline{A}^{(n)}\underline{M}^{(n)} = \underline{A}^{(n)}_D = \begin{bmatrix} a_1^{(n)} & & & \\ & a_2^{(n)} & & \\ & & \ddots & \\ & & & a_{v_n}^{(n)} \end{bmatrix} \tag{3.234}$$

The matrices $\underline{B}$ and $\underline{A}$ are now simultaneously diagonal.

The modal matrix gives the new eigenstates as

$$|b_n, a_\lambda\rangle = \sum_{\nu=1}^{v_n} \mathcal{M}^{(n)}_{\nu\lambda}|b_{n,\nu}\rangle \tag{3.235}$$

One of the basic challenges in quantum mechanics is to find a *maximal set of mutually commuting observables* for any problem. In the central-force problem, for example, we used

$$[\hat{H}, \hat{\mathbf{L}}^2] = [\hat{H}, \hat{L}_z] = [\hat{\mathbf{L}}^2, \hat{L}_z] = 0 \qquad ; \text{ central-force} \qquad (3.236)$$

The solutions to the Schrödinger equation are simultaneous eigenstates of these quantities.

$$\psi_{nlm}(r, \theta, \phi) = R_{nl}(r) Y_{lm}(\theta, \phi) \qquad ; \text{ solutions} \qquad (3.237)$$

We conclude this chapter on formal developments with a summary of the postulates of quantum mechanics.

3.12 Quantum Mechanics Postulates

Here we *summarize* the quantum mechanics postulates arrived at in the previous discussion. They are formulated in the abstract Hilbert space.

(1) There is a state vector $|\Psi(t)\rangle$ that provides a complete dynamical description of a system;

(2) An observable F is represented by a linear hermitian operator $\hat{F}$;

(3) The operators obey canonical commutation relations, in particular

$$[\hat{p}, \hat{x}] = \frac{\hbar}{i} \qquad (3.238)$$

(4) The dynamics is given by the Schrödinger equation

$$i\hbar \frac{\partial}{\partial t} |\Psi(t)\rangle = \hat{H} |\Psi(t)\rangle \qquad (3.239)$$

(5) The eigenstates of a linear hermitian operator form a complete set[48]

$$\hat{F} |f_n\rangle = f_n |f_n\rangle \qquad ; n = 1, 2, \cdots, \infty$$

$$\sum_n |f_n\rangle \langle f_n| = \hat{1} \qquad (3.240)$$

(6) Measurement postulate:

   (a) A precise measurement of F must yield one of the eigenvalues f_n;

   (b) If the state vector is normalized, then the probability of observing an eigenvalue f_n at the time t is $|\langle f_n | \Psi(t)\rangle|^2$;

---

[48] Compare Prob. 3.9.

(c) A measurement $f' \leq f \leq f''$ at time t_0 reduces the state vector to

$$|\Psi(t_0)\rangle' = \frac{\sum_n' a_{f_n}(t_0)|f_n\rangle}{\left(\sum_n' |a_{f_n}(t_0)|^2\right)^{1/2}} \qquad ; \text{ where } f' \leq f_n \leq f'' \quad (3.241)$$

Through his many years in physics, the author has found this to be a complete and essential set of postulates for the implementation of quantum mechanics.

PART 2
Applications of Quantum Mechanics

Chapter 4

Approximation Methods for Bound States

In chapter 2, we solved a few problems exactly. Usually, one is not in a position to do this. It is extremely valuable to have approximation techniques that allow one to get close to the solution for *any* problem. We start with the *Rayleigh-Ritz variational method*, an extension to quantum mechanics of a classical technique.[1]

4.1 Variational Method

Consider a stationary-state solution to the Schrödinger equation

$$\Psi = e^{-iEt/\hbar}\,\phi$$

$$H\phi = E\phi \qquad \text{; time-independent Schrödinger equation}$$
$$\text{eigenvalue equation} \qquad (4.1)$$

We are interested in finding approximate solutions to this stationary-state eigenvalue equation.

Let ψ be *any* wave function satisfying the appropriate boundary conditions for the problem, including the requirement that the hamiltonian H be hermitian with respect to ψ. Consider the following *functional* of ψ [2]

$$\mathcal{E} \equiv \frac{\int \psi^\star H\psi \, d\tau}{\int \psi^\star \psi \, d\tau} \qquad (4.2)$$

This energy functional can be analyzed as follows:

---

[1]See [Fetter and Walecka (2003a)].

[2]A *functional* is a function of a function. Here the volume element $d\tau$ (and later the coordinate x) are of dimension appropriate to the problem at hand.

(1) The quantity $\mathcal{E}$ is real, since H is hermitian

$$\mathcal{E} = \frac{\int (H\psi)^\star \psi \, d\tau}{\int \psi^\star \psi \, d\tau} = \mathcal{E}^\star \qquad ; \text{ real} \qquad (4.3)$$

(2) Expand ψ in terms of the complete set of *exact eigenfunctions* for the problem[3]

$$\psi(x) = \sum_{E_n} a_{E_n} \phi_{E_n}(x) \qquad ; \text{ completeness}$$

$$H\phi_{E_n}(x) = E_n \phi_{E_n}(x) \qquad ; \text{ exact eigenfunctions} \quad (4.4)$$

Here the $\phi_{E_n}(x)$ are complicated and unknown. Substitute this expansion into Eq. (4.2), and use the orthonormality of the eigenfunctions

$$\mathcal{E} = \frac{\sum_{E_n} |a_{E_n}|^2 E_n}{\sum_{E_n} |a_{E_n}|^2} \qquad ; \text{ orthonormality} \qquad (4.5)$$

Now make use of the following inequality

$$\sum_{E_n} |a_{E_n}|^2 E_n \geq E_0 \sum_{E_n} |a_{E_n}|^2 \qquad (4.6)$$

where E_0 is the *exact lowest eigenvalue*. Substitution into Eq. (4.5) then gives

$$\mathcal{E} \geq E_0 \qquad ; \text{ exact } E_0 \qquad (4.7)$$

- The energy functional is greater than the exact lowest eigenvalue for *any* ψ satisfying the boundary conditions;
- $\mathcal{E}$ evidently possesses a *minimum value*;
- If the system is non-degenerate, the equality in Eq. (4.6) holds when

$$a_{E_0} = 1 \qquad ; \text{ non-degenerate}$$

$$a_{E_n} = 0 \qquad ; E_n > E_0 \qquad (4.8)$$

(3) Let us approach the problem another way. Consider that ψ that *minimizes the energy functional* $\mathcal{E}$. This is the basic problem in the *calculus of variations*. To solve it, let

$$\psi = \psi_0 + \lambda\eta(x) \equiv \psi_0 + \delta\psi(x) \qquad ; \text{ still satisfies B.C.}$$

$$\psi^\star = \psi_0^\star + \lambda\eta^\star(x) \equiv \psi_0^\star + \delta\psi^\star(x) \qquad (4.9)$$

---

[3]Here $\{E_n\} = (E_0, E_1, E_2, \cdots)$ is an ordered set of eigenvalues, some of which may be equal.

Here

- ψ_0 is that ψ which minimizes $\mathcal{E}$;
- λ is a real *infinitesimal*;
- $\eta(x)$ is an *arbitrary, complex function* that still satisfies the boundary conditions.

Now compute

$$\mathcal{E}(\lambda) = \frac{\int (\psi_0 + \lambda\eta)^\star H (\psi_0 + \lambda\eta) \, d\tau}{\int (\psi_0 + \lambda\eta)^\star (\psi_0 + \lambda\eta) \, d\tau} \tag{4.10}$$

Since ψ_0 is the solution to the problem, one has

$$\left. \frac{d\mathcal{E}(\lambda)}{d\lambda} \right|_{\lambda=0} = 0 \tag{4.11}$$

Our previous analysis indicates that this is indeed a minimum.
Re-define

$$\mathcal{E} \equiv \mathcal{E}(0) = \frac{\int \psi_0^\star H \psi_0 \, d\tau}{\int |\psi_0|^2 \, d\tau} \tag{4.12}$$

Expansion of Eq. (4.10) in powers of λ then gives

$$\mathcal{E}(\lambda) = \mathcal{E} + \frac{\lambda}{\int |\psi_0|^2 \, d\tau} \int [\eta^\star(H - \mathcal{E})\psi_0 + \psi_0^\star(H - \mathcal{E})\eta] \, d\tau + O(\lambda^2) \tag{4.13}$$

The terms in $\mathcal{E}$ in the integral come from the expansion of the denominator in Eq. (4.10). The minimum condition in Eq. (4.11) now implies

$$\int \{\eta^\star(H - \mathcal{E})\psi_0 + [(H - \mathcal{E})\psi_0]^\star\eta\} \, d\tau = 0 \tag{4.14}$$

Here the hermiticity of H, and the reality of $\mathcal{E}$, have been used to rewrite the second term. This relation must be true for any arbitrary, complex $\eta(x)$. First take η to be real, and then imaginary. This gives

$$\begin{aligned}
(H - \mathcal{E})\psi_0 + [(H - \mathcal{E})\psi_0]^\star &= 0 \qquad ; \eta(x) \text{ real and arbitrary} \\
(H - \mathcal{E})\psi_0 - [(H - \mathcal{E})\psi_0]^\star &= 0 \qquad ; \eta(x) \text{ imaginary and arbitrary}
\end{aligned} \tag{4.15}$$

Hence

$$(H - \mathcal{E})\psi_0 = 0 \qquad ; \text{ for stationary } \mathcal{E}(\lambda) \tag{4.16}$$

Several comments:

- *The Euler-Lagrange equation for this variational problem is just the time-independent Schrödinger equation.*[4] This gives the Schrödinger equation from a variational principle. The energy functional is made stationary [see Eq. (4.11) and Fig. 4.1] by those functions satisfying the time-independent Schrödinger equation;

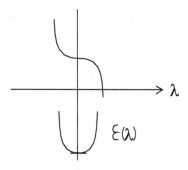

Fig. 4.1   Stationary points of the energy functional $\mathcal{E}(\lambda)$ where $[d\mathcal{E}(\lambda)/d\lambda]_{\lambda=0} = 0$. The ground state corresponds to a *minimum* of the energy functional, as illustrated in the bottom curve. The top curve illustrates a point of inflection.

- The ground state corresponds to a *minimum* of the energy functional;
- If we *achieve* the lowest eigenvalue $\mathcal{E} = E_0$, then by Eq. (4.8)

$$\psi_0 = \phi_{E_0} \qquad \text{; minimum achieved} \qquad (4.17)$$

- First-order changes around the exact ground state produce only *second-order* changes in the eigenvalue

$$\psi(x) = \phi_{E_0}(x) + \lambda\eta(x) \qquad \text{; first-order correction}$$
$$\mathcal{E}(\lambda) = E_0 + O(\lambda^2) \qquad \text{; second-order correction} \qquad (4.18)$$

This implies that crude wave functions can give very good eigenvalues with the variational principle;

- It is convenient to formalize the above calculation and carry it out in a *variational language*.[5] In this language, Eqs. (4.9) read

$$\psi = \psi_0 + \delta\psi(x) \qquad ; \ \psi^\star = \psi_0^\star + \delta\psi^\star(x) \qquad (4.19)$$

---

[4]The second Euler-Lagrange equation is just the complex conjugate of the first.
[5]Same calculation, different language.

If the energy functional is required to be stationary under these variations, then one has the *variational principle*

$$\delta\mathcal{E} = 0 \qquad ; \text{ variational principle} \qquad (4.20)$$

Since the energy functional is of the form $\mathcal{E} = A/B$, the first-order variation becomes

$$\delta\mathcal{E} = \frac{\delta A}{B} - \frac{A}{B^2}\delta B = \frac{1}{B}(\delta A - \mathcal{E}\delta B) = 0 \qquad ; \mathcal{E} = \frac{A}{B} \qquad (4.21)$$

Identification with Eq. (4.2) leads to the first-order result

$$\int d\tau \left[\psi_0^\star H \delta\psi + \delta\psi^\star H \psi_0 - \mathcal{E}\left(\psi_0^\star \delta\psi + \delta\psi^\star \psi_0\right)\right] = 0 \qquad (4.22)$$

Since H is hermitian, this is re-written as

$$\int d\tau \left\{\delta\psi^\star(H - \mathcal{E})\psi_0 + [(H - \mathcal{E})\psi_0]^\star \delta\psi\right\} = 0 \qquad (4.23)$$

Since $(\delta\psi, \delta\psi^\star)$ are arbitrary independent variations, one concludes that the Euler-Lagrange equation for this variational problem again reproduces the stationary-state Schrödinger equation[6]

$$(H - \mathcal{E})\psi_0 = 0 \qquad ; \text{ Euler-Lagrange eqn} \qquad (4.24)$$

- So far the discussion has focused on the ground state. What about the *excited states?* For the first excited state, introduce the following function into the energy functional

$$\psi_1(x) \equiv \psi(x) - \phi_{E_0}(x)\int \phi_{E_0}^\star(y)\psi(y)\, d\tau \qquad (4.25)$$

This subtracts off the projection of ψ_1 on ϕ_{E_0}, and in Eq. (4.4)

$$a_{E_0} = \int \phi_{E_0}^\star(y)\psi_1(y)\, d\tau = 0$$

$$\psi_1(x) = \sum_{E_n \geq E_1} a_{E_n} \phi_{E_n}(x) \qquad (4.26)$$

The argument in Eqs. (4.5)–(4.7) then leads to

$$\mathcal{E}_1 = \frac{\sum_{E_n \geq E_1} |a_{E_n}|^2 E_n}{\sum_{E_n \geq E_1} |a_{E_n}|^2} \geq E_1 \qquad (4.27)$$

---

[6]First take a real $\delta\psi = \delta\psi^\star$, and then an imaginary $\delta\psi = -\delta\psi^\star$. The second Euler-Lagrange equation is again just the complex conjugate of the first.

In this subspace of functions orthogonal to ϕ_{E_0}, the exact first-excited-state energy provide a lower bound on the energy functional, and all the previous arguments can now be invoked:

- Minimize $\mathcal{E}$ in the subspace (we know that $\mathcal{E}$ has an absolute minimum there);
- If the minimum is achieved, then $(H - \mathcal{E})\psi_1 = 0$, and we have a solution to the Schrödinger equation;
- It cannot be ϕ_{E_0}, and therefore if must be ϕ_{E_1} with eigenvalue E_1.

A generalization of this argument leads to the *Nth*-order result

$$\psi_N(x) \equiv \psi(x) - \sum_{E_n < E_N} \phi_{E_n}(x) \int \phi_{E_n}^{\star}(y)\psi(y)\,d\tau$$

$$\mathcal{E} \geq E_N \tag{4.28}$$

The *catch*, of course, is that one has to have determined the previous ϕ_{E_n} to implement this procedure. This gets more and more difficult the higher one goes in the spectrum.[7]

4.1.1   *Application: Ground State in Exponential Potential*

As an example, take a non-trivial problem where we know the exact answer. Consider the *s*-wave ground-state in the following exponential potential

$$V = -V_0 e^{-r/a}$$

$$\frac{8ma^2}{\hbar^2}V_0 = Z_{11}^2 = (3.8318)^2 \qquad ;\text{ first zero of } J_1(\rho) \tag{4.29}$$

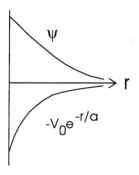

Fig. 4.2   Exponential potential $V = -V_0 e^{-r/a}$ and trial *s*-wave wave function $\psi = [(2\alpha)^3/8\pi]^{1/2}e^{-\alpha r}$. Here $8ma^2 V_0/\hbar^2 = Z_{11}^2$ where Z_{11} is the first zero of $J_1(\rho)$.

---

[7]Convergence is improved by building any symmetries in the problem into $\psi_N(x)$.

As a variational wave function, take

$$\psi(r) = \left[\frac{(2\alpha)^3}{8\pi}\right]^{1/2} e^{-\alpha r} \qquad \text{; trial wave function} \quad (4.30)$$

This wave function has the following properties:

- This is a hydrogenic wave function, and we can use α as a *variational parameter*;
- $\psi(r)$ satisfies the boundary conditions appropriate to the problem;
- It is normalized, since[8]

$$\int |\psi|^2 d^3r = \frac{(2\alpha)^3}{8\pi} 4\pi \int_0^\infty e^{-2\alpha r} r^2 dr = 1 \qquad (4.31)$$

The energy functional is then given by

$$\begin{aligned} .\mathcal{E} &= \int \psi^* \left[-\frac{\hbar^2}{2m}\nabla^2 + V(r)\right] \psi \, d^3r \\ &= \int \left[\frac{\hbar^2}{2m}|\nabla\psi|^2 + V(r)|\psi|^2\right] d^3r \end{aligned} \qquad (4.32)$$

We have used the fact that $\mathbf{p} = (\hbar/i)\nabla$ is hermitian in arriving at the second line.[9] It is clear from the second form that the kinetic energy must be played off against the potential energy in the minimization of the energy functional.

Call $t \equiv r/a$. The second of Eqs. (4.32) then becomes

$$\mathcal{E} = \frac{(2\alpha a)^3}{2} \int_0^\infty t^2 \, dt \left[\frac{\hbar^2}{2ma^2}\left|\frac{d}{dt}e^{-\alpha at}\right|^2 - \frac{\hbar^2 Z_{11}^2}{8ma^2}e^{-(2\alpha a+1)t}\right] \quad (4.33)$$

Now call $x \equiv 2\alpha a$, and again make use of the normalization integral

$$\mathcal{E}(x) = \frac{\hbar^2}{8ma^2}\left[\frac{x^3}{2}\left(x^2\frac{2}{x^3} - \frac{2}{(1+x)^3}Z_{11}^2\right)\right] \qquad (4.34)$$

Thus

$$\mathcal{E}(x) = \frac{\hbar^2}{8ma^2}\left[x^2 - \left(\frac{x}{1+x}\right)^3 Z_{11}^2\right] \equiv \frac{\hbar^2}{8ma^2}f(x) \qquad (4.35)$$

The function $f(x)$ is plotted in Fig. 4.3.

---

[8] Use $\int_0^\infty t^n e^{-t} \, dt = \Gamma(n+1) = n!$.
[9] See Prob. 4.1(a).

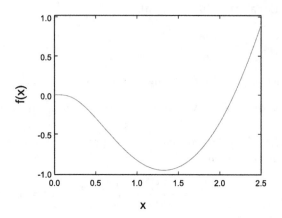

Fig. 4.3　The function $f(x)$ in Eq. (4.35) with $Z_{11} = 3.8318$.

This calculation provides an upper bound to the exact ground-state energy for any $x = 2\alpha a$. One then obtains the best estimate by *minimizing $f(x)$ with respect to x.* A simple numerical calculation gives [see Prob. 4.1(b)]

$$x_{\min} = 1.323 \cdots$$
$$f(x_{\min}) = -0.962 \cdots \tag{4.36}$$

Hence our best variational estimate for the ground-state energy in this problem is

$$\mathcal{E}_{\min} = \frac{\hbar^2}{8ma^2}[-0.962 \cdots] \qquad ; \text{ variational} \tag{4.37}$$

The exact eigenvalue condition with this potential was obtained in Prob. 2.25 as

$$J_\nu\left(\left[\frac{8ma^2V_0}{\hbar^2}\right]^{1/2}\right) = 0 \qquad ; \nu = \left[\frac{8ma^2|E|}{\hbar^2}\right]^{1/2} \tag{4.38}$$

Thus the exact lowest eigenvalue is obtained from $J_\nu(Z_{11}) = 0$, or $\nu = 1$

$$\frac{8ma^2}{\hbar^2}|E| = 1$$
$$E_0 = \frac{\hbar^2}{8ma^2}[-1.000 \cdots] \quad ; \text{ exact} \tag{4.39}$$

One gets within 3.8% of the exact answer, which indeed lies below the variational estimate, with this simple one-parameter calculation.

4.2 Perturbation Theory

As another approximation technique, we examine *perturbation theory*. Suppose we are faced with the stationary-state eigenvalue equation

$$H\psi = E\psi \qquad \text{; want this} \qquad (4.40)$$

but it is too difficult to solve directly. Suppose we can, however, solve a closely related problem

$$H_0\phi_n = E_n^0\phi_n \qquad \text{; know this} \qquad (4.41)$$

where the hamiltonian H differs from H_0 by the *perturbation* H'

$$H = H_0 + H' \qquad \text{; perturbation } H' \qquad (4.42)$$

The goal is to express the solutions to the first problem in terms of those to the second. We start by deriving some exact results.

4.2.1 *Some Exact Results*

Use completeness to expand[10]

$$\psi(x) = \sum_n a_n\phi_n(x) \qquad \text{; completeness} \qquad (4.43)$$

Then re-write Eq. (4.40) as

$$(E - H_0)\psi(x) = H'\psi(x) \qquad (4.44)$$

Now substitute the expansion in Eq. (4.43) on the l.h.s., and use Eq. (4.41)

$$\sum_m a_m(E - E_m^0)\phi_m(x) = H'\psi(x) \qquad (4.45)$$

Take $\int \phi_n^\star(x)\,d\tau$ on this equation, and use the orthonormality of the eigenfunctions

$$a_n(E - E_n^0) = \int \phi_n^\star(x)H'\psi(x)\,d\tau \equiv \langle\phi_n|H'|\psi\rangle \qquad (4.46)$$

---

[10] Again, the dimensions of the coordinate x and volume element $d\tau$ are appropriate to the problem at hand. Note also the familiar definition in the last of Eqs. (4.46).

Hence

$$a_n = \frac{\langle \phi_n | H' | \psi \rangle}{E - E_n^0} \qquad ; E \neq E_n^0 \qquad (4.47)$$

It is assumed here that the exact eigenvalue is unequal to any of the un-perturbed eigenvalues.[11] Equation (4.43) now reads

$$\psi(x) = \sum_n \phi_n(x) \frac{\langle \phi_n | H' | \psi \rangle}{E - E_n^0} \qquad (4.48)$$

This is an *exact integral equation for* $\psi(x)$, as ψ appears in the integral on the r.h.s.

To obtain some insight into Eq. (4.48), let us re-write this result. Define the *Green's function* by

$$G_E(\mathbf{x}, \mathbf{y}) \equiv \sum_n \frac{\phi_n(\mathbf{x}) \phi_n^\star(\mathbf{y})}{E - E_n^0} \qquad ; \text{Green's function} \qquad (4.49)$$

Some comments:

- Here, because it makes a nice connection to later work, we specialize to three dimensions;
- What equation does this Green's function satsfy? Compute

$$(H_0 - E) G_E(\mathbf{x}, \mathbf{y}) = \sum_n \frac{E_n^0 - E}{E - E_n^0} \phi_n(\mathbf{x}) \phi_n^\star(\mathbf{y})$$

$$= -\sum_n \phi_n(\mathbf{x}) \phi_n^\star(\mathbf{y}) \qquad (4.50)$$

The use of completeness then gives the Dirac delta-function

$$(H_0 - E) G_E(\mathbf{x}, \mathbf{y}) = -\delta^{(3)}(\mathbf{x} - \mathbf{y}) \qquad ; \text{Green's function} \qquad (4.51)$$

- Note that we have not yet explicitly determined $G_E(\mathbf{x}, \mathbf{y})$ in Eq. (4.49), since it *depends on the exact eigenvalue E*;
- This Green's function can be used to write the integral Eq. (4.48) in a more transparent fashion

$$\psi(\mathbf{x}) = \int G_E(\mathbf{x}, \mathbf{y}) H' \psi(\mathbf{y}) \, d^3 y \qquad ; \text{exact integral equation for } \psi$$

$$\text{with corresponding } E \quad (4.52)$$

---

[11] For the situation where this condition is not satisfied, see Prob. 4.17.

• One can check this result by applying $(H_0 - E)$ and using Eq. (4.51)

$$(H_0 - E)\psi(\mathbf{x}) = -\int \delta^{(3)}(\mathbf{x} - \mathbf{y})H'\psi(\mathbf{y}) \, d^3y = -H'\psi(\mathbf{x}) \quad (4.53)$$

This reproduces Eq. (4.44).

Equation (4.48) presents a *homogeneous* condition on $\psi(\mathbf{x})$, and the wave function can be *normalized* in any fashion we choose, provided the expectation value of a physical operator is always calculated as

$$\langle O \rangle \equiv \frac{\langle \psi|O|\psi \rangle}{\langle \psi|\psi \rangle} \qquad ; \text{expectation value} \quad (4.54)$$

It is most convenient to *choose the following norm for* ψ

$$\langle \phi_n|\psi \rangle = 1 \qquad ; \text{choice of norm for } \psi \quad (4.55)$$

Here $\langle \phi_n|\psi \rangle$ is the overlap with one particular $\phi_n(x)$. Equations (4.48) and (4.55) then take the form

$$E - E_n^0 = \langle \phi_n|H'|\psi \rangle$$
$$\psi(x) = \phi_n(x) + \sum_{m \neq n} \phi_m(x)\frac{\langle \phi_m|H'|\psi \rangle}{E - E_m^0} \quad (4.56)$$

This is still an exact set of equations for any ψ and corresponding E! It is an exact re-expression of the solution to the time-independent Schrödinger equation.

The reader may not feel that much progress has been made, since we are still faced with an integral equation for ψ, where the kernal depends on the exact eigenvalue E. However, the equations have now been cast in a form such that $\psi \to \phi_n$ as $H' \to 0$, and we are in a perfect position to now *iterate these equations in H'* and generate perturbation theory.

4.2.2 *Non-Degenerate Perturbation Theory*

Suppose we start with $\psi = \phi_n$ and $E = E_n^0$, and then turn on H'. One can proceed to make a consistent expansion of Eqs. (4.56) in powers of H'.

(1) *Lowest Order:* When $H' = 0$, one has[12]

$$\psi_n^{(0)}(x) = \phi_n(x) \qquad ; \text{lowest order}$$
$$E_n^{(0)} = E_n^0 \quad (4.57)$$

---

[12] We use the notation $(\psi_n^{(p)}, E_n^{(p)})$ where p is the order.

(2) *First Order:* To first order in H', one finds

$$\psi_n^{(1)}(x) = \phi_n(x) + \sum_{m \neq n} \phi_m(x) \frac{\langle \phi_m | H' | \phi_n \rangle}{E_n^0 - E_m^0} \qquad ; \text{ first order}$$

$$E_n^{(1)} - E_n^0 = \langle \phi_n | H' | \phi_n \rangle \qquad\qquad (4.58)$$

Several comments:

- Here we have made use of the fact that the spectrum of H_0 is *non-degenerate*

$$E_n^0 \neq E_m^0 \quad \text{if} \quad n \neq m \qquad ; \text{ non-degenerate} \qquad (4.59)$$

  This allows us to use the lowest-order result $E = E_n^0$ in the energy denominator of the correction to the wave function that is already proportional to H' in Eqs. (4.56);

- We can then similarly use the lowest-order expression $\psi = \phi_n$ to compute that correction;

- Since the energy shift in Eqs. (4.56) is already of order H', one can compute it using the lowest-order expression $\psi = \phi_n$;

- *The first-order energy shift is then just the expectation value of the perturbation in the unperturbed state;*

- We had to make sure that the correction to the wave function was also truly of order H' in order to obtain this expression for the energy shift;

- These are *important results.*

(3) *Second Order:* One must both iterate, and expand the energy denominator consistently, in order to obtain the second-order wave function $\psi_n^{(2)}$. Everything is well-defined if the system is non-degenerate and the condition in Eq. (4.59) is satisfied.[13]
As above, one only needs $\psi_n^{(1)}$ to obtain $E_n^{(2)}$. The use of the first of Eqs. (4.58) in the first of Eqs. (4.56) gives the second-order energy shift in non-degenerate perturbation theory as

$$E_n^{(2)} - E_n^0 = \langle \phi_n | H' | \phi_n \rangle + \sum_{m \neq n} \frac{|\langle \phi_m | H' | \phi_n \rangle|^2}{E_n^0 - E_m^0} \qquad ; \text{ second order}$$

$$(4.60)$$

It was essential to establish that the correction to the wave function was also truly of $O(H'^2)$ to arrive at this result.

---

[13] See Prob. 4.4.

4.2.3 *Degenerate Perturbation Theory–I*

We now generalize the previous discussion to the case where the spectrum of H_0 is degenerate, but *the perturbation is diagonal in the degenerate subspace* (see Fig. 4.4). For the subspace under consideration

$$\langle \phi_m | H' | \phi_n \rangle = \delta_{mn} \langle \phi_n | H' | \phi_n \rangle \qquad ; \text{ all } (m, n) \text{ with } E_m^0 = E_n^0 \quad (4.61)$$

The perturbation matrix in the given degenerate subspace is thus assumed to have the structure

$$H'_{mn} = \begin{pmatrix} \checkmark & & \\ & \checkmark & \\ & & \checkmark \end{pmatrix} \qquad (4.62)$$

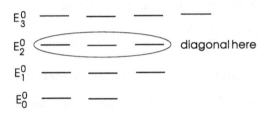

Fig. 4.4 With degenerate perturbation theory–I, the perturbation is diagonal in the subspace of states degenerate with ϕ_n. Illustrated here is the subspace with energy E_2^0.

We make the assumption that a perturbation series exists, so that $\psi = \phi_n + O(H')$, and the second Eqs. (4.56) can then be iterated in *two steps*:

(1) First, iterate the wave function ψ in the matrix element $\langle \phi_n | H' | \psi \rangle$, while retaining the exact eigenvalue E in the denominator

$$\psi = \phi_n + \sum_{m \neq n} \phi_m \frac{\langle \phi_m | H' | \phi_n \rangle}{E - E_m^0} + O(H'^2) \qquad (4.63)$$

The condition in Eq. (4.61) now *eliminates all offending terms in the sum*, that is, any contribution from the other states with $E_m^0 = E_n^0$

$$\psi = \phi_n + \sum_{m \neq n} \phi_m \frac{\langle \phi_m | H' | \phi_n \rangle}{E - E_m^0} + \cdots \qquad ; \text{ Now } E_m^0 \neq E_n^0 \quad (4.64)$$

(2) Next, insert the corresponding iteration of the first of Eqs. (4.56) for the eigenvalue

$$E = E_n^0 + \langle \phi_n | H' | \phi_n \rangle + O(H'^2) \qquad (4.65)$$

With the assumption that the energy shifts are small compared to the unperturbed spacing, Eq. (4.64) then becomes

$$\psi = \phi_n + \sum_{m \neq n} \phi_m \frac{\langle \phi_m | H' | \phi_n \rangle}{E_n^0 - E_m^0} + O(H'^2) \qquad ; E_m^0 \neq E_n^0 \quad (4.66)$$

Now the corrections to the wave function are indeed of order H', and the theory makes sense. There is no problem with vanishing energy denominators, since the vanishing numerators provide protection.

In *summary*, in degenerate perturbation theory–I, where the perturbation is diagonal in the degenerate subspace, one has through $O(H')$

$$\psi_n^{(1)} = \phi_n + \sum_{m \neq n} \phi_m \frac{\langle \phi_m | H' | \phi_n \rangle}{E_n^0 - E_m^0} \qquad ; E_m^0 \neq E_n^0 \qquad (4.67)$$

$$E_n^{(1)} = E_n^0 + \langle \phi_n | H' | \phi_n \rangle \qquad ; \text{ degenerate perturbation theory–I}$$

The first-order energy shift is exactly the same result we had before!

4.2.4   *Applications*

We proceed to discuss several applications of this analysis.

4.2.4.1   *Oscillator with $H' = \beta x^2$ Perturbation*

As an example, we again choose a problem where the exact answer is known. Consider the simple harmonic oscillator with a perturbation (see Fig. 4.5)

$$H' = \beta x^2 \qquad ; \text{ perturbation} \qquad (4.68)$$

The unperturbed hamiltonian and eigenvalues are given by

$$H_0 = \frac{p^2}{2m} + \frac{1}{2} m \omega_0^2 x^2$$

$$E_n^0 = \hbar \omega_0 \left(n + \frac{1}{2} \right) \qquad ; n = 0, 1, 2, \cdots, \infty \qquad (4.69)$$

Since there is no degeneracy in the spectrum, this is an example of *non-degenerate* perturbation theory.

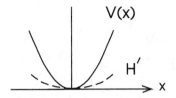

Fig. 4.5   Harmonic oscillator with $H' = \beta x^2$ perturbation.

The dimensions can be taken out by writing

$$H' \equiv \lambda \left(\frac{1}{2}m\omega_0^2\right) x^2 \tag{4.70}$$

The matrix element of the perturbation is

$$\langle n|H'|n\rangle = \frac{\lambda}{2}m\omega_0^2 \langle n|x^2|n\rangle \tag{4.71}$$

This is readily calculated using completeness and previous oscillator results in Eqs. (2.98)–(2.99)

$$
\begin{aligned}
\langle n|x^2|n\rangle &= \sum_m \langle n|x|m\rangle\langle m|x|n\rangle \\
&= \langle n|x|n-1\rangle\langle n-1|x|n\rangle + \langle n|x|n+1\rangle\langle n+1|x|n\rangle \\
&= \frac{\hbar}{2m\omega_0}[n + (n+1)]
\end{aligned} \tag{4.72}
$$

Hence

$$\langle n|H'|n\rangle = \frac{\lambda}{2}\hbar\omega_0 \left(n + \frac{1}{2}\right) \tag{4.73}$$

First-order non-degenerate perturbation theory then gives the energy shift

$$E_n^{(1)} = E_n^0 \left(1 + \frac{1}{2}\lambda\right) \qquad ; \text{first-order shift} \tag{4.74}$$

The exact answer in this case is obtained from the full hamiltonian

$$H = H_0 + H' = \frac{p^2}{2m} + \frac{1}{2}m\omega_0^2(1 + \lambda)x^2 \tag{4.75}$$

The perturbation just shifts the oscillator frequency to $\omega = \omega_0(1 + \lambda)^{1/2}$, and the exact eigenvalues are

$$E_n = \hbar\omega \left(n + \frac{1}{2}\right) = \hbar\omega_0(1 + \lambda)^{1/2} \left(n + \frac{1}{2}\right) \tag{4.76}$$

Hence

$$E_n = E_n^0 \, (1 + \lambda)^{1/2}$$

$$= E_n^0 \left(1 + \frac{1}{2}\lambda - \frac{1}{8}\lambda^2 + \cdots \right) \qquad ; \text{ exact answer} \qquad (4.77)$$

The first-order correction is exactly reproduced using perturbation theory.[14] With the aid of the exact solution in Eq. (4.77) one can say interesting things about perturbation theory in this case:

- Perturbation theory provides a power series in the coupling constant λ;
- The function $(1 + \lambda)^{1/2}$ has a branch point at $\lambda = -1$, where the net oscillator potential in Eq. (4.75) vanishes;
- The radius of convergence of the power series is thus $R = 1$ (see Fig. 4.6);

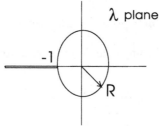

Fig. 4.6   Circle of convergence of perturbation series for oscillator with $H' = \lambda m \omega_0^2 x^2 / 2$.

- Perturbation theory only makes sense within this circle of convergence.[15]

4.2.4.2   *Ground State of the He Atom*

The He atom consists of two electrons moving in the Coulomb field of a nucleus with $Z = 2$ (Fig. 4.7). The hamiltonian is

$$H = T(1) + T(2) + V(1) + V(2) + V(1,2)$$

$$= -\frac{\hbar^2}{2m}(\nabla_1^2 + \nabla_2^2) - \frac{2e^2}{r_1} - \frac{2e^2}{r_2} + \frac{e^2}{|\mathbf{r}_1 - \mathbf{r}_2|} \qquad (4.78)$$

---

[14]For the second-order correction, see Prob. 4.6.

[15]It is not surprising that the character of the solution changes entirely here when $\lambda = -1$, and the confining potential disappears. In most applications, the radius of convergence of the perturbation series is unknown. There are examples where the series is divergent for *any* coupling (see [Bender and Wu (1969)]), and a finite set of terms in the series only provides an asymptotic approximation to the correct answer.

This presents a three-body problem, where we assume the nucleus is heavy and fixed. The hamiltonian can be written

$$H = H_0 + H'$$
$$H' \equiv \frac{e^2}{|\mathbf{r}_1 - \mathbf{r}_2|} \qquad (4.79)$$

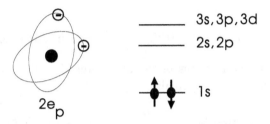

Fig. 4.7 The He atom consists of two electrons moving in the Coulomb field of a nucleus with charge $2e_p$. The ground state of H_0, which neglects the Coulomb interaction between the electrons, is $\Phi_0 = \phi_{100}(1)\phi_{100}(2)$, where the electrons have opposite spins.

One wants to solve the time-independent Schrödinger equation

$$H\psi(\mathbf{r}_1, \mathbf{r}_2) = E\psi(\mathbf{r}_1, \mathbf{r}_2) \qquad ; \text{S-eqn} \qquad (4.80)$$

Here the wave function depends on *two* coordinates. The corresponding probability of finding one electron in the volume element d^3r_1 about $\mathbf{r}_1$, and the second electron in the volume element d^3r_2 about $\mathbf{r}_2$ is

$$|\psi(\mathbf{r}_1, \mathbf{r}_2)|^2 d^3r_1 d^3r_2 \qquad ; \text{probability} \qquad (4.81)$$

We carry out a perturbation calculation on the ground state of H_0, where each electron occupies the single-particle state $\phi_{100}(\mathbf{r})$ of the hydrogen-like atom (see Fig. 4.7)[16]

$$\Phi_0 = \phi_{100}(1)\phi_{100}(2) \qquad ; \text{ground state of } H_0 \qquad (4.82)$$

The electrons can exist in the same spatial state if they have opposite spins. The goal is to compute the first-order energy shift due to the Coulomb repulsion between the electrons

$$E_0^{(1)} = E_0^0 + \langle \Phi_0 | H' | \Phi_0 \rangle \qquad ; \text{perturbation theory} \qquad (4.83)$$

---

[16]This is just separation of variables where $H_0 = H_0(1) + H_0(2)$, and $\Phi_0 = \phi_0(1)\phi_0(2)$. The notation in Eq. (4.82) is ϕ_{nlm}.

We use the hydrogen wave functions for a nucleus with charge Z from Prob. 2.22

$$\phi_{100}(\mathbf{r}) = \frac{1}{\sqrt{8\pi}}\left(\frac{2Z}{a_0}\right)^{3/2} e^{-Zr/a_0} \tag{4.84}$$

Here $a_0 = \hbar^2/m_e e^2$ is the Bohr radius. For reasons that will soon become apparent, we will compute Eq. (4.83) through the fully equivalent expression

$$E_0^{(1)} = \langle \Phi_0 | H | \Phi_0 \rangle \qquad \text{; perturbation theory} \tag{4.85}$$

and leave the result in terms of the parameter Z in the wave functions. These wave functions are normalized for any Z[17]

$$\int |\phi_{100}|^2 d^3r = \frac{1}{8\pi}\left(\frac{2Z}{a_0}\right)^3 \int r^2\, dr d\Omega\, e^{-2Zr/a_0}$$
$$= \frac{1}{2}\int_0^\infty \rho^2 e^{-\rho}\, d\rho = 1 \tag{4.86}$$

The terms required with the hamiltonian of Eq. (4.78) are as follows:

(1) The one-body kinetic energy is evaluated with the aid of a partial integration

$$\langle \phi_{100} | -\frac{\hbar^2}{2m}\nabla^2 | \phi_{100} \rangle = \frac{\hbar^2}{2m}\int d^3r |\nabla\phi_{100}|^2$$
$$= \frac{\hbar^2}{2m}\frac{1}{8\pi}\left(\frac{2Z}{a_0}\right)^3 \int d^3r\, e^{-2Zr/a_0}\left(\frac{Z}{a_0}\right)^2$$
$$= \frac{1}{2}\frac{Z^2 e^2}{a_0} \tag{4.87}$$

(2) The expectation value of the one-body Coulomb potential energy is

$$\langle \phi_{100} | -\frac{2e^2}{r} | \phi_{100} \rangle = -2e^2 \frac{1}{8\pi}\left(\frac{2Z}{a_0}\right)^3 \int d^3r \frac{1}{r} e^{-2Zr/a_0}$$
$$= -2e^2\left(\frac{2Z}{a_0}\right)\frac{1}{2}$$
$$= -\frac{2Ze^2}{a_0} \tag{4.88}$$

---

[17]We again use $\int_0^\infty t^n e^{-t}\, dt = \Gamma(n+1) = n!$.

(3) The matrix element of the perturbation is obtained as

$$\langle \Phi_0 | H' | \Phi_0 \rangle = \frac{1}{(8\pi)^2} \left(\frac{2Z}{a_0} \right)^6 \int d^3 r_1 \int d^3 r_2 \; e^{-2Zr_1/a_0} e^{-2Zr_2/a_0} \frac{e^2}{|\mathbf{r}_1 - \mathbf{r}_2|} \tag{4.89}$$

The angular integrals are done as follows:

- Make use of the generating function for the Legendre polynomials in Eq. (2.66)

$$\frac{1}{|\mathbf{r}_1 - \mathbf{r}_2|} = \sum_{l=0}^{\infty} \frac{r_<^l}{r_>^{l+1}} P_l(\cos \theta_{12}) \tag{4.90}$$

- Fix $\mathbf{r}_1$, and take this to define the z-axis. Then $\cos \theta_{12} = \cos \theta_2$;
- Use the orthogonality of the Legendre polynomials, with $P_0(\cos \theta_2) = 1$

$$\int d\Omega_2 P_l(\cos \theta_2) P_0(\cos \theta_2) = 4\pi \delta_{l0} \tag{4.91}$$

- The final angular integral then gives $\int d\Omega_1 = 4\pi$.

With the definition $\rho \equiv 2Zr/a_0$, Eq. (4.89) now becomes[18]

$$\langle \Phi_0 | H' | \Phi_0 \rangle = \frac{1}{4} \frac{2Z}{a_0} e^2 \int_0^\infty \rho_1^2 d\rho_1 \, e^{-\rho_1} \int_0^\infty \rho_2^2 d\rho_2 \, e^{-\rho_2} \left(\frac{1}{\rho_>} \right) \tag{4.92}$$

(4) A collection of these results gives[19]

$$\langle \Phi_0 | H | \Phi_0 \rangle = \frac{e^2}{a_0} \left[Z^2 - 4Z + IZ \right]$$

$$I \equiv \frac{1}{2} \int_0^\infty \rho_1^2 d\rho_1 \, e^{-\rho_1} \int_0^\infty \rho_2^2 d\rho_2 \, e^{-\rho_2} \left(\frac{1}{\rho_>} \right) \tag{4.93}$$

We claim the remaining definite integral is given by

$$I = \frac{5}{8} \tag{4.94}$$

---

[18] Compare Prob. 4.7. Recall $\rho_> = \rho_1$ if $\rho_1 > \rho_2$, and $\rho_> = \rho_2$ if $\rho_2 > \rho_1$.
[19] Note that there are *two* one-body kinetic and potential energy terms.

This is demonstrated as follows.[20] Write out I

$$I = \frac{1}{2} \int_0^\infty \rho_1^2 d\rho_1 \, e^{-\rho_1} \left[\int_{\rho_1}^\infty \rho^2 d\rho \, e^{-\rho} \frac{1}{\rho} + \frac{1}{\rho_1} \int_0^{\rho_1} \rho^2 d\rho \, e^{-\rho} \right]$$

$$\equiv \frac{1}{2} \int_0^\infty \rho_1^2 d\rho_1 \, e^{-\rho_1} \left[I_A + I_B \right] \tag{4.95}$$

The first integral I_A is evaluated by introducing a parameter λ, where one is instructed to set $\lambda = 1$ at the end of the calculation

$$I_A = \int_{\rho_1}^\infty \rho e^{-\rho} \, d\rho = -\frac{d}{d\lambda} \int_{\rho_1}^\infty e^{-\lambda\rho} \, d\rho = -\frac{d}{d\lambda} \left(\frac{1}{\lambda} e^{-\lambda\rho_1} \right)$$

$$= \frac{1}{\lambda^2} e^{-\lambda\rho_1} + \frac{\rho_1}{\lambda} e^{-\lambda\rho_1} \qquad\qquad ; \text{ set } \lambda = 1 \tag{4.96}$$

Thus one obtains

$$I_A = (1 + \rho_1) e^{-\rho_1} \tag{4.97}$$

The second integral I_B is similarly given by

$$\rho_1 I_B = \int_0^{\rho_1} \rho^2 e^{-\rho} \, d\rho = \frac{d^2}{d\lambda^2} \int_0^{\rho_1} e^{-\lambda\rho} \, d\rho = -\frac{d^2}{d\lambda^2} \frac{1}{\lambda} (e^{-\lambda\rho_1} - 1)$$

$$= \frac{d}{d\lambda} \left[\frac{1}{\lambda^2} (e^{-\lambda\rho_1} - 1) + \frac{\rho_1}{\lambda} e^{-\lambda\rho_1} \right]$$

$$= -\frac{2}{\lambda^3} (e^{-\lambda\rho_1} - 1) - \frac{\rho_1}{\lambda^2} e^{-\lambda\rho_1} - \frac{\rho_1}{\lambda^2} e^{-\lambda\rho_1} - \frac{\rho_1^2}{\lambda} e^{-\lambda\rho_1}$$

$$; \text{ set } \lambda = 1 \tag{4.98}$$

Hence

$$I_B = \frac{1}{\rho_1} \left[2 - 2e^{-\rho_1} - 2\rho_1 e^{-\rho_1} - \rho_1^2 e^{-\rho_1} \right] \tag{4.99}$$

A combination with Eq. (4.97), and insertion into Eq. (4.95), then gives[21]

$$I = \frac{1}{2} \int_0^\infty \rho e^{-\rho} \, d\rho \left[2 - 2e^{-\rho} - \rho e^{-\rho} \right]$$

$$= \frac{1}{2} \left[2 - \frac{2}{4} - \frac{2}{8} \right] = \frac{5}{8} \tag{4.100}$$

This establishes the result in Eq. (4.94).

---

[20] The result is sufficiently important that we provide a detailed evaluation of I.
[21] Again, use $\int_0^\infty \rho^n e^{-\rho} \, d\rho = n!$.

With the insertion of Eq. (4.94), Eqs. (4.93) and (4.85) become

$$E_0^{(1)} = \langle \Phi_0 | H | \Phi_0 \rangle = \frac{e^2}{a_0} \left[Z^2 - 4Z + \frac{5}{8}Z \right] \tag{4.101}$$

We proceed to discuss this result:

- For the He atom $Z = 2$, and the first-order perturbation theory result for the ground-state energy is

$$E_0^{(1)} = \frac{e^2}{a_0} \left[4 - 8 + \frac{5}{4} \right] = \frac{e^2}{a_0} \left[-\frac{11}{4} \right]$$
$$= \frac{e^2}{a_0} [-2.75] \qquad ; \text{ perturbation theory} \tag{4.102}$$

Note the convergence of the terms for the one-body potential energy, the one-body kinetic energy, and the two-body Coulomb energy in the first of Eqs. (4.102);

- The reason for evaluating the first-order perturbation theory result through Eq. (4.85), while leaving Z as a parameter in the wave functions, now becomes clear. We are evaluating the expectation value of the hamiltonian with normalized wave functions, and in this form, *the first-order perturbation theory result is also a variational calculation!* One can now treat Z as a *variational parameter*, and a better upper bound on the ground-state energy is obtained by *minimizing with respect to Z*

$$\frac{dE_0^{(1)}}{dZ} = 0 \qquad ; \text{ for minimum}$$
$$\implies \quad 2Z^\star - 4 + \frac{5}{8} = 0$$
$$Z^\star = \frac{27}{16} \tag{4.103}$$

This gives

$$\left[E_0^{(1)} \right]_{\min} = \frac{e^2}{a_0} \left[-\left(\frac{27}{16} \right)^2 \right]$$
$$= \frac{e^2}{a_0} [-2.847 \cdots] \quad ; \text{ variational} \tag{4.104}$$

Note that $Z^\star < 2$, reflecting a shielding of the nuclear charge by the electrons;

- The experimental number for the minimum energy to remove the two electrons from atomic He is

$$E_0 = \frac{e^2}{a_0}[-2.904\cdots] \qquad ; \text{experiment} \quad (4.105)$$

Perturbation theory is good to 5.3%, while the variational estimate lies within 2% of the exact ground-state energy;

- It is possible to get extremely accurate variational answers for E_0 for this problem by using more and more complicated wave functions. For example, with a determinant of $O(1078)$ [Pekeris (1959)] obtains[22]

$$E_0 = \frac{e^2}{a_0}[-2.903, 724, 375\cdots] \qquad ; \text{Pekeris} \quad (4.106)$$

The theoretical error is estimated to be in the last digit, while the experimental error lies in the seventh digit. When relativistic QED corrections are included in the calculation, theory and experiment agree to approximately one part in 2×10^6 for the ground-state energy of He.

4.2.4.3 *Hydrogen Spectrum with Finite Nuclei*

The nucleus has a finite size, which is small compared to the size of the atom.[23] The charge distribution is spread out over the nucleus, so the real nuclear Coulomb potential is actually (see Fig. 4.8)

$$V_C(r) = -Ze^2 \int \frac{\rho_N(\mathbf{r'})}{|\mathbf{r} - \mathbf{r'}|} d^3r' \qquad ; \text{Coulomb potential} \quad (4.107)$$

Here $\rho_N(\mathbf{r})$ is the nuclear charge distribution, with the following properties:

- We take $\rho_N(r)$ to be spherically symmetric, which will necessarily be the case for spin-zero nuclei;
- Since the total nuclear charge Z has been extracted, the nuclear charge distribution here is normalized

$$\int \rho_N(r) d^3r = 1 \qquad ; \text{normalized} \quad (4.108)$$

The unperturbed hamiltonian for hydrogen-like atoms is

$$H_0 = -\frac{\hbar^2}{2m}\nabla^2 - \frac{Ze^2}{r} \qquad (4.109)$$

---

[22]For an interesting historical perspective here, see [Koutschan and Zeilberger (2010)].
[23]This includes the proton, which is the nucleus of the hydrogen atom.

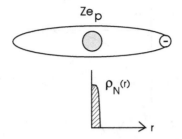

Fig. 4.8 The nuclear charge distribution in hydrogen-like atoms has a finite spatial distribution given by $\rho_N(r)$. The Coulomb potential is actually given by Eq. (4.107).

The *perturbation* for the atom with the extended charge distribution is then

$$H' = \delta V(r) = -Ze^2 \int \rho_N(r')d^3r' \left[\frac{1}{|\mathbf{r} - \mathbf{r}'|} - \frac{1}{r} \right] \qquad (4.110)$$

This gives the correct total hamiltonian as

$$H = H_0 + H' \qquad (4.111)$$

A repetition of the arguments in Eqs. (4.90)–(4.91) leads to

$$\delta V(r) = -Ze^2 \int \rho_N(r')d^3r' \left[\frac{1}{r_>} - \frac{1}{r} \right] \qquad (4.112)$$

We observe the following:

- If the radial coordinate r lies *outside* the nucleus so that $r_> = r$, then the perturbation vanishes

$$\delta V(r) = 0 \qquad ; r_> = r \qquad (4.113)$$

  This is a reflection of the fact that a spherically symmetric charge distribution acts as a point source if one is outside of it. As a consequence, *the perturbation is only non-zero over the nucleus!* (See Fig. 4.9.)
- If the matrix element of the perturbation $\langle nlm|\delta V(r)|nlm \rangle$ is to be non-zero, there must be some overlap of the atomic wave functions with the nucleus, that is, with the origin. Hence *it is only the atomic s-states whose energy is shifted.*

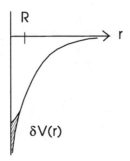

Fig. 4.9   The modification of the Coulomb potential $\delta V(r)$ is only non-zero over the interior of the nucleus.

The only non-zero matrix elements of the perturbation within a degenerate subspace in Fig. 2.15 are therefore

$$\delta E_{n00}^{\text{f.s.}} \equiv \langle n00|\delta V(r)|n00\rangle$$

$$= -Ze^2 \int \rho_N(r')d^3r' \int |\phi_{n00}(r)|^2 d^3r \left[\frac{1}{r_>} - \frac{1}{r}\right] \quad (4.114)$$

We are interested here in the case where the nuclear size is much less than the atomic size

$$\mathcal{R}(\text{nuclear size}) \ll \frac{a_0}{Z}(\text{atomic size}) \quad (4.115)$$

In this case, the atomic wave function can be taken to be constant over the nucleus, and Eq. (4.114) becomes

$$\delta E_{n00}^{\text{f.s.}} \approx -Ze^2|\phi_{n00}(0)|^2 \int \rho_N(r')d^3r' \int_0^{r'} d^3r \left[\frac{1}{r'} - \frac{1}{r}\right] \quad (4.116)$$

Now

$$\int \rho_N(r')d^3r' \int_0^{r'} d^3r \left[\frac{1}{r'} - \frac{1}{r}\right] = \int \rho_N(r')d^3r' \left[4\pi \left(\frac{1}{3}r'^2 - \frac{1}{2}r'^2\right)\right]$$

$$= -\frac{4\pi}{6} \int \rho_N(r')r'^2 \, d^3r'$$

$$= -\frac{2\pi}{3}\langle r^2\rangle_N \qquad ; \text{ mean-square radius}$$

$$(4.117)$$

Here $\langle r^2 \rangle_N$ is the *mean-square radius of the nuclear charge distribution.* Thus Eq. (4.114) becomes

$$\delta E_{n00}^{\text{f.s.}} = \frac{2\pi}{3} Z\alpha\hbar c |\phi_{n00}(0)|^2 \langle r^2 \rangle_N \tag{4.118}$$

where $\alpha = e^2/\hbar c = 1/137.0$ is the fine-structure constant.

We make several comments on this result:

(1) The first few levels of the hydrogen-like spectrum of Fig. 2.15 are shown again in Fig. 4.10, with the degenerate subspaces explicitly indicated.

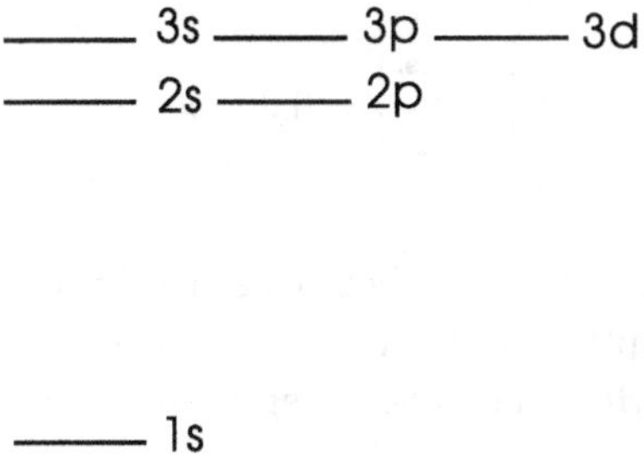

Fig. 4.10 First few levels of the hydrogen-like spectrum of Fig. 2.15, with degenerate subspaces explicitly indicated.

We have here an example of degenerate perturbation theory–I. The only non-zero matrix element in the degenerate subspaces is the diagonal matrix element between the s-states, and hence the perturbation is diagonal in the degenerate subspaces;

(2) *It is only the energies of the s-states that are shifted, and they are shifted up.* This upward shift arises because one does not have the full attractive $-Ze^2/r$ potential at the origin (see Fig. 4.9);

(3) From Eq. (2.298), one has for the probability density at the origin

$$|\phi_{n00}(0)|^2 = \frac{1}{\pi} \left(\frac{Z\alpha}{n} \right)^3 \left(\frac{mc}{\hbar} \right)^3 \tag{4.119}$$

For hydrogen-like atoms, the unperturbed energy is

$$E_n^0 = -\frac{1}{2} \left(\frac{Z\alpha}{n} \right)^2 mc^2 \tag{4.120}$$

Hence the magnitude of the fractional level shift of the s-states is

$$\left|\frac{\delta E_{n00}^{\text{f.s.}}}{E_n^0}\right| = \frac{4}{3}\frac{Z^2\alpha^2}{n}\frac{\langle r^2\rangle_N}{(\hbar/mc)^2} \tag{4.121}$$

(4) For the $1s$ level in the hydrogen atom

$$\langle r^2\rangle_p = \left(0.8\times 10^{-13}\,\text{cm}\right)^2$$
$$(\hbar/m_e c)^2 = \left(3.862\times 10^{-11}\,\text{cm}\right)^2 \tag{4.122}$$

The first number is the mean-square charge radius of the proton as measured through electron scattering by [Hofstadter (1956)].[24] This gives

$$\left|\frac{\delta E_{1s}^{\text{f.s.}}}{E_{1s}^0}\right| = 0.305\times 10^{-9} \tag{4.123}$$

This is a *very small number;*

(5) In a muonic atom, formed when a negatively charged muon stops in matter and is captured into an atomic orbit, the situation is significantly different. Here the ratio of lepton masses is

$$\frac{m_\mu}{m_e} = \frac{105.66}{0.511} \approx 207 \tag{4.124}$$

This has the following consequences:

- The Bohr radius for the muon is

$$(a_0)_\mu \equiv \frac{\hbar^2}{m_\mu e^2} = \frac{m_e}{m_\mu}a_0 = \frac{a_0}{207} \tag{4.125}$$

As a result, the muon sits well *inside* the electron cloud in the low-lying levels of the muonic atom (see Fig. 4.11).

Fig. 4.11   The low-lying states of the muonic atom lie well inside of the electron cloud and exhibit a hydrogen-like spectrum.

---

[24]See [Walecka (2001)].

One really has a hydrogen-like atom for the muon with a nuclear charge Z and lepton mass m_μ;[25]

- One picks up a factor of Z^2 in the energy shift in Eq. (4.121) for heavier atoms;

- The nuclear mean-square charge radius is often quoted in terms of the radius R of the equivalent uniform distribution, and a simple calculation gives

$$\langle r^2 \rangle_{\text{uniform}} = \frac{\int_0^R r^4 \, dr}{\int_0^R r^2 \, dr} = \frac{3}{5} R^2 \qquad (4.126)$$

The observed values for nuclei with nucleon number A can be parametrized as

$$R \approx 1.2 \, A^{1/3} \times 10^{-13} \, \text{cm} \qquad ; \text{ experiment} \qquad (4.127)$$

- For $^{32}_{16}$S, this gives

$$\left| \frac{\delta E_{1s}^{\text{f.s.}}}{E_{1s}^0} \right| = 0.045 \qquad (4.128)$$

This is a shift that is *easily seen*;

- In fact, the size of the nuclear charge distribution exhibited in Eq. (4.127) was first established through muonic X-rays by [Fitch and Rainwater (1953)].

4.2.4.4 *Uniform Magnetic Field—Zeeman Effect*

The non-relativistic hamiltonian for a particle with mass m_0 and charge e moving in a central potential $V(r)$, with additional external electromagnetic fields $(\mathbf{B}, \mathbf{E})$, is[26]

$$H = \frac{1}{2m_0} \left[\mathbf{p} - \frac{e}{c} \mathbf{A}(\mathbf{r}, t) \right]^2 + V(r) + e\Phi(\mathbf{r}, t) \qquad (4.129)$$

The fields are related to the electromagnetic potentials by

$$\mathbf{B} = \boldsymbol{\nabla} \times \mathbf{A}$$
$$\mathbf{E} = -\boldsymbol{\nabla}\Phi - \frac{1}{c} \frac{\partial \mathbf{A}}{\partial t} \qquad (4.130)$$

This hamiltonian has the following properties:

---

[25]The corresponding energy spectrum of the muonic atom is obtained from Eq. (4.120) as $[E_n^0]_\mu = 207[E_n^0]_e$.

[26]See Vol. I and chapter 7. We again use c.g.s. units here.

- It is the correct classical H;
- It provides a linear, hermitian hamiltonian;
- As demonstrated in appendix B, Ehrenfest's theorem gives the classical equations of motion in the correct limit;
- The resulting Schrödinger equation is gauge invariant.[27]

The hamiltonian in Eq. (4.129) is written out as

$$H = H_0 + H'$$
$$= H_0 - \frac{e}{2m_0 c}\left(\mathbf{p}\cdot\mathbf{A} + \mathbf{A}\cdot\mathbf{p}\right) + \frac{e^2}{2m_0 c^2}\mathbf{A}^2 + e\Phi \cdot \qquad (4.131)$$

Canonical quantization in the coordinate representation gives

$$\mathbf{p} = \frac{\hbar}{i}\boldsymbol{\nabla} \qquad \text{; canonical quantization} \qquad (4.132)$$
$$\text{; coordinate representation}$$

Consider the case where the one-electron atom sits is an additional constant uniform magnetic field $\mathbf{B}$ (Fig. 4.12).

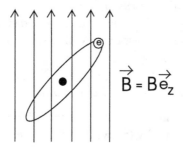

$$\vec{B} = B\vec{e}_z$$

Fig. 4.12   One-electron atom in a constant uniform magnetic field $\mathbf{B} = B\mathbf{e}_z$, where $\mathbf{e}_z$ is a unit vector in the z-direction.

Take the potentials to be

$$\mathbf{A}(\mathbf{r}) = \frac{1}{2}\mathbf{B}\times\mathbf{r} \qquad ; \ \Phi = 0 \qquad (4.133)$$

It is readily verified that this gives the correct magnetic field

$$\boldsymbol{\nabla}\times\mathbf{A} = \frac{1}{2}\boldsymbol{\nabla}\times(\mathbf{B}\times\mathbf{r}) = \frac{1}{2}\left[\mathbf{B}(\boldsymbol{\nabla}\cdot\mathbf{r}) - (\mathbf{B}\cdot\boldsymbol{\nabla})\mathbf{r}\right] = \mathbf{B} \qquad (4.134)$$

---

[27]See Prob. 4.9.

Now substitute Eqs. (4.133) into Eqs. (4.131), and use some simple vector manipulations

$$H' = -\frac{e}{2m_0c}\left[\mathbf{p}\cdot\frac{1}{2}(\mathbf{B}\times\mathbf{r}) + \frac{1}{2}(\mathbf{B}\times\mathbf{r})\cdot\mathbf{p}\right] + \frac{e^2}{2m_0c^2}\frac{1}{4}(\mathbf{B}\times\mathbf{r})\cdot(\mathbf{B}\times\mathbf{r})$$

$$= -\frac{e}{2m_0c}\mathbf{B}\cdot(\mathbf{r}\times\mathbf{p}) + \frac{e^2}{8m_0c^2}\mathbf{B}\cdot[\mathbf{r}\times(\mathbf{B}\times\mathbf{r})]$$

$$= -\frac{e}{2m_0c}\mathbf{B}\cdot(\mathbf{r}\times\mathbf{p}) + \frac{e^2}{8m_0c^2}\left[\mathbf{B}^2\mathbf{r}^2 - (\mathbf{B}\cdot\mathbf{r})^2\right] \tag{4.135}$$

The angular momentum of the particle is identified as

$$\boldsymbol{\mathcal{L}} = \mathbf{r}\times\mathbf{p} = \hbar\mathbf{L} \qquad ; \text{ angular momentum} \tag{4.136}$$

Hence the perturbation for a particle of charge e moving in a central potential $V(r)$ with an additional constant uniform magnetic field $\mathbf{B}$ becomes

$$H' = -\frac{e\hbar}{2m_0c}\mathbf{B}\cdot\mathbf{L} + \frac{e^2}{8m_0c^2}\left[\mathbf{B}^2\mathbf{r}^2 - (\mathbf{B}\cdot\mathbf{r})^2\right] \tag{4.137}$$

If the magnetic field is taken to define the z-direction as in Fig. 4.12, and one writes $\mathbf{r} = (x, y, z)$, then

$$H' = -\frac{e\hbar}{2m_0c}BL_z + \frac{e^2B^2}{8m_0c^2}(x^2 + y^2) \qquad ; \mathbf{B} = B\mathbf{e}_z \tag{4.138}$$

Consider an electron in an atom. The electron charge is negative, so

$$e = -|e| \qquad ; \text{ electron} \tag{4.139}$$

The *Bohr magneton* is defined by[28]

$$\mu_0 \equiv \frac{|e|\hbar}{2m_ec} \qquad ; \text{ Bohr magneton}$$
$$= 9.274 \times 10^{-21} \text{ erg/Gauss}$$
$$= 5.789 \times 10^{-9} \text{ eV/Gauss} \tag{4.140}$$

Now assume *weak fields*, where $B \to 0$.[29] The perturbation in Eq. (4.138) then becomes

$$H' = \mu_0 BL_z \qquad ; B \to 0 \tag{4.141}$$

---

[28] Note 1.602×10^{-12} erg $= 1$ eV.
[29] The precise criterion is discussed below.

The general matrix element of this perturbation between the eigenstates ϕ_{nlm} in a central field is readily calculated

$$\langle \phi_{n'l'm'}|H'|\phi_{nlm}\rangle = \mu_0 B \langle \phi_{n'l'm'}|L_z|\phi_{nlm}\rangle \qquad (4.142)$$

Recall that in a central field

$$\phi_{nlm} = R_{nl}(r)Y_{lm}(\theta, \phi) \qquad\qquad ; \text{ central field}$$

$$L_z Y_{lm}(\theta, \phi) = \frac{1}{i}\frac{\partial}{\partial \phi}Y_{lm}(\theta, \phi) = mY_{lm}(\theta, \phi) \qquad (4.143)$$

Therefore

$$\langle n'l'm'|H'|nlm\rangle = \mu_0 Bm\, \delta_{n'n}\delta_{l'l}\delta_{m'm} \qquad (4.144)$$

Several comments:

- This provides an example of degenerate perturbation theory–I. Consider the hydrogen spectrum in Fig. 4.10, which exhibits maximum degeneracy. The perturbation is diagonal in the degenerate subspaces;
- The energy shift of the levels is then given by

$$\delta E_{nlm} = \mu_0 Bm \qquad ; \text{ normal Zeeman effect} \qquad (4.145)$$

This is the *normal Zeeman effect*. It is illustrated in Fig. 4.13.

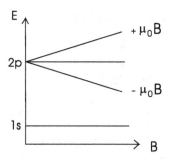

Fig. 4.13   Normal Zeeman effect for the 2p-level, where $E_{2pm}^{(1)} = E_{2p}^0 + \mu_0 Bm$.

- The energy shift in *independent of the radial wave function $R_{nl}(r)$*;
- It depends on m, and the slope as a function of B gives μ_0;
- There is no spin in the analysis here, and the normal Zeeman effect is only applicable to something like a π^--atom, where the pion is spinless;

- The result in Eq. (4.145) is more general than first-order perturbation theory, for with Eq. (4.144), we have actually *solved this problem exactly!* The full weak-field hamiltonian is

$$H = H_0 + \mu_0 B L_z \qquad ; \text{ weak fields} \qquad (4.146)$$

The eigenfunctions of H_0 in Eqs. (4.143) are again *eigenfunctions of the full hamiltonian*

$$H\phi_{nlm} = (E_{nl}^0 + \mu_0 B m)\phi_{nlm} \qquad ; \text{ eigenfunctions} \qquad (4.147)$$

- It remains to examine the *criterion for weak fields*. From Eq. (4.138), the condition that the second term can be neglected is

$$\frac{e^2 B^2}{8 m_0 c^2}\langle x^2 + y^2 \rangle \ll \frac{|e|\hbar}{2 m_0 c}B|m| \qquad ; \text{ weak fields} \qquad (4.148)$$

The l.h.s. can be estimated for hydrogen-like atoms as

$$\langle x^2 + y^2 \rangle \sim \langle r^2 \rangle \sim \left(\frac{n a_0}{Z}\right)^2 = \left(\frac{n}{Z\alpha}\frac{\hbar}{m_0 c}\right)^2 \qquad (4.149)$$

The weak-field condition then becomes

$$\frac{1}{2}\frac{\mu_0 B}{m_0 c^2} \ll \frac{Z^2 \alpha^2}{n^2}|m| \qquad ; \text{ weak fields} \qquad (4.150)$$

This is always satisfied for small enough B, with any finite $|m|$.[30]

4.2.4.5 *Uniform Electric Field—Stark Effect*

Consider next a particle with mass m and charge e on a spring with spring constant k aligned along the x-axis, and let the equilibrium position be at $x = 0$ (see Fig. 4.14). Assume an additional constant uniform electric field $\mathcal{E} = \mathcal{E}\mathbf{e}_x$, which can be described with the potentials

$$\Phi = -\mathcal{E}x \qquad ; \mathbf{A} = 0 \qquad (4.151)$$

From Eq. (4.131), the hamiltonian for this system is

$$H = H_0 - e\mathcal{E}x$$
$$H_0 = \frac{p^2}{2m} + \frac{1}{2}kx^2 \qquad ; k = m\omega_0^2 \qquad (4.152)$$

---

[30]Note that if the orbit is large enough (n large enough), the condition will eventually be violated for any B.

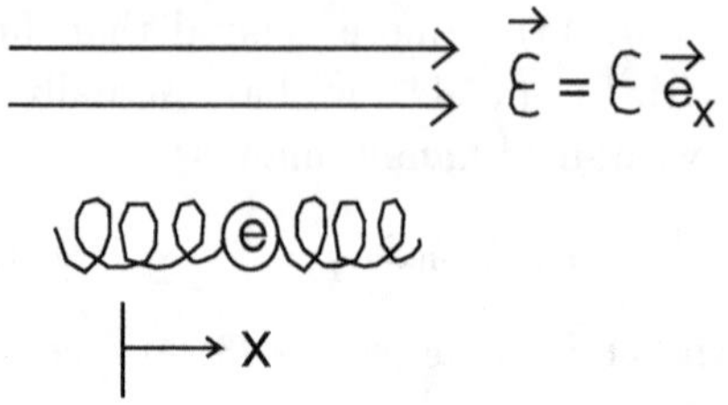

Fig. 4.14 Particle of charge e on a spring lying along the x-axis in a uniform electric field $\boldsymbol{\mathcal{E}} = \mathcal{E}\mathbf{e}_x$, where $\mathbf{e}_x$ is a unit vector in the x-direction. The spring constant is k, and the equilibrium position is $x = 0$.

The spectrum of the simple harmonic oscillator is non-degenerate, and we will do non-degenerate perturbation theory with the perturbation

$$H' = -e\mathcal{E}x \tag{4.153}$$

There is no diagonal matrix element of the coordinate x, so that[31]

$$\langle n|H'|n\rangle = -e\mathcal{E}\langle n|x|n\rangle = 0 \tag{4.154}$$

One must therefore go to second-order perturbation theory to get the energy shift, and from Eq. (4.60)

$$\delta E_n \equiv E_n^{(2)} - E_n^{(0)} = \sum_{m \neq n} \frac{|\langle m|H'|n\rangle|^2}{E_n^0 - E_m^0}$$

$$= e^2\mathcal{E}^2 \sum_{m \neq n} \frac{|\langle m|x|n\rangle|^2}{E_n^0 - E_m^0} \tag{4.155}$$

With the simple harmonic oscillator, the coordinate x connects only to the states with $m = n \pm 1$, and it follows from Eqs. (2.98)–(2.99) that

$$\sum_{m \neq n} \frac{|\langle m|x|n\rangle|^2}{E_n^0 - E_m^0} = \frac{|\langle n-1|x|n\rangle|^2}{\hbar\omega_0} + \frac{|\langle n+1|x|n\rangle|^2}{-\hbar\omega_0}$$

$$= \frac{1}{\hbar\omega_0}\frac{\hbar}{2m\omega_0}[n - (n+1)] = -\frac{1}{2m\omega_0^2} \tag{4.156}$$

Since $m\omega_0^2 = k$, this gives

$$\delta E_n = -\frac{e^2\mathcal{E}^2}{2k} \qquad ; \text{ 2nd-order energy shift} \tag{4.157}$$

We make several comments:

------

[31] This follows from parity considerations.

- The shift in energy in a uniform electric field is the *Stark effect*;
- The electric *polarizability* of a system is defined in general by

$$\delta E_n \equiv -\frac{1}{2}\alpha\mathcal{E}^2 \quad ; \text{ polarizability } \alpha \quad (4.158)$$

It follows that here

$$\alpha = \frac{e^2}{k} \quad ; \text{ oscillator} \quad (4.159)$$

- The induced *dipole moment* of a system is given in terms of the polarizability by

$$d = \alpha\mathcal{E} \quad ; \text{ induced dipole moment} \quad (4.160)$$

This relation can be verified in the present case through the use of the new wave functions

$$\psi_n^{(1)}(x) = \phi_n(x) + \sum_{m\neq n} \frac{\phi_m(x)\langle\phi_m|H'|\phi_n\rangle}{E_n^0 - E_m^0} \quad (4.161)$$

The induced dipole moment is computed from this wave function as

$$d = \langle\psi_n^{(1)}|ex|\psi_n^{(1)}\rangle \quad (4.162)$$

Through first-order in H', this gives[32]

$$d = \sum_{m\neq n} \left[\frac{\langle\phi_n|ex|\phi_m\rangle\langle\phi_m|H'|\phi_n\rangle}{E_n^0 - E_m^0} + \frac{\langle\phi_n|H'|\phi_m\rangle\langle\phi_m|ex|\phi_n\rangle}{E_n^0 - E_m^0} \right]$$

$$= -\frac{2}{\mathcal{E}} \sum_{m\neq n} \frac{|\langle m|H'|n\rangle|^2}{E_n^0 - E_m^0} = -\frac{2}{\mathcal{E}} \delta E_n \quad (4.163)$$

Hence from Eq. (4.157)

$$d = \frac{e^2}{k}\mathcal{E} = \alpha\mathcal{E} \quad ; \text{ oscillator} \quad (4.164)$$

which verifies the result in Eq. (4.160).

---

[32]Note that $\langle\phi_n|ex|\phi_n\rangle = 0$; recall also Prob. 4.3.

- This problem can actually be *solved exactly*. The hamiltonian in Eq. (4.152) can be re-written be completing the square

$$H = \frac{p^2}{2m} + \frac{1}{2}kx^2 - e\mathcal{E}x$$

$$= \frac{p^2}{2m} + \frac{1}{2}k\left(x - \frac{e\mathcal{E}}{k}\right)^2 - \frac{e^2\mathcal{E}^2}{2k} \qquad (4.165)$$

Now shift the coordinate by the constant displacement $e\mathcal{E}/k$

$$\xi \equiv x - \frac{e\mathcal{E}}{k} \qquad ; \text{ shifted coordinate}$$

$$p_x = \frac{\hbar}{i}\frac{\partial}{\partial x} = \frac{\hbar}{i}\frac{\partial}{\partial \xi} = p_\xi \qquad (4.166)$$

Then

$$H = H_0 - \frac{e^2\mathcal{E}^2}{2k}$$

$$H_0 = \frac{p_\xi^2}{2m} + \frac{1}{2}k\xi^2 \qquad ; \text{ again just s.h.o.} \qquad (4.167)$$

Here H_0 is again just the hamiltonian of the simple harmonic oscillator. The exact solution to this problem is thus given by

$$H_0\phi_n(\xi) = \hbar\omega_0\left(n + \frac{1}{2}\right)\phi_n(\xi) \qquad ; \text{ s.h.o.} \qquad (4.168)$$

$$H\phi_n(\xi) = \left[\hbar\omega_0\left(n + \frac{1}{2}\right) - \frac{e^2\mathcal{E}^2}{2k}\right]\phi_n(\xi)$$

$$E_n = \hbar\omega_0\left(n + \frac{1}{2}\right) - \frac{e^2\mathcal{E}^2}{2k} \qquad ; \text{ exact eigenvalues}$$

The second-order perturbation theory result gives the exact answer for the energy eigenvalues in this case![33]

4.2.5 *Degenerate Perturbation Theory–II*

Consider a degenerate subspace, as illustrated in Fig. 4.4,

$$H_0\phi_{n,\nu} = E_n^0\phi_{n,\nu} \qquad ; \nu = 1, 2, \cdots, p$$

$$p\text{-fold degenerate} \qquad (4.169)$$

---

[33]Since $\xi = x - e\mathcal{E}/k$, the field $\mathcal{E}$ enters in a complicated fashion in the wave functions.

The level is here p-fold degenerate. Now apply a perturbation H'.

- If the perturbation is diagonal in the degenerate subspace, we have the previously-discussed case of degenerate perturbation theory–I;
- Suppose the perturbation is *not* diagonal in this subspace. We observe that any new orthonormal linear combination of the degenerate eigenstates is again an eigenstate of H_0 with eigenvalue E_n^0

$$\sum_{\nu=1}^{p} c_\nu^{(s)} \phi_{n,\nu} \equiv \chi_s \qquad ; s = 1, 2, \cdots, p$$

$$H_0 \chi_s = E_n^0 \chi_s \qquad (4.170)$$

Let us try to choose the χ_s so that H' *is* again diagonal in the subspace

$$\langle \chi_s | H' | \chi_t \rangle = \langle \chi_s | H' | \chi_s \rangle \delta_{st} \qquad (4.171)$$

If this can be accomplished, the problem is reduced to degenerate perturbation theory–I! To see this, simply start with the degenerate set $\{\chi_1, \cdots, \chi_p\}$ instead of the $\{\phi_{n,1}, \cdots, \phi_{n,p}\}$. A repetition of the previous arguments then gives

$$\psi_s^{(1)} = \chi_s + \sum_{m \neq s} \frac{\phi_m \langle \phi_m | H' | \chi_s \rangle}{E_n^0 - E_m^0} \qquad ; E_m^0 \neq E_n^0$$

$$E_s^{(1)} = E_n^0 + \langle \chi_s | H' | \chi_s \rangle \qquad (4.172)$$

Since H_0 is already diagonal, the condition in Eq. (4.171) is equivalent to

$$\langle \chi_s | H | \chi_t \rangle = E_s^{(1)} \delta_{st} \qquad (4.173)$$

Can one find a set of coefficients and wave functions in Eqs. (4.170) that satisfy the condition in Eq. (4.173)? The answer is *yes*. *This is just the basic problem in matrix mechanics, as analyzed previously.* One proceeds as follows in the degenerate subspace:

(1) Determine the eigenvalues $E_{(1)}, \cdots, E_{(p)}$ by setting

$$\det \left[\underline{H} - E \underline{I} \right] = 0 \qquad ; \text{gives } E_{(1)}, \cdots, E_{(p)} \qquad (4.174)$$

(2) Solve the resulting set of linear equations for each $E_{(s)}$ to determine the coefficients $\underline{c}^{(s)}$

$$[\underline{H} - E_{(s)}\underline{I}][\underline{c}^{(s)}] = 0 \qquad ; \text{ gives } \underline{c}^{(s)} = \begin{pmatrix} c_1^{(s)} \\ \vdots \\ c_p^{(s)} \end{pmatrix} \qquad (4.175)$$

(3) Construct the modal matrix

$$M_{st} \equiv c_s^{(t)} \qquad ; \ (s,t) = 1, 2, \cdots, p \qquad (4.176)$$

In matrix notation, this is

$$\underline{M} = \begin{bmatrix} \underline{c}^{(1)} & \cdots & \underline{c}^{(p)} \\ \downarrow & & \downarrow \end{bmatrix} \qquad (4.177)$$

Then

$$\underline{M}^\dagger \underline{M} = \underline{I}$$

$$\underline{M}^\dagger \underline{H} \underline{M} = \underline{H}_D = \begin{pmatrix} E_{(1)} & & \\ & \ddots & \\ & & E_{(p)} \end{pmatrix} \qquad (4.178)$$

The hamiltonian has been diagonalized in this subspace.[34]
(4) The new wave functions are given by

$$\chi_s = \sum_{\nu=1}^{p} M_{\nu s} \phi_{n,\nu} = \sum_{\nu=1}^{p} c_\nu^{(s)} \phi_{n,\nu} \qquad ; \ s = 1, 2, \cdots, p \qquad (4.179)$$

They are orthonormal.

4.2.6 *Application: Linear Stark Effect in Hydrogen*

As an example, consider atomic hydrogen in a constant uniform electric field oriented in the z-direction (Fig. 4.15). The perturbation is

$$H' = e\Phi = -e\mathcal{E}z \qquad ; \ \boldsymbol{\mathcal{E}} = \mathcal{E}\mathbf{e}_z \qquad (4.180)$$

This can be re-written in spherical coordinates as (see Fig. 2.6)

$$H' = -e\mathcal{E}r\cos\theta = -e\mathcal{E}r \left(\frac{4\pi}{3}\right)^{1/2} Y_{10}(\theta, \phi) \qquad (4.181)$$

---

[34]In the previous notation $E_{(s)} = \langle \chi_s | H | \chi_s \rangle = E_s^{(1)}$.

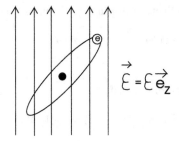

Fig. 4.15   One-electron atom in a constant uniform electric field $\boldsymbol{\mathcal{E}} = \mathcal{E}\mathbf{e}_z$.

We focus on the behavior of the first four degenerate excited states with energy E_1^0, as illustrated in Fig. 4.16.   The wave functions $\psi_{nlm}(r, \theta, \phi)$ for

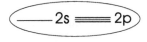

Fig. 4.16   Four states in the degenerate subspace with energy E_1^0 in hydrogen.

these states are given in Probs. 2.14 and 2.22

$$\psi_{200}(r, \theta, \phi) = \left(\frac{Z}{2a_0}\right)^{3/2} (2 - \rho)e^{-\rho/2}\,Y_{00}(\theta, \phi) \qquad ; \rho = \frac{Zr}{a_0}$$

$$\psi_{21m}(r, \theta, \phi) = \frac{1}{\sqrt{3}}\left(\frac{Z}{2a_0}\right)^{3/2} \rho e^{-\rho/2}\,Y_{1m}(\theta, \phi) \qquad (4.182)$$

We proceed to construct the hamiltonian matrix $\langle i|H|j\rangle$ for these four degenerate states; it is *not* diagonal.

There are 16 matrix elements of H_{ij} in this case. Fortunately, simple symmetry arguments reduce the problem to the calculation of just one of them![35]

- The behavior of the integrand under spatial reflection eliminates many

---

[35]Remember that $H_{ji} = H_{ij}^\star$, since H is hermitian.

of them. Spatial reflection implies $\mathbf{r} \to -\mathbf{r}$, or (see Fig. 2.6)

$$\theta \to \pi - \theta, \qquad \phi \to \phi + \pi, \qquad r \to r \qquad ; \text{ spatial reflection}$$
$$(4.183)$$

The spherical harmonics satisfy[36]

$$Y_{lm}(\pi - \theta, \phi + \pi) = (-1)^l Y_{lm}(\theta, \phi) \qquad (4.184)$$

Under spatial reflection, the overall wave function then behaves as

$$\psi_{nlm}(-\mathbf{r}) = (-1)^l \psi_{nlm}(\mathbf{r}) \qquad ; \text{ parity} \qquad (4.185)$$

This is the *parity* of the wave function. Now consider the matrix element

$$\int d^3r \, \psi^\star_{n'l'm'}(\mathbf{r}) \, z \, \psi_{nlm}(\mathbf{r}) \equiv \langle n'l'm' | z | nlm \rangle \qquad (4.186)$$

Change variables to $\mathbf{r}' = -\mathbf{r}$. Then[37]

$$\int d^3r' = \int d^3r \qquad ; \, \mathbf{r}' = -\mathbf{r} \qquad (4.187)$$

Hence

$$\langle n'l'm' | z | nlm \rangle = (-1)^{1+l+l'} \langle n'l'm' | z | nlm \rangle \qquad (4.188)$$
$$\implies \qquad 1 + l + l' \text{ an even integer} \qquad ; \text{ parity}$$

This is a *parity selection rule*. It implies the perturbation in Eq. (4.180) only connects the s- and p-states in Fig. 4.16, and this reduces the number of required matrix elements to three;

- Now consider the angular integrals, using the form of the perturbation in Eq. (4.181). Observe that

$$\int d\Omega \, Y^\star_{1m}(\theta, \phi) Y_{10}(\theta, \phi) Y_{00} = \frac{1}{\sqrt{4\pi}} \int d\Omega \, Y^\star_{1m}(\theta, \phi) Y_{10}(\theta, \phi) \qquad (4.189)$$

$$= \frac{1}{\sqrt{4\pi}} \delta_{m0} \qquad ; \text{ angular momentum}$$

This is an *angular-momentum selection rule*. It implies that the perturbation only connects the s-state to the p-state with $m = 0$, and hence the only remaining non-zero matrix element is $\langle 210 | H' | 200 \rangle$!

---

[36] See Prob. 4.11.
[37] Use $\int_{-\infty}^\infty dx = \int_{-\infty}^\infty dx'$, etc.

With the use of the wave functions in Eqs. (4.182), and the perturbation in Eqs. (4.181), the required matrix element becomes

$$\langle 210|H'|200\rangle = \frac{1}{8\sqrt{3}}\left[-e\mathcal{E}\frac{a_0}{Z}\left(\frac{4\pi}{3}\right)^{1/2}\right]\frac{1}{\sqrt{4\pi}}\int_0^\infty e^{-\rho}\rho^2(2-\rho)\rho^2\,d\rho$$

$$= -\frac{e\mathcal{E}a_0}{24Z}(2\cdot 4! - 5!) \tag{4.190}$$

Thus the required matrix element is

$$\langle 210|H'|200\rangle = \frac{3e\mathcal{E}a_0}{Z} \equiv \lambda \tag{4.191}$$

We now proceed to *diagonalize* the hamiltonian matrix in the degenerate subspace, which, since it only mixes the two states $|200\rangle$ and $|210\rangle$, takes the form

$$\underline{H} = \begin{pmatrix} E^0 & \lambda \\ \lambda & E^0 \end{pmatrix} \tag{4.192}$$

We proceed through the appropriate steps:

(1) The eigenvalues are obtained from

$$\det\begin{vmatrix} E^0 - E & \lambda \\ \lambda & E^0 - E \end{vmatrix} = 0$$

$$(E - E^0)^2 - \lambda^2 = 0 \tag{4.193}$$

Therefore

$$E_\pm = E^0 \pm \lambda \qquad ; \text{ eigenvalues} \tag{4.194}$$

(2) The wave-function coefficients corresponding to the eigenvalue E_+ satisfy

$$\begin{pmatrix} -\lambda & \lambda \\ \lambda & -\lambda \end{pmatrix}\begin{pmatrix} c_1^{(+)} \\ c_2^{(+)} \end{pmatrix} = 0 \tag{4.195}$$

Thus the normalized coefficients in this case are[38]

$$c_1^{(+)} = c_2^{(+)} = \frac{1}{\sqrt{2}} \tag{4.196}$$

---

[38] With an appropriate choice of overall phase.

Similarly, the wave-function coefficients corresponding to the eigenvalue E_- satisfy

$$\begin{pmatrix} \lambda & \lambda \\ \lambda & \lambda \end{pmatrix} \begin{pmatrix} c_1^{(-)} \\ c_2^{(-)} \end{pmatrix} = 0 \qquad (4.197)$$

Thus the normalized coefficients in this case are

$$c_1^{(-)} = -c_2^{(-)} = \frac{1}{\sqrt{2}} \qquad (4.198)$$

(3) The modal matrix for this problem is therefore

$$\underline{M} = \frac{1}{\sqrt{2}} \begin{pmatrix} 1 & 1 \\ 1 & -1 \end{pmatrix} \qquad ; \text{ modal matrix} \quad (4.199)$$

(4) The new eigenstates that diagonalize the hamiltonian in the degenerate subspace of the first excited states with energy E_1^0 for the hydrogen atom (Fig. 4.16), in a uniform electric field $\boldsymbol{\mathcal{E}} = \mathcal{E} \mathbf{e}_z$ (Fig. 4.15), are thus

$$\chi_\pm(\mathbf{r}) = \frac{1}{\sqrt{2}} [\psi_{200}(\mathbf{r}) \pm \psi_{210}(\mathbf{r})] \qquad ; \text{ eigenstates}$$

$$E_\pm = E_1^0 \pm \frac{3ea_0}{Z} \mathcal{E} \qquad ; \text{ eigenvalues} \qquad (4.200)$$

Recall that in hydrogen, the electron charge is negative [see Eq. (4.139)]. We proceed to discuss these results:

- This is the *linear Stark effect* in the degenerate subspace with energy E_1^0 in hydrogen. The shift of the two levels that mix is linear in the applied electric field $\mathcal{E}$ (Fig. 4.17).

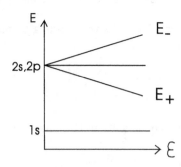

Fig. 4.17   Linear Stark effect in the degenerate subspace with energy E_1^0 in hydrogen.

In the previous non-degenerate result in Eq. (4.157), the shift went as $\mathcal{E}^2$;

- Let us calculate the *electric dipole moment* in the two mixed states $\chi_\pm$. The charge displacement is

$$\langle \chi_\pm | z | \chi_\pm \rangle = -\frac{1}{e\mathcal{E}} \langle \chi_\pm | H' | \chi_\pm \rangle = \mp \frac{3a_0}{Z} \qquad (4.201)$$

The result is *independent of $\mathcal{E}$*. These two states have a *permanent electric dipole moment!* (See Fig. 4.18.)

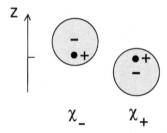

Fig. 4.18 Charge displacement in the mixed states $\chi_\pm$ obtained in Eq. (4.201). Compare Prob. 4.12.

- The character of the eigenstates in the presence of the field is here determined *entirely by H'*. The unperturbed hamiltonian no longer controls the situation. Here the splittings are *not* small compared to the unperturbed spacings, since the unperturbed levels are *degenerate*.

Chapter 5

Scattering Theory

So far, the discussion has focused on bound states with discrete eigenvalues and localized wave functions. With finite potentials, there are additional continuum solutions to the Schrödinger equation, and these are analyzed with *scattering theory*.[1]

5.1 Formulation of Problem

We proceed to formulate the scattering problem.

- The analysis is concerned with continuum solutions to the time-independent Schrödinger equation in three dimensions;
- It will be assumed that the potential $V(r)$ is spherically symmetric and of finite range

$$rV(r) \to 0 \qquad ; r \to \infty \qquad (5.1)$$

- The discussion is based on *scattering states*, which are stationary states where the physics is contained in the *ratio of probability fluxes* (see Vol. I). If an experiment is performed over and over again with many particles, these ratios give the ratios of *count rates*;
- The time-independent Schrödinger equation is

$$\left[-\frac{\hbar^2}{2m} \nabla^2 + V(r) \right] \psi(\mathbf{r}) = E\psi(\mathbf{r}) \qquad ; \text{ time-independent S-eqn} \quad (5.2)$$

Here the analysis applies to either

(1) A particle of mass m moving in a fixed potential $V(r)$; or

---

[1] See, for example, [Newton (1982); Goldberger and Watson (2004)].

(2) A two-body problem in the center-of-mass (C-M) system where (see appendix A)

    – The coordinate $\mathbf{r} = \mathbf{r}_1 - \mathbf{r}_2$ is the relative coordinate
    – The mass m is the reduced mass $\mu = m_1 m_2/(m_1 + m_2)$.

- In contrast to the discrete eigenvalues of the bound states, in the scattering states the initial energy is *continuous and specified*. It is determined, for example, by the energy of a beam coming out of an accelerator.

$$E = E_{\text{inc}} = \frac{\hbar^2 k^2}{2m} \qquad ; \text{given} \qquad (5.3)$$

The Schrödinger equation then takes the form

$$(\nabla^2 + k^2)\psi(\mathbf{r}) = v(r)\psi(\mathbf{r}) \qquad ; v \equiv \frac{2m}{\hbar^2} V \qquad (5.4)$$

5.1.1 Scattering Boundary Conditions

The scattering state is now illustrated in Fig. 5.1.

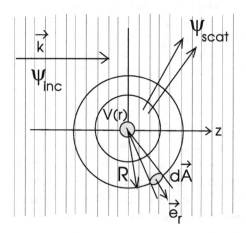

Fig. 5.1 Scattering state with a central potential $V(r)$. There is an incident wave $\sqrt{V}\,\psi_{\text{inc}} = e^{i\mathbf{k}\cdot\mathbf{r}}$ where $\mathbf{k} = k\mathbf{e}_z$ defines the z-axis. There is an additional scattered wave, which far away from the origin takes the form $\sqrt{V}\,\psi_{\text{scat}} = f(\theta, \phi)e^{ikr}/r$. $d\mathbf{A} = \mathbf{e}_r dA$ is a small element of surface area on a large sphere of radius R surrounding the scatterer. Here $(\mathbf{e}_z, \mathbf{e}_r)$ are unit vectors.

Far away from the scatterer on the left in Fig. 5.1, where $V(r) = 0$, one sends in an *incident wave* of given energy $E_{\rm inc}$ and momentum $\mathbf{p} = \hbar\mathbf{k}$

$$\psi(\mathbf{r}) = \psi_{\rm inc}(\mathbf{r}) = \frac{1}{\sqrt{V}} e^{i\mathbf{k}\cdot\mathbf{r}} \qquad ; \text{ incident wave} \qquad (5.5)$$

We take $\mathbf{k} = k\mathbf{e}_z$ to define the z-axis (see Fig. 2.6).

When the potential is included, there is an additional *scattered wave*, and everywhere in space the total wave function takes the form

$$\psi(\mathbf{r}) = \psi_{\rm inc}(\mathbf{r}) + \psi_{\rm scatt}(\mathbf{r}) \qquad ; \text{ total wave function} \qquad (5.6)$$

Far away from the scatterer, the scattered wave function takes the form of an outgoing spherical wave emanating from the scatterer, and asymptotically

$$\psi(\mathbf{r}) \to \frac{1}{\sqrt{V}} \left[e^{i\mathbf{k}\cdot\mathbf{r}} + f(\theta,\phi)\frac{e^{ikr}}{r} \right] \qquad ; \ r \to \infty \qquad (5.7)$$

Four comments:[2]

- The total wave function is the sum of the incident wave and the outgoing scattered wave (see Vol. I);
- The factor of $1/r$ in the outgoing scattered wave is required to produce a finite outgoing radial flux;
- The outgoing wave has the same wavenumber k as the incident wave for elastic scattering;
- The quantity $f(\theta,\phi)$ is known as the *scattering amplitude*.

5.1.2 *Cross Section*

Suppose one has an incident *beam of particles*, with an incident particle flux of

$$\mathcal{I}_{\rm inc} \equiv \text{\# of particles crossing unit area} \perp \text{beam/sec}$$

$$\qquad ; \text{ incident particle flux} \qquad (5.8)$$

The *cross section* $d\sigma$ is defined as follows. If a particle goes through a little element of transverse area $d\sigma$, it leads to an *event*. In this case the event is elastic scattering into a solid angle $d\Omega$ (see Fig. 5.2). Hence

$$\mathcal{I}_{\rm inc}\, d\sigma = \text{\# of particles scattered into solid angle } d\Omega/\text{sec} \qquad (5.9)$$

---

[2]We shall later show that our solution to the scattering problem produces an asymptotic scattered wave of this form (see also Prob. 5.1). In elastic scattering, the outgoing particles again have an energy $E_{\rm inc}$.

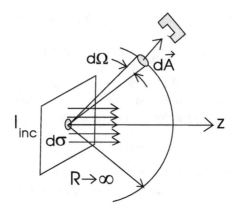

Fig. 5.2   All those particles in the incident beam passing through the little element of transverse area $d\sigma$ get scattered into the solid angle $d\Omega = dA/R^2$.

The cross section is therefore given by

$$d\sigma = \frac{\text{particle flux through dA}}{\text{incident particle flux}} \qquad ; \text{ cross section} \qquad (5.10)$$

Here the solid angle is $d\Omega = dA/R^2$, where R is the radius of a large sphere far away from the scatterer (Fig. 5.2). It is evident from this discussion that the cross section is an *experimental quantity* directly accessed through a scattering experiment.

In the form in Eq. (5.10), as a ratio of fluxes, the cross section can also be obtained theoretically as the ratio of quantum-mechanical one-body probability fluxes.[3] *If **S** is the probability current, then*

$$d\sigma = \frac{I_{\text{scatt}}}{I_{\text{inc}}} = \frac{\mathbf{S}_{\text{scat}} \cdot d\mathbf{A}}{\mathbf{S}_{\text{inc}} \cdot \mathbf{e}_z} \qquad (5.11)$$

As before, the probability current is given by

$$\mathbf{S} = \frac{\hbar}{2im}[\psi^\star \boldsymbol{\nabla}\psi - (\boldsymbol{\nabla}\psi)^\star\psi] \qquad ; \text{ probability current} \quad (5.12)$$

The insertion of ψ_{inc} from Eq. (5.5) gives

$$\mathbf{S}_{\text{inc}} = \frac{\hbar}{2im}[\psi_{\text{inc}}^\star \boldsymbol{\nabla}\psi_{\text{inc}} - (\boldsymbol{\nabla}\psi_{\text{inc}})^\star\psi_{\text{inc}}]$$
$$= \frac{1}{V}\frac{\hbar\mathbf{k}}{m} \qquad\qquad ; \psi_{\text{inc}} = \frac{1}{\sqrt{V}}e^{i\mathbf{k}\cdot\mathbf{r}} \quad (5.13)$$

---

[3] See Vol. I.

The incident probability flux in Fig. 5.2 is therefore[4]

$$I_{\text{inc}} = \mathbf{S}_{\text{inc}} \cdot \mathbf{e}_z = \frac{1}{V}\frac{\hbar k}{m} \qquad ; \text{ incident flux} \quad (5.14)$$

The scattered probability flux through the surface element $d\mathbf{A} = \mathbf{e}_r R^2 d\Omega$ at large R in Fig. 5.2 is

$$\mathbf{S}_{\text{scat}} \cdot d\mathbf{A} = \frac{\hbar}{2im}\left[\psi_{\text{scat}}^\star \frac{\partial \psi_{\text{scat}}}{\partial r} - \left(\frac{\partial \psi_{\text{scat}}}{\partial r}\right)^\star \psi_{\text{scat}}\right]_R R^2 d\Omega$$
$$; \text{ scattered flux} \quad (5.15)$$

It is assumed here that the detector in Fig. 5.2 shields out the incident beam and one only observes the scattered beam coming from the target. Now from Eq. (5.7), for large r

$$\psi_{\text{scat}}(\mathbf{r}) = \frac{1}{\sqrt{V}}\left[f(\theta,\phi)\frac{e^{ikr}}{r}\right] \qquad ; r \to \infty \qquad (5.16)$$

It follows that the scattered probability flux is given for large R by[5]

$$I_{\text{scat}} = \mathbf{S}_{\text{scat}} \cdot d\mathbf{A} = \frac{1}{V}\frac{\hbar k}{m}|f(\theta,\phi)|^2 \frac{1}{R^2}R^2 d\Omega$$
$$= \frac{1}{V}\frac{\hbar k}{m}|f(\theta,\phi)|^2 d\Omega \qquad ; \text{ scattered flux} \quad (5.17)$$

Note the following:

- The factors of R^2 *cancel* in this expression, and the result for I_{scat} is simply proportional to the element of solid angle $d\Omega$;
- The ratio of scattered flux to incident flux is *independent of the overall norm of the wave function* $1/V$.

The differential cross section is given by the ratio of the expressions in Eqs. (5.17) and (5.14)

$$d\sigma = |f(\theta,\phi)|^2 d\Omega$$
$$\frac{d\sigma}{d\Omega} = |f(\theta,\phi)|^2 \qquad ; \text{ differential cross section} \quad (5.18)$$

The differential cross section $d\sigma/d\Omega$ is given by the absolute square of the scattering amplitude $|f(\theta,\phi)|^2$.

---

[4]This expression has a nice hydrodynamic interpretation as $I_{\text{inc}} = \rho v$, where $\rho = 1/V$ is the particle density and $v = \hbar k/m$ is the particle velocity.

[5]As $R \to \infty$, one only needs the radial derivative of the exponential in Eq. (5.16).

The goal is thus to solve the Schrödinger Eq. (5.4) everywhere in space, with a wave function of the form in Eq. (5.6), and then obtain the scattering amplitude $f(\theta, \phi)$ from the asymptotic expression in Eq. (5.7).

5.2   Scattering Green's Function

The time-independent Schrödinger Eq. (5.4) is solved for the scattering state with the aid of the *scattering Green's function*. Define the Green's function through the following inhomogeneous differential equation with a Dirac delta-function source[6]

$$(\nabla^2 + k^2)G_k(\mathbf{x}, \mathbf{y}) = -\delta^{(3)}(\mathbf{x} - \mathbf{y}) \qquad ; \text{Green's function} \quad (5.19)$$

Then the scattering solution to the Schrödinger Eq. (5.4) can be written as

$$\psi(\mathbf{x}) = \psi_{\text{inc}}(\mathbf{x}) - \int d^3y\, G_k(\mathbf{x}, \mathbf{y})v(\mathbf{y})\psi(\mathbf{y}) \qquad ; \text{scattering state} \quad (5.20)$$

Here $\psi_{\text{inc}}(\mathbf{x})$ is a solution to the homogeneous equation[7]

$$(\nabla^2 + k^2)\psi_{\text{inc}}(\mathbf{x}) = 0 \qquad ; \text{incident wave} \quad (5.21)$$

The solution in Eq. (5.20) can be verified by computing

$$(\nabla^2 + k^2)\psi(\mathbf{x}) = (+) \int d^3y\, \delta^{(3)}(\mathbf{x} - \mathbf{y})v(\mathbf{y})\psi(\mathbf{y}) = v(x)\psi(\mathbf{x}) \quad (5.22)$$

The Green's function can be constructed from plane waves as

$$G_k(\mathbf{x}, \mathbf{y}) = \int \frac{d^3t}{(2\pi)^3} \frac{e^{it\cdot(\mathbf{x}-\mathbf{y})}}{t^2 - k^2} \qquad (5.23)$$

This is immediately checked by calculating

$$(\nabla^2 + k^2)G_k(\mathbf{x}, \mathbf{y}) = \int \frac{d^3t}{(2\pi)^3} \left[\frac{k^2 - t^2}{t^2 - k^2}\right] e^{it\cdot(\mathbf{x}-\mathbf{y})}$$
$$= -\delta^{(3)}(\mathbf{x} - \mathbf{y}) \qquad (5.24)$$

A little thought, however, leads one to realize that the expression in Eq. (5.23) is not well-defined until we decide how to handle the singularities in the integrand at $t^2 = k^2$. This shall be done by specifying the

---

[6]Recall the arguments for the bound-state Green's function in Eqs. (4.50)– (4.53); in these equations $\nabla^2 = \nabla_x^2$.

[7]This is known as the *scalar Helmholtz equation.*

integration *contour*, with different contours resulting in Green's functions satisfying different *boundary conditions*.

To see this, define[8]

$$G_k^{(\pm)}(\mathbf{x}, \mathbf{y}) \equiv \int \frac{d^3t}{(2\pi)^3} \frac{e^{it \cdot (\mathbf{x} - \mathbf{y})}}{t^2 - (k \pm i\eta)^2} \qquad ; \eta \to 0^+$$

$$= \int \frac{d^3t}{(2\pi)^3} \frac{e^{it \cdot (\mathbf{x} - \mathbf{y})}}{t^2 - k^2 \mp i\eta} \tag{5.25}$$

Here η is a small positive quantity, and the limit $\eta \to 0$ is implied, in which case the second equality also holds. Let us evaluate this integral. Write it out as

$$G_k^{(\pm)}(\mathbf{x}, \mathbf{y}) = \frac{1}{8\pi^3} \int_0^\infty \frac{t^2 \, dt}{t^2 - (k \pm i\eta)^2} \int_0^{2\pi} d\phi \int_{-1}^1 d\mu \, e^{it|\mathbf{x} - \mathbf{y}|\mu} \tag{5.26}$$

$$= \frac{1}{4\pi^2} \int_0^\infty \frac{t^2 \, dt}{t^2 - (k \pm i\eta)^2} \frac{1}{it|\mathbf{x} - \mathbf{y}|} \left[e^{it|\mathbf{x} - \mathbf{y}|} - e^{-it|\mathbf{x} - \mathbf{y}|} \right]$$

Here we have used $\mu \equiv \cos\theta$. Now introduce $t \equiv -u$ in the last integral

$$G_k^{(\pm)}(\mathbf{x}, \mathbf{y}) = \frac{1}{4\pi^2 i |\mathbf{x} - \mathbf{y}|} \times$$

$$\left[\int_0^\infty \frac{t \, dt}{t^2 - (k \pm i\eta)^2} e^{it|\mathbf{x} - \mathbf{y}|} - \int_0^{-\infty} \frac{u \, du}{u^2 - (k \pm i\eta)^2} e^{iu|\mathbf{x} - \mathbf{y}|} \right] \tag{5.27}$$

Hence

$$G_k^{(\pm)}(\mathbf{x}, \mathbf{y}) = \frac{1}{4\pi^2 i |\mathbf{x} - \mathbf{y}|} \int_{-\infty}^\infty \frac{t \, dt}{t^2 - (k \pm i\eta)^2} e^{it|\mathbf{x} - \mathbf{y}|} \tag{5.28}$$

This can be converted to a closed-contour integral in the complex t-plane by adding the contribution from a large semi-circle of radius R in the upper-half t-plane, where there is exponential damping (see Fig. 5.3).

$$G_k^{(\pm)}(\mathbf{x}, \mathbf{y}) = \frac{1}{4\pi^2 i |\mathbf{x} - \mathbf{y}|} \oint_C \frac{t \, dt}{t^2 - (k \pm i\eta)^2} e^{it|\mathbf{x} - \mathbf{y}|} \qquad ; R \to \infty \tag{5.29}$$

The contribution from this semi-circle vanishes as $R \to \infty$.

The function $G_k^{(+)}(\mathbf{x}, \mathbf{y})$ has singularities at $t = \pm(k + i\eta)$, and only the pole at $t = k + i\eta$ lies within the contour C. One picks up only the

---

[8] The elements of complex analysis required here are detailed in appendix A of [Fetter and Walecka (2003a)] (see also appendix B of [Walecka (2010)]). In fact, this calculation of the scattering Green's function appears in both of those references. Problem 5.2 takes the reader through an alternate derivation of $G_k^{(\pm)}(\mathbf{x}, \mathbf{y})$.

residue of the pole at this singularity, and contour integration evaluates the integral as

$$G_k^{(+)}(\mathbf{x}, \mathbf{y}) = \frac{1}{4\pi^2 i |\mathbf{x} - \mathbf{y}|} 2\pi i \operatorname{Res} \left[\frac{t}{t^2 - (k + i\eta)^2} e^{it|\mathbf{x}-\mathbf{y}|} \right]_{t=k+i\eta}$$

$$= \frac{1}{4\pi^2 i |\mathbf{x} - \mathbf{y}|} 2\pi i \left[\frac{k}{2k} e^{ik|\mathbf{x}-\mathbf{y}|} \right] \tag{5.30}$$

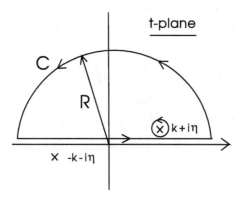

Fig. 5.3   Contour C and location of singularities at $t = \pm(k + i\eta)$ in the complex t-plane for the evaluation of the scattering Green's functions $G_k^{(+)}(\mathbf{x}, \mathbf{y})$. The radius of the semi-circle is R, and its contribution to the integral vanishes as $R \to \infty$ (see Prob. 5.3).

Thus

$$G_k^{(+)}(\mathbf{x}, \mathbf{y}) = \frac{e^{ik|\mathbf{x}-\mathbf{y}|}}{4\pi|\mathbf{x} - \mathbf{y}|} \qquad ; \text{ scattering Green's function} \tag{5.31}$$

This is the *scattering Green's function for the scalar Helmholtz equation with outgoing-wave boundary conditions.*[9]

A parallel calculation gives[10]

$$G_k^{(-)}(\mathbf{x}, \mathbf{y}) = \frac{e^{-ik|\mathbf{x}-\mathbf{y}|}}{4\pi|\mathbf{x} - \mathbf{y}|} \tag{5.32}$$

These Green's functions depend only on the vector difference $\mathbf{x} - \mathbf{y}$.

---

[9]See [Fetter and Walecka (2003a)].

[10]See Prob. 5.3. This is the scattering Green's function with *incoming-wave* boundary conditions. Its use is discussed, for example, in [Walecka (2010)].

5.3 Integral Equation

With the aid of the scattering Green's function, an *exact integral equation* can be constructed for the scattering state with outgoing-wave boundary conditions

$$\Psi_{\mathbf{k}}^{(+)}(\mathbf{x}) = \frac{1}{\sqrt{V}}e^{i\mathbf{k}\cdot\mathbf{x}} - \int d^3y\, G_k^{(+)}(\mathbf{x}-\mathbf{y})v(y)\Psi_{\mathbf{k}}^{(+)}(\mathbf{y}) \qquad (5.33)$$

The (irrelevant) overall norm of $1/V$ can be extracted by writing

$$\Psi_{\mathbf{k}}^{(+)}(\mathbf{x}) \equiv \frac{1}{\sqrt{V}}\psi_{\mathbf{k}}^{(+)}(\mathbf{x})$$

$$\psi_{\mathbf{k}}^{(+)}(\mathbf{x}) = e^{i\mathbf{k}\cdot\mathbf{x}} - \int d^3y\, G_k^{(+)}(\mathbf{x}-\mathbf{y})v(y)\psi_{\mathbf{k}}^{(+)}(\mathbf{y})$$

$$; \text{ exact integral equation} \qquad (5.34)$$

One can now show that this integral equation has the correct scattering boundary conditions *built in*. We observe that

- The vector $\mathbf{y}$ in the integral is confined to the region of the potential $v(y)$, which is of finite extent [see Eq. (5.1)];[11]
- The scattering boundary condition in Eq. (5.7) is applicable to the asymptotic regime far away from the scatterer where $|\mathbf{x}| \to \infty$. Note that here $|\mathbf{x}| = x \equiv r$, where r is usual radial coordinate in three dimensions;
- In this asymptotic regime, the magnitude of the vector difference can be expanded as

$$|\mathbf{x}-\mathbf{y}| = (x^2 - 2\mathbf{x}\cdot\mathbf{y} + y^2)^{1/2}$$

$$= x\left(1 - \frac{2\mathbf{x}\cdot\mathbf{y}}{x^2} + \frac{y^2}{x^2}\right)^{1/2}$$

$$\to x - \mathbf{e}_x\cdot\mathbf{y} + O\left(\frac{y^2}{x}\right) \qquad ; |\mathbf{x}| \to \infty \qquad (5.35)$$

Here $\mathbf{x} = x\mathbf{e}_x$, where $\mathbf{e}_x$ is a unit vector in the direction of observation. The situation is illustrated in Fig. 5.4.

Now define $k\mathbf{e}_x$ as the final wave vector $\mathbf{k}'$

$$k\mathbf{e}_x \equiv \mathbf{k}' \qquad (5.36)$$

---

[11]Recall from Eq. (5.4) that $v(y) \equiv 2mV(y)/\hbar^2$.

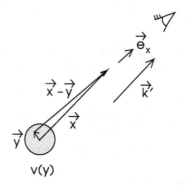

Fig. 5.4   Vector configuration where $|\mathbf{x}| \to \infty$ while $\mathbf{y}$ is confined to the potential. Here $\mathbf{x} = x\mathbf{e}_x$ where $\mathbf{e}_x$ is a unit vector in the direction of observation, and the final wave vector is defined by $\mathbf{k}' \equiv k\mathbf{e}_x$. Also here $|\mathbf{x}| = x \equiv r$ is the usual radial coordinate in three dimensions.

With the use of Eq. (5.35), the Green's function appearing in Eq. (5.34) then takes the asymptotic form

$$G_k^{(+)}(\mathbf{x} - \mathbf{y}) \to \frac{1}{4\pi x} e^{ikx} e^{-i\mathbf{k}' \cdot \mathbf{y}} \qquad\qquad ; \ x \to \infty \qquad (5.37)$$
$$; \ y \text{ in potential}$$

Here we are keeping finite corrections to the *phase* in the scattered wave, while neglecting corrections of $O(y/x)$ in its *amplitude*. The integral Eq. (5.34) then takes the asymptotic form

$$\psi_{\mathbf{k}}^{(+)}(\mathbf{x}) \to e^{i\mathbf{k}\cdot\mathbf{x}} + f(\theta, \phi)\frac{e^{ikx}}{x} \qquad\qquad ; \ x \to \infty$$
$$f(\theta, \phi) \equiv -\frac{1}{4\pi} \int d^3y \, e^{-i\mathbf{k}'\cdot\mathbf{y}} \, v(y)\psi_{\mathbf{k}}^{(+)}(\mathbf{y}) \qquad\qquad (5.38)$$

We proceed to discuss these results:

- Equation (5.34) is an *exact integral equation* for the scattering wave function $\psi_{\mathbf{k}}^{(+)}(\mathbf{x})$;
- The correct scattering boundary condition in Eq. (5.38) has been *built into* the Green's function. Recall that here $|\mathbf{x}| = x \equiv r$.
- Note that the asymptotic form of the scattering state in Eq. (5.38) comes out of the Green's function; it does not have to be assumed;[12]

---

[12]In this manner, one can extend scattering theory to any number of dimensions.

- The second of Eqs. (5.38) provides an *exact expression for the scattering amplitude* $f(\theta, \phi)$, and hence through the differential cross section in Eq. (5.18), a *direct connection with experiment*;
- One has to first solve for the scattering state $\psi_{\mathbf{k}}^{(+)}(\mathbf{x})$ everywhere, in particular *in the potential* $v(y)$, and then use it to compute $f(\theta, \phi)$;
- Note that $\mathbf{k}$ is the initial wave vector, and $\mathbf{k}'$ is the final wave vector in the scattering process (see Fig. 5.5).

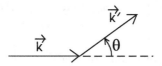

Fig. 5.5 Wave vectors in the scattering process. The initial and final particle momenta are $\mathbf{p} = \hbar\mathbf{k}$ and $\mathbf{p}' = \hbar\mathbf{k}'$, respectively. Here θ is the scattering angle.

One of the great advantages of formulating a problem through an inhomogeneous integral equation as in Eq. (5.34), is that the expression can be immediately *iterated*, which we proceed to do.[13]

5.4 Born's First Approximation

In order to calculate the scattering amplitude in Eq. (5.38), one needs to know the full scattering wave function $\psi_{\mathbf{k}}^{(+)}(\mathbf{y})$ in the region of the potential $v(y)$. The first approximation in Eq. (5.34) is to neglect the effect of the potential *on* the wave function and simply replace (see Fig. 5.6)

$$\psi_{\mathbf{k}}^{(+)}(\mathbf{y}) \approx \psi_{\mathrm{inc}}(\mathbf{y}) = e^{i\mathbf{k}\cdot\mathbf{y}} \qquad ; \text{ Born Approximation} \quad (5.39)$$

in the integral in Eq. (5.38). This gives

$$f_{\mathrm{BA}}(\mathbf{k}', \mathbf{k}) = -\frac{1}{4\pi} \int d^3y\, e^{-i\mathbf{q}\cdot\mathbf{y}}\, v(y) \qquad ; \mathbf{q} \equiv \mathbf{k}' - \mathbf{k}$$

$$\equiv -\frac{1}{4\pi} \tilde{v}(\mathbf{q}) \qquad\qquad (5.40)$$

Some comments:

- This is known as Born's first approximation for the scattering amplitude, or more simply, *Born approximation*;

---

[13] Compare Prob. 5.4.

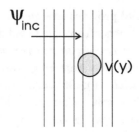

Fig. 5.6   First iteration of the scattering integral Eq. (5.34) assumes $\psi_{\mathbf{k}}^{(+)} = \psi_{inc}$ in the region of the potential $v(y)$.

- Here $\mathbf{q}$ is the "momentum transfer" in the scattering process (see Fig. 5.7)[14]

$$\mathbf{q} \equiv \mathbf{k}' - \mathbf{k} \qquad ; \text{ momentum transfer} \quad (5.41)$$

Since in elastic scattering $|\mathbf{k}'| = |\mathbf{k}|$, the square of this quantity is given by

$$q^2 = 2k^2(1 - \cos\theta) = 4k^2 \sin^2\frac{\theta}{2} \qquad (5.42)$$

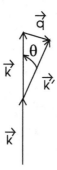

Fig. 5.7   The momentum transfer $\mathbf{q} = \mathbf{k}' - \mathbf{k}$ in the scattering process.

- With the insertion of the definition of v in Eq. (5.4), and the use of the

---

[14]The actual momentum transfer is $\hbar\mathbf{q}$.

reduced mass μ for two-body scattering, one has

$$f_{\mathrm{BA}}(\mathbf{q}) = -\frac{2\mu}{4\pi\hbar^2} \int d^3y\, e^{-i\mathbf{q}\cdot\mathbf{y}}\, V(y) \qquad ; \text{Born Approximation}$$

$$= -\frac{2\mu}{4\pi\hbar^2}\, \tilde{V}(\mathbf{q}) \tag{5.43}$$

We observe that $\tilde{V}(\mathbf{q})$ is just the *three-dimensional Fourier transform of the two-body potential;*

- This is an *important result.*[15]

5.4.1 *Validity*

Let us examine the validity of the Born approximation in Eq. (5.39):

- It assumes that the scattered wave is negligible with respect to the incident wave in the region of the potential

$$\frac{|\psi_{\text{scatt}}|}{|\psi_{\text{inc}}|} \ll 1 \qquad ; \text{in potential} \tag{5.44}$$

- The first approximation for the scattered wave in Eq. (5.34) is obtained by again making the approximation in Eq. (5.39) in the integral

$$\psi_{\text{scatt}}^{(1)}(\mathbf{x}) = -\int d^3y\, G_k^{(+)}(\mathbf{x}-\mathbf{y}) v(y)\psi_{\text{inc}}(\mathbf{y})$$

$$= -\int d^3y\, G_k^{(+)}(\mathbf{x}-\mathbf{y}) v(y) e^{i\mathbf{k}\cdot\mathbf{y}} \tag{5.45}$$

- The scattered wave is required in the region of the potential, that is, for $x \approx 0$

$$\psi_{\text{scatt}}^{(1)}(0) = -\frac{1}{4\pi} \int d^3y\, \frac{e^{iky}}{y} v(y) e^{i\mathbf{k}\cdot\mathbf{y}} \tag{5.46}$$

- The angular integral is carried out as in Eq. (5.26)[16]

$$\int d\Omega_y\, e^{i\mathbf{k}\cdot\mathbf{y}} = \int_0^{2\pi} d\phi \int_{-1}^1 d\mu\, e^{iky\mu} = 4\pi\frac{\sin ky}{ky} = 4\pi j_0(ky) \tag{5.47}$$

---

[15]Legend has it that in order to help get some feel for quantum mechanics, Bethe simply computed the Fourier transform of every potential he could think of!

[16]Here $j_0(ky)$ is a spherical Bessel function. This is a result worth remembering.

Hence

$$\psi_{\text{scatt}}^{(1)}(0) = -\int_0^\infty y^2 dy \, \frac{e^{iky}}{y} \frac{\sin ky}{ky} v(y)$$

$$= -\frac{1}{2ik} \int_0^\infty dy \, (e^{2iky} - 1)v(y) \tag{5.48}$$

- A *sufficient condition* for the validity of Born approximation is therefore[17]

$$\left| \frac{1}{2ik} \int_0^\infty dy \, (e^{2iky} - 1)v(y) \right| \ll 1 \qquad \text{; sufficient condition} \tag{5.49}$$

Three comments:

(1) This may not be a *necessary* condition. It is possible that the higher-order corrections go into a pure phase, as with the Coulomb potential, and an overall phase leaves the differential cross section unchanged;

(2) At *low energy*, where $k \to 0$, one can expand the exponential and the condition in Eq. (5.49) becomes

$$\left| \int_0^\infty y dy \, v(y) \right| \ll 1 \qquad \text{; low energy} \tag{5.50}$$

This condition depends on the integrated strength of the potential. Note that by our assumption in Eq. (5.1), this integral exists;

(3) At *high energy*, where $k \to \infty$, the first term in Eq. (5.49) gives a rapidly oscillating integrand, which leads to a vanishing integral, and the condition in Eq. (5.49) reduces to

$$\left| \frac{1}{2ik} \int_0^\infty dy \, v(y) \right| \ll 1 \qquad \text{; high energy} \tag{5.51}$$

Thus, if the potential is integrable, Born approximation *always* holds at high energy in non-relativistic potential scattering.

5.4.2   *Applications*

We discuss several applications of the above results.

---

[17]Note that $\psi_{\text{inc}}(0) = 1$.

5.4.2.1 *Square-Well Potential*

Consider the "square-well potential", with either sign for V_0 (see Fig. 5.8),

$$V(r) = V_0 \theta(R - r) \qquad\qquad \text{; square-well potential} \qquad (5.52)$$

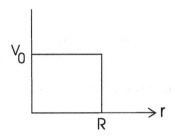

Fig. 5.8   The square-well potential, with either sign for V_0.

The Fourier transform of this potential is evaluated with the aid of Eq. (5.47) as[18]

$$\int d^3x\, e^{-i\mathbf{q}\cdot\mathbf{x}}\, V(x) = 4\pi V_0 \int_0^R x^2 dx\, j_0(qx) = \frac{4\pi V_0}{q^3} \int_0^{qR} \rho^2 d\rho\, j_0(\rho)$$

$$= \frac{4\pi V_0}{q^3}(qR)^2 j_1(qR) \qquad (5.53)$$

Hence

$$\tilde{V}_{\text{SW}}(\mathbf{q}) = \frac{4\pi R^3}{3} V_0 \left[\frac{3 j_1(qR)}{qR}\right] \qquad (5.54)$$

The factor in front is the volume of the potential, and the last factor $F(qR) \equiv 3 j_1(qR)/qR$ is known as the *form factor*[19]

$$F(qR) \equiv \frac{3 j_1(qR)}{qR} \qquad \text{; form factor}$$

$$F(0) = 1 \qquad (5.55)$$

The scattering amplitude and differential cross section in Born approxima-

---

[18]We use $\int \rho^2 d\rho\, j_0(\rho) = \rho^2 j_1(\rho)$ (see, for example, [Schiff (1968)]).
[19]Note $j_1(\rho) \to \rho/3$ as $\rho \to 0$.

tion are now given by Eqs. (5.43) and (5.18) as

$$f_{BA}(\mathbf{q}) = -\frac{2\mu}{\hbar^2}\frac{R^3 V_0}{3}F(qR) \qquad ; \text{square-well potential}$$

$$\left(\frac{d\sigma}{d\Omega}\right)_{BA} = \left|\frac{V_0}{\hbar^2/2\mu R^2}\right|^2 \frac{R^2}{9}|F(qR)|^2 \qquad (5.56)$$

Several comments:

- The differential cross section depends on the momentum transfer only through qR, and this dependence is given by the absolute square of the form factor;
- The quantity $|F(qR)|^2$ is plotted in Fig. 5.9;

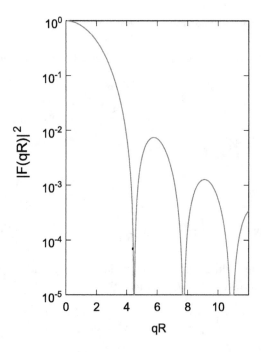

Fig. 5.9 Absolute square of the form factor $F(qR) \equiv 3j_1(qR)/qR$ for scattering from a square-well potential in Born approximation. Note $F(0) = 1$. The first zero occurs at $qR = X_{11} = 4.493$.

- One sees a *diffraction pattern*;

- The first zero of the diffraction pattern occurs at $j_1(X_{11}) = 0$, which is $X_{11} = (1.430)\pi = 4.493;$[20]
- The differential cross section at $qR = 0$ provides the overall scale

$$\left(\frac{d\sigma}{d\Omega}\right)_{\text{BA}} = \left|\frac{V_0}{\hbar^2/2\mu R^2}\right|^2 \frac{R^2}{9} \quad ; \quad qR = 0 \tag{5.57}$$

- The differential cross section is independent of the *sign* of V_0;
- If the Born approximation is valid, one can invert the Fourier transform in Eq. (5.40), and use the scattering amplitude to determine the spatial structure of the potential itself!

$$v(x) = \int \frac{d^3q}{(2\pi)^3} e^{i\mathbf{q}\cdot\mathbf{x}} \left[-4\pi f_{\text{BA}}(\mathbf{q})\right] \tag{5.58}$$

The only catch is that one has to know the scattering amplitude, not just the cross section, and this must be known for all $\mathbf{q}$;

- This analysis gives a qualitative picture of the scattering of intermediate energy hadrons ($N, \pi, \text{etc.}$) from nuclei through the nuclear optical potential.

5.4.2.2 *Yukawa Potential*

Consider the Yukawa potential, with either sign for V_0 (see Fig. 5.10),[21]

$$V(r) = V_0 \frac{e^{-\lambda r}}{r} \quad ; \text{ Yukawa potential} \tag{5.59}$$

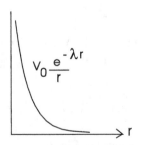

Fig. 5.10 The Yukawa potential, with either sign for V_0.

---

[20]See, for example, [Morse and Feshbach (1953)] p. 1576.
[21]Here V_0 has the dimensions $[EL]$.

The Fourier transform is

$$\tilde{V}_{\text{Yukawa}}(\mathbf{q}) = \int d^3x\, e^{-i\mathbf{q}\cdot\mathbf{x}}\, V(x)$$

$$= 4\pi V_0 \int_0^\infty x^2 dx\, j_0(qx) \frac{e^{-\lambda x}}{x} \qquad (5.60)$$

where the angular integrals are evaluated using Eq. (5.47).[22] The remaining integral is evaluated as

$$\tilde{V}_{\text{Yukawa}}(q) = \frac{4\pi V_0}{2i} \int_0^\infty x^2 dx\, \frac{(e^{iqx} - e^{-iqx})}{qx} \frac{e^{-\lambda x}}{x}$$

$$= \frac{4\pi V_0}{2iq} \int_0^\infty dx\, \left(e^{iqx - \lambda x} - e^{-iqx - \lambda x} \right)$$

$$= \frac{4\pi V_0}{2iq} \left[-\frac{1}{iq - \lambda} + \frac{1}{-iq - \lambda} \right] \qquad (5.61)$$

Thus

$$\tilde{V}_{\text{Yukawa}}(q) = \frac{4\pi V_0}{q^2 + \lambda^2} \qquad (5.62)$$

The scattering amplitude in Born approximation follows from Eqs. (5.43) as

$$f_{\text{BA}}(q) = -\frac{2\mu}{\hbar^2} \frac{V_0}{\lambda^2} F\left(\frac{q}{\lambda}\right) \qquad ; \text{ Yukawa potential}$$

$$F\left(\frac{q}{\lambda}\right) \equiv \frac{1}{1 + q^2/\lambda^2} \qquad (5.63)$$

Here $F(q/\lambda)$ is the form factor for the Yukawa potential. The corresponding differential cross section is given by

$$\left(\frac{d\sigma}{d\Omega}\right)_{\text{BA}} = \left|\frac{2\mu V_0}{\hbar^2 \lambda^2}\right|^2 \left|F\left(\frac{q}{\lambda}\right)\right|^2 \qquad ; \text{ Yukawa potential} \quad (5.64)$$

The quantity $|F(q/\lambda)|^2$ is shown in Fig. 5.11.

In contrast to the situation with the square well, this is a monotonically decreasing function of q^2.

---

[22]Note that the three-dimensional Fourier transform of $f(|\mathbf{x}|)$ depends only on $|\mathbf{q}|$.

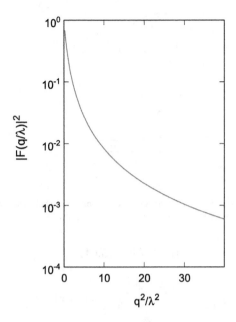

Fig. 5.11 Absolute square of the form factor $F(q/\lambda) \equiv 1/(1 + q^2/\lambda^2)$ for scattering from a Yukawa potential in Born approximation. Note $F(0) = 1$.

5.4.2.3 *Scattering of Charged Particles From Atoms*

Consider the scattering of a particle with charge $Z_2 e_p$ (for example, an α-particle $^4\mathrm{He}^{++}$, a muon $\mu^{\pm}$, *etc.*) from a neutral atom with nuclear charge $Z_1 e_p$ in Born approximation (see Fig. 5.12).

We observe the following:

- Close to the nucleus, one sees the point Coulomb potential $Z_1 Z_2 e^2/r$;
- Let $a = O(a_0/Z_1)$ characterize the size of the atom. Outside of the neutral atom, for $r > a$, the potential vanishes since the atomic electrons shield the nuclear charge;
- Let us *model* the Coulomb potential for an atom by

$$V(r) = \frac{Z_1 Z_2 e^2}{r} e^{-r/a} \qquad \text{; model potential} \quad (5.65)$$

With an appropriate identification of constants, the cross section in

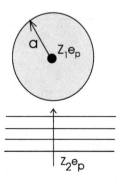

Fig. 5.12 Scattering of a particle with charge $Z_2 e_p$ from a neutral atom with nuclear charge $Z_1 e_p$ in Born approximation. Here a is a measure of the size of the atom.

Born approximation is now just that of Eq. (5.64)

$$\left(\frac{d\sigma}{d\Omega}\right)_{\text{BA}} = \left| \frac{2\mu Z_1 Z_2 e^2}{\hbar^2} \frac{1}{q^2 + (1/a)^2} \right|^2 \tag{5.66}$$

If the momentum transfer is large enough, so that

$$qa = 2ka \sin\frac{\theta}{2} \gg 1 \tag{5.67}$$

then

$$\left(\frac{d\sigma}{d\Omega}\right)_{\text{BA}} = \left| \frac{2\mu Z_1 Z_2 e^2}{\hbar^2 q^2} \right|^2 \qquad ; \, qa \gg 1 \tag{5.68}$$

This is re-written as

$$\left(\frac{d\sigma}{d\Omega}\right)_{\text{BA}} = \left| \frac{Z_1 Z_2 e^2}{4E_{\text{inc}} \sin^2 \theta/2} \right|^2 \qquad ; \, E_{\text{inc}} = \frac{\hbar^2 k^2}{2\mu} \tag{5.69}$$

A few comments:

- This is precisely the *Rutherford cross section* (see Vol. I) in the C-M system;
- If $qa \gg 1$, the scattered object sees just the point nuclear Coulomb potential;
- Born approximation evidently holds for small Z_1, Z_2, and high k;
- The Rutherford cross section is actually an *exact result* for the point Coulomb potential (see [Schiff (1968)]).

5.5 Higher-Order Born Approximations

The exact equations for the scattering state and scattering amplitude are

$$\psi_{\mathbf{k}}^{(+)}(\mathbf{x}) = e^{i\mathbf{k}\cdot\mathbf{x}} - \int d^3y\, G_k^{(+)}(\mathbf{x} - \mathbf{y})v(y)\psi_{\mathbf{k}}^{(+)}(\mathbf{y})$$

$$-4\pi f(\mathbf{k}', \mathbf{k}) = \int d^3y\, e^{-i\mathbf{k}'\cdot\mathbf{y}}\, v(y)\psi_{\mathbf{k}}^{(+)}(\mathbf{y}) \qquad (5.70)$$

The scattering Green's function with the correct outgoing-wave boundary condition is

$$G_k^{(+)}(\mathbf{x} - \mathbf{y}) = \int \frac{d^3t}{(2\pi)^3} \frac{e^{it\cdot(\mathbf{x}-\mathbf{y})}}{t^2 - (k + i\eta)^2} = \frac{e^{ik|\mathbf{x}-\mathbf{y}|}}{4\pi|\mathbf{x} - \mathbf{y}|} \qquad (5.71)$$

We proceed to *iterate* the inhomogeneous integral equation for $\psi_{\mathbf{k}}^{(+)}(\mathbf{x})$ by repeatedly substituting the r.h.s. into the integral for the scattered wave.[23] When the result is substituted into the scattering amplitude, the following series is obtained

$$-4\pi f(\mathbf{k}', \mathbf{k}) = \int d^3x\, e^{-i\mathbf{k}'\cdot\mathbf{x}} v(x) e^{i\mathbf{k}\cdot\mathbf{x}} - \qquad (5.72)$$

$$\int d^3x \int d^3y\, e^{-i\mathbf{k}'\cdot\mathbf{x}} v(x) G_k^{(+)}(\mathbf{x} - \mathbf{y})v(y) e^{i\mathbf{k}\cdot\mathbf{y}} + \cdots$$

The first term on the r.h.s. is just Born's first approximation ("Born approximation"). The remaining terms give the higher Born approximations, each characterized by the number of times the potential v appears in the amplitude. We proceed to analyze the general term in this series.

5.5.1 *Diagrammatic Analysis*

The structure of the series in Eq. (5.72) is illustrated in Fig. 5.13. The diagrams and accompanying rules are simply a means of keeping track of the terms in this series.

To compute a_n, the nth-order term in the series for $-4\pi f(\mathbf{k}', \mathbf{k})$:[24]

(1) Connect the n points $\{\mathbf{x}_1, \mathbf{x}_2, \cdots, \mathbf{x}_n\}$ with a directed line segment running from $\mathbf{x}_n$ to $\mathbf{x}_1$;

---

[23]See Prob. 5.4.

[24]There are n factors of v in nth order, and $-4\pi f(\mathbf{k}', \mathbf{k}) = \sum_{n=1}^{\infty} a_n$.

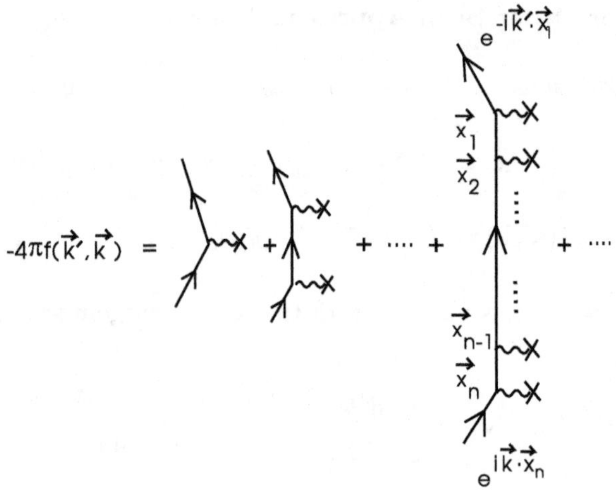

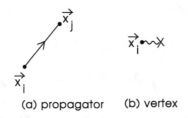

Fig. 5.13   Diagrammatic analysis of the Born series for the scattering amplitude.

(a) propagator   (b) vertex

Fig. 5.14   Components (a) propagator, and (b) vertex, in the diagrammatic analysis.

(2) For the *propagator*, the directed line segment running from the point $\mathbf{x}_i$ to the point $\mathbf{x}_j$ [see Fig. 5.14(a)], include a factor

$$(-1)\, G_k^{(+)}(\mathbf{x}_j - \mathbf{x}_i) \qquad ; \text{propagator} \qquad (5.73)$$

(3) For each *vertex* [see Fig. 5.14(b)], the interaction at $\mathbf{x}_i$, include a factor

$$v(x_i) \qquad ; \text{vertex} \qquad (5.74)$$

(4) Add the *external lines*, and include factors of

$$e^{i\mathbf{k}\cdot\mathbf{x}_n} \qquad ; \text{incoming line at } \mathbf{x}_n$$
$$e^{-i\mathbf{k}'\cdot\mathbf{x}_1} \qquad ; \text{outgoing line at } \mathbf{x}_1 \qquad (5.75)$$

(5) Integrate over all internal space points $d^3x_1\, d^3x_2 \cdots d^3x_n$.

5.6   Partial-Wave Analysis

The scattering problem is now well formulated. The goal here is to make use of the spherical symmetry of $v(x)$ and separate variables into radial and angular parts, as done with the differential Schrödinger equation, only now the starting point will be the *integral equation* in the first of Eqs. (5.70).

5.6.1   *Two Formulae From Analysis*

Two formulae from classical analysis will be employed:

(1) The *plane-wave expansion* is[25]

$$e^{i\mathbf{k}\cdot\mathbf{x}} = \sum_{l=0}^{\infty} (2l+1)i^l j_l(kx) P_l(\cos\theta_{kx}) \quad ; \text{ plane-wave expansion} \quad (5.76)$$

Here $j_l(kx)$ is a spherical Bessel function, and $P_l(\cos\theta_{kx})$ is a Legendre polynomial. The relevant variables are illustrated in Fig. 5.15.

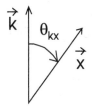

Fig. 5.15   Quantities in the plane-wave expansion.

(2) The *addition theorem* for spherical harmonics is[26]

$$P_l(\cos\theta_{12}) = \frac{4\pi}{2l+1} \sum_{m=-l}^{l} Y_{lm}(\Omega_1) Y_{lm}^\star(\Omega_2) \quad ; \text{ addition theorem} \quad (5.77)$$

Here $Y_{lm}(\Omega) \equiv Y_{lm}(\theta,\phi)$ is a spherical harmonic, and the relevant variables are shown in Fig. 5.16. This is a very useful formula since it *separates* the dependence on the angles in Ω_1 and Ω_2.

---

[25]See [Schiff (1968)] p. 119, or [Fetter and Walecka (2003a)] Prob. D.7.
[26]See [Edmonds (1974)] Sec. 4.6.6.

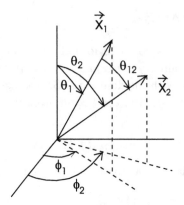

Fig. 5.16   Quantities in the addition theorem.

A combination of these two results gives the separated plane-wave expansion

$$e^{i\mathbf{k}\cdot\mathbf{x}} = \sum_{l=0}^{\infty} \sum_{m=-l}^{l} 4\pi i^l j_l(kx) Y_{lm}(\Omega_k) Y_{lm}^{\star}(\Omega_x) \tag{5.78}$$

5.6.2   Green's Function

The scattering Green's function has the following expansion, derived below,

$$\frac{e^{ik|\mathbf{x}-\mathbf{y}|}}{4\pi|\mathbf{x}-\mathbf{y}|} = ik \sum_{l=0}^{\infty} \left(\frac{2l+1}{4\pi} \right) j_l(kx_<) h_l^{(1)}(kx_>) P_l(\cos\theta_{xy})$$

$$= ik \sum_{l=0}^{\infty} \sum_{m=-l}^{l} j_l(kx_<) h_l^{(1)}(kx_>) Y_{lm}(\Omega_x) Y_{lm}^{\star}(\Omega_y) \tag{5.79}$$

Here the *spherical Hankel functions* are defined by

$$h_l^{(1)}(\rho) \equiv j_l(\rho) + i n_l(\rho) \qquad ; \text{ spherical Hankel functions}$$

$$h_l^{(2)}(\rho) \equiv j_l(\rho) - i n_l(\rho) \tag{5.80}$$

They have the following asymptotic behavior for large ρ [27]

$$h_l^{(1)}(\rho) \to \frac{1}{\rho} e^{i[\rho - (l+1)\pi/2]} \qquad ; \; \rho \to \infty$$

$$h_l^{(2)}(\rho) \to \frac{1}{\rho} e^{-i[\rho - (l+1)\pi/2]} \qquad (5.81)$$

The following symmetry properties will also be employed

$$j_l(-\rho) = (-1)^l j_l(\rho)$$
$$n_l(-\rho) = (-1)^{l+1} n_l(\rho)$$
$$\implies h_l^{(2)}(-\rho) = (-1)^l h_l^{(1)}(\rho) \qquad (5.82)$$

We proceed to the *derivation* of Eq. (5.79).

The distance $|\mathbf{x} - \mathbf{y}|$ is given by

$$|\mathbf{x} - \mathbf{y}| = (x^2 + y^2 - 2xy \cos\theta_{xy})^{1/2} \qquad (5.83)$$

The angle dependence of $G_k^{(+)}(\mathbf{x} - \mathbf{y})$ can then be expanded in the complete set of Legendre polynomials

$$G_k^{(+)}(\mathbf{x} - \mathbf{y}) = \sum_{l=0}^{\infty} \left(\frac{2l+1}{4\pi}\right) G_l^{(+)}(x, y; k) P_l(\cos\theta_{xy})$$

$$= \sum_{l=0}^{\infty} G_l^{(+)}(x, y; k) Y_{lm}(\Omega_x) Y_{lm}^{\star}(\Omega_y) \qquad (5.84)$$

where the second line follows from the use of the addition theorem.

Now use the integral expression for the Green's function in Eq. (5.71), and the expansion of the plane wave in Eq. (5.78) (twice)

$$G_k^{(+)}(\mathbf{x} - \mathbf{y}) = \int \frac{d^3 t}{(2\pi)^3} \frac{e^{i\mathbf{t}\cdot(\mathbf{x}-\mathbf{y})}}{t^2 - (k+i\eta)^2} \qquad (5.85)$$

$$= \sum_{lm} \sum_{l'm'} \int_0^{\infty} \frac{t^2 dt}{(2\pi)^3} \frac{1}{t^2 - (k+i\eta)^2} \int d\Omega_t \times$$

$$\left[4\pi i^l j_l(tx) Y_{lm}(\Omega_x) Y_{lm}^{\star}(\Omega_t)\right] \left[4\pi i^{l'} j_{l'}(ty) Y_{l'm'}(\Omega_y) Y_{l'm'}^{\star}(\Omega_t)\right]^{\star}$$

The angular integration $\int d\Omega_t$ can now be carried out, and the orthonormality of the spherical harmonics gives

$$\int d\Omega_t\, Y_{lm}^{\star}(\Omega_t) Y_{l'm'}(\Omega_t) = \delta_{ll'} \delta_{mm'} \qquad (5.86)$$

---

[27] See, for example, [Schiff (1968)].

Thus

$$G_k^{(+)}(\mathbf{x} - \mathbf{y}) = \sum_{lm} \frac{2}{\pi} \int_0^\infty t^2 dt\, \frac{j_l(tx)j_l(ty)}{t^2 - (k + i\eta)^2} Y_{lm}(\Omega_x)Y_{lm}^*(\Omega_y) \qquad (5.87)$$

Comparison with Eq. (5.84) allows us to identify

$$G_l^{(+)}(x, y; k) = \frac{2}{\pi} \int_0^\infty t^2 dt\, \frac{j_l(tx)j_l(ty)}{t^2 - (k + i\eta)^2} \qquad (5.88)$$

The use of the symmetry properties in Eqs. (5.82) permit this to be re-written as

$$G_l^{(+)}(x, y; k) = \frac{1}{\pi} \int_{-\infty}^\infty t^2 dt\, \frac{j_l(tx)j_l(ty)}{t^2 - (k + i\eta)^2} \qquad (5.89)$$

This integral can be evaluated through contour integration. Assume $x > y$, and write

$$G_l^{(+)} = \frac{1}{2\pi} \int_{-\infty}^\infty \frac{t^2 dt\, j_l(ty)}{t^2 - (k + i\eta)^2} [h_l^{(1)}(tx) + h_l^{(2)}(tx)] \qquad (5.90)$$

Since $x > y$, the exponentials in $h_l(tx)$ dominate those in $j_l(ty)$ for large arguments.[28] With the use of the asymptotic expressions in Eqs. (5.81), one sees that the first integral can be completed with a large semi-circle in the *upper-1/2* t-plane, and the second with a large semi-circle in the *lower-1/2* t-plane. Thus (see Fig. 5.17)

$$G_l^{(+)} = \frac{1}{2\pi} \oint_{C_1} \frac{t^2 dt\, j_l(ty)h_l^{(1)}(tx)}{t^2 - (k + i\eta)^2} + \frac{1}{2\pi} \oint_{C_2} \frac{t^2 dt\, j_l(ty)h_l^{(2)}(tx)}{t^2 - (k + i\eta)^2} \qquad (5.91)$$

The integrands are analytic except for the poles indicated in Fig. 5.17, and evaluation with residues gives

$$G_l^{(+)} = \frac{2\pi i}{2\pi} \left[\frac{k^2}{2k} j_l(ky)h_l^{(1)}(kx) - \frac{(-k)^2}{2(-k)} j_l(-ky)h_l^{(2)}(-kx) \right] \qquad (5.92)$$

The use of the symmetry properties in Eqs. (5.82) reduces this to

$$G_l^{(+)}(x, y; k) = ik j_l(ky)h_l^{(1)}(kx) \qquad ; x > y \qquad (5.93)$$

---

[28]Note $2j_l(\rho) = h_l^{(1)}(\rho) + h_l^{(2)}(\rho)$, and hence

$$2\rho j_l(\rho) \to e^{i[\rho - (l+1)\pi/2]} + e^{-i[\rho - (l+1)\pi/2]} = 2\cos[\rho - (l+1)\pi/2] \qquad ; \rho \to \infty$$

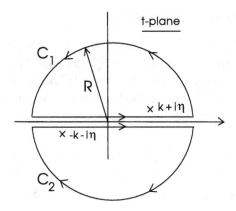

Fig. 5.17   Contours of integration for the Green's function $G_l^{(+)}$ in Eq. (5.91).

If $x < y$, one just reverses their role in evaluating the integral in Eq. (5.89)

$$G_l^{(+)}(x, y; k) = ikj_l(kx)h_l^{(1)}(ky) \qquad ; x < y \qquad (5.94)$$

These results reproduce the expression for the Green's function in Eq. (5.79).

5.6.3   *Separation of Variables in Scattering Equation*

The goal now is to make use of the spherical symmetry of the potential $v(x)$ and separate variables in the scattering equation

$$\psi_{\mathbf{k}}^{(+)}(\mathbf{x}) = e^{i\mathbf{k}\cdot\mathbf{x}} - \int d^3y\, G_k^{(+)}(\mathbf{x} - \mathbf{y})v(y)\psi_{\mathbf{k}}^{(+)}(\mathbf{y}) \qquad (5.95)$$

The vector configuration is shown again in Fig. 5.18.

Fig. 5.18   Quantities in the scattering equation.

The separation will be accomplished through a series of steps:

(1) What variables can $\psi_{\mathbf{k}}^{(+)}(\mathbf{x})$ depend on? The system must be invariant under *overall* rotations, and hence $\psi_{\mathbf{k}}^{(+)}(\mathbf{x})$ must be a scalar. As such, it can only depend on the combinations $\{\mathbf{x}^2, \mathbf{k}^2, \mathbf{k} \cdot \mathbf{x}\}$, or equivalently on $\{x, k, \cos\theta_{kx}\}$.[29] The angle dependence can then be expanded in the complete set of Legendre polynomials[30]

$$\psi_{\mathbf{k}}^{(+)}(\mathbf{x}) = \sum_{l=0}^{\infty}(2l+1)i^l\psi_l^{(+)}(x;k)P_l(\cos\theta_{kx}) \qquad (5.96)$$

(2) Substitute this expression into the integral equation, and make use of

- the plane-wave expansion in Eq. (5.78);
- the Green's function expansion in Eq. (5.84);
- the addition theorem in Eq. (5.77).

If the incident wave is taken to the l.h.s., the result of these substitutions is

$$\sum_{l=0}^{\infty}(2l+1)i^l[\psi_l^{(+)}(x;k) - j_l(kx)]P_l(\cos\theta_{kx}) = \qquad (5.97)$$

$$-\sum_{lm}\sum_{l'm'}\int y^2\,dy\,d\Omega_y\,[G_l^{(+)}(x,y;k)Y_{lm}(\Omega_x)Y_{lm}^{\star}(\Omega_y)]v(y)\times$$

$$[4\pi i^{l'}\psi_{l'}^{(+)}(y;k)Y_{l'm'}(\Omega_y)Y_{l'm'}^{\star}(\Omega_k)]$$

(3) Now use the orthonormality of the spherical harmonics

$$\int d\Omega_y\,Y_{lm}^{\star}(\Omega_y)Y_{l'm'}(\Omega_y) = \delta_{ll'}\delta_{mm'} \qquad (5.98)$$

(4) Then use the addition theorem again

$$4\pi\sum_m Y_{lm}(\Omega_x)Y_{lm}^{\star}(\Omega_k) = (2l+1)P_l(\cos\theta_{kx}) \qquad (5.99)$$

(5) Finally, the coefficients of the $P_l(\cos\theta_{kx})$ can be equated to give

$$\psi_l^{(+)}(x;k) = j_l(kx) - \int_0^{\infty}y^2\,dy\,G_l^{(+)}(x,y;k)v(y)\psi_l^{(+)}(y;k) \qquad (5.100)$$

---

[29]This is true, for example, for each term in the iterated series in Eq. (5.72).

[30]Alternatively, one can take this as an *ansatz*, substitute it into the integral equation, and see if it separates.

The Green's function $G_l^{(+)}(x, y; k)$ is given by Eqs. (5.93)–(5.94) as

$$G_l^{(+)}(x, y; k) = ik j_l(ky) h_l^{(1)}(kx) \qquad ; x > y$$
$$= ik j_l(kx) h_l^{(1)}(ky) \qquad ; x < y \qquad (5.101)$$

A few comments:

- The original three-dimensional integral equation for $\psi_{\mathbf{k}}^{(+)}(\mathbf{x})$ in the co-ordinate $\mathbf{x}$ has been reduced to a one-dimensional integral equation for $\psi_l^{(+)}(x; k)$, in the radial coordinate $|\mathbf{x}| = x = r$;
- The equations *decouple*, and there is one equation for each l;
- It is the spherical symmetry of $v(x)$ that permits this reduction;[31]
- The full scattering state is recovered through Eq. (5.96);
- This separation is equivalent to the separation $\psi_{lm}(\mathbf{r}) = R(r) Y_{lm}(\theta, \phi)$ in the differential Schrödinger equation (see below). The advantage here is that the *scattering boundary conditions have now been built into the radial integral Eq. (5.100)*.

5.6.4 *Scattering Boundary Conditions*

The asymptotic behavior of $\psi_l^{(+)}(x; k)$ as $x \to \infty$ is obtained through the use of Eqs. (5.81), and the linear combinations in Eqs. (5.80). If y is confined to the potential, then the first of Eqs. (5.101) is applicable, and from Eq. (5.100)

$$\psi_l^{(+)}(x; k) \to \frac{1}{2kx} \left\{ e^{i[kx - (l+1)\pi/2]} + e^{-i[kx - (l+1)\pi/2]} \right\} - \qquad (5.102)$$
$$\frac{ik}{kx} e^{i[kx - (l+1)\pi/2]} \int_0^\infty y^2 \, dy \, j_l(ky) v(y) \psi_l^{(+)}(y; k) \qquad ; x \to \infty$$

The incident plane wave has both an *incoming and outgoing radial wave*, arising from $j_l(kx)$, while the scattered wave has only an *outgoing radial wave* arising from $h_l^{(1)}(kx)$ in this asymptotic regime.[32]

The expression in Eq. (5.102) can be re-cast in the form

$$\psi_l^{(+)}(x; k) \to \frac{1}{2kx} \left\{ S_l(k) e^{i[kx - (l+1)\pi/2]} + e^{-i[kx - (l+1)\pi/2]} \right\} \qquad ; x \to \infty$$
$$S_l(k) \equiv 1 - 2ik \int_0^\infty y^2 \, dy \, j_l(ky) v(y) \psi_l^{(+)}(y; k) \qquad (5.103)$$

---

[31] See the angular integration in step (3).
[32] Recall Fig. 5.1.

In the absence of scattering, $S_l(k) = 1$.

The *scattering amplitude* is obtained from the asymptotic form of the full scattering state $\psi_{\mathbf{k}}^{(+)}(\mathbf{x})$ in the first of Eqs. (5.38). The scattering state is, in turn, constructed from the partial-wave amplitudes $\psi_l^{(+)}(x; k)$ in Eq. (5.96). The asymptotic form of the scattered radial wave in Eq. (5.102) allows us to write

$$\psi_{\mathbf{k}}^{(+)}(\mathbf{x}) \to e^{i\mathbf{k}\cdot\mathbf{x}} + \frac{e^{ikx}}{x} \sum_{l=0}^{\infty} (2l+1)i^l P_l(\cos\theta_{kx}) \times \qquad (5.104)$$

$$e^{-i(l+1)\pi/2}[-i \int_0^{\infty} y^2 \, dy \, j_l(ky)v(y)\psi_l^{(+)}(y; k)] \qquad ; \, x \to \infty$$

Hence the scattering amplitude is identified as[33]

$$f(\theta,\phi) = -\sum_{l=0}^{\infty}(2l+1)P_l(\cos\theta_{kx}) \int_0^{\infty} y^2 \, dy \, j_l(ky)v(y)\psi_l^{(+)}(y; k) \quad (5.105)$$

With the definition of $S_l(k)$ in Eqs. (5.103), this is re-written as

$$f(k,\theta) = \sum_{l=0}^{\infty}(2l+1)\left[\frac{S_l(k)-1}{2ik}\right]P_l(\cos\theta) \qquad ; \text{ scattering amplitude}$$

$$(5.106)$$

Here the scattering angle θ_{kx} in Fig. 5.18 is simply denoted by θ, and the energy dependence is explicitly exhibited in $f(k,\theta)$. There is no ϕ-dependence in the scattering amplitude for a spherically symmetric potential.

5.6.5 *Phase Shifts*

We now claim that *conservation of probability* implies that with only elastic scattering, the partial-wave scattering amplitudes $S_l(k)$ must have unit modulus[34]

$$|S_l(k)| = 1$$
$$\implies \quad S_l(k) \equiv e^{2i\delta_l(k)} \qquad ; \text{ phase shift} \quad (5.107)$$

The second line defines the *phase shift*.

Thus the entire effect of the scattering is to shift the phase of the outgoing radial wave in the asymptotic regime in Eq. (5.103).

---

[33]Note that $e^{-i(l+1)\pi/2} = (-i)^{l+1}$.

[34]This is often referred to as the condition of *unitarity*.

We proceed to a proof of this result. Surround the scatterer with a large sphere of radius R as in Fig. 5.1. For stationary states with no sources or sinks of probability, the integrated probability density ρ is independent of time

$$I \equiv \frac{d}{dt} \int_V \rho \, d^3x = \int_V \left(\frac{\partial \rho}{\partial t} \right) d^3x = 0 \qquad (5.108)$$

Use the continuity equation and Gauss' law to re-write this as

$$I = - \int_V (\nabla \cdot \mathbf{S}) \, d^3x = - \int_A \mathbf{S} \cdot d\mathbf{A} \qquad (5.109)$$

With $d\mathbf{A} = \mathbf{e}_r R^2 d\Omega$, this becomes

$$I = -\frac{\hbar}{2im} \int_A R^2 \, d\Omega \left[\psi^* \frac{\partial \psi}{\partial r} - \left(\frac{\partial \psi}{\partial r} \right)^* \psi \right]_R \qquad ; \ |\mathbf{x}| = x = r \quad (5.110)$$

Now do the following:[35]

(1) Substitute the asymptotic form of the full scattering state $\psi_{\mathbf{k}}^{(+)}(\mathbf{x})$ obtained from the partial-wave expansion in Eq. (5.96) and the asymptotic behavior of the radial waves in Eq. (5.103)

$$\psi_{\mathbf{k}}^{(+)}(\mathbf{x}) \to \sum_{l=0}^{\infty} (2l+1) i^l \frac{1}{2kx} (S_l e^{i\varphi} + e^{-i\varphi}) P_l(\cos\theta_{kx}) \qquad ; \ x \to \infty$$

$$\varphi \equiv kx + \frac{\pi}{2}(l+1) \qquad (5.111)$$

(2) Use the orthogonality of the Legendre polynomials

$$\int d\Omega_x \, P_l(\cos\theta_{kx}) P_{l'}(\cos\theta_{kx}) = \frac{4\pi}{2l+1} \delta_{ll'} \qquad (5.112)$$

The result is

$$I = -4\pi R^2 \frac{\hbar k}{2m} \sum_{l=0}^{\infty} \frac{(2l+1)}{4k^2 R^2} [(S_l e^{i\varphi} + e^{-i\varphi})^* (S_l e^{i\varphi} - e^{-i\varphi}) +$$

$$(S_l e^{i\varphi} - e^{-i\varphi})^* (S_l e^{i\varphi} + e^{-i\varphi})] \quad (5.113)$$

It follows from Eqs. (5.108) and (5.113) that

$$\frac{d}{dt} \int_V \rho \, d^3x = -\frac{\hbar k}{m} \frac{\pi}{k^2} \sum_{l=0}^{\infty} (2l+1)[\,|S_l|^2 - 1\,] = 0 \qquad (5.114)$$

; no sources or sinks

---

[35] Recall our notation that $|\mathbf{x}| = x = r$; remember also that the radius $R \to \infty$.

(3) The condition in Eq. (5.107) is clearly *sufficient* to satisfy this relation. It is also *necessary*. To see this, we invoke our old friend the *superposition axiom*. Consider the linear combination of two distinct scattering states of the same energy

$$\psi = \alpha\psi_{\mathbf{k}_1}^{(+)}(\mathbf{x}) + \beta\psi_{\mathbf{k}_2}^{(+)}(\mathbf{x}) \qquad ; \; |\mathbf{k}_1| = |\mathbf{k}_2| = k \quad (5.115)$$

The vector configuration is indicated in Fig. 5.19.

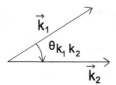

Fig. 5.19   Quantities in the superposed scattering state. Here $|\mathbf{k}_1| = |\mathbf{k}_2| = k$.

Now repeat the previous calculation, using instead of Eq. (5.112)[36]

$$\int d\Omega_x P_l(\cos\theta_{k_1 x}) P_{l'}(\cos\theta_{k_2 x}) = \frac{(4\pi)^2}{(2l+1)(2l'+1)} \sum_m \sum_{m'} Y_{lm}(\Omega_{k_1}) \times$$

$$\left[\int d\Omega_x Y_{lm}^\star(\Omega_x) Y_{l'm'}(\Omega_x)\right] Y_{l'm'}^\star(\Omega_{k_2})$$

$$= \frac{4\pi}{2l+1}\delta_{ll'}\, P_l(\cos\theta_{k_1 k_2}) \qquad (5.116)$$

The new result is then

$$\frac{d}{dt}\int_V \rho\, d^3x = -\frac{\hbar k}{m}\frac{\pi}{k^2}\sum_{l=0}^{\infty}(2l+1)[\,|S_l|^2 - 1]\times \qquad (5.117)$$

$$[\,|\alpha|^2 + |\beta|^2 + 2\mathrm{Re}(\alpha^\star\beta)\,P_l(\cos\theta_{k_1 k_2})] = 0$$

$$; \text{ no sources or sinks}$$

(4) Now take $\int d\Omega_{k_2} P_l(\cos\theta_{k_1 k_2})$ on this expression, and use Eq. (5.112). One concludes that each term in the series must vanish, or[37]

$$|S_l|^2 - 1 = 0 \qquad (5.118)$$

This was the stated goal.

---

[36] Once again, this result follows from judicious use of the addition theorem and the orthonormality of the spherical harmonics.

[37] By assumption, $\mathrm{Re}(\alpha^\star\beta) \neq 0$ and $|\alpha + \beta|^2 \neq 0$.

A combination of Eqs. (5.103), (5.106), and (5.107) gives our result for the partial-wave decomposition of the scattering amplitude expressed in terms of the phase shifts

$$\frac{S_l(k) - 1}{2ik} = \frac{e^{i\delta_l(k)} \sin \delta_l(k)}{k} = -\int_0^\infty y^2 dy \, j_l(ky) v(y) \psi_l^{(+)}(y; k)$$

$$f(k, \theta) = \sum_{l=0}^\infty (2l + 1) \frac{e^{i\delta_l(k)} \sin \delta_l(k)}{k} P_l(\cos \theta) \qquad (5.119)$$

Here the partial-wave scattering state $\psi_l^{(+)}(x; k)$ satisfies the radial integral Eq. (5.100), with the Green's function of Eqs. (5.101), and the wave vectors in the scattering process are illustrated in Fig. 5.5. These are *exact* results.

5.6.5.1 *Born Approximation*

Born approximation again consists of replacing the full partial-wave scattering state in the region of the potential by the incident wave in Eq. (5.100) (see Fig. 5.20)

$$\psi_l^{(+)}(x; k) = j_l(kx) \qquad ; \text{ Born approximation} \qquad (5.120)$$

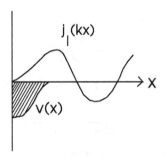

Fig. 5.20 Born approximation for the partial-wave scattering state $\psi_l^{(+)}(x; k)$ replacing it by the incident wave $j_l(kx)$ in the region of the potential.

Born approximation holds for sufficiently weak potentials $v(x)$. It is clear from the first of Eqs. (5.119), that in this limit the phase shift is *also* small. For a consistent expansion, the phase shift δ_l must also be expanded in v, and the leading term gives the *Born approximation for the phase shift*

$$\frac{\delta_l^{BA}(k)}{k} = -\int_0^\infty y^2 dy \, j_l^2(ky) v(y) \qquad ; \text{ Born approximation} \qquad (5.121)$$

At long wavelength, the expansions in Eqs. (2.210) allow this relation to be written as

$$\delta_l^{\mathrm{BA}}(k) \to -\frac{k^{2l+1}}{[(2l+1)!!]^2} \int_0^\infty y^{2l+2}\, dy\, v(y) \qquad ; \; k \to 0 \qquad (5.122)$$

A few comments:

- If the potential $v(x)$ has a *finite range* a, then

$$\delta_l^{\mathrm{BA}}(k) \sim (ka)^{2l+1} \qquad ; \; ka \to 0 \qquad (5.123)$$

- The angular momentum barrier keeps the high-l states out of $v(x)$;
- *Only a few partial waves contribute to the scattering as $ka \to 0$;*
- Although derived in Born approximation, we proceed to show that this is a *general result.*

5.6.5.2 *Long Wavelength*

The general long-wavelength behavior is extracted as follows:

- The behavior of the spherical Bessel and Hankel functions as the argument $\rho \to 0$ is obtained from Eqs. (2.210) and (5.80)

$$j_l(\rho) \to \frac{\rho^l}{(2l+1)!!} \qquad ; \; \rho \to 0$$

$$h_l^{(1)}(\rho) \to -i\frac{(2l-1)!!}{\rho^{l+1}} \qquad (5.124)$$

The Green's function in Eqs. (5.101) in the same limit behaves as[38]

$$G_l^{(+)}(x,y;k) \to \frac{1}{2l+1}\frac{1}{x_>}\left(\frac{x_<}{x_>}\right)^l \qquad ; \; k \to 0 \qquad (5.125)$$

This expression is independent of k;
- Now define

$$\psi_l^{(+)}(x;k) \equiv k^l \phi_l^{(+)}(x) \qquad (5.126)$$

Substitute Eqs. (5.124)–(5.126) into Eq. (5.100), and divide by k^l,

$$\phi_l^{(+)}(x) = \frac{x^l}{(2l+1)!!} - \frac{1}{2l+1}\int_0^\infty y^2 dy\left[\frac{1}{x_>}\left(\frac{x_<}{x_>}\right)^l\right]v(y)\phi_l^{(+)}(y)$$

$$\qquad ; \; k \to 0 \qquad (5.127)$$

---

[38]Compare Prob. 5.8.

The k-dependence has now been scaled out of this integral equation;

- The corresponding behavior of the phase shifts in this limit follows from the first of Eqs. (5.119)[39]

$$\delta_l(k) \to -\frac{k^{2l+1}}{(2l+1)!!} \int_0^\infty y^{l+2} dy \, v(y)\phi_l^{(+)}(y) \qquad ; k \to 0 \qquad (5.128)$$

- Thus if the potential is of *finite range*, so that all the integrals are well-defined, then the phase shift $\delta_l(k)$ goes as k^{2l+1} as $k \to 0$. This is the claimed result.

5.6.6 *Differential Equation*

The advantage of working with an integral equation in scattering theory is that we are able to build the scattering boundary conditions into the Green's function at the outset. Let us now convert the problem *back* to the differential Schrödinger equation, and investigate the implications. Our starting point in scattering theory was Eq. (5.4)

$$(\nabla^2 + k^2)\psi_{\mathbf{k}}^{(+)}(\mathbf{x}) = v(x)\psi_{\mathbf{k}}^{(+)}(\mathbf{x}) \qquad ; v \equiv \frac{2\mu}{\hbar^2}V \quad (5.129)$$

It was shown in chapter 2 that for any radial function $f(r)$

$$\nabla^2 P_l(\cos\theta)f(r) = P_l(\cos\theta)\left[\frac{1}{r}\frac{d^2}{dr^2}r - \frac{l(l+1)}{r^2}\right]f(r) \qquad (5.130)$$

Now substitute the partial-wave expansion of $\psi_{\mathbf{k}}^{(+)}(\mathbf{x})$ in Eq. (5.96) into Eq. (5.129), use Eq. (5.130), and invoke the linear independence of the Legendre polynomials. One concludes that the partial-wave amplitudes satisfy the following radial differential equation[40]

$$\left[\frac{1}{r}\frac{d^2}{dr^2}r - \frac{l(l+1)}{r^2} + [k^2 - v(r)]\right]\psi_l^{(+)}(r;k) = 0 \qquad (5.131)$$

The proper form of the radial wave functions far from the scattering center is given by Eqs. (5.103) and (5.107)

$$\psi_l^{(+)}(r;k) \to \frac{1}{2kr}\left\{e^{2i\delta_l(k)}e^{i[kr-(l+1)\pi/2]} + e^{-i[kr-(l+1)\pi/2]}\right\} \qquad (5.132)$$

$$= \frac{e^{i\delta_l(k)}}{kr}\cos\left[kr - (l+1)\pi/2 + \delta_l(k)\right] \quad ; r \to \infty$$

---

[39]Note that $\delta_l(k) \to 0$ as $k \to 0$.

[40]Recall that $|\mathbf{x}| = x = r$. Also, for generality, here we write $v = 2\mu V/\hbar^2$ where μ is the reduced mass (with a fixed target, $\mu = m$).

In *summary*:

- The time-independent scattering state is given by Eq. (5.96)[41]

$$\psi_{\mathbf{k}}^{(+)}(\mathbf{x}) = \sum_{l=0}^{\infty}(2l+1)i^l\psi_l^{(+)}(r;k)P_l(\cos\theta) \qquad (5.133)$$

where $\cos\theta = \cos\theta_{kx}$, and the partial-wave amplitudes satisfy the radial differential Eq. (5.131);
- The scattering boundary conditions in the first of Eqs. (5.38) are built in, provided the radial amplitudes have the asymptotic phase-shifted behavior in Eq. (5.132);
- The scattering amplitude is then given in terms of the phase shifts by Eq. (5.119)

$$f(k,\theta) = \sum_{l=0}^{\infty}(2l+1)\frac{e^{i\delta_l(k)}\sin\delta_l(k)}{k}P_l(\cos\theta) \qquad (5.134)$$

- In the asymptotic regime where $r \to \infty$, one can re-write Eq. (5.132) as follows

$$\psi_l^{(+)}(r;k) \to \frac{e^{i\delta_l(k)}}{kr}\cos\left[kr - (l+1)\pi/2 + \delta_l(k)\right] \qquad ; r \to \infty$$

$$= \frac{e^{i\delta_l(k)}}{kr}\{\cos\left[kr - (l+1)\pi/2\right]\cos\delta_l(k) - \sin\left[kr - (l+1)\pi/2\right]\sin\delta_l(k)\}$$

$$= e^{i\delta_l(k)}[j_l(kr)\cos\delta_l(k) - n_l(kr)\sin\delta_l(k)] \qquad (5.135)$$

Hence

$$\psi_l^{(+)}(r;k) \to e^{i\delta_l(k)}\cos\delta_l(k)[j_l(kr) - \tan\delta_l(k)n_l(kr)] \qquad ; r \to \infty$$

$$(5.136)$$

This is the result given in [Schiff (1968)].[42] Expressed in this form, in the asymptotic regime one has the appropriate linear combinations of the regular and irregular radial solutions to the free scalar Helmholtz equation.

---

[41] Recall Fig. 5.1.
[42] See [Schiff (1968)] p. 119.

5.6.6.1 *Problem Solving*

The solution to a three-dimensional scattering problem with a central potential through the differential Schrödinger equation involves the following steps:

(1) Construct the general solution to the radial Schrödinger Eq. (5.131) for each l;
(2) Match all the boundary conditions in the region of the potential;
(3) Identify the phase shift from the relative admixture of the regular and irregular solutions in Eq. (5.136) as $r \to \infty$;
(4) Use the phase shifts to calculate the scattering amplitude in Eq. (5.134) and the differential cross section in Eq. (5.18);
(5) Normalize each partial-wave amplitude as indicated in Eq. (5.136);[43]
(6) The full scattering state, with the correct asymptotic scattering boundary conditions, is then given by the sum in Eq. (5.133).

5.6.6.2 *Scattering from a Hard Sphere*

To see how this works, consider the scattering from a *hard sphere* (see Fig. 5.21).

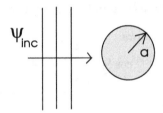

Fig. 5.21   Scattering from a hard sphere.

The general solution to the Schrödinger Eq. (5.131) outside of the sphere is given in terms of the fundamental system of solutions $\{j_l(kr), n_l(kr)\}$ to the radial part of the scalar Helmholtz equation

$$\psi_l^{(+)}(r; k) = \alpha j_l(kr) + \beta n_l(kr) \qquad ; r \geq a \qquad (5.137)$$

The boundary condition for a hard sphere is that this wave function must

---

[43]Note that the calculation of the phase shifts, and hence the scattering amplitude and differential cross section, is independent of this choice of normalization!

vanish at the wall

$$\psi_l^{(+)}(a; k) = 0 \qquad ; \text{ hard sphere} \qquad (5.138)$$

It follows that

$$\psi_l^{(+)}(r; k) = \alpha \left[j_l(kr) - \frac{j_l(ka)}{n_l(ka)} n_l(kr) \right] \qquad ; r \geq a \qquad (5.139)$$

A comparison with Eq. (5.136) identifies the phase shift through

$$\tan \delta_l(k) = \frac{j_l(ka)}{n_l(ka)} \qquad ; \text{ hard sphere} \qquad (5.140)$$

This provides an *exact solution for the phase shifts for the hard sphere for all l and all k*. The overall normalization (and phase!) of the partial waves is now determined according to Eq. (5.136) as

$$\alpha(k) = e^{i\delta_l(k)} \cos \delta_l(k) \qquad ; \text{ choice of norm} \qquad (5.141)$$

It is evident from Eq. (5.128) that at very low energies, it is only the *s*-waves with $l = 0$ that contribute to the scattering. For *s*-waves

$$j_0(\rho) = \frac{\sin \rho}{\rho} \qquad ; n_0(\rho) = -\frac{\cos \rho}{\rho} \qquad (5.142)$$

The equations for the phase shift and the wave function in this case become

$$\tan \delta_0(k) = -\tan ka$$
$$\psi_0^{(+)}(r; k) = e^{i\delta_0(k)} \cos \delta_0(k) \left[\frac{\sin kr}{kr} + \tan \delta_0(k) \frac{\cos kr}{kr} \right]$$
$$\qquad ; \text{ s-waves} \qquad (5.143)$$

The $l = 0$ phase shift is evidently

$$\delta_0(k) = -ka \qquad ; \text{ s-waves} \qquad (5.144)$$

Take out the $1/r$ dependence, and define the wave function $u_0(r)$ by[44]

$$\psi_0^{(+)}(r; k) \equiv e^{i\delta_0(k)} \frac{u_0(r)}{kr} \qquad (5.145)$$

Then everywhere outside of the hard sphere, the phase-shifted *s*-wave $u_0(r)$ has the form

$$u_0(r) = \sin(kr + \delta_0) = \sin k(r - a) \qquad ; \text{ s-waves} \qquad (5.146)$$

---

[44]In keeping with previous usage, we now suppress the k-dependence in $u_0(r)$.

This result could have been derived immediately from Eq. (2.201), and it has a nice, simple interpretation, as illustrated in Fig. 5.22.[45]

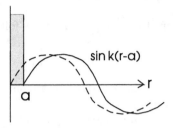

Fig. 5.22 Phase-shifted s-wave $u_0(r) = \sin k(r - a) = \sin (kr + \delta_0)$ for scattering from a hard sphere.

The numerical evaluation of the differential cross section for a hard sphere as a function of energy is left as an exercise (see Probs. 5.10–5.11 and Fig. 11.6).

5.6.6.3 *S-Wave Scattering from Attractive Square-Well*

As a second example, consider low-energy s-wave scattering from an attractive square-well potential (Fig. 5.23). Here

$$v_0 = \frac{2\mu}{\hbar^2} V_0 \tag{5.147}$$

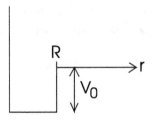

Fig. 5.23 Attractive square-well potential.

The wave function $u_0(r)$ defined in Eq. (5.145) again satisfies the s-wave

---

[45]Note, again, that the determination of the phase shift here is independent of the normalization of $u_0(r)$.

Schrödinger equation

$$\frac{d^2 u_0(r)}{dr^2} + [k^2 - v(r)]u_0(r) = 0 \quad ; \text{s-waves} \tag{5.148}$$

The appropriate solutions to this equation are as follows:

(1) *Outside the potential, for $r \geq R$:*

$$u_0(r) = \alpha \sin kr + \beta \cos kr \equiv \bar{\alpha} \sin (kr + \delta_0) \tag{5.149}$$

This incorporates the correct scattering boundary condition and defines the phase shift as $r \to \infty$;

(2) *Inside the potential, for $0 \leq r \leq R$:*

$$u_0(r) = A \sin \kappa r + B \cos \kappa r \quad ; \kappa^2 = v_0 + k^2 \tag{5.150}$$

The boundary condition at the origin is $u_0(0) = 0$, so that the origin does not become a point source of probability. This implies that $B = 0$

$$u_0(r) = A \sin \kappa r \quad ; u_0(0) = 0 \tag{5.151}$$

The boundary conditions at $r = R$ are that both the wave function and its derivative must be continuous there

$$\begin{aligned} u_0|_+ &= u_0|_- \quad ; r = R \\ u_0'|_+ &= u_0'|_- \end{aligned} \tag{5.152}$$

Substitute Eqs. (5.149) and (5.151), and take the *ratio* of the two results. This elimates the normalization constants and results in the relation

$$k \cot [kR + \delta_0(k)] = \kappa \cot (\kappa R) \quad ; \text{determines } \delta_0 \tag{5.153}$$

Several comments:

- Equation (5.153) determines the s-wave phase shift $\delta_0(k)$ for the attractive square-well potential at all k;
- It follows from Eqs. (5.143) and (5.145) that the correct scattering normalization is $\bar{\alpha} = 1$, so that Eq. (5.149) becomes

$$u_0(r) = \sin (kr + \delta_0) \quad ; \text{normalized} \tag{5.154}$$

- Take the limit $k \to 0$ of Eq. (5.153), and define the *s-wave scattering length* by

$$\delta_0(k) \equiv -ka \qquad ; \ k \to 0 \qquad (5.155)$$
$$; \text{ scattering length } a$$

Then Eq. (5.153) becomes

$$\frac{1}{R-a} = \sqrt{v_0} \cot\left(\sqrt{v_0}R\right) \qquad ; \ k \to 0 \qquad (5.156)$$

This implies

$$a = R\left[1 - \frac{1}{\sqrt{v_0}R\cot\left(\sqrt{v_0}R\right)}\right] \qquad ; \text{ scattering length} \quad (5.157)$$

- Recall that with *s*-wave scattering, the scattering amplitude and cross section at zero energy are[46]

$$f(k,\theta) = \frac{e^{i\delta_0(k)}\sin\delta_0(k)}{k} \to -a \qquad ; \ k \to 0$$
$$\frac{d\sigma}{d\Omega} = \frac{\sin^2\delta_0(k)}{k^2} \to |a|^2 \qquad (5.158)$$

- In contrast to hard-sphere scattering, where $a = a_{\text{hard-sphere}}$, *the scattering length a in Eq. (5.157) can take any value $-\infty \leq a \leq \infty$!*
- Outside the potential, it follows from Eq. (5.154) that at zero energy the wave function $u_0(r)$ takes the form

$$\frac{u_0(r)}{k} \to r - a \qquad ; \ k \to 0 \qquad (5.159)$$

The scattering length therefore has an interpretation as the place where this asymptotic form of $u_0(r)/k$ *extrapolates back and crosses the r-axis* (see Fig. 5.24). Note that even though the physical values of r lie in the interval $0 \leq r \leq \infty$, the extrapolation back of the asymptotic form in Eq. (5.159) can cross the *r*-axis *anywhere*;

- *Effective-Range Expansion.* The energy-dependence of the low-energy *s*-wave phase shift is provided by the effective-range expansion. The expression for $\delta_0(k)$ in Eq. (5.153) can be re-written as

$$k\left[\frac{\cos\left(kR\right)\cos\delta_0 - \sin\left(kR\right)\sin\delta_0}{\sin\left(kR\right)\cos\delta_0 + \cos\left(kR\right)\sin\delta_0}\right] = \kappa\cot\left(\kappa R\right) \qquad (5.160)$$

---

[46]Here $(2l+1)P_l = 1$ for $l = 0$.

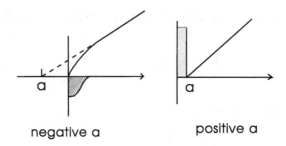

negative a   positive a

Fig. 5.24 Behavior of the true wave function $u_0(r)/k$ at zero energy, and its extrapolated value $r - a$, for negative and positive scattering lengths a. (Compare Prob. 5.13.)

This, in turn, can be re-arranged to read

$$k \cot \delta_0 \left[\cos (kR) - (\kappa/k) \sin (kR) \cot (\kappa R) \right] =$$
$$\kappa \cot (\kappa R) \cos (kR) + k \sin (kR) \qquad (5.161)$$

which gives

$$k \cot \delta_0(k) = \frac{\kappa \cot (\kappa R) + k \tan (kR)}{1 - (\kappa/k) \tan (kR) \cot (\kappa R)} \quad ; \; \kappa^2 = v_0 + k^2 \quad (5.162)$$

This is an *exact* expression for $k \cot \delta_0(k)$ for the attractive square-well potential at all energies. It is evident that the r.h.s. of this expression is a function of k^2, and at *low energy*, it can be expanded as a power series in k^2

$$k \cot \delta_0(k) = -\frac{1}{a} + \frac{1}{2} r_0 k^2 + \cdots$$
$$; \text{ effective-range expansion} \quad (5.163)$$

This is the *effective-range expansion*:

- The first term on the r.h.s. is the negative reciprocal of the *scattering length*;
- The parameter r_0 in the second term is the *effective range*;
- The expansion is a useful one here, provided the dimensionless parameters in Eq. (5.162) satisfy $kR \ll 1$ and $k^2/v_0 \ll 1$.
- It is evident that low-energy scattering measurements can then determine two parameters of the potential, the scattering length a and the effective range r_0.[47]

---

[47] Compare Prob. 5.14.

- To verify the first term in Eq. (5.163), let $k \to 0$ in Eq. (5.162)

$$k \cot \delta_0(k) \to \frac{\sqrt{v_0} \cot (\sqrt{v_0} R)}{1 - \sqrt{v_0} R \cot (\sqrt{v_0} R)} \qquad ; k \to \infty$$

$$= -\frac{1}{a} \qquad (5.164)$$

The last equality follows from Eq. (5.157);
- The effective-range expansion gives $k \cot \delta_0(k)$. For the low-energy s-wave scattering amplitude one actually needs[48]

$$f_0(k) = \frac{e^{i\delta_0(k)} \sin \delta_0(k)}{k} = \frac{\sin \delta_0(k)}{ke^{-i\delta_0(k)}} = \frac{1}{k \cot \delta_0(k) - ik} \qquad (5.165)$$

The low-energy cross section then takes the form

$$\frac{d\sigma}{d\Omega} = \frac{1}{(-1/a + r_0 k^2/2)^2 + k^2} \qquad ; k \to 0 \qquad (5.166)$$

- The next term in the scattering amplitude comes from p-wave scattering

$$f_1(k) = 3 \cos \theta \frac{e^{i\delta_1(k)} \sin \delta_1(k)}{k}$$

$$\to \cos \theta \frac{(-ka_1)^3}{k} \qquad ; k \to 0 \qquad (5.167)$$

Here a_1 is the p-wave scattering length.[49] Two comments:

- This term does not contribute at $\theta = 90°$;
- This term does not contribute to $O(k^2)$ in the integrated cross section $\sigma = \int d\Omega \, (d\sigma/d\Omega)$.

5.6.7 *Effective-Range Theory*

In Eq. (5.162), we solved explicitly for $k \cot \delta_0(k)$ for the attractive square-well potential. Effective-range theory, due to [Schwinger (1947); Bethe (1949)], derives a general expression for the effective range r_0, valid for *any* potential.

Let $w(r, k)$ be a particular solution to the s-wave Schrödinger equation

$$\frac{d^2 w(r, k)}{dr^2} + [k^2 - v(r)]w(r, k) = 0 \qquad ; \text{S-eqn} \qquad (5.168)$$

---

[48]Note that Im $[1/f_0(k)] = -k$ is determined by unitarity.
[49]The p-wave scattering length is here defined by $\delta_1(k) \to (-ka_1)^3/3$ as $k \to 0$.

This solution is chosen to satisfy all the boundary conditions in the region of the potential and to match onto the following asymptotic form, here chosen for convenience,

$$w(r, k) \to \frac{\sin[kr + \delta_0(k)]}{\sin \delta_0(k)} \equiv \phi(r, k) \qquad ; r \to \infty \qquad (5.169)$$

Then the effective range is given by the following integral of a difference between the actual solution and its asymptotic form defined above

$$\frac{1}{2} r_0 = \int_0^\infty dr[\phi^2(r, 0) - w^2(r, 0)] \qquad ; \text{ effective range} \quad (5.170)$$

The situation is illustrated in Fig. 5.25.

The derivation of Eq. (5.170) goes as follows:

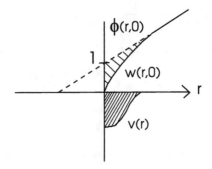

Fig. 5.25   Behavior of the true wave function $w(r, k)$, and its extrapolated asymptotic value $\phi(r, k) = \sin(kr + \delta_0)/\sin \delta_0$, at zero energy in effective range-theory. Note that $\phi(r, 0) = 1 - r/a$, and here the scattering length $a < 0$.

(1) The asymptotic solution $\phi(r, k)$ satisfies the following equation *everywhere*

$$\frac{d^2\phi(r, k)}{dr^2} + k^2\phi(r, k) = 0 \qquad ; \text{ all } r \geq 0 \qquad (5.171)$$

(2) Take the $k \to 0$ limit of Eqs. (5.168) and (5.171)

$$\frac{d^2w(r, 0)}{dr^2} - v(r)w(r, 0) = 0$$

$$\frac{d^2\phi(r, 0)}{dr^2} = 0 \qquad (5.172)$$

It follows from its definition in Eqs. (5.169) that

$$\phi(r,0) = 1 - \frac{r}{a} \tag{5.173}$$

where a is the scattering length.

(3) Label the l.h.s. of Eqs. (5.168), (5.171), and (5.172) by (A, B, C, D) respectively, and take the following combination

$$\int_0^\infty dr \left\{ [w(r,0)(A) - w(r,k)(C)] - [\phi(r,0)(B) - \phi(r,k)(D)] \right\} = 0 \tag{5.174}$$

Now observe that:

- The potential drops out of the first term in square brackets;
- The second term in square brackets cancels the first term in the region outside the potential;
- The terms in k^2 can be taken to the r.h.s. of this relation;
- The second derivatives can then be re-arranged to give[50]

$$\int_0^\infty dr \frac{d}{dr} \left\{ \left[w(r,0)\frac{d}{dr}w(r,k) - w(r,k)\frac{d}{dr}w(r,0) \right] - \tag{5.175}$$
$$\left[\phi(r,0)\frac{d}{dr}\phi(r,k) - \phi(r,k)\frac{d}{dr}\phi(r,0) \right] \right\} =$$
$$k^2 \int_0^\infty dr \left[\phi(r,0)\phi(r,k) - w(r,0)w(r,k) \right]$$

Now further observe the following:

- The integrand on the l.h.s. is a perfect differential;
- There is no contribution to the integral from the upper limit, since $w = \phi$ there;
- The exact solution vanishes at the origin, so $w(0,k) = w(0,0) = 0$;
- It follows from Eq. (5.169) that

$$\phi(0,k) = \phi(0,0) = 1$$
$$\phi'(0,k) = k \cot \delta_0(k)$$
$$\phi'(0,0) = -\frac{1}{a} \tag{5.176}$$

Here a is the scattering length.

---

[50] As in the derivation of Green's theorem.

Thus Eq. (5.175) becomes

$$k \cot \delta_0(k) = -\frac{1}{a} + \frac{1}{2} r_0 k^2$$

$$\frac{1}{2} r_0 \equiv \int_0^\infty dr \, [\phi(r,0)\phi(r,k) - w(r,0)w(r,k)] \quad (5.177)$$

This is an *exact* relation. The $k \to 0$ limit of the second line gives the result quoted in Eq. (5.170). Note that the integral in Eq. (5.170) only gets a contribution from the region where $w(r,0)$ differs from its asymptotic value $\phi(r,0)$, that is, over the region of the potential (see Fig. 5.25). Hence the name "effective range".[51]

5.7   Phase Shift at High Energy–The WKB Approximation

In the preceding sections, we analyzed the scattering problem at *low energy*. Here the angular momentum barrier implies that with a finite-range potential, only s-wave scattering is important. We now go to the other extreme of *high energy*, where the scattering amplitude receives a contribution from many partial waves.

Let us go back to the partial-wave radial differential equation. Define[52]

$$\psi_l^{(+)}(r; k) \equiv \frac{u_l^{(+)}(r)}{kr} \quad (5.178)$$

Then $u_l^{(+)}(r)$ satisfies

$$\left\{ \frac{d^2}{dr^2} + k^2 - \left[v(r) + \frac{l(l+1)}{r^2} \right] \right\} u_l^{(+)}(r) = 0 \quad (5.179)$$

This presents a one-dimensional problem in an *effective potential*

$$v_{\text{eff}}(r) \equiv v(r) + \frac{l(l+1)}{r^2} \quad ; \text{effective potential} \quad (5.180)$$

The additional term represents the angular momentum barrier.

We are interested in the solution to Eq. (5.179) when the parameter k^2, representing the incident energy, becomes very large. The leading term in the solution to this problem, obtained when $k^2 \gg v_{\text{eff}}(r)$, is $u_l^{(+)}(r) \sim e^{\pm ikr}$. This can be readily improved within the framework of the *WKB*

---

[51] For applications of effective-range theory, see Probs. 5.15–5.16.

[52] Once again, we conventionally suppress the k-dependence in $u_l^{(+)}(r)$.

approximation [Wentzel (1926); Kramers (1926); Brillouin (1926)], where the wavenumber in the solution becomes a function of position. We look for solutions to Eq. (5.179) of the form

$$u_l^{(+)}(r) = e^{\pm ik\phi(r)} \qquad ; \text{ still general} \qquad (5.181)$$

In this form, no generality has yet been sacrificed. It follows that

$$\frac{d}{dr} u_l^{(+)}(r) = \pm ik\phi'(r)e^{\pm ik\phi(r)}$$

$$\frac{d^2}{dr^2} u_l^{(+)}(r) = -k^2 \left[\phi'(r)\right]^2 e^{\pm ik\phi(r)} \pm ik\phi''(r)e^{\pm ik\phi(r)} \qquad (5.182)$$

The expression in Eq. (5.181) will then yield a solution to Eq. (5.179) provided that

$$\left\{\pm ik\phi''(r) - k^2 \left[\phi'(r)\right]^2 + \left[k^2 - v_{\text{eff}}(r)\right]\right\} e^{\pm ik\phi(r)} = 0 \qquad (5.183)$$

Now keep just the leading term in k from the first two terms, and assume

$$\left| \frac{\phi''(r)}{k\left[\phi'(r)\right]^2} \right| \ll 1 \qquad (5.184)$$

Equation (5.183) then reduces to[53]

$$[\phi'(r)]^2 = 1 - \frac{v_{\text{eff}}(r)}{k^2}$$

$$\implies \quad \frac{d\phi(r)}{dr} = \left[1 - \frac{v_{\text{eff}}(r)}{k^2}\right]^{1/2} \qquad ; \text{ new S-eqn} \qquad (5.185)$$

This is now the new high-energy Schrödinger equation!

Let us investigate the conditions under which the inequality in Eq. (5.184) holds. Substitution of Eq. (5.185) gives

$$\left| \frac{\phi''(r)}{k\left[\phi'(r)\right]^2} \right| = \frac{1}{2k^3} \left| \frac{dv_{\text{eff}}(r)}{dr} \right| \left[1 - \frac{v_{\text{eff}}(r)}{k^2}\right]^{-3/2} \ll 1 \qquad (5.186)$$

This relation is clearly satisfied if

(1) The *potential varies slowly over a wavelength*, so that

$$\frac{1}{k^3} \left| \frac{dv_{\text{eff}}(r)}{dr} \right| \ll 1 \qquad ; \text{ slowly varying} \qquad (5.187)$$

---

[53]One cannot neglect $v_{\text{eff}}(r)$ with respect to k^2, since $v_{\text{eff}}(r)$ blows up as $r \to 0$ due to the angular momentum barrier (see Fig. 5.26).

(2) The *energy is well above the potential*

$$\left|\frac{v_{\text{eff}}(r)}{k^2}\right| \ll 1 \qquad ; k^2 \text{ above } v_{\text{eff}} \qquad (5.188)$$

These conditions are clearly satisfied in the indicated region in Fig. 5.26.

The first integral of Eq. (5.185) immediately solves for $\phi(r)$ as[54]

$$\phi(r) = \int_{r_0}^{r} dr \left[1 - \frac{v_{\text{eff}}(r)}{k^2}\right]^{1/2} + \phi(r_0) \qquad (5.189)$$

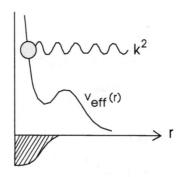

Fig. 5.26 Region where the conditions in Eqs. (5.187)–(5.188) are satisfied for a high-energy wave in the effective potential $v_{\text{eff}}(r)$.

Now add and subtract 1 in the integrand to produce an integral that converges for large r

$$\phi(r) = r - r_0 + \int_{r_0}^{r} dr \left\{\left[1 - \frac{v_{\text{eff}}(r)}{k^2}\right]^{1/2} - 1\right\} + \phi(r_0) \qquad (5.190)$$

The general solution for $u_l^{(+)}(r)$ under the condition in Eq. (5.184) is then given by

$$u_l^{(+)}(r) = ae^{ik\phi(r)} + be^{-ik\phi(r)} \qquad ; \text{general solution} \qquad (5.191)$$

---

[54]As a solution to an ordinary first-order differential equation, this relation holds for any r_0 for which the integral exists; we subsequently identify r_0 with the classical turning point.

The condition in Eq. (5.188) is clearly *violated* at the *classical turning point* r_0 where (see Fig. 5.27)

$$k^2 = v_{\text{eff}}(r_0) \qquad ; \text{ turning point } r_0 \qquad (5.192)$$

It is therefore necessary to make *some further approximation at* r_0.

As a first attempt, let us demand that the wave function *vanish* at the classical turning point[55]

$$u_l^{(+)}(r_0) = 0 \qquad ; \text{ first try} \qquad (5.193)$$

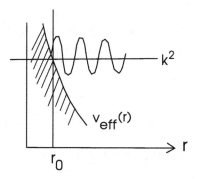

Fig. 5.27 The classical turning point where $v_{\text{eff}}(r_0) = k^2$.

The imposition of this boundary condition on Eq. (5.191) gives

$$u_l^{(+)}(r_0) = a e^{ik\phi(r_0)} + b e^{-ik\phi(r_0)} = 0$$

$$\implies \quad \frac{b}{a} = -\frac{e^{ik\phi(r_0)}}{e^{-ik\phi(r_0)}} \qquad (5.194)$$

Hence $u_l^{(+)}(r)$ in Eq. (5.191) then takes the form

$$u_l^{(+)}(r) = \bar{a} \left\{ e^{ik[\phi(r)-\phi(r_0)]} - e^{-ik[\phi(r)-\phi(r_0)]} \right\} \qquad (5.195)$$

As in Eq. (5.132), the phase shift $\delta_l(k)$ is identified from the asymptotic

---

[55]We know from Vol. I that demanding that the wave function vanish in a classically forbidden region where $k^2 < v_{\text{eff}}$ leads to the incorrect boundary condition. Here there will evidently be some barrier penetration and leaking of the wave function into the forbidden region in Fig. 5.27, which must be taken into account. (Compare Prob. 5.17.)

form of the radial solution[56]

$$2u_l^{(+)}(r) \rightarrow \left\{ e^{2i\delta_l(k)} e^{i[kr-(l+1)\pi/2]} + e^{-i[kr-(l+1)\pi/2]} \right\} \quad ; r \rightarrow \infty \quad (5.196)$$

From Eqs. (5.195) and (5.190)

$$u_l^{(+)}(r) = \bar{a} \left\{ \exp\left[ik\left(r - r_0 + \int_{r_0}^r dr \left\{ \left[1 - \frac{v_{\text{eff}}(r)}{k^2}\right]^{1/2} - 1\right\}\right)\right] - \right.$$
$$\left. \exp\left[-ik\left(r - r_0 + \int_{r_0}^r dr \left\{ \left[1 - \frac{v_{\text{eff}}(r)}{k^2}\right]^{1/2} - 1\right\}\right)\right]\right\}$$
$$(5.197)$$

Now take the asymptotic ratio of the amplitude of the outgoing radial wave to that of the incoming radial wave, and identify with Eq. (5.196)

$$\exp\left[2i\delta_l(k) - i(l+1)\pi\right] =$$
$$- \text{Lim}_{r\to\infty} \exp\left[2ik\left(-r_0 + \int_{r_0}^r dr \left\{ \left[1 - \frac{v_{\text{eff}}(r)}{k^2}\right]^{1/2} - 1\right\}\right)\right] \quad (5.198)$$

Hence the expression for the phase shift under the assumed boundary condition in Eq. (5.193) is[57]

$$\delta_l(k) = l\frac{\pi}{2} - kr_0 + k\int_{r_0}^\infty dr \left\{ \left[1 - \frac{v_{\text{eff}}(r)}{k^2}\right]^{1/2} - 1\right\} \quad (5.199)$$

In this relation

$$v_{\text{eff}}(r) = v(r) + \frac{l(l+1)}{r^2}$$
$$v_{\text{eff}}(r_0) = k^2 \quad (5.200)$$

We can now ask, "Have we treated the classical turning point correctly?" The answer is, of course, no. For example, we know the phase shift must *vanish* if there is no potential $v(r)$, and as readily shown, one does not recover this limit. It is relatively simple, however, to *modify* the result in Eq. (5.199) so that the phase shift *does* vanish in the absence of an interaction, and we proceed to do so.

---

[56] Note Eq. (5.178).
[57] Recall $e^{-i\pi} = -1$.

We note that the above analysis is valid for large k and *large l*, where many partial waves contribute to the scattering. In this regime, it is appropriate to make the following replacements in Eqs. (5.199)–(5.200)

$$\frac{l(l+1)}{r^2} \to \frac{(l+1/2)^2}{r^2} \qquad ; \text{large } l$$

$$l\frac{\pi}{2} \to \left(l+\frac{1}{2}\right)\frac{\pi}{2} \qquad\qquad (5.201)$$

This gives the *WKB approximation for the phase shift*

$$\delta_l^{\text{WKB}}(k) \equiv \left(l+\frac{1}{2}\right)\frac{\pi}{2} - kr_0 + k\int_{r_0}^{\infty} dr \left\{ \left[1 - \frac{\tilde{v}_{\text{eff}}(r)}{k^2}\right]^{1/2} - 1 \right\}$$

$$\tilde{v}_{\text{eff}}(r) \equiv v(r) + \frac{(l+1/2)^2}{r^2}$$

$$\tilde{v}_{\text{eff}}(r_0) \equiv k^2 \qquad\qquad ; \text{WKB approximation} \quad (5.202)$$

Now consider the limit where $v(r) = 0$, so that there is no interaction and hence no scattering. Define

$$kr \equiv \rho$$

$$(kr_0)^2 = \left(l+\frac{1}{2}\right)^2 \equiv \rho_0^2 \qquad ; v(r) = 0 \quad (5.203)$$

Then, when $v(r) = 0$,

$$\delta_l^{\text{WKB}}(k) = \rho_0\left(\frac{\pi}{2} - 1\right) + \mathcal{I} \qquad ; v(r) = 0$$

$$\mathcal{I} \equiv \int_{\rho_0}^{\infty} d\rho \left\{ \left[1 - \left(\frac{\rho_0}{\rho}\right)^2\right]^{1/2} - 1 \right\} \qquad\qquad (5.204)$$

We claim the definite integral is

$$\mathcal{I} = \rho_0\left(1 - \frac{\pi}{2}\right) \qquad\qquad (5.205)$$

Hence, when $v(r) = 0$,

$$\delta_l^{\text{WKB}}(k) = 0 \qquad ; v(r) = 0 \quad (5.206)$$

Thus the WKB phase shift vanishes in the absence of an interaction, as it should.

To establish Eq. (5.205), introduce[58]

$$\frac{\rho_0}{\rho} \equiv \sin\theta \qquad\qquad ; \quad -\frac{\rho_0}{\rho^2}\,d\rho = \cos\theta\,d\theta \qquad (5.207)$$

The integral $\mathcal{I}$ then takes the form

$$\mathcal{I} = \int_{\pi/2}^{0} \left(-\frac{\rho^2}{\rho_0}\cos\theta\,d\theta\right)(\cos\theta - 1)$$

$$= \rho_0 \int_{0}^{\pi/2} \frac{\cos\theta}{\sin^2\theta}(\cos\theta - 1)\,d\theta$$

$$= -\rho_0 \int_{0}^{\pi/2} \frac{\cos\theta}{1 + \cos\theta}\,d\theta \qquad (5.208)$$

This integral can be done by introducing half-angles

$$\mathcal{I} = -\rho_0 \int_{0}^{\pi/2} \frac{\cos^2\theta/2 - \sin^2\theta/2}{2\cos^2\theta/2}\,d\theta$$

$$= -\rho_0 \int_{0}^{\pi/2} \left[1 - \frac{1}{2\cos^2\theta/2}\right]d\theta$$

$$= -\rho_0 \frac{\pi}{2} + \rho_0 \int_{0}^{\pi/4} dx\,\sec^2 x \qquad\qquad ; \quad x \equiv \frac{\theta}{2}$$

$$= -\rho_0 \frac{\pi}{2} + \rho_0 \tan x \Big|_{0}^{\pi/4}$$

$$= \rho_0 \left(1 - \frac{\pi}{2}\right) \qquad (5.209)$$

This is the result quoted in Eq. (5.205).

Three comments:

- The fact that $\delta_l^{\text{WKB}}(k)$ now vanishes when $v(r) = 0$ serves as the justification for the modifications introduced in obtaining Eqs. (5.202);
- *The WKB expression for the phase shift in Eqs. (5.202) provides a remarkably effective approximation at high k;*[59]
- The WKB approximation plays a useful role, for example, in analyzing heavy-ion reactions through recognition of those partial waves whose turning point reaches the nuclear surface.

---

[58]This is a sufficiently important result that we carry out the integral in detail.
[59]See, for example, [Donnelly, Dubach, and Walecka (1974)]. For more applications of the WKB approximation, including bound states, see [Brack and Bhaduri (1997)].

5.8 High-Energy Eikonal Approximation

Instead of pursuing applications of $\delta_l^{\text{WKB}}(k)$, we make a further high-energy simplification to obtain the *eikonal approximation* for the scattering amplitude.[60]

Consider classical orbit theory (see Fig. 5.28). The magnitude of the angular momentum is given by

$$|\mathcal{L}| = |\mathbf{r} \times \mathbf{p}| = rp \sin\theta = bp \tag{5.210}$$

- Here b is the *impact parameter*;
- r_0 is the distance of closest approach.

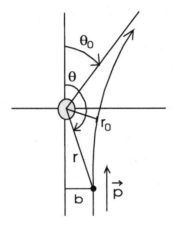

Fig. 5.28 Classical orbit theory. A particle with momentum $\mathbf{p}$ at (r, θ) is scattered through a scattering angle θ_0. Here $b = r \sin\theta$ is the impact parameter, and r_0 is the distance of closest approach.

The quantum mechanical version of Eq. (5.210) becomes

$$\hbar\left(l + \frac{1}{2}\right) = \hbar k b$$

$$\implies \quad \left(l + \frac{1}{2}\right) = kb \qquad ; \text{ defines } b \tag{5.211}$$

This relation *defines* the impact parameter $b(k, l)$ in quantum mechanics.

---

[60]This is often referred to as the "Glauber approximation" [Glauber (1959)].

Insert this last relation in $\delta_l^{\text{WKB}}(k)$ in Eq. (5.202)

$$\delta_l^{\text{WKB}}(k) = kb\frac{\pi}{2} - kr_0 + k\int_{r_0}^{\infty} dr\left\{\left[1 - \frac{b^2}{r^2} - \frac{v(r)}{k^2}\right]^{1/2} - 1\right\}$$

$$= kb\frac{\pi}{2} - kr_0 + k\int_{r_0}^{\infty} dr\left\{\left(1 - \frac{b^2}{r^2}\right)^{1/2}\left[1 - \left(\frac{r^2}{r^2 - b^2}\right)\frac{v(r)}{k^2}\right]^{1/2} - 1\right\}$$

$$\tag{5.212}$$

When expressed in terms of the impact parameter b,

$$\tilde{v}_{\text{eff}}(r) = v(r) + \frac{(kb)^2}{r^2}$$

$$\tilde{v}_{\text{eff}}(r_0) = v(r_0) + \frac{(kb)^2}{r_0^2} = k^2 \qquad ; \text{ turning point} \qquad (5.213)$$

Now let $v(r)/k^2 \to 0$. Note this is $v(r)$ and *not* $\tilde{v}_{\text{eff}}(r)$. With a non-singular potential, it will alway be true that $v(r)/k^2 \to 0$ at high energy. Then

(1) From the second of Eqs. (5.213)

$$r_0 = \frac{b}{[1 - v(r_0)/k^2]^{1/2}} = b\left[1 + \frac{v(r_0)}{2k^2} + \cdots\right] \qquad (5.214)$$

An iteration of this relation gives

$$r_0 = b\left[1 + \frac{v(b)}{2k^2} + \cdots\right] \qquad (5.215)$$

Note that $r_0 \to b$ as $k^2 \to \infty$ (see Fig. 5.28);

(2) As $v(r)/k^2 \to 0$, Eq. (5.212) then takes the form

$$\delta_l^{\text{WKB}}(k) \to kb\frac{\pi}{2} - kb\left[1 + \frac{v(b)}{2k^2}\right] + \qquad (5.216)$$

$$k\int_{b[1+v(b)/2k^2]}^{\infty} dr\left[\left(1 - \frac{b^2}{r^2}\right)^{1/2} - 1 - \left(\frac{r^2}{r^2 - b^2}\right)^{1/2}\frac{v(r)}{2k^2}\right] + \cdots$$

Now expand the integral of the first two terms about its lower limit

$$\int_{b+\epsilon}^{\infty} dr\, f(r) = \int_{b}^{\infty} dr\, f(r) - \epsilon f(b) + \cdots \qquad (5.217)$$

Then Eq. (5.216) becomes

$$\delta_l^{\text{WKB}}(k) \to kb\left(\frac{\pi}{2} - 1\right) + k\int_b^\infty dr\left[\left(1 - \frac{b^2}{r^2}\right)^{1/2} - 1\right]$$

$$-kb\frac{v(b)}{2k^2} - kb\frac{v(b)}{2k^2}\left[\left(1 - \frac{b^2}{r^2}\right)^{1/2} - 1\right]_{r=b}$$

$$-\frac{1}{2k}\int_b^\infty dr\left(\frac{r^2}{r^2 - b^2}\right)^{1/2} v(r) + O\left(\frac{1}{k^3}\right) \qquad (5.218)$$

Now

- The two terms on the r.h.s. in the first line cancel by the analysis in Eqs. (5.204)–(5.206);
- The two terms in the second line also cancel.

Thus one arrives at the simple high-energy expression

$$\delta_l^{\text{WKB}}(k) \to -\frac{1}{2k}\int_b^\infty dr\left(\frac{r^2}{r^2 - b^2}\right)^{1/2} v(r) \qquad ; \quad \frac{v(r)}{k^2} \to 0 \quad (5.219)$$

(3) The integral is to be carried out at fixed impact parameter b. Change variables in the integral to (see Fig. 5.29)

$$r^2 \equiv z^2 + b^2$$
$$r\,dr = z\,dz \qquad (5.220)$$

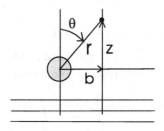

Fig. 5.29   Coordinates (r, θ) and (b, z) on the straight-line eikonal trajectory.

Then

$$\left(\frac{r^2}{r^2 - b^2}\right)^{1/2} dr = \frac{r\,dr}{z} = dz \qquad (5.221)$$

Hence as $v(r)/k^2 \to 0$, the WKB phase shift takes the form

$$\delta_l^{\text{WKB}}(k) \to \delta_l^{\text{Glauber}}(k) \qquad\qquad ; \frac{v(r)}{k^2} \to 0$$

$$= -\frac{1}{2k} \int_0^\infty dz\, v\left(\sqrt{b^2 + z^2}\right) \qquad (5.222)$$

This expression is now a function of the energy and impact parameter (k, b), and the angular momentum l enters only through the impact parameter $b = (l+1/2)/k$ [see Eq. (5.211)]. The integral is immediately extended to negative z to give

$$\delta_l^{\text{Glauber}}(k) = -\frac{1}{4k} \int_{-\infty}^\infty dz\, v\left(\sqrt{b^2 + z^2}\right) \qquad (5.223)$$

To get the high-energy limit of the phase shift $\delta_l^{\text{Glauber}}(k)$, one simply integrates through the potential on a straight-line eikonal trajectory at an impact parameter b.

(4) The partial-wave scattering amplitude $S_l(k)$ is now given by[61]

$$S_l(k) = e^{2i\delta_l(k)} \to e^{i\chi(b,k)}$$

$$i\chi(b, k) = 2i\delta_l^{\text{Glauber}}(k) = -\frac{i}{2k} \int_{-\infty}^\infty dz\, v\left(\sqrt{b^2 + z^2}\right) \quad (5.224)$$

(5) The scattering amplitude is then given by

$$f(k, \theta) = \sum_{l=0}^\infty (2l + 1) P_l(\cos\theta) \left[\frac{S_l(k) - 1}{2ik}\right] \qquad (5.225)$$

This expression can be analyzed as follows:

- Since many partial waves contribute at high energy, one can make use of Eq. (5.211) to replace the sum over l by an integral over b

$$\frac{1}{k} \sum_l \to \int db \qquad (5.226)$$

- Write

$$\frac{2l + 1}{2ik} = \frac{l + 1/2}{ik} = \frac{b}{i} \qquad (5.227)$$

---

[61] This result can be derived directly, without going to radial coordinates and discussing the turning point (see [Schiff (1968)] p. 339, also §34); however, one then misses out on the extremely useful WKB approximation for the phase shift $\delta_l^{\text{WKB}}(k)$.

- Use *Heine's relation*, which states that as $l \to \infty$[62]

$$P_l\left(1 - \frac{z^2}{2l^2}\right) \to J_0(z) \qquad ; l \to \infty \qquad (5.228)$$

$$\text{Heine's relation}$$

Here $J_0(z)$ is the cylindrical Bessel function of order zero;
- Recall that for elastic scattering, the square of the momentum transfer is

$$q^2 = 2k^2(1 - \cos\theta) \qquad (5.229)$$

Therefore

$$\cos\theta = 1 - \frac{q^2}{2k^2} = 1 - \frac{q^2 b^2}{2(l + 1/2)^2}$$

$$\to 1 - \frac{q^2 b^2}{2l^2} \qquad ; l \to \infty \qquad (5.230)$$

Hence at a given qb, Eq. (5.228) can be re-written as

$$P_l(\cos\theta) = J_0(qb) \qquad ; l \to \infty \qquad (5.231)$$

A combination of these results allows the high-energy scattering amplitude in Eqs. (5.224)–(5.225) to be written as

$$f(k, \theta) \to \frac{k}{i} \int_0^\infty b\,db\, J_0(qb)\left[e^{i\chi(b,k)} - 1\right] \qquad ; \frac{v(r)}{k^2} \to 0$$

$$i\chi(b, k) = -\frac{i}{2k} \int_{-\infty}^\infty dz\, v\left(\sqrt{b^2 + z^2}\right) \qquad (5.232)$$

Three comments:

- The integration in the scattering amplitude goes over all *impact parameters* $b = (l + 1/2)/k$;
- The only dependence on the scattering angle is through the argument of the cylindrical Bessel function, where

$$qb = 2kb \sin\frac{\theta}{2} \qquad (5.233)$$

- The result for the phase $\chi(k, b)$ in Eqs. (5.232) is *exact* as $k \to \infty$.

---

[62]See [Whittaker and Watson (1969)] §17.4.

This result can be re-written by making use of the following integral representation of $J_0(z)$[63]

$$J_0(z) = \frac{1}{2\pi} \int_0^{2\pi} d\phi \, e^{-iz\cos\phi} \tag{5.234}$$

Introduce a plane transverse to the incident direction (see Fig. 5.30). For very large $\mathbf{k}$, the scattering occurs primarily at small scattering angle θ. Under these conditions, the momentum transfer $\mathbf{q}$ lies in this plane

$$\mathbf{q} \approx \mathbf{q}_\perp \tag{5.235}$$

Consider a cylindrical eikonal tube surrounding the scatterer. The impact parameter vector $\mathbf{b}$ to the tube also lies in the transverse plane, and it makes an angle ϕ_b with $\mathbf{q}$. Therefore

$$\mathbf{b} \cdot \mathbf{q} = bq\cos\phi_b \tag{5.236}$$

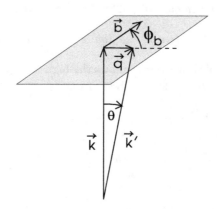

Fig. 5.30 Plane transverse to the incident direction. Here the incident $\mathbf{k}$ is large, the scattering takes place primarily at small angles, and the momentum transfer $\mathbf{q} \approx \mathbf{q}_\perp$. The impact parameter $\mathbf{b}$ lies in this plane, and it makes an angle ϕ_b with $\mathbf{q}$.

Hence, from Eq. (5.234)

$$J_0(qb) = \frac{1}{2\pi} \int_0^{2\pi} d\phi_b \, e^{-i\mathbf{q}\cdot\mathbf{b}} \tag{5.237}$$

---

[63] See, for example, [Fetter and Walecka (2003a)].

The two-dimensional area in the transverse plane is given by

$$bdbd\phi_b = d^{(2)}b \tag{5.238}$$

Equations (5.232) then take the form

$$f(k, \theta) = \frac{k}{2\pi i} \int_0^\infty d^{(2)}b\, e^{-i\mathbf{q}\cdot\mathbf{b}} \left[e^{i\chi(b,k)} - 1 \right] \quad ; \text{2-D Fourier transform}$$

$$i\chi(b, k) = -\frac{i}{2k} \int_{-\infty}^\infty dz\, v\left(\sqrt{b^2 + z^2} \right) \tag{5.239}$$

The scattering amplitude is now a *two-dimensional Fourier transform*, in the transverse plane, of the straight-line eikonal pattern $[e^{i\chi(b,k)} - 1]$ produced by the scattering potential.

5.8.1 *Integrated Elastic Cross Section*

Consider the total elastic cross section, obtained by integrating the differential cross section over solid angle

$$\sigma_{\text{elastic}} = \int d\Omega \left(\frac{d\sigma}{d\Omega} \right) \tag{5.240}$$

It will be assumed here that the differential cross section falls off sufficiently fast with q^2 that all the significant contribution to the integral comes from $\mathbf{q}$ lying in the transverse plane. One can then make use of the following (see Fig. 5.31)

$$q^2 = 2k^2(1 - \cos\theta)$$

$$d\cos\theta = -\frac{dq^2}{2k^2} = -\frac{q\,dq}{k^2} \tag{5.241}$$

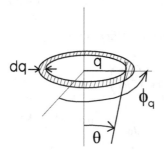

Fig. 5.31   Coordinates (q, ϕ_q) in the transverse plane. The ring thickness is dq.

It follows that

$$
\begin{aligned}
\sigma_{\text{elastic}} &= \int_{-1}^{1} d\cos\theta \int_{0}^{2\pi} d\phi_q \left(\frac{d\sigma}{d\Omega}\right) \\
&= \int_{0}^{2k} \frac{q\,dq}{k^2} \int_{0}^{2\pi} d\phi_q \left(\frac{d\sigma}{d\Omega}\right) \\
&= \frac{1}{k^2} \int d^{(2)}q \left(\frac{d\sigma}{d\Omega}\right) \qquad\qquad ; \ k\to\infty \qquad (5.242)
\end{aligned}
$$

Now use the completeness relation for the two-dimensional plane waves

$$
\int d^{(2)}q\, e^{-i\mathbf{q}\cdot\mathbf{b}}\, e^{i\mathbf{q}\cdot\mathbf{b}'} = (2\pi)^2 \delta^{(2)}(\mathbf{b}-\mathbf{b}') \qquad (5.243)
$$

Then, when Eq. (5.239) is substituted into Eq. (5.242), one obtains the following expression for the integrated cross section

$$
\sigma_{\text{elastic}} = \int d^{(2)}b \left| e^{i\chi(b,k)} - 1 \right|^2 \qquad ; \ \text{integrated cross section} \qquad (5.244)
$$

This is just the integral over impact parameters in the transverse plane of the absolute square of the straight-line eikonal pattern $[e^{i\chi(b,k)} - 1]$ produced by the scattering potential.

Since the differential cross section is independent of the angle ϕ_q, one can re-write Eq. (5.242) as

$$
\sigma_{\text{elastic}} = \frac{\pi}{k^2} \int_{0}^{\infty} dq^2 \left(\frac{d\sigma}{d\Omega}\right) \qquad (5.245)
$$

This relation permits the identification of

$$
\begin{aligned}
\frac{d\sigma}{dq^2} &= \frac{\pi}{k^2}\left(\frac{d\sigma}{d\Omega}\right) \\
&= \frac{1}{4\pi} \left| \int_{0}^{\infty} d^{(2)}b\, e^{-i\mathbf{q}\cdot\mathbf{b}} \left[e^{i\chi(b,k)} - 1 \right] \right|^2 \qquad (5.246)
\end{aligned}
$$

The differential cross section is now the ratio of two quantities $d\sigma$ and dq^2, both of which are defined in the transverse plane.

5.8.2 *Complex Potentials*

Let us generalize the discussion to *complex potentials*, which permits a description of the absorption of the incident beam by the target.[64] Write

$$v(r) = v_R(r) - iv_I(r) \qquad ; \text{complex potential} \quad (5.247)$$

Then in Eqs. (5.239)

$$i\chi(b,k) = -\frac{i}{2k} \int_{-\infty}^{\infty} dz\, v_R(\sqrt{b^2 + z^2}) - \frac{1}{2k} \int_{-\infty}^{\infty} dz\, v_I(\sqrt{b^2 + z^2})$$

$$\equiv i\chi_R(b,k) - \eta(b,k) \qquad (5.248)$$

Hence

$$e^{i\chi(b,k)} = e^{i\chi_R(b,k)}\, e^{-\eta(b,k)} \qquad (5.249)$$

The optical theorem from Probs. 5.7 and 5.20 states that

$$\text{Im}\, f_{\text{el}}(k,0) = \frac{k}{4\pi}\sigma_{\text{total}} \qquad ; \text{optical theorem} \quad (5.250)$$

where σ_{total} is the *total* cross section, now including an absorptive part. From Eq. (5.239)

$$\text{Im}\, f_{\text{el}}(k,0) = \frac{k}{2\pi} \int d^{(2)}b\, \text{Im}\, \frac{1}{i}\left[e^{i\chi(b,k)} - 1\right] \qquad (5.251)$$

Therefore, within the eikonal approximation, the total cross section is given by

$$\sigma_{\text{total}} = \int d^{(2)}b\, 2\,\text{Re}\left[1 - e^{i\chi(b,k)}\right] \qquad ; \text{total cross section} \quad (5.252)$$

The *reaction* cross section σ_{reaction} is defined to include everything else but elastic scattering

$$\sigma_{\text{total}} \equiv \sigma_{\text{elastic}} + \sigma_{\text{reaction}} \qquad (5.253)$$

The integrated elastic cross section is given in Eq. (5.244). Now use

$$2\,\text{Re}\left[1 - e^{\chi(b,k)}\right] - \left|1 - e^{i\chi(b,k)}\right|^2 = 1 - \left|e^{i\chi(b,k)}\right|^2 \qquad (5.254)$$

It follows that

$$\sigma_{\text{reaction}} = \int d^{(2)}b\left[1 - \left|e^{i\chi(b,k)}\right|^2\right] \qquad (5.255)$$

---

[64]Compare Probs. 5.7, 5.20, and 5.24. It is assumed here that $v_I(r) \geq 0$.

5.8.2.1   *Example: Black Disc*

As an example, consider the scattering from a *black disc* (see Fig. 5.32), which is described in the following way

$$e^{i\chi(b,k)} = 0 \qquad ; \ b < R \qquad ; \text{ complete absorption}$$
$$e^{i\chi(b,k)} = 1 \qquad ; \ b > R \qquad ; \text{ no phase shift} \qquad (5.256)$$

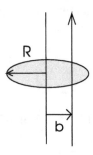

Fig. 5.32   Scattering from a black disc of radius R in the eikonal approximation.

To evaluate Eq. (5.246), one requires[65]

$$2\pi \int_0^R b\,db\,J_0(qb) = \frac{2\pi}{q^2} \int_0^{qR} z J_0(z)\,dz$$
$$= \frac{2\pi}{q^2}(qR)J_1(qR) \qquad (5.257)$$

Thus the two-dimensional Fourier transform of a circle is

$$\int_0^R d^{(2)}b\, e^{-i\mathbf{q}\cdot\mathbf{b}} = \pi R^2 \left[\frac{J_1(qR)}{qR/2} \right] \qquad (5.258)$$

The differential cross section for elastic scattering from a black disc then follows from Eq. (5.246) as

$$\left(\frac{d\sigma}{dq^2} \right)_{\text{el}} = \frac{\pi R^4}{4} \left| \frac{J_1(qR)}{qR/2} \right|^2 \qquad (5.259)$$

This cross section is plotted as a function of qR in Fig. 5.33.

Several comments:

[65] Use $\int z J_0(z)dz = z J_1(z)$; see [Fetter and Walecka (2003a)]. Recall that $J_1(z) \to z/2$ as $z \to 0$.

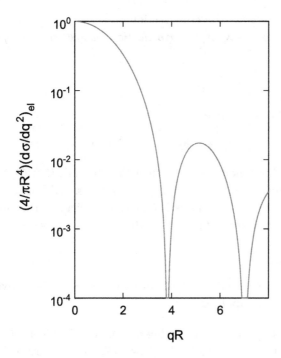

Fig. 5.33 Cross section $(4/\pi R^4)(d\sigma/dq^2)_{el} = |J_1(qR)/(qR/2)|^2$ for scattering from a black disc of radius R in the eikonal approximation. The first minimum occurs at $qR = Z_{11} = 3.832$, and the second at $Z_{21} = 7.016$.

- This result is due to [Bethe and Placzek (1937)];
- The cross section depends on the momentum transfer only through the combination qR;
- This is the diffraction pattern of a circular aperture in classical optics (compare Prob. 5.19);
- Note the magnification in Fig. 5.33—small values of R are studied with large values of q (see Prob. 5.5);
- It follows from Eqs. (5.244), (5.255), and (5.256) that

$$\sigma_{\text{elastic}} = \sigma_{\text{reaction}} = \int_0^R d^{(2)}b$$

$$= \pi R^2 \qquad \text{; black disc} \qquad (5.260)$$

Each is just the *geometrical* cross section;

- Note that there is always *diffractive elastic (shadow) scattering behind the black disc!*[66]

5.8.2.2 An Extension

Note that in the above example

$$\chi(b, k) \to \chi(b) \qquad \qquad \text{; independent of } k \qquad (5.261)$$

More generally, if this condition holds, then Eqs. (5.244), (5.246), (5.252), and (5.255) all depend only on *transverse quantities*. Although derived within the framework of non-relativistic quantum mechanics, this now provides a *relativistically invariant* picture of the scattering process.

5.8.2.3 Applications

To give the reader some feel for what can be done with this analysis, we very briefly present two applications:

(1) An application to p-p elastic scattering at a center-of-momentum energy of $E_{CM} = 53.2\,\text{GeV}$ is shown in Fig. 5.34, taken from [Leith (1974)].

- The data are from an Intersecting Storage Ring (ISR) experiment at CERN (the data are referenced in [Leith (1974)]);[67]
- The calculation is done using the optical-model approach of Durand and Lipes (see Probs. 5.23–5.24);
- A good fit at high energies is obtained if the nucleon radius *increases* very slowly with $(E_{CM})^2 \equiv s$, so that $R^2 = R_0^2 + \bar{\alpha}\ln(s/s_0)$.

(2) A second application to elastic p-^{16}O scattering at an incident proton momentum of $p_{inc} = 1.7\,\text{Gev}/c$ is shown in Fig. 5.35, taken from [Czyz and Maximon (1969)].

- The data are from Brookhaven (the data are referenced in [Czyz and Maximon (1969)]);
- The calculation labeled (B) is done with the high-energy Glauber optical potential (see Prob. 5.24);
- The calculation (C) is the result obtained by just using the experimental oxygen and proton form factors in the optical limit;

---

[66]After all, we are doing wave mechanics!

[67]Relativistically, the center-of-mass (C-M) becomes the center-of-momentum.

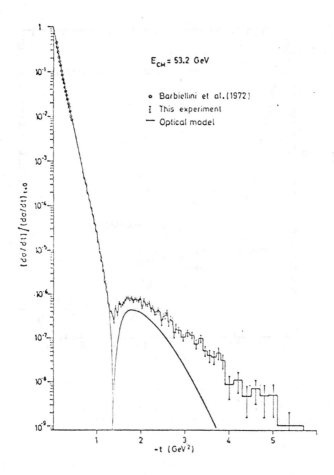

Fig. 5.34   Fit to ISR p-p elastic differential scattering cross section using the optical model approach of Durand and Lipes [Leith (1974)]. Here $-t \equiv (\hbar c \mathbf{q})^2$.

- The calculation labeled (A) goes further and utilizes Glauber multiple scattering theory based on a picture of oxygen as composed of individual nucleons (see [Glauber (1959); Walecka (2004)]).

5.9   Nuclear Reactions

To conclude this chapter on scattering theory, we provide a brief overview of the theory of *nuclear reactions*. Most of this material is taken from [Blatt

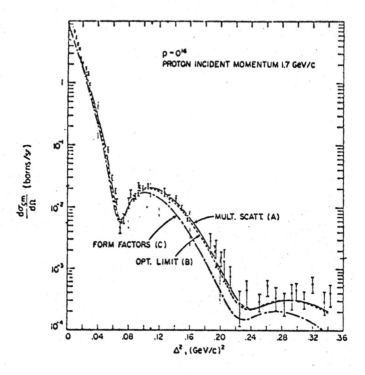

Fig. 5.35   Proton-^{16}O elastic scattering cross sections at $p_{\text{inc}} = 1.7\,\text{Gev}/c$ [Czyz and Maximon (1969)]. Here $\Delta^2 \equiv (\hbar\mathbf{q})^2$, and $1\,\text{barn} \equiv 10^{-24}\,\text{cm}^2$.

and Weisskopf (1952)].

5.9.1   *Reaction Cross Section*

The asymptotic form of the partial-wave scattering state is given by the first of Eqs. (5.103) as

$$\psi_l^{(+)}(r;k) \to \frac{1}{2kr}\left\{S_l(k)e^{i[kr-(l+1)\pi/2]} + e^{-i[kr-(l+1)\pi/2]}\right\}$$
$$;\ r \to \infty \qquad (5.262)$$

Consider a large sphere A surrounding the scatterer (Fig. 5.36). The net incoming probability flux through A in the scattering state was calculated previously in Eqs. (5.109) and (5.114)

$$I_{\text{incoming}} = -\int_A \mathbf{S} \cdot d\mathbf{A} = \frac{\hbar k}{m} \frac{\pi}{k^2} \sum_{l=0}^{\infty} (2l+1) \left[1 - |S_l(k)|^2 \right]$$

$$; \text{ incoming flux} \quad (5.263)$$

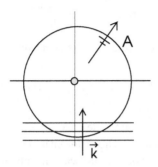

Fig. 5.36   A large sphere A surrounding the scatterer through which the net incoming probability flux in the scattering state is calculated. The scattering state for the incident channel has both an incident and elastically scattered wave.

There are now two possibilities, depending on whether or not the target acts as a *probability sink* for the incident channel

$$I_{\text{incoming}} = 0 \qquad ; \text{ no probability sink}$$
$$> 0 \qquad ; \text{ probability sink at target} \quad (5.264)$$

Here the incident wave and incident probability flux are given by

$$\psi_{\text{inc}} = e^{i\mathbf{k} \cdot \mathbf{x}}$$
$$I_0 = \frac{\hbar k}{m} \qquad ; \text{ incident flux} \quad (5.265)$$

Now recall the previous discussion of the cross section in Eqs. (5.8)–(5.11). With a beam of particles, the *reaction* cross section is defined in terms of the number of particles disappearing from the incident channel per unit time

$$I_0 \, \sigma_{\text{reaction}} \equiv \# \text{ of particles disappearing from incident channel/unit time}$$
$$= I_{\text{incoming}} \quad (5.266)$$

Hence the reaction cross section is a ratio of particle fluxes

$$\sigma_{\text{reaction}} = \frac{\mathcal{I}_{\text{incoming}}}{\mathcal{I}_0} \qquad (5.267)$$

As previously, this can again be written as the ratio of one-body probability fluxes

$$\sigma_{\text{reaction}} = \frac{I_{\text{incoming}}}{I_0} \qquad (5.268)$$

Thus from Eqs. (5.263) and (5.265), the reaction cross section is given by

$$\sigma_{\text{reaction}} = \frac{\pi}{k^2} \sum_{l=0}^{\infty} (2l+1) \left[1 - |S_l(k)|^2\right] \quad ; \text{ reaction cross section} \qquad (5.269)$$

This vanishes if $|S_l(k)| = 1$, as is the case for purely elastic scattering.

From the previous analysis, the differential cross section for elastic scattering is given in terms of the partial-wave amplitudes by

$$\left(\frac{d\sigma}{d\Omega}\right)_{\text{el}} = |f_{\text{el}}(k, \theta)|^2 \qquad ; \text{ differential cross section}$$

$$f_{\text{el}}(k, \theta) = \sum_{l=0}^{\infty} (2l+1) \left[\frac{S_l(k) - 1}{2ik}\right] P_l(\cos\theta) \qquad (5.270)$$

It follows that the integrated elastic cross section is then given by

$$\sigma_{\text{elastic}} = \int d\Omega \left(\frac{d\sigma}{d\Omega}\right)_{\text{el}}$$

$$= \frac{\pi}{k^2} \sum_{l=0}^{\infty} (2l+1) |S_l(k) - 1|^2 \quad ; \text{ elastic cross section} \qquad (5.271)$$

The *total* cross section is the sum of the elastic and reaction contributions

$$\sigma_{\text{total}} = \sigma_{\text{elastic}} + \sigma_{\text{reaction}}$$

$$= \frac{\pi}{k^2} \sum_{l=0}^{\infty} (2l+1) 2\,\text{Re}[1 - S_l(k)] \quad ; \text{ total cross section} \qquad (5.272)$$

Several comments:

- All of these quantities are obtained from the scattering state in the incident channel, which has both an incident and elastically scattered wave;

- The optical theorem follows directly from these results (see Prob. 5.20)

$$\sigma_{\text{total}} = \frac{4\pi}{k}\text{Im}\, f_{\text{el}}(k, 0) \tag{5.273}$$

- Since the reaction cross section is non-negative for a probability sink, the partial-wave amplitudes satisfy the following unitarity relation

$$|S_l(k)|^2 \le 1 \tag{5.274}$$

The result for each partial wave is again obtained with the aid of the superposition principle (see Prob. 5.20);

- These results agree with those obtained in the high-energy eikonal (Glauber) approximation with an absorptive complex potential, where [see Eq. (5.249)]

$$S_l(k) \to e^{i\chi(b,k)} = e^{i\chi_R(b,k)}\, e^{-\eta(b,k)} \tag{5.275}$$

- These results can be given a geometrical interpretation (see Fig. 5.37).

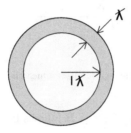

Fig. 5.37 Geometrical interpretation of the maximum reaction cross section for a given l. The relevant quantities are defined in Eqs. (5.276)–(5.278).

Define the reduced wavelength and corresponding impact parameter by

$$\lambdabar \equiv \frac{1}{k} = \frac{\lambda}{2\pi} \qquad ; \; b \approx l\lambdabar \tag{5.276}$$

The following area can then be associated with each impact parameter

$$a(l) = \pi b^2 = \pi l^2 \lambdabar^2 \tag{5.277}$$

If the partial-wave reaction cross section corresponds to the absorption of everything in the ring between l and $l+1$ in Fig. 5.37, then

$$\sigma^l_{\text{reaction}} = a(l+1) - a(l) = \frac{\pi}{k^2}[(l+1)^2 - l^2]$$

$$= \frac{\pi}{k^2}(2l+1) \qquad\qquad \text{; complete absorption} \qquad (5.278)$$

This is the result obtained with $|S_l(k)| = 0$ in Eq. (5.269).

5.9.2   *Reactions with S-Wave Neutrons*

Consider low-energy, s-wave neutron reactions on a nuclear target. The problem will be analyzed in the following manner:[68]

- It is assumed that the nucleus has a well-defined radius R, outside of which the neutron has no interaction (see Fig. 5.38).

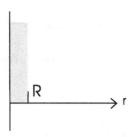

Fig. 5.38   Radius R for the nucleus, outside of which the neutron's interaction vanishes.

The $l = 0$ wave function for $r \geq R$ follows from the first of Eqs. (5.103)

$$u_{\text{out}}(r) = \frac{\bar{A}}{2ik}\left[S_0(k)e^{ikr} - e^{-ikr}\right] \qquad (5.279)$$

- We work in terms of the *logarithmic derivative* of the wave function at the radius R

$$f_0(k) \equiv \left[\frac{1}{u_0}\left(r\frac{du_0}{dr}\right)\right]_R \qquad\qquad \text{; logarithmic derivative}$$

$$= ikR\left[\frac{S_0(k)e^{ikR} + e^{-ikR}}{S_0(k)e^{ikR} - e^{-ikR}}\right] \qquad (5.280)$$

This quantity is independent of the norm of the wave function.[69] Equa-

---

[68] This discussion of nuclear reactions is based on [Blatt and Weisskopf (1952)].
[69] Recall the argument in Eqs. (5.152)–(5.153).

tion (5.280) is readily inverted to give

$$S_0(k) = e^{-2i\rho} \left[\frac{f_0(k) + i\rho}{f_0(k) - i\rho} \right] \qquad ; \rho \equiv kR \qquad (5.281)$$

- Whatever happens *inside* the nucleus, the wave function and its derivative, and hence $f_0(k)$, must be continuous across the boundary;
- If $f_0(k)$ is *real*, then $|S_0(k)| = 1$ and there is no reaction. Thus we write

$$f_0(k) = \mathrm{Re} f_0(k) + i \,\mathrm{Im} f_0(k)$$

$$\sigma^0_{\text{elastic}} = \frac{\pi}{k^2} \left| e^{2i\rho} - 1 - \frac{2i\rho}{f_0(k) - i\rho} \right|^2$$

$$\sigma^0_{\text{reaction}} = \frac{\pi}{k^2} \left[\frac{-4\rho \mathrm{Im} f_0(k)}{[\mathrm{Re} f_0(k)]^2 + [\mathrm{Im} f_0(k) - \rho]^2} \right] \qquad (5.282)$$

The last two results are the s-wave parts of Eqs. (5.271) and (5.269).

5.9.2.1 *Model of Interior Behavior*

The external physics has now been dealt with correctly. To proceed, we have to provide a description of the nuclear interior and calculate the corresponding logarithmic derivative $f_0(k)$. We make two simple models:

(1) Assume first that there is complete absorption inside the target. Assume that just inside the nuclear surface there is only an *incoming* wave, with a substantial nuclear wavenumber κ (see Fig. 5.39)

$$u_{\text{inside}} = Ce^{-i\kappa r} \qquad ; \text{ pure incoming wave} \qquad (5.283)$$

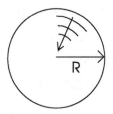

Fig. 5.39 Model of pure incoming wave with wavenumber κ in the nuclear interior.

The corresponding logarithmic derivative is readily calculated as

$$f_0(k) = -i\kappa R \qquad (5.284)$$

The reaction cross section in this case becomes

$$\sigma^0_{\text{reaction}} = \frac{\pi}{k^2}\left[\frac{4k\kappa}{(\kappa+k)^2}\right]$$

$$\rightarrow \frac{4\pi\hbar}{mv}\frac{1}{\kappa} \qquad ; k \rightarrow 0 \qquad (5.285)$$

Here v is the incident neutron velocity, and this gives the celebrated $1/v$–*law* for the capture of low-energy neutrons;

(2) Let us extend the model. Assume that just inside the nuclear surface there is both an incoming wave and also a returning phase-shifted, but unabsorbed, outgoing wave (see Fig. 5.40)

$$u_{\text{inside}} = \bar{C}\left[e^{-i\kappa r} + e^{2i\delta_i}e^{i\kappa r}\right]$$

$$= C\cos\left(\kappa r + \delta_i\right) \qquad (5.286)$$

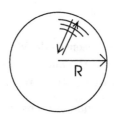

Fig. 5.40   Extended model with both an incoming wave and a phase-shifted outgoing wave in the nuclear interior; both have wavenumber κ.

The corresponding logarithmic derivative is

$$f_0(k) = -\kappa R\tan\left(\kappa R + \delta_i\right) \qquad (5.287)$$

Suppose the parameters are such that the logarithmic derivative *vanishes* for some incident energy E_0

$$f_0(E_0) = 0 \qquad ; \text{some } E_0 \qquad (5.288)$$

Make a Taylor series expansion of $f_0(E)$ about his value

$$f_0(E) = f_0(E_0) + (E - E_0)f_0'(E_0) + \cdots$$

$$\equiv -\frac{2\rho}{\Gamma_0}(E - E_0) + \cdots \qquad (5.289)$$

Here the second line defines the quantity Γ_0.

Substitution of Eq. (5.289) into the second of Eqs. (5.282) gives

$$\sigma^0_{\text{elastic}} = \frac{\pi}{k^2} \left| e^{2ikR} - 1 + \frac{i\Gamma_0}{E - E_0 + i\Gamma_0/2} \right|^2 \qquad (5.290)$$

- The first two terms in the amplitude here represent *potential scattering* from a hard sphere. To see this, recall that for a hard sphere $\delta_0 = -kR$, and therefore

$$\left| 1 - S_0(k) \right|^2 = \left| e^{2ikR} - 1 \right|^2 \qquad ; \text{ hard sphere } (5.291)$$

- The last term in the amplitude in Eq. (5.290) represents a *Breit-Wigner resonance*. If $kR \ll 1$, then near resonance Eq. (5.290) takes the form

$$\sigma^0_{\text{elastic}} = \frac{4\pi}{k^2} \left[\frac{(\Gamma_0/2)^2}{(E - E_0)^2 + (\Gamma_0/2)^2} \right] \qquad ; kR \ll 1 \qquad (5.292)$$

This resonance structure is sketched in Fig. 5.41.

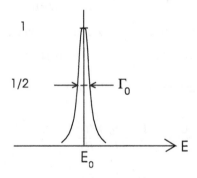

Fig. 5.41   Breit-Wigner resonance structure $(\Gamma_0^2/4) \left[(E - E_0)^2 + \Gamma_0^2/4 \right]^{-1}$. The quantity Γ_0 is known as the *half-width*.

The quantity Γ_0 is referred to as the *half-width*; it is the full width at half-maximum of the resonance curve.

A multitude of low-energy s-wave neutron resonances have been observed, many with very narrow widths.

The neutron wave function outside of the nucleus takes the form[70]

$$u_{\text{out}}(r) = A \sin [kr + \delta_0(k)] \qquad ; \text{ outside nucleus}$$

(5.293)

The logarithmic derivative at the nuclear radius is therefore

$$f_0(k) = kR \cot [kR + \delta_0(k)]$$
$$\approx kR \cot \delta_0(k) \qquad ; kR \ll 1 \qquad (5.294)$$

where the last relation holds at low energy. Hence at resonance, when $f_0 = 0$, the low-energy s-wave phase shift is $\pi/2$

$$\delta_0 = \frac{\pi}{2} \qquad ; f_0 = 0 \qquad (5.295)$$

It follows from Eqs. (5.293)–(5.294) that at the nuclear surface the outside wave function satisfies

$$|u_{\text{out}}(R)|^2 = |A|^2 \frac{1}{1 + (f_0/kR)^2} \qquad (5.296)$$

Similarly, from Eqs. (5.286)–(5.287), the inside wave function satisfies

$$|u_{\text{in}}(R)|^2 = |C|^2 \frac{1}{1 + (f_0/\kappa R)^2} \qquad (5.297)$$

Continuity of the wave function at the nuclear surface demands that

$$|u_{\text{out}}(R)|^2 = |u_{\text{in}}(R)|^2 \qquad (5.298)$$

Thus the ratio of squares of amplitudes is[71]

$$\frac{|C|^2}{|A|^2} = \frac{1 + (f_0/\kappa R)^2}{1 + (f_0/kR)^2}$$
$$\approx \frac{1}{1 + (f_0/kR)^2} \qquad ; \frac{k}{\kappa} \to 0 \qquad (5.299)$$

Hence, with the aid of Eq. (5.289),

$$\frac{|C|^2}{|A|^2} = \frac{\Gamma_0^2/4}{(E - E_0)^2 + \Gamma_0^2/4} \qquad (5.300)$$

---

[70]Recall Eq. (5.154).

[71]Note that $f_0/kR \approx -2(E - E_0)/\Gamma_0$, and $f_0/\kappa R \approx (k/\kappa)f_0/kR$. The quantity $(f_0/\kappa R)^2$ is thus negligible as $k/\kappa \to 0$.

The situation is sketched in Fig. 5.42. *Off resonance*, very little wave function gets into the nucleus, while *on resonance*, it all penetrates.

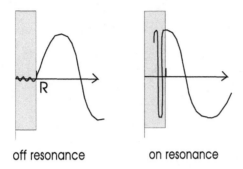

off resonance   on resonance

Fig. 5.42   Square of the $l = 0$ wave function $u(r)$ inside and outside the nucleus both off-resonance, where the wave function is excluded from the nucleus, and on-resonance, where there is a phase shift of $\delta_0 = \pi/2$ and the wave function penetrates the nuclear interior.

A thorough treatise on nuclear reactions can be found in [Feshbach (1991)].

Chapter 6

Time-Dependent Perturbation Theory

So far, we have discussed *stationary states*

$$\Psi(x,t) = \psi(x)e^{-iEt/\hbar} \qquad ; \text{ stationary state} \qquad (6.1)$$

These are solutions to the Schrödinger equation with time-independent densities. We have considered both bound states and scattering states, where particles get scattered or absorbed. Now we will talk about *transitions* between stationary states caused by some perturbation H_1. Examples here include both photoemission and photoabsorption, and inelastic scattering where the target is excited to a discrete state (see Fig. 6.1).

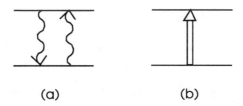

(a)   (b)

Fig. 6.1   Examples of transitions between stationary states caused by some perturbation H_1: (a) photoemission and photoabsorption; (b) inelastic scattering between discrete states of the target.

We proceed to the general analysis of this problem.[1]

---

[1] For continuity and completeness, some repetition of the material in appendix I of Vol. I is included here.

245

6.1   General Analysis

The time-dependent Schrödinger equation is[2]

$$i\hbar\frac{\partial}{\partial t}\Psi(x,t) = H\Psi(x,t) \qquad ; \text{S-eqn} \qquad (6.2)$$

The hamiltonian now consists of two parts

$$H = H_0 + H_1 \qquad ; \text{perturbation } H_1 \quad (6.3)$$

The unperturbed hamiltonian H_0 is just one whose effects we have been studying, and the perturbation H_1 may, or may not, have an explicit time dependence

$$
\begin{aligned}
H_1 &= H_1(t) \qquad ; \text{explicit time-dependence} \\
&= H_1 \qquad\quad ; \text{time-independent}
\end{aligned} \qquad (6.4)
$$

The eigenstates of H_0, the stationary states in the absence of H_1, will be denoted by

$$H_0\phi_n(x) = E_n\phi_n(x) \qquad ; n = 0, 1, 2, \cdots, \infty \quad (6.5)$$

At any time t, completeness of the eigenfunctions of H_0 allows one to expand the solution to the full Schrödinger equation as

$$\Psi(x,t) = \sum_n c_n(t)\phi_n(x)e^{-iE_nt/\hbar} \qquad ; \text{completeness} \qquad (6.6)$$

We make three observations:

(1) The full state is *normalized*, and the orthonormality of the eigenstates of H_0 then implies

$$\langle\Psi|\Psi\rangle = \sum_n |c_n(t)|^2 = 1 \qquad ; \text{normalization} \qquad (6.7)$$

(2) The quantity $|\langle\phi_n|\Psi\rangle|^2$ is similarly given by

$$|\langle\phi_n|\Psi\rangle|^2 = \left|c_n(t)e^{-iE_nt/\hbar}\right|^2 = |c_n(t)|^2 \qquad (6.8)$$

From the general principles of quantum mechanics, this is the *probability of finding the system in the state ϕ_n at the time t*;

---

[2]Once again, x represents the coordinate appropriate to the dimensions of the problem.

(3) The rapidly varying time-dependent phase $e^{-iE_n t/\hbar}$ has been explicitly *extracted* in the definition of the coefficients $c_n(t)$ in Eq. (6.6).

Now substitute the expansion in Eq. (6.6) into the Schrödinger Eq. (6.2)

$$\sum_n \left[i\hbar \frac{dc_n(t)}{dt} + E_n c_n(t) \right] \phi_n(x) e^{-iE_n t/\hbar}$$

$$= \sum_n c_n(t) \left(H_0 + H_1 \right) \phi_n(x) e^{-iE_n t/\hbar}$$

$$= \sum_n c_n(t) \left(E_n + H_1 \right) \phi_n(x) e^{-iE_n t/\hbar} \qquad (6.9)$$

A cancellation of terms gives

$$\sum_n i\hbar \frac{dc_n(t)}{dt} \phi_n(x) e^{-iE_n t/\hbar} = \sum_n c_n(t) H_1 \phi_n(x) e^{-iE_n t/\hbar} \qquad (6.10)$$

The orthonormality of the eigenfunctions $\phi_n(x)$ then leads to

$$i\hbar \frac{dc_n(t)}{dt} = \sum_m \langle \phi_n | H_1 | \phi_m \rangle c_m(t) e^{i(E_n - E_m)t/\hbar} \qquad (6.11)$$

$$; \ n = 0, 1, 2, \cdots, \infty$$

$$; \ \text{exact coupled eqns}$$

This is an infinite set of coupled, linear, first-order differential equations in the time for the coefficients $c_n(t)$. The coefficients in these equations $\langle \phi_n | H_1 | \phi_m \rangle$ are the matrix elements of the perturbation taken between eigenstates of H_0. These equations are exact.

Equations (6.11) can be analyzed in the following manner:

(1) *Suppose $H_1 = 0$:* In this case the equations reduce to

$$\frac{dc_n(t)}{dt} = 0 \qquad ; \ H_1 = 0 \qquad (6.12)$$

The coefficients are constants independent of the time

$$c_n(t) = c_n(t_0) \equiv c_n e^{iE_n t_0/\hbar} \qquad ; \ \text{time-independent} \qquad (6.13)$$

Here the constant $c_n(t_0)$ depends on just how the system is prepared, and the last relation defines c_n. It follows that the full solution to the Schrödinger equation in this case is given by

$$\Psi(x, t) = \sum_n c_n(t_0) \phi_n(x) e^{-iE_n t/\hbar} = \sum_n c_n \phi_n(x) e^{-iE_n(t-t_0)/\hbar} \qquad (6.14)$$

Hence

$$\Psi(x,t) = \sum_n c_n \phi_n(x) e^{-iE_n(t-t_0)/\hbar} \qquad ; H_1 = 0$$

$$\Psi(x,t_0) = \sum_n c_n \phi_n(x) \tag{6.15}$$

This is simply the general solution to the time-dependent Schrödinger equation, constructed by superposition, that reduces to the initial value $\Psi(x,t_0)$ at time t_0.[3]

(2) *Consider $H_1 \neq 0$*: Suppose H_1 is now active, and the system is prepared in some arbitrary state $\Psi_i(x,t_0)$ at time $t = t_0$

$$\Psi_i(x,t_0) = \sum_n c_n^{(i)}(t_0) \phi_n(x) e^{-iE_n t_0/\hbar} \qquad ; \text{ prepared initial state} \tag{6.16}$$

Then

- If H_1 is absent, the previous analysis in Eq. (6.14) says that at a future time

$$\Psi_i(x,t) = \sum_n c_n^{(i)}(t_0) \phi_n(x) e^{-iE_n t/\hbar} \qquad ; H_1 = 0 \tag{6.17}$$

- If H_1 is active, then from Eq. (6.6)

$$\Psi_i(x,t) = \sum_n c_n^{(i)}(t) \phi_n(x) e^{-iE_n t/\hbar} \qquad ; H_1 \neq 0 \tag{6.18}$$

The whole effect of H_1 now is to replace the constant coefficients $c_n^{(i)}(t_0)$ by the time-dependent coefficients $c_n^{(i)}(t)$. To compute the *difference* of these coefficients, just perform $\int_{t_0}^t dt$ on Eq. (6.11)

$$c_n^{(i)}(t) - c_n^{(i)}(t_0) = -\frac{i}{\hbar} \sum_m \int_{t_0}^t dt \, \langle \phi_n | H_1 | \phi_m \rangle c_m^{(i)}(t) e^{i(E_n - E_m)t/\hbar}$$

$$; \text{ coupled integral eqns} \tag{6.19}$$

The differential equations for the time-dependent coefficients $c_n^{(i)}(t)$ have now been converted to an infinite set of coupled, linear *integral equations*, which offers two important advantages:

- The *initial conditions* are now built in

$$c_n^{(i)}(t)\Big|_{t=t_0} = c_n^{(i)}(t_0) \qquad ; \text{ initial conditions} \tag{6.20}$$

---

[3] Compare Prob. 2.35.

- As previously, the integral equations provide a most convenient framework for constructing an *iterated* solution to the problem as a power series in H_1.

We note that Eqs. (6.19) are still *exact*.[4]

6.2 Dirac Perturbation Theory

Assume the transition rate is weak, H_1 is small, and iterate Eqs. (6.19). Since the right-hand-side is explicitly of order H_1, then to first order in H_1 one can use $c_m^{(i)}(t_0)$ *in the integral*

$$c_m^{(i)}(t) = c_m^{(i)}(t_0) + O(H_1)$$
$$c_m^{(i)}(t) \approx c_m^{(i)}(t_0) \qquad ; \text{ on r.h.s. to } O(H_1) \qquad (6.21)$$

Now suppose the system is initially prepared in a given state ϕ_i at time t_0, so that the initial expansion coefficients reduce to

$$c_m^{(i)}(t_0) = c_i(t_0)\delta_{mi} \qquad ; \text{ prepared initial state} \qquad (6.22)$$

Since the initial state is normalized

$$|c_i(t_0)|^2 = 1 \qquad ; \text{ normalization} \qquad (6.23)$$

In the absence of a perturbation, this state simply propagates as

$$\Psi_i(x,t) = c_i(t_0)\phi_i(x)e^{-iE_it/\hbar} \qquad ; H_1 = 0 \qquad (6.24)$$

If H_1 is present, and one wants the coefficient $c_n^{(i)}(t)$ for a state ϕ_n *different* from the initial state ϕ_i, then from Eqs. (6.19)

$$c_n^{(i)}(t) \approx -\frac{i}{\hbar}c_i(t_0)\int_{t_0}^t dt\,\langle\phi_n|H_1|\phi_i\rangle e^{i(E_n-E_i)t/\hbar} \qquad ; n \neq i \qquad (6.25)$$

If one is interested in the *probability that the system is in a specific final state* $\phi_f \neq \phi_i$ *at the time* t, then from Eqs. (6.8) and (6.25)[5]

$$P_{fi}(t,t_0) = |c_f^{(i)}(t)|^2 = \left|-\frac{i}{\hbar}\int_{t_0}^t dt\,\langle\phi_f|H_1|\phi_i\rangle e^{i(E_f-E_i)t/\hbar}\right|^2$$
$$; \text{ transition probability } ; f \neq i \qquad (6.26)$$

Two comments:

---

[4]Remember the r.h.s. of Eq. (6.19) involves the *time-integral* over $c_m^{(i)}(t)$ from t_0 to t.
[5]Note also Eq. (6.23).

- Given the presence of H_1, this is the probability that if one starts in the eigenstate ϕ_i of H_0 at time t_0, one will be in the state ϕ_f at time t (see Fig. 6.2);

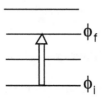

Fig. 6.2   Transition from eigenstate ϕ_i of H_0, to state ϕ_f, under the perturbation H_1.

- The amplitude is exact to $O(H_1)$, and Eq. (6.26) constitutes first-order time-dependent (or *Dirac*) perturbation theory.

We present an example with a time-dependent $H_1(t)$, which provides a model for the energy loss of a charged particle passing through matter, or for a nuclear reaction through Coulomb excitation.[6]

6.2.1   *Example with $H_1(t)$*

A heavy charged particle moves by a target with a velocity **v** in an undeviated, straight-line trajectory at an impact parameter b (see Fig. 6.3).

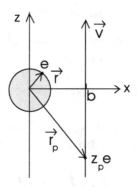

Fig. 6.3   A heavy charged particle with charge $z_p e$ moves by a target with a velocity **v** in an undeviated, straight-line trajectory at an impact parameter b. It interacts through the Coulomb interaction with another charged particle at position **r** in the target. We use unit vectors $(\mathbf{e}_x, \mathbf{e}_z)$ to indicate the x- and z-directions.

---

[6]For another example, see Prob. 6.2.

The position of the projectile is given as a function of time by

$$\mathbf{r}_p(t) = b\mathbf{e}_x + vt\mathbf{e}_z \qquad (6.27)$$

The projectile interacts with another charged particle with charge e at a position $\mathbf{r}$ in the target through the Coulomb interaction

$$H_1(t) = \frac{z_p e^2}{|\mathbf{r}_p(t) - \mathbf{r}|} \qquad ; \text{ Coulomb potential} \quad (6.28)$$

The target particle's motion is thus perturbed by an interaction with a specified time dependence. The interaction at the origin, for example, has the time-dependence sketched in Fig. 6.4. The target particle gets a

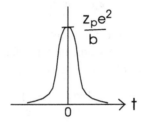

Fig. 6.4 Potential pulse at the origin felt by the target particle.

transverse kick from the electric field, which is effectively of the form

$$\boldsymbol{\mathcal{E}}(t) \approx \mathcal{E}(t)\mathbf{e}_x \qquad ; \text{ effective field} \qquad (6.29)$$

Model the target as a three-dimensional simple harmonic oscillator with the following hamiltonian and ground-state

$$H_0 = \frac{\mathbf{p}^2}{2m} + \frac{1}{2}m\omega_0^2 \mathbf{r}^2 \qquad ; \text{ target hamiltonian}$$
$$\phi_i(\mathbf{r}) = u_0(x)u_0(y)u_0(z) \qquad ; \text{ ground state} \qquad (6.30)$$

Although the target is excited to other states, consider the transition where the target has one quantum of excitation in the $\mathbf{e}_x$-direction (Fig. 6.5).

$$\phi_f(\mathbf{r}) = u_1(x)u_0(y)u_0(z) \qquad ; \text{ excited state} \qquad (6.31)$$

In order to simplify the calculation, make the dipole approximation where

$$r \ll r_p \,^7$$

$$H_1(t) \approx z_p e^2 \left[\frac{1}{r_p} + \frac{\mathbf{r}_p \cdot \mathbf{r}}{r_p^3} + \cdots \right] \qquad ; \text{ dipole approximation}$$

$$r/r_p \ll 1 \qquad (6.32)$$

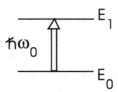

Fig. 6.5   Target excited to oscillator state with one quantum in the $\mathbf{e}_x$-direction.

Since the target states are orthogonal, the term in $1/r_p$ in Eq. (6.32) does not contribute to the transition matrix element of $H_1(t)$, and from Eq. (2.98) the transition dipole moment is

$$\langle \phi_f | \mathbf{r} | \phi_i \rangle = \langle \phi_f | x\mathbf{e}_x + y\mathbf{e}_y + z\mathbf{e}_z | \phi_i \rangle$$

$$= \mathbf{e}_x \left(\frac{\hbar}{2m\omega_0} \right)^{1/2} \qquad (6.33)$$

Recall from Prob. 2.26 that the oscillator parameter is defined by

$$\frac{\hbar^2}{mb_{\mathrm{osc}}^2} = \hbar\omega_0 \qquad (6.34)$$

The transition matrix element of the perturbation is thus given by

$$\langle \phi_f | H_1(t) | \phi_i \rangle = \frac{z_p e^2}{r_p^3} (\mathbf{e}_x \cdot \mathbf{r}_p) \left(\frac{\hbar}{2m\omega_0} \right)^{1/2}$$

$$= \frac{z_p e^2 b}{(b^2 + v^2 t^2)^{3/2}} \left(\frac{b_{\mathrm{osc}}^2}{2} \right)^{1/2} \qquad (6.35)$$

The transition probability as the time runs from $t_0 = -\infty$ to $t = +\infty$, and the projectile passes by the target at an impact parameter b, then follows from Eq. (6.26) as

$$P_{fi}(\infty, -\infty; b) = \left(\frac{z_p e^2}{\hbar} \right)^2 \frac{b_{\mathrm{osc}}^2}{2} \left| \int_{-\infty}^{\infty} dt \, \frac{b e^{i\omega_0 t}}{(b^2 + v^2 t^2)^{3/2}} \right|^2 \qquad (6.36)$$

---

[7]Use $|\mathbf{r}_p - \mathbf{r}|^{-1} = (r_p^2 - 2\mathbf{r}_p \cdot \mathbf{r} + r^2)^{-1/2} = 1/r_p + \mathbf{r}_p \cdot \mathbf{r}/r_p^3 + \cdots$.

The integral is the time Fourier transform of the pulse with respect to the frequency $(E_1 - E_0)/\hbar = \omega_0$. The integral is re-written as

$$\int_{-\infty}^{\infty} dt \frac{be^{i\omega_0 t}}{(b^2 + v^2 t^2)^{3/2}} = -\frac{d}{db} \int_{-\infty}^{\infty} dt \frac{e^{i\omega_0 t}}{(b^2 + v^2 t^2)^{1/2}} \tag{6.37}$$

$$= -\frac{1}{v} \frac{d}{db} \int_{-\infty}^{\infty} d\tau \frac{e^{i(\omega_0 b/v)\tau}}{(1 + \tau^2)^{1/2}} \qquad ; \tau \equiv \frac{vt}{b}$$

The expression in Eq. (6.36) can therefore be put into a standard form

$$P_{fi}(\infty, -\infty; b) = \left(\frac{z_p e^2}{\hbar v}\right)^2 2b_{\mathrm{osc}}^2 \left(\frac{\omega_0}{v}\right)^2 \left|-\frac{d}{d\alpha} K_0(\alpha)\right|^2 \qquad ; \alpha \equiv \frac{\omega_0 b}{v} \tag{6.38}$$

Here $\alpha \equiv \omega_0 b/v$, and $K_0(\alpha)$ is given by the integral

$$K_0(\alpha) = \frac{1}{2} \int_{-\infty}^{\infty} d\tau \frac{e^{i\alpha\tau}}{(1 + \tau^2)^{1/2}} = \int_0^{\infty} d\tau \frac{\cos(\alpha\tau)}{(1 + \tau^2)^{1/2}} \tag{6.39}$$

This is a Hankel function of imaginary argument, defined by[8]

$$K_m(z) \equiv \frac{\pi}{2} i^{m+1} H_m^{(1)}(iz) = \frac{\pi}{2} i^{m+1} [J_m(iz) + iN_m(iz)] \tag{6.40}$$

It has the following properties

$$-\frac{d}{d\alpha} K_0(\alpha) = K_1(\alpha) \longrightarrow \left(\frac{\pi}{2\alpha}\right)^{1/2} e^{-\alpha} \qquad ; \alpha \to \infty$$

$$\longrightarrow \frac{1}{\alpha} \qquad ; \alpha \to 0 \tag{6.41}$$

6.2.1.1 *Inelastic Cross Section*

Let the incident particle flux be $\mathcal{I}_0$. Now surround the target with a ring of radius b, and thickness db, in the transverse plane (see Fig. 6.6). The *inelastic cross section* is then defined by

$$\mathcal{I}_0 \, d\sigma_{fi}(b) = (\# \text{ transitions } i \to f)/(\text{unit time})$$

$$= \mathcal{I}_0 \, 2\pi b \, db \, P_{fi}(\infty, -\infty; b) \tag{6.42}$$

Here $\mathcal{I}_0 \, 2\pi b \, db$ is the number of particles passing through the ring per unit time, and $P_{fi}(\infty, -\infty; b)$ is the previously calculated transition probability.

---

[8]See [Morse and Feshbach (1953)] pp. 1321–1323 for the definition and properties. See also Prob. 6.1.

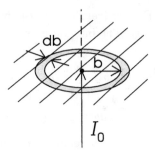

Fig. 6.6   Ring of radius b, and thickness db, surrounding the target in the transverse plane. The incident particle flux is $\mathcal{I}_0$.

It follows that the inelastic cross section is given by

$$d\sigma_{fi}(b) = 2\pi b\, db\, P_{fi}(\infty, -\infty; b) \qquad ; \text{ inelastic cross section}$$

$$\sigma_{fi} = 2\pi \int b\, db\, P_{fi}(\infty, -\infty; b) \qquad (6.43)$$

Several comments:

(1) It is clear from Eqs. (6.38) and (6.41) that most of the contribution to the integrated cross section comes from $\alpha \lesssim 1$, or $b \lesssim v/\omega_0$; otherwise, there are not enough Fourier components in the pulse, and the transition probability falls off as $e^{-2\alpha}$.[9] In the regime $\alpha \lesssim 1$

$$P_{fi}(\infty, -\infty; b) \approx \left(\frac{z_p e^2}{\hbar v}\right)^2 2\left(\frac{b_{\text{osc}}}{b}\right)^2 \qquad ; b \lesssim \frac{v}{\omega_0} \qquad (6.44)$$

(2) The previous analysis is only valid if the projectile is *outside* of the target, or $b \gtrsim b_{\text{osc}}$.[10] The calculated integrated inelastic cross section in Eqs. (6.43) thus takes the form

$$\sigma_{fi} = 4\pi \left(\frac{z_p e^2}{\hbar v}\right)^2 b_{\text{osc}}^2 \int_{b_{\text{min}}}^{b_{\text{max}}} \frac{db}{b}$$

$$= 4\pi \left(\frac{z_p e^2}{\hbar v}\right)^2 b_{\text{osc}}^2 \ln\left(\frac{b_{\text{max}}}{b_{\text{min}}}\right) \qquad (6.45)$$

---

[9]If the effective angular velocity of the projectile is $\omega_{\text{eff}} = v/b$, then the condition for sufficient Fourier components in the pulse for the excitation is $\omega_{\text{eff}} > \omega_0$, or $b < v/\omega_0$.

[10]Note one must always have $P_{fi}(\infty, -\infty; b) \leq 1$.

Substitution of the above values for the limits gives the integrated inelastic cross section as

$$\sigma_{fi} = 4\pi \left(\frac{z_p e^2}{\hbar v}\right)^2 b_{\text{osc}}^2 \ln\left(\frac{v}{\omega_0 b_{\text{osc}}}\right) \tag{6.46}$$

The logarithm is not particularly sensitive to the limits;

(3) This calculation provides a simple version of the Bohr energy-loss formula for charged particles in matter, where

$$\sigma_{fi} \sim \frac{z_p^2}{v^2} \qquad ; \text{ Bohr energy-loss formula} \tag{6.47}$$

Here z_p is the charge of the projectile, and v is its velocity.

6.2.2 H_1 Independent of Time

Consider next the situation where H_1 is independent of time. An example is provided by a two-body collision where the particles interact through a time-independent potential

$$H_1 = V(|\mathbf{x}_a - \mathbf{x}_b|) \qquad ; \text{ time-independent potential} \tag{6.48}$$

and one of the particles is promoted to an excited state in the collision process (see Fig. 6.7).

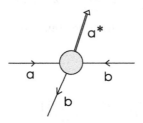

Fig. 6.7 Inelastic collision where the particles interact through the time-independent potential $H_1 = V(|\mathbf{x}_a - \mathbf{x}_b|)$, and one of the particles is promoted to an excited state in the collision process.

The transition probability is again given to leading order in H_1 by Eq. (6.26)

$$P_{fi}(t, t_0) = \left| -\frac{i}{\hbar} \int_{t_0}^{t} dt \, \langle \phi_f | H_1 | \phi_i \rangle e^{i(E_f - E_i)t/\hbar} \right|^2 \tag{6.49}$$

Define the difference of the initial and final total energies, and the total time interval over which H_1 acts, as

$$\frac{E_f - E_i}{\hbar} \equiv \omega \qquad ; \text{total energies}$$

$$t - t_0 \equiv T \qquad ; \text{total time interval} \qquad (6.50)$$

The integral in Eq. (6.49) then becomes

$$\int_{t_0}^{t} dt\, e^{i\omega t} = \frac{1}{i\omega} \left[e^{i\omega t} - e^{i\omega t_0} \right]$$

$$= \frac{e^{i\omega t_0}}{i\omega} \left[e^{i\omega T} - 1 \right]$$

$$= e^{i\omega t_0}\, e^{i\omega T/2} \left[\frac{2\sin(\omega T/2)}{\omega} \right] \qquad (6.51)$$

The transition probability in Eq. (6.49) thus takes the form

$$P_{fi}(T) = \frac{1}{\hbar^2} |\langle \phi_f | H_1 | \phi_i \rangle|^2 \left[\frac{\sin^2(\omega T/2)}{(\omega/2)^2} \right] \qquad (6.52)$$

The expectation is that the transition probability here will be proportional to T, the total time over which the perturbation acts. Therefore it is sensible to define the *transition rate* through

$$\text{transition rate} \equiv (\text{transition probability})/(\text{total time}) \qquad (6.53)$$

Thus the transition rate, which is the transition probability per unit time, is given by

$$R_{fi} \equiv \frac{P_{fi}(T)}{T} \qquad ; \text{transition rate}$$

$$= \frac{1}{\hbar^2} |\langle \phi_f | H_1 | \phi_i \rangle|^2 \left[T \frac{\sin^2(\omega T/2)}{(\omega T/2)^2} \right] \qquad (6.54)$$

This quantity should be independent of T, for large T.

Consider the large-T limit of the function $f_T(\omega)$ appearing in this expresssion

$$f_T(\omega) \equiv T \frac{\sin^2(\omega T/2)}{(\omega T/2)^2} \qquad (6.55)$$

This function has the following properties (see Fig. 6.8):

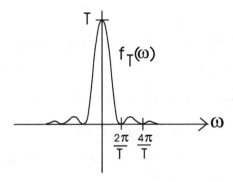

Fig. 6.8 Sketch of the function $f_T(\omega)$ in Eq. (6.55).

- The height at the origin is T, which goes to infinity as $T \to \infty$

$$f_T(0) = T$$
$$\to \infty \qquad ; T \to \infty \qquad (6.56)$$

- At any finite ω, this function goes to zero as $T \to \infty$

$$f_T(\omega) \to 0 \qquad ; T \to \infty \qquad ; \omega \neq 0 \qquad (6.57)$$

Hence $f_T(\omega)$ gets all its contribution at the origin in the large-T limit;[11]
- Consider the following integral over $f_T(\omega)$

$$\int_{-\infty}^{\infty} d\omega \, f_T(\omega) = 2 \int_{-\infty}^{\infty} dx \, \frac{\sin^2 x}{x^2} \qquad ; x \equiv \frac{\omega T}{2}$$
$$= 2\pi \qquad (6.58)$$

This relation holds for all T;
- Thus in the large-T limit, $f_T(\omega)$ has all the properties of a *Dirac delta-function*[12]

$$f_T(\omega) \to 2\pi\delta(\omega) \qquad ; T \to \infty$$
$$= 2\pi\hbar \, \delta(E_f - E_i) \qquad (6.59)$$

In the large-T limit, the transition rate in Eq. (6.54) therefore takes the

---

[11]Note that $\omega = 0$ is *energy conservation*! See Prob. 6.3.
[12]Recall $\delta(ax) = |a|^{-1}\delta(x)$.

form

$$R_{fi} = \frac{P_{fi}}{T} = \frac{2\pi}{\hbar}|\langle\phi_f|H_1|\phi_i\rangle|^2\delta(E_f - E_i) \quad ; T \to \infty$$

$$; \text{ transition rate} \quad (6.60)$$

Note that, as expected, this ratio is finite as $T \to \infty$.[13]

6.2.2.1 *Particle in Continuum in Final State*

As an application, consider the case where there is one particle in the continuum in the final state (see Fig. 6.9).

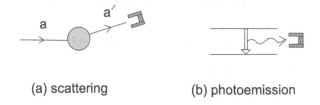

$$\text{(a) scattering} \qquad\qquad \text{(b) photoemission}$$

Fig. 6.9　Examples of transitions with one particle in the continuum in the final state: (a) Scattering, both elastic and inelastic; (b) Photoemission. The detector of the final particle, which plays a central role in the analysis, is indicated.

We start from the expression for the transition rate in Eq. (6.54), where $f_T(\omega)$ is defined in Eq. (6.55) and sketched in Fig. 6.8. The present calculation is then carried out in sufficient detail that one can identify and analyze subsequent approximations made along the way.

Put the whole system in a box of volume $\Omega = L^3$ (Fig. 6.10). The final wave functions are then

$$\phi_f = \frac{1}{\sqrt{\Omega}}e^{i\mathbf{k}_f \cdot \mathbf{x}} \qquad (6.61)$$

---

[13]We are still assuming the perturbation is weak enough so that in the integral in Eqs. (6.19) and (6.25)

$$c_i(t) \approx c_i(t_0) \qquad ; |c_i(t_0)|^2 = 1$$

The initial state must not be significantly depleted, even for large T. Thus the $T \to \infty$ limit is, in fact, rather subtle. One actually assumes that

• $P_{fi}(T) \ll 1$, which implies that $|T\langle f|H_1|i\rangle/\hbar| \ll 1$;

• $\omega T \gg 1$, which implies that $|T(E_f - E_i)/\hbar| \gg 1$.

We shall return shortly to a discussion of the consistency of these assumptions.

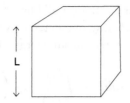

Fig. 6.10 Quantization in a big box of volume $\Omega = L^3$.

With periodic boundary conditions in the box, the wavenumbers are given by

$$k_j = \frac{2\pi n_j}{L} \qquad ; n_j = 0, \pm 1, \pm 2, \cdots, \pm\infty$$
$$; j = x, y, z \qquad (6.62)$$

These states lie very close together in **k**-space as $L \to \infty$ (see Fig. 6.11).

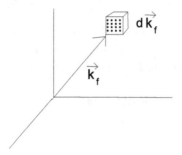

Fig. 6.11 States in the small volume d^3k_f in **k**-space in a big box of volume $\Omega = L^3$.

In any experiment, one measures the transition rate into that *group* of states around $\mathbf{k}_f$ that get into the detector. The number of these states in a differentially small volume d^3k_f is[14]

$$dn_f = dn_x dn_y dn_z = \left(\frac{L}{2\pi}\right)^3 dk_x dk_y dk_z = \frac{\Omega}{(2\pi)^3} d^3k_f \qquad (6.63)$$

Call the sum of the transition rates over these states $\sum'_f R_{fi}$. Then what gets into the detector is

$$\sum_f{}' R_{fi} = R_{fi} dn_f \qquad ; \text{into detector} \quad (6.64)$$

[14]Recall the arguments in Vol. I.

Now change variables, and sum over all final states allowed by the *energy resolution of the experiment*. Say one detects everything in a range $E_f = E_i \pm \Delta$, where E_i is the prepared initial energy. Then

$$\sum_f{}'' R_{fi} = \int R_{fi} dn_f = \int_{E_i - \Delta}^{E_i + \Delta} R_{fi} \left(\frac{\partial n_f}{\partial E_f} \right) dE_f \qquad (6.65)$$

Substitution of the expression for the transition rate in Eq. (6.54) then gives

$$\sum_f{}'' R_{fi} = \int_{E_i - \Delta}^{E_i + \Delta} \left[\frac{2\pi}{\hbar} |\langle \phi_f | H_1 | \phi_i \rangle|^2 \left(\frac{\partial n_f}{\partial E_f} \right) \right] \left[\frac{1}{2\pi\hbar} f_T \left(\frac{E_f - E_i}{\hbar} \right) dE_f \right] \qquad (6.66)$$

Define the following function of E_f

$$\mathcal{R}_{fi}(E_f) \equiv \frac{2\pi}{\hbar} |\langle \phi_f | H_1 | \phi_i \rangle|^2 \left(\frac{\partial n_f}{\partial E_f} \right) \qquad (6.67)$$

It is clear from our previous discussion that $f_T(\omega)$ is strongly peaked about $\omega = 0$ (energy conservation). Therefore, one can make the following replacement in the integral in Eq. (6.66)

$$\mathcal{R}_{fi}(E_f) \approx \mathcal{R}_{fi}(E_i) \qquad (6.68)$$

It follows that

$$\sum_f{}'' R_{fi} \approx \mathcal{R}_{fi}(E_i) \int_{E_i - \Delta}^{E_i + \Delta} \frac{dE_f}{2\pi\hbar} f_T \left(\frac{E_f - E_i}{\hbar} \right)$$

$$= \mathcal{R}_{fi}(E_i) \frac{1}{2\pi} \int_{-\Delta/\hbar}^{\Delta/\hbar} d\omega \, f_T(\omega) \qquad (6.69)$$

Substitution of the definition of $f_T(\omega)$ in Eq. (6.55) gives

$$\frac{1}{2\pi} \int_{-\Delta/\hbar}^{\Delta/\hbar} d\omega \, f_T(\omega) = \frac{1}{2\pi} \int_{-\Delta/\hbar}^{\Delta/\hbar} T d\omega \left[\frac{\sin^2(\omega T/2)}{(\omega T/2)^2} \right] \qquad (6.70)$$

Define

$$x \equiv \frac{\omega T}{2} = \frac{(E_f - E_i)T}{2\hbar} \qquad (6.71)$$

Then

$$\frac{1}{2\pi} \int_{-\Delta/\hbar}^{\Delta/\hbar} d\omega \, f_T(\omega) = \frac{1}{\pi} \int_{-T\Delta/2\hbar}^{T\Delta/2\hbar} dx \, \frac{\sin^2 x}{x^2}$$

$$\approx \frac{1}{\pi} \int_{-\infty}^{\infty} dx \, \frac{\sin^2 x}{x^2} = 1 \qquad (6.72)$$

The last relation holds provided one lets the perturbation act long enough so that

$$T\Delta \gg \hbar \qquad (6.73)$$

The net result of this analysis is the following expression for the transition rate, where there is one particle in the continuum in the final state,

$$\int R_{fi} dn_f = \mathcal{R}_{fi}(E_i) = \frac{2\pi}{\hbar} \left[|\langle \phi_f | H_1 | \phi_i \rangle|^2 \left(\frac{\partial n_f}{\partial E_f} \right) \right]_{E_f = E_i}$$

$$; \text{ Fermi's Golden Rule} \qquad (6.74)$$

This is *Fermi's Golden Rule* for the transition rate. It is an extremely useful expression, as we shall see. This result is fully equivalent to the use of Eq. (6.60)

$$R_{fi} = \frac{2\pi}{\hbar} |\langle \phi_f | H_1 | \phi_i \rangle|^2 \delta(E_f - E_i) \qquad (6.75)$$

from the outset in Eq. (6.65).

Let us consider the *validity* of the approximations made along the way:

(1) In arriving at the starting expression in Eq. (6.54), it was assumed that the initial state is not depleted

$$|c_i(t)|^2 \approx |c_i(t_0)|^2 \qquad ; \text{ initial state not depleted} \qquad (6.76)$$

This is equivalent to the condition[15]

$$P_{fi}(T) \ll 1 \qquad (6.77)$$

From the amplitude in Eq. (6.52) evaluated at $\omega = 0$, this condition becomes

$$|T\langle \phi_f | H_1 | \phi_i \rangle| \ll \hbar \qquad (6.78)$$

The time T cannot be *too long*, or the initial state will be depopulated;

---

[15]This must remain true when summed over $f \neq i$.

(2) To estimate the validity of the approximation in Eq. (6.68), expand

$$\mathcal{R}_{fi}(E_f) = \mathcal{R}_{fi}(E_i) + (E_f - E_i)\mathcal{R}'_{fi}(E_i) + \frac{1}{2}(E_f - E_i)^2\mathcal{R}''_{fi}(E_i) + \cdots$$
(6.79)

The mean value $\langle (E_f - E_i) \rangle$ vanishes with the distribution in Fig. 6.8, and hence the validity of Eq. (6.68) depends on the condition

$$\left| \langle (E_f - E_i)^2 \rangle \frac{1}{2}\mathcal{R}''_{fi}(E_i) \right| \ll \mathcal{R}_{fi}(E_i)$$
(6.80)

Here, from the above,

$$\langle (E_f - E_i)^2 \rangle = \frac{1}{\pi} \int_{-T\Delta/2\hbar}^{T\Delta/2\hbar} dx \left(\frac{2\hbar x}{T} \right)^2 \frac{\sin^2 x}{x^2}$$

$$= \frac{1}{\pi} \left(\frac{2\hbar}{T} \right)^2 \frac{T\Delta}{2\hbar} = \frac{2}{\pi} \left(\frac{\hbar\Delta}{T} \right)$$
(6.81)

Hence Eq. (6.80) becomes[16]

$$\left| \frac{\hbar\Delta}{T} \mathcal{R}''_{fi}(E_i) \right| \ll \mathcal{R}_{fi}(E_i)$$
(6.82)

The transition rate cannot vary too much within the energy resolution of the experiment;

(3) The condition in Eq. (6.73) is

$$T\Delta \gg \hbar$$
(6.83)

The time interval cannot be *too short*, or one does not observe energy conservation within the resolution.

A combination of these criteria indicates that the following conditions must be respected for Eq. (6.74) to hold

$$\text{Max} \left\{ \frac{\hbar}{\Delta}, \left| \frac{\hbar\Delta\mathcal{R}''_{fi}(E_i)}{\mathcal{R}_{fi}(E_i)} \right| \right\} \ll T \ll \frac{\hbar}{|\langle \phi_f | H_1 | \phi_i \rangle|}$$
(6.84)

This is a consistent set of conditions, which can always be satisfied in the limit $H_1 \to 0$.

If the conditions in Eqs. (6.84) are satisfied, then Eq. (6.74) provides an expression for the transition rate that is correct to leading order in H_1. We shall later consider contributions to the transition rate that are of higher

---

[16]To within numerical factors.

order in H_1, but for now, we focus on some applications of the result in Eq. (6.74).

6.2.2.2 *Relation to Previous Scattering Theory*

As a first application, we recover the previous result for elastic scattering obtained with time-independent scattering theory. The calculation is carried out systematically through a series of steps:

(1) The hamiltonian in a static central potential is

$$H = \frac{\mathbf{p}^2}{2m} + V(x) = H_0 + H_1 \tag{6.85}$$

where $x = |\mathbf{x}|$. The initial and final scattering states in a big box with periodic boundary conditions are (see Fig. 6.11)

$$\phi_i = \frac{1}{\sqrt{\Omega}} e^{i\mathbf{k_i} \cdot \mathbf{x}} \qquad ; \phi_f = \frac{1}{\sqrt{\Omega}} e^{i\mathbf{k_f} \cdot \mathbf{x}} \tag{6.86}$$

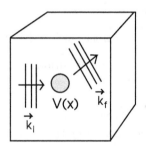

Fig. 6.12 Interaction $H_1 = V(x)$, and the initial and final states of Eqs. (6.86) in a big box of volume $\Omega = L^3$ with periodic boundary conditions. Here $x = |\mathbf{x}|$.

The transition is from $\mathbf{k}_i$ to $\mathbf{k}_f$ (see Fig. 6.13), and the transition matrix element of H_1 is given by

$$\langle \phi_f | H_1 | \phi_i \rangle = \frac{1}{\Omega} \int d^3x \, e^{-i\mathbf{k_f} \cdot \mathbf{x}} \, V(x) e^{i\mathbf{k_i} \cdot \mathbf{x}} \qquad ; \text{transition M.E.} \tag{6.87}$$

(2) The number of final states in the volume element d^3k_f in momentum space is

$$dn_f = \frac{\Omega}{(2\pi)^3} d^3 k_f = \frac{\Omega}{(2\pi)^3} k_f^2 d\Omega_f dk_f \tag{6.88}$$

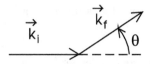

Fig. 6.13   Transition from $\mathbf{k}_i$ to $\mathbf{k}_f$.

(3) The final energy is

$$E_f = \frac{\hbar^2 k_f^2}{2m} \tag{6.89}$$

Hence the derivative $\partial E_f / \partial k_f$ is[17]

$$\frac{\partial E_f}{\partial k_f} = \frac{\hbar^2 k_f}{m} \tag{6.90}$$

Thus the *density of states* is given by

$$\frac{\partial n_f}{\partial E_f} = \frac{\Omega}{(2\pi)^3} k_f^2 \, d\Omega_f \, \frac{m}{\hbar^2 k_f} \qquad ; \text{ density of states} \tag{6.91}$$

(4) The *cross section* for the transition $i \to f$ is given in general by

$$I_{\text{inc}} d\sigma_{fi} = \text{transition rate } i \to f \tag{6.92}$$

In this expression

- I_{inc} is the *incident probability flux*;
- The r.h.s. is the *probability of the transition $i \to f$ per unit time.*

The use of Eq. (6.74) for the transition rate gives the following expression for the cross section obtained with the use of the Golden Rule

$$d\sigma_{fi} = \frac{\mathcal{R}_{fi}(E_i)}{I_{\text{inc}}} \qquad ; \text{ cross section} \tag{6.93}$$

(5) With $\phi_{\text{inc}} = \phi_i$ in Eq. (6.86), the incident flux is given by Eq. (5.14)

$$I_{\text{inc}} = \frac{1}{\Omega} \frac{\hbar k_i}{m} = \rho v_i \qquad ; \text{ incident flux} \tag{6.94}$$

(6) At energy conservation $E_f = E_i$, which implies

$$|\mathbf{k}_f| = |\mathbf{k}_i| \equiv k \qquad ; \text{ energy conservation} \tag{6.95}$$

---

[17]We use a partial derivative here to indicate that it is only the length of the final vector dk_f that is being varied. Note that $\partial k_f / \partial E_f = (\partial E_f / \partial k_f)^{-1}$.

Substitution into Eq. (6.74) then gives the cross section as

$$
d\sigma_{fi} = \frac{\mathcal{R}_{fi}(E_i)}{I_{\text{inc}}}
$$

$$
= \left[\frac{2\pi}{\hbar} \left| \frac{1}{\Omega} \tilde{V}(\mathbf{q}) \right|^2 \right] \left[\frac{\Omega k^2 d\Omega_f}{(2\pi)^3} \frac{m}{\hbar^2 k} \right] \left[\frac{1}{\hbar k/\Omega m} \right] \tag{6.96}
$$

Several comments:

- The first factor is $(2\pi/\hbar)|\langle\phi_f|H_1|\phi_i\rangle|^2$, the second $(\partial n_f/\partial E_f)$, and the last $1/I_{\text{inc}}$;
- The artificial quantization volume Ω *disappears from this relation, as it must!*
- Here $\mathbf{q} \equiv \mathbf{k}_f - \mathbf{k}_i$ is the momentum transfer, and $\tilde{V}(\mathbf{q})$ is the Fourier transform of the potential, exactly as in Eqs. (5.43);
- A cancellation of factors then reduces Eq. (6.96) to

$$
\left(\frac{d\sigma}{d\Omega} \right)_{\text{el}} = \left| -\frac{1}{4\pi} \frac{2m}{\hbar^2} \tilde{V}(\mathbf{q}) \right|^2 \quad ; \text{ Born approximation} \tag{6.97}
$$

This is just the Born approximation result in Eq. (5.40);[18]
- This is an important calculation. Not only does it reproduce the result for elastic scattering obtained with time-independent scattering theory, but, in addition, it provides an extended framework within which to calculate the cross section for *any inelastic process.*

6.2.2.3 *Scattering of Charged Lepton from Atom*

As a second application, consider the Coulomb scattering of a charged lepton $l^\pm$, for example $(e^\pm, \mu^\pm)$, from an atom. This now includes inelastic scattering, where the atom is promoted to an excited state

$$
l^\pm + (e^-, Ze_p) \to l^\pm + (e^-, Ze_p)^\star \tag{6.98}
$$

The incident lepton here is non-relativistic, with energy $\hbar^2 k_i^2/2m_l$. Leptons have only electromagnetic, and the much weaker, weak interactions. The scattering configuration is illustrated in Fig. 6.14. The hamiltonian for this

---

[18]The sign of the amplitude is conventional; recall $v \equiv 2mV/\hbar^2$.

problem, incorporating the Coulomb interactions, is[19]

$$H_0 = \frac{\mathbf{p}_l^2}{2m_l} + \left[\frac{\mathbf{p}_e^2}{2m_e} - \frac{Ze^2}{r_e} \right]$$

$$H_1 = z_l \left[Ze^2 \int d^3r' \frac{\rho_N(r')}{|\mathbf{r}_l - \mathbf{r}'|} - \frac{e^2}{|\mathbf{r}_l - \mathbf{r}_e|} \right] \qquad (6.99)$$

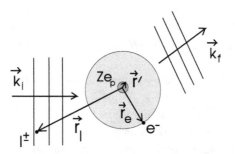

Fig. 6.14   Configuration for scattering of a charged lepton from an atom. The scattered lepton goes from $\mathbf{k}_i \to \mathbf{k}_f$, and the target atom goes from $\psi_i \to \psi_f$. The particles interact through the Coulomb interaction [see Eqs. (6.99)]. The charge on the nucleus is Ze_p; it is distributed according to $\rho_N(r')$ with $\int d^3r' \, \rho_N(r') = 1$. The charge on the lepton is $z_l e_p$ with $z_l = \pm 1$.

We now repeat our systematic calculation of the cross section:

(1) The atom starts in the ground state $\psi_i(\mathbf{r}_e)$ (see Fig. 6.15) and makes a transition to another state $\psi_f(\mathbf{r}_e)$. The initial and final wave functions and energies for the two-body system of atomic electron and incident lepton are therefore[20]

$$\phi_i = \psi_i(\mathbf{r}_e) \frac{1}{\sqrt{\Omega}} e^{i\mathbf{k}_i \cdot \mathbf{r}_l} \qquad ; \ E_i = \varepsilon_i^0 + \frac{\hbar^2 k_i^2}{2m_l}$$

$$\phi_f = \psi_f(\mathbf{r}_e) \frac{1}{\sqrt{\Omega}} e^{i\mathbf{k}_f \cdot \mathbf{r}_l} \qquad ; \ E_f = \varepsilon_f^0 + \frac{\hbar^2 k_f^2}{2m_l} \qquad (6.100)$$

We assume a hydrogen-like target atom with wave functions $\psi_{nlm}(\mathbf{r}_e)$ and corresponding energies

$$\varepsilon_n^0 = -\frac{Z^2 \alpha^2 m_e c^2}{2n^2} \qquad ; \ \text{H-like atom} \qquad (6.101)$$

---

[19]Here we neglect the effect of finite nuclear size on the atomic electron (see §4.2.4.3).
[20]This is just separation of variables in the unperturbed two-body problem.

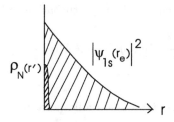

Fig. 6.15   Sketch of the initial target densities.

(2) The matrix element of the perturbation is now given by

$$\langle \phi_f | H_1 | \phi_i \rangle = \frac{z_l e^2}{\Omega} \int d^3 r_e \int d^3 r_l \, e^{-i\mathbf{k}_f \cdot \mathbf{r}_l} \, \psi_f^\star(\mathbf{r}_e) \times$$

$$\left[Z \int d^3 r' \frac{\rho_N(r')}{|\mathbf{r}_l - \mathbf{r}'|} - \frac{1}{|\mathbf{r}_l - \mathbf{r}_e|} \right] \psi_i(\mathbf{r}_e) \, e^{i\mathbf{k}_f \cdot \mathbf{r}_l} \qquad (6.102)$$

The integral over $d^3 r_l$ can be carried out by introducing $\mathbf{t} \equiv \mathbf{r}_l - \mathbf{r}$, with $d^3 r_l = d^3 t$,

$$\int d^3 r_l \, e^{-i\mathbf{q} \cdot \mathbf{r}_l} \frac{1}{|\mathbf{r}_l - \mathbf{r}|} = \int d^3 t \, \frac{1}{t} e^{-i\mathbf{q} \cdot \mathbf{t}} e^{-i\mathbf{q} \cdot \mathbf{r}}$$

$$= \frac{4\pi}{q^2} e^{-i\mathbf{q} \cdot \mathbf{r}} \qquad ; \, \mathbf{q} = \mathbf{k}_f - \mathbf{k}_i \qquad (6.103)$$

Here we have made use of the $\lambda \to 0$ limit of the Fourier transform of the Yukawa potential in Eqs. (5.60)–(5.62)

$$\int d^3 t \, \frac{e^{-\lambda t}}{t} e^{-i\mathbf{q} \cdot \mathbf{t}} = \frac{4\pi}{q^2 + \lambda^2} \qquad (6.104)$$

Introduce the following *form factors*

$$F_N(q) \equiv \int d^3 r' \, \rho_N(r') \, e^{-i\mathbf{q} \cdot \mathbf{r}'} \qquad ; \text{ elastic nuclear f.f.}$$

$$\mathcal{F}_{fi}(\mathbf{q}) \equiv \int d^3 r_e \, \psi_f^\star(\mathbf{r}_e) \psi_i(\mathbf{r}_e) \, e^{-i\mathbf{q} \cdot \mathbf{r}_e} \qquad ; \text{ atomic f.f.} \qquad (6.105)$$

The second relation holds for both elastic and inelastic scattering. The orthonormality of the atomic wave functions gives

$$\int d^3 r_e \, \psi_f^\star(\mathbf{r}_e) \psi_i(\mathbf{r}_e) = \delta_{fi} \qquad (6.106)$$

Thus the matrix element of the perturbation in Eq. (6.102) becomes

$$\langle \phi_f | H_1 | \phi_i \rangle = \frac{z_l}{\Omega} \frac{4\pi e^2}{q^2} [Z F_N(q) \delta_{fi} - \mathcal{F}_{fi}(\mathbf{q})] \quad ; \text{ matrix element of } H_1$$

$$(6.107)$$

(3) The *density of final states* again follows from

$$dn_f = \frac{\Omega}{(2\pi)^3} d^3 k_f$$

$$\frac{\partial n_f}{\partial E_f} = \frac{\Omega}{(2\pi)^3} k_f^2 d\Omega_f \left(\frac{\partial k_f}{\partial E_f} \right) \qquad (6.108)$$

The final energy is given by the last of Eqs. (6.100)

$$E_f = \varepsilon_f^0 + \frac{\hbar^2 k_f^2}{2m_l} \qquad (6.109)$$

Since ε_f^0 is constant, it follows that

$$\left(\frac{\partial E_f}{\partial k_f} \right) = \frac{\hbar^2 k_f}{m_l} \qquad (6.110)$$

Hence the density of states is given by

$$\frac{\partial n_f}{\partial E_f} = \frac{\Omega}{(2\pi)^3} k_f^2 d\Omega_f \left(\frac{m_l}{\hbar^2 k_f} \right) \qquad ; \text{ density of states} \qquad (6.111)$$

Energy conservation relates the initial and final wave numbers according to

$$\varepsilon_f^0 + \frac{\hbar^2 k_f^2}{2m_l} = \varepsilon_i^0 + \frac{\hbar^2 k_i^2}{2m_l} \qquad ; \text{ energy conservation} \qquad (6.112)$$

(4) The incident flux follows exactly as in Eq. (6.94)

$$I_{\text{inc}} = \frac{1}{\Omega} \left(\frac{\hbar k_i}{m_l} \right) \qquad ; \text{ incident flux} \qquad (6.113)$$

(5) The cross section is now obtained from the relation in Eq. (6.93)

$$d\sigma_{fi} = \frac{\mathcal{R}_{fi}(E_i)}{I_{\text{inc}}} \qquad (6.114)$$

Substitution of the above results into Eq. (6.74) then gives

$$d\sigma_{fi} = \frac{2\pi}{\hbar} \left| \frac{z_l}{\Omega} \frac{4\pi e^2}{q^2} [ZF_N(q)\delta_{fi} - \mathcal{F}_{fi}(\mathbf{q})] \right|^2 \times$$
$$\left[\frac{\Omega}{(2\pi)^3} k_f^2 d\Omega_f \left(\frac{m_l}{\hbar^2 k_f} \right) \right] \left[\frac{1}{\hbar k_i / \Omega m_l} \right] \qquad (6.115)$$

A cancellation and combination of terms then yields

$$\frac{d\sigma_{fi}}{d\Omega} = \frac{4}{q^4} \left(\frac{e^2}{\hbar c} \right)^2 \left(\frac{m_l c}{\hbar} \right)^2 \left(\frac{k_f}{k_i} \right) |ZF_N(q)\delta_{fi} - \mathcal{F}_{fi}(\mathbf{q})|^2 \qquad (6.116)$$

This is a *general expression* for the cross section to lowest order in the fine-structure constant $\alpha = e^2/\hbar c = 1/137$.

We discuss a few applications of the result in Eq. (6.116):

(1) *Elastic scattering from H-atom*: Consider elastic scattering from the hydrogen atom where $Z = 1$, and assume a point proton.[21] From Eq. (4.84), the ground-state wave function for the H-atom is

$$\psi_{1s}(r) = \frac{1}{\sqrt{8\pi}} \left(\frac{2Z}{a_0} \right)^{3/2} e^{-Zr/a_0} \qquad (6.117)$$

The elastic atomic form factor $\mathcal{F}_{ii} \equiv \mathcal{F}_{\text{el}}$ is then given by

$$\mathcal{F}_{\text{el}}(q) = \int d^3r \, |\psi_{1s}(r)|^2 \, e^{-i\mathbf{q}\cdot\mathbf{r}}$$
$$= \frac{1}{8\pi} \left(\frac{2Z}{a_0} \right)^3 \int d^3r \, e^{-2Zr/a_0} \, e^{-i\mathbf{q}\cdot\mathbf{r}} \qquad (6.118)$$

The integral is evaluated through the derivative of Eq. (6.104)

$$-\frac{d}{d\lambda} \int d^3t \, \frac{e^{-\lambda t}}{t} e^{-i\mathbf{q}\cdot\mathbf{t}} = \int d^3t \, e^{-\lambda t} e^{-i\mathbf{q}\cdot\mathbf{t}}$$
$$= \frac{8\pi\lambda}{(q^2 + \lambda^2)^2} \qquad (6.119)$$

Hence

$$\mathcal{F}_{\text{el}}(q) = \frac{(2Z/a_0)^4}{[q^2 + (2Z/a_0)^2]^2}$$
$$= \frac{1}{(1 + q^2 a_0^2 / 4Z^2)^2} \qquad ; \text{ elastic f.f. for H-atom} \qquad (6.120)$$

---

[21] For elastic scattering $\delta_{fi} = 1$, and for a point proton $F_N(q) = 1$.

Recall that for elastic scattering

$$q_{el}^2 = 4k^2 \sin^2 \frac{\theta}{2} \qquad ; E_0 = \frac{\hbar^2 k^2}{2m_l}$$

$$\sigma_R = \left| \frac{e^2}{4E_0 \sin^2 \theta/2} \right|^2 \qquad ; \text{Rutherford cross section} \qquad (6.121)$$

For elastic scattering of a charged lepton from the hydrogen atom with $Z = 1$ and a point nucleus, Eq. (6.116) thus becomes

$$\left(\frac{d\sigma}{d\Omega} \right)_{el} = \sigma_R \left[1 - \frac{1}{(1 + q^2 a_0^2/4)^2} \right]^2 \qquad ; \text{H-atom} \qquad (6.122)$$

Here σ_R is the Rutherford cross section in Eqs. (6.121). We observe the following:

- Since the atom is *neutral*, the cross section vanishes as $qa_0 \to 0$;
- For $qa_0 \gg 1$, the cross section is just that for scattering from the point proton;
- This atomic Coulomb cross section passes smoothly from one limit to the other.

(2) *Elastic scattering from H-like atom at large q*: Consider elastic scattering at large momentum transfers from H-like atoms with a nuclear charge Z. In this case

- If $qa_0/Z \gg 1$, the atomic form factor in Eq. (6.120) vanishes;
- One then observes scattering from the *charge distribution of the nucleus*.

Equation (6.116) now becomes

$$\left(\frac{d\sigma}{d\Omega} \right)_{el} = Z^2 \sigma_R |F_N(q)|^2 \qquad (6.123)$$

Here $F_N(q)$ is the nuclear form factor, which for a spherically symmetric charge distribution takes the form[22]

$$F_N(q) = \int d^3 r \, \frac{\sin qr}{qr} \rho_N(r) \qquad ; \text{nuclear charge f.f.} \qquad (6.124)$$

The nuclear Coulomb cross section now exhibits a diffraction structure similar to that shown in Fig. 5.9.

---

[22]Recall Eq. (5.47). Although illustrating the principles, this calculation becomes much more realistic with the scattering of relativistic electrons (see [Walecka (2001)]).

(3) *Inelastic scattering from H-atom*: Consider inelastic scattering from the hydrogen atom, with transitions as illustrated in Fig. 6.16.[23]

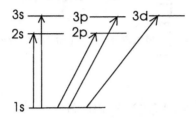

Fig. 6.16   Inelastic transitions in hydrogen.

The energy transfer to the target is

$$\Delta\varepsilon = \frac{\hbar^2 k_i^2}{2m_l} - \frac{\hbar^2 k_f^2}{2m_l} = \varepsilon_f^0 - \varepsilon_i^0 \qquad ; \text{ energy transfer} \quad (6.125)$$

The inelastic atomic form factor is

$$\mathcal{F}_{nlm,1s}(\mathbf{q}) = \int d^3 r \, R_{nl}(r) Y_{lm}^*(\Omega_r) R_{1s}(r) Y_{00}(\Omega_r) \, e^{-i\mathbf{q}\cdot\mathbf{r}} \quad (6.126)$$

Here, for convenience. we choose the $\mathbf{q}$-axis as the z-axis for the quantization of the target angular momentum, and all the m values now refer to this as the $\mathbf{e}_z$-axis (see Fig. 6.17).

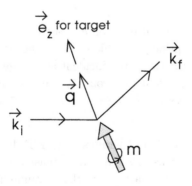

Fig. 6.17   Target quantization axis for inelastic scattering.

---

[23] For inelastic scattering $\delta_{fi} = 0$.

Now insert the plane-wave expansion[24]

$$e^{-i\mathbf{q}\cdot\mathbf{r}} = \sum_{l=0}^{\infty}(2l+1)(-i)^l j_l(qr)P_l(\cos\theta_r)$$

$$= \sum_{l=0}^{\infty}[4\pi(2l+1)]^{1/2}(-i)^l j_l(qr)Y_{l0}(\cos\theta_r) \qquad (6.127)$$

Use

$$\int d\Omega_r Y_{lm}^*(\Omega_r)Y_{l'0}(\Omega_r)Y_{00} = \frac{1}{\sqrt{4\pi}}\delta_{ll'}\delta_{m0} \qquad (6.128)$$

The Coulomb inelastic atomic form factor is then given by

$$\mathcal{F}_{nlm,1s}(\mathbf{q}) = \delta_{m0}(-i)^l(2l+1)^{1/2}\int_0^{\infty} r^2 dr\, R_{nl}(r)j_l(qr)R_{1s}(r)$$
$$\text{; inelastic atomic f.f.} \qquad (6.129)$$

A few comments:

- These inelastic form factors start as q^l, which helps to order them. The oscillation of the integrand for large q then drives them to zero;
- The analytic and numerical calculations for a few typical cases are assigned as Probs. 6.4–6.6;
- With degenerate final states, as is Fig. 6.16, *one must sum over everything that gets into the detector*;
- There are some additional complications if the incident lepton is an $e^{\pm}$:
  - With an incident electron e^-, one must incorporate the Pauli exclusion principle. This leads to the possibility of exchange scattering between the incident and target particles;
  - With an incident positron e^+, one must include the annihilation process $e^+ + e^- \to \gamma^* \to e^+ + e^-$ involving the target electron.

An additional application of Fermi's Golden Rule to direct nuclear reactions is covered in Probs. 6.7–6.8. An extension of Dirac perturbation theory to the case with two particles in the continuum in the final state is covered in Prob. 6.9.

---

[24]Recall $P_l(\cos\theta) = [4\pi/(2l+1)]^{1/2}Y_{l0}(\Omega)$. Note the subsequent selection rule δ_{m0} in Eq. (6.129).

6.2.3 Interaction with External Electromagnetic Fields

As another application of time-dependent perturbation theory, consider the interaction of a charged quantum system, say an electron in an atom, with a classical, external electromagnetic field.

6.2.3.1 Formulation of Problem

The Schrödinger equation and hamiltonian for this problem are[25]

$$i\hbar\frac{\partial}{\partial t}\Psi = H\Psi$$

$$H = \frac{1}{2m}\left[\mathbf{p} - \frac{e}{c}\mathbf{A}(\mathbf{x}, t)\right]^2 + e\Phi(\mathbf{x}, t) + V(\mathbf{x}) \qquad (6.130)$$

Here, to satisfy the canonical commutation relations in the coordinate representation,

$$\mathbf{p} = \frac{\hbar}{i}\boldsymbol{\nabla} \qquad ; \text{ coordinate rep} \qquad (6.131)$$

The electromagnetic fields, characterized by the potentials $(\mathbf{A}, \Phi)$, have a specified time dependence, and are evaluated at the position of the system. The fields themselves are related to the potentials by

$$\mathbf{B} = \boldsymbol{\nabla} \times \mathbf{A}$$

$$\boldsymbol{\mathcal{E}} = -\boldsymbol{\nabla}\Phi - \frac{1}{c}\frac{\partial \mathbf{A}}{\partial t} \qquad (6.132)$$

6.2.3.2 Gauge Invariance

Let $\lambda(\mathbf{x}, t)$ be any differentiable scalar function of space-time. The electromagnetic fields are *unchanged* under a gauge transformation, which replaces the potentials by

$$\mathbf{A}_\lambda \equiv \mathbf{A} + \boldsymbol{\nabla}\lambda \qquad ; \text{ gauge transformation}$$

$$\Phi_\lambda \equiv \Phi - \frac{1}{c}\frac{\partial \lambda}{\partial t} \qquad (6.133)$$

Since the physical fields $(\boldsymbol{\mathcal{E}}, \mathbf{B})$ are unchanged, this gauge transformation *cannot change the physics.* What happens to the Schrödinger equation?

- The Schrödinger equation is formulated in terms of the *potentials* $(\mathbf{A}, \Phi)$;

---

[25]See appendix B. The corresponding conserved probability current is constructed in Prob. 6.11.

- This formulation reproduces the correct classical limit for particle motion in the electromagnetic field through Ehrenfest's theorem;[26]
- Recall that one has invariance under *unitary transformations* in quantum mechanics

$$\Psi_U = U\Psi \qquad ; \text{ unitary transformation}$$
$$O_U = UOU^{-1} \qquad\qquad (6.134)$$

Consider that unitary transformation that changes the phase of the wave function everywhere in space-time by $e\lambda(\mathbf{x}, t)/\hbar c$

$$\Psi_\lambda \equiv U_\lambda\Psi = e^{ie\lambda(\mathbf{x},t)/\hbar c}\Psi \qquad ; \text{ local change of phase} \qquad (6.135)$$

Then

— The quantity $\mathbf{p}\Psi$ becomes

$$\frac{\hbar}{i}\boldsymbol{\nabla}\left(e^{-ie\lambda/\hbar c}\Psi_\lambda\right) = e^{-ie\lambda/\hbar c}\left(\frac{\hbar}{i}\boldsymbol{\nabla} - \frac{e}{c}\boldsymbol{\nabla}\lambda\right)\Psi_\lambda \qquad (6.136)$$

— A repetition of this step gives $\mathbf{p}^2\Psi$;
— The time derivative $i\hbar\partial\Psi/\partial t$ is given by

$$i\hbar\frac{\partial}{\partial t}\left(e^{-ie\lambda/\hbar c}\Psi_\lambda\right) = e^{-ie\lambda/\hbar c}\left(i\hbar\frac{\partial}{\partial t} + \frac{e}{c}\frac{\partial\lambda}{\partial t}\right)\Psi_\lambda \qquad (6.137)$$

It follows from Eq. (6.130) that the Schrödinger equation becomes

$$\left[\frac{1}{2m}\left(\mathbf{p} - \frac{e}{c}\mathbf{A} - \frac{e}{c}\boldsymbol{\nabla}\lambda\right)^2 + V - \left(i\hbar\frac{\partial}{\partial t} - e\Phi + \frac{e}{c}\frac{\partial\lambda}{\partial t}\right)\right]\Psi_\lambda = 0 \qquad (6.138)$$

The gauge-transformed potentials are immediately identified from Eqs. (6.133), and thus

$$\left[\frac{1}{2m}\left(\mathbf{p} - \frac{e}{c}\mathbf{A}_\lambda\right)^2 + V - \left(i\hbar\frac{\partial}{\partial t} - e\Phi_\lambda\right)\right]\Psi_\lambda = 0 \qquad (6.139)$$

One concludes that

> *A gauge transformation of the electromagnetic potentials is equivalent to a unitary transformation that changes the phase of the system's wave function at each point in space-time.*

Such a unitary transformation leaves the physics invariant.

---

[26]See appendix B.

• This argument can be *turned around*. Suppose one demands gauge invariance from the outset. From the above analysis, this is achieved by making the following *gauge-invariant replacements* in the classical hamiltonian for the system

$$\mathbf{p} \to \mathbf{p} - \frac{e}{c}\mathbf{A} \qquad ; \text{ gauge-invariant replacements}$$
$$H \to H + e\Phi \qquad \text{classical H} \qquad (6.140)$$

This reproduces the hamiltonian in Eq. (6.130).

6.2.3.3 *External Radiation Field*

sources outside

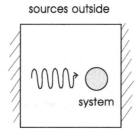

system

Fig. 6.18   Interaction of a quantum system with an external (specified) radiation field.

Suppose the quantum system is subjected to a specified *radiation field* produced by sources external to the active volume. Maxwell's equations in the source-free region read

$$\Box\boldsymbol{\mathcal{E}} = \Box\mathbf{B} = 0 \qquad ; \text{ Maxwell's equations}$$
$$\boldsymbol{\nabla} \cdot \boldsymbol{\mathcal{E}} = \boldsymbol{\nabla} \cdot \mathbf{B} = 0 \qquad \text{no sources} \qquad (6.141)$$

A transverse gauge for the radiation field can be chosen, satisfying

$$\boldsymbol{\nabla} \cdot \mathbf{A} = 0 \qquad ; \text{ radiation field}$$
$$\Phi_\gamma = 0 \qquad (6.142)$$

For monochromatic radiation, this vector potential can be explicitly constructed as follows. Let $\mathbf{A}_0$ be a constant vector transverse to the wave vector $\mathbf{k}$. Then take (see Fig. 6.19)

$$\mathbf{A}(\mathbf{x}, t) = \mathbf{A}_0 \left[e^{i\mathbf{k}\cdot\mathbf{x} - i\omega t} + e^{-i\mathbf{k}\cdot\mathbf{x} + i\omega t} \right] \qquad ; \text{real}$$
$$; \ \omega = kc \qquad (6.143)$$

We impose the condition that the classical fields must be *real*, and also the dispersion relation for radiation $\omega = |\mathbf{k}|c$.

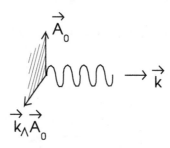

Fig. 6.19   Vector potential for the external (specified) radiation field. We later write $\mathbf{A}_0 = A_0 \mathbf{e}_s$, where $(\mathbf{e}_1, \mathbf{e}_2)$ form an orthonormal set of unit vectors transverse to $\mathbf{k}$.

The corresponding electric and magnetic components of the radiation field are then

$$\boldsymbol{\mathcal{E}}_\gamma = -\frac{1}{c}\frac{\partial \mathbf{A}}{\partial t} = \frac{i\omega}{c}\mathbf{A}_0 \left[e^{i\mathbf{k}\cdot\mathbf{x} - i\omega t} - e^{-i\mathbf{k}\cdot\mathbf{x} + i\omega t} \right]$$
$$\mathbf{B} = \boldsymbol{\nabla} \times \mathbf{A} = i\mathbf{k} \times \mathbf{A}_0 \left[e^{i\mathbf{k}\cdot\mathbf{x} - i\omega t} - e^{-i\mathbf{k}\cdot\mathbf{x} + i\omega t} \right] \qquad (6.144)$$

The *energy flux* in the radiation field is given by the Poynting vector

$$\mathbf{S}_\gamma = \frac{c}{4\pi}\left(\boldsymbol{\mathcal{E}}_\gamma \times \mathbf{B} \right) \qquad ; \text{Poynting vector} \qquad (6.145)$$

Thus

$$\mathbf{S}_\gamma = \frac{\omega}{4\pi}|\mathbf{A}_0|^2 \mathbf{k} \left[2 - e^{2i\mathbf{k}\cdot\mathbf{x} - 2i\omega t} - e^{-2i\mathbf{k}\cdot\mathbf{x} + 2i\omega t} \right] \qquad (6.146)$$

The time-average of this quantity is

$$\overline{\langle \mathbf{S}_\gamma \rangle} = \frac{\omega}{2\pi}|\mathbf{A}_0|^2 \mathbf{k} \qquad ; \text{time-average} \qquad (6.147)$$

This can be used to define an *incident photon flux* in the $\mathbf{k}$-direction, in terms of the number of photons of energy $\hbar\omega$ per unit area per unit time

$$I_{\text{inc}} = \frac{\overline{\langle S_\gamma \rangle}}{\hbar\omega} = \frac{k}{2\pi\hbar}|\mathbf{A}_0|^2 \qquad ; \text{incident photon flux} \qquad (6.148)$$

6.2.3.4 *Transition Rate*

Divide the hamiltonian into H_0 and H_1[27]

$$H_0 = \frac{\mathbf{p}^2}{2m} + V(\mathbf{x}) + e\Phi(\mathbf{x})$$

$$H_1 = -\frac{e}{2mc}\left[\mathbf{p}\cdot\mathbf{A}(\mathbf{x},t) + \mathbf{A}(\mathbf{x},t)\cdot\mathbf{p}\right] + \frac{e^2}{2mc^2}\left[\mathbf{A}(\mathbf{x},t)\right]^2 \quad (6.149)$$

The transition matrix element of the perturbation is then given to lowest order in e by[28]

$$\langle\phi_f|H_1(t)|\phi_i\rangle = -\frac{e}{2mc}\int d^3x\,\phi_f^\star(\mathbf{x})\left[\mathbf{p}\cdot\mathbf{A}(\mathbf{x},t) + \mathbf{A}(\mathbf{x},t)\cdot\mathbf{p}\right]\phi_i(\mathbf{x})$$

$$= -\frac{e}{2mc}\int d^3x\,\left\{\phi_f^\star(\mathbf{x})\mathbf{p}\phi_i(\mathbf{x}) + \left[\mathbf{p}\phi_f(\mathbf{x})\right]^\star\phi_i(\mathbf{x})\right\}\cdot\mathbf{A}(\mathbf{x},t)$$

$$; \text{ to order } e \quad (6.150)$$

The second line follows from the hermiticity of $\mathbf{p}$. This expression can be re-written as

$$\langle\phi_f|H_1(t)|\phi_i\rangle = -\frac{e}{c}\int d^3x\,\mathbf{S}_{fi}(\mathbf{x})\cdot\mathbf{A}(\mathbf{x},t)$$

$$\mathbf{S}_{fi}(\mathbf{x}) = \frac{\hbar}{2im}\left\{\phi_f^\star(\mathbf{x})\boldsymbol{\nabla}\phi_i(\mathbf{x}) - \left[\boldsymbol{\nabla}\phi_f(\mathbf{x})\right]^\star\phi_i(\mathbf{x})\right\}$$

$$; \text{ transition probability current} \quad (6.151)$$

The last line identifies the *transition probability current*.

Substitution of the vector potential in Eq. (6.143) exhibits the harmonic time dependence of the transition matrix element of the perturbation

$$\langle\phi_f|H_1(t)|\phi_i\rangle = \langle\phi_f|H_1|\phi_i\rangle_{\text{abs}}\,e^{-i\omega t} + \langle\phi_f|H_1|\phi_i\rangle_{\text{em}}\,e^{i\omega t} \quad (6.152)$$

Here the absorption and emission matrix elements (see below) are given by

$$\langle\phi_f|H_1|\phi_i\rangle_{\text{abs}} = -\frac{e}{c}\int d^3x\,\mathbf{S}_{fi}(\mathbf{x})\cdot\mathbf{A}_0\,e^{i\mathbf{k}\cdot\mathbf{x}}$$

$$\langle\phi_f|H_1|\phi_i\rangle_{\text{em}} = -\frac{e}{c}\int d^3x\,\mathbf{S}_{fi}(\mathbf{x})\cdot\mathbf{A}_0\,e^{-i\mathbf{k}\cdot\mathbf{x}} \quad (6.153)$$

---

[27]We include in H_0 the possibility of an electrostatic potential of the form $\Phi(\mathbf{x})$.

[28]The dimensionless expansion parameter here is $\alpha = e^2/\hbar c = 1/137.0$, where α is the fine-structure constant.

The transition rate is given in Dirac perturbation theory by Eq. (6.49) and the first of Eqs. (6.54)

$$R_{fi} = \frac{1}{T} \left| -\frac{i}{\hbar} \int_{t_0}^{t} dt \, \langle \phi_f | H_1(t) | \phi_i \rangle e^{i(E_f - E_i)t/\hbar} \right| \quad ; \text{ transition rate} \quad (6.154)$$

Now substitute Eq. (6.152) with its harmonic time dependence, and repeat the previous arguments on the time integrals. As $T = t - t_0 \to \infty$, one obtains $\delta(E_f - E_i + \hbar\omega)$ from the absolute square of the emission term, and $\delta(E_f - E_i - \hbar\omega)$ from the absolute square of the absorption term (see Fig. 6.20).

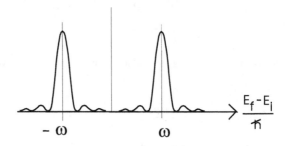

Fig. 6.20   The two contributions to the transition rate in Eq. (6.154).

Since there is no overlap of the two terms, they contribute incoherently to the rate. *Only one term, or the other will contribute in the limit $T \to \infty$ for a transition from a given initial state ϕ_i, with a given E_i, to a distinct final state ϕ_f (see Fig. 6.21).*

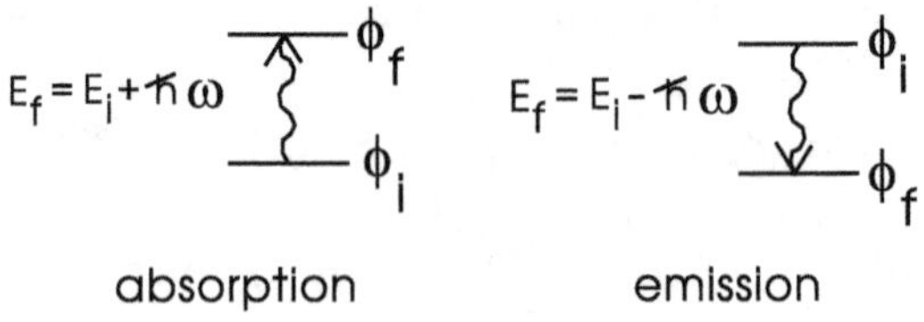

$$E_f = E_i + \hbar\omega \qquad \phi_f \qquad\qquad E_f = E_i - \hbar\omega \qquad \phi_i$$

$$\phi_i \qquad\qquad\qquad \phi_f$$

absorption   emission

Fig. 6.21   Absorption and emission of radiation from a given initial state ϕ_i to distinct final states ϕ_f.

The transition rates corresponding to the two cases then take the form

$$R_{fi} = \frac{2\pi}{\hbar} |\langle \phi_f | H_1 | \phi_i \rangle_{\text{abs}}|^2 \delta(E_f - E_i - \hbar\omega) \qquad ; \text{ absorption}$$

$$= \frac{2\pi}{\hbar} |\langle \phi_f | H_1 | \phi_i \rangle_{\text{em}}|^2 \delta(E_f - E_i + \hbar\omega) \qquad ; \text{ emission} \qquad (6.155)$$

These expressions have the same meaning as before; they must still be summed over the appropriate final states.

A *photon cross section* can be obtained by using the above rates and the incident photon flux in Eq. (6.148)

$$d\sigma_{fi} = \frac{R_{fi}}{I_{\text{inc}}} \qquad ; \text{ photon cross section} \qquad (6.156)$$

$$= \frac{(2\pi)^2}{k} \frac{1}{|\mathbf{A}_0|^2} |\langle \phi_f | H_1 | \phi_i \rangle_{\text{abs}}|^2 \delta(E_f - E_i - \hbar\omega) \qquad ; \text{ absorption}$$

$$= \frac{(2\pi)^2}{k} \frac{1}{|\mathbf{A}_0|^2} |\langle \phi_f | H_1 | \phi_i \rangle_{\text{em}}|^2 \delta(E_f - E_i + \hbar\omega) \qquad ; \text{ emission}$$

Three comments:

- These expressions must still be summed over the appropriate final states, as before;
- The field strength $|\mathbf{A}_0|^2$ cancels from these ratios.
- Emission induced by an external radiation field is known as *stimulated emission*.[29]

6.2.3.5 *Transition Matrix Element*

We proceed to analyze the transition matrix elements in Eqs. (6.153), where the transition current is given by the second of Eqs. (6.151).

(1) The vector potential for the radiation field is *transverse*, which implies (see Fig. 6.19)

$$\boldsymbol{\nabla} \cdot \left(\mathbf{A}_0 \, e^{\pm i \mathbf{k} \cdot \mathbf{x}} \right) = 0 \qquad ; \text{ transverse} \qquad (6.157)$$

The momentum $\mathbf{p}$ can then be partially integrated to produce

$$\langle \phi_f | H_1 | \phi_i \rangle_{\text{abs}} = -\frac{e}{mc} \int d^3x \, \phi_f^\star(\mathbf{x}) \mathbf{p} \phi_i(\mathbf{x}) \cdot \mathbf{A}_0 \, e^{i\mathbf{k} \cdot \mathbf{x}}$$

$$\langle \phi_f | H_1 | \phi_i \rangle_{\text{em}} = -\frac{e}{mc} \int d^3x \, \phi_f^\star(\mathbf{x}) \mathbf{p} \phi_i(\mathbf{x}) \cdot \mathbf{A}_0 \, e^{-i\mathbf{k} \cdot \mathbf{x}} \qquad (6.158)$$

---

[29] As opposed to *spontaneous* photoemission, discussed in the next chapter.

(2) *Dipole Approximation*: The size of an atomic system is characterized by

$$R \sim \frac{a_0}{Z} = \frac{1}{Z} \frac{\hbar^2}{m_e e^2} \qquad ; \text{ atomic size} \qquad (6.159)$$

Suppose the wavelength of the radiation is large compared to the size of the system

$$kR = \frac{2\pi R}{\lambda} = \frac{\omega R}{c} = \frac{\omega}{c} \frac{\hbar^2}{m_e e^2} \frac{1}{Z} \ll 1 \qquad (6.160)$$

This inequality is re-written as

$$\frac{\hbar\omega}{m_e c^2} \ll Z\alpha$$

$$\frac{\hbar\omega}{0.51 \times 10^6 \, \text{eV}} \ll \frac{Z}{137} \qquad ; \text{ condition } R \ll \lambda \qquad (6.161)$$

If this condition holds, one can make the *dipole approximation* (see Fig. 6.22)

$$e^{\pm i \mathbf{k} \cdot \mathbf{x}} \approx 1 \qquad ; \text{ dipole approximation} \qquad (6.162)$$

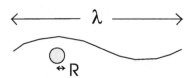

Fig. 6.22   Condition for dipole approximation $R \ll \lambda$.

In the dipole approximation, the transition matrix elements for emission and absorption take a common form

$$\langle \phi_f | H_1 | \phi_i \rangle = -\frac{e}{mc} \mathbf{A}_0 \cdot \langle \phi_f | \mathbf{p} | \phi_i \rangle \qquad ; \text{ emission and absorption}$$

$$(6.163)$$

(3) The transition matrix element of the momentum operator can be manipulated as follows. Recall

$$H_0 = \frac{\mathbf{p}^2}{2m} + V(\mathbf{x}) + e\Phi(\mathbf{x}) \qquad (6.164)$$

It follows, as in Ehrenfest's theorem, that

$$\mathbf{p} = \frac{im}{\hbar}[H_0, \mathbf{x}] = m\left(\frac{d\mathbf{x}}{dt}\right)_{\text{op}} \tag{6.165}$$

This is readily demonstrated in terms of components by writing[30]

$$[H_0, x_i] = \frac{1}{2m}[p_j p_j, x_i] = \frac{1}{2m}\left(p_j[p_j, x_i] + [p_j, x_i]p_j\right)$$
$$= \frac{1}{m}p_j\left(\frac{\hbar}{i}\delta_{ij}\right) \tag{6.166}$$

It follows from Eq. (6.165) that for the matrix elements of $\mathbf{p}$

$$\langle\phi_f|\mathbf{p}|\phi_i\rangle = \frac{im}{\hbar}\langle\phi_f|[H_0, \mathbf{x}]|\phi_i\rangle$$
$$= \frac{im}{\hbar}(E_f - E_i)\langle\phi_f|\mathbf{x}|\phi_i\rangle \tag{6.167}$$

The last equality holds since the initial and final states are eigenstates of H_0. Since $E_f - E_i = \hbar\omega$ for absorption, and $E_f - E_i = -\hbar\omega$ for emission

$$\langle\phi_f|H_1|\phi_i\rangle_{\text{abs}} = -\frac{ie\omega}{c}\mathbf{A}_0 \cdot \langle\phi_f|\mathbf{x}|\phi_i\rangle \qquad ;\text{ absorption}$$
$$\langle\phi_f|H_1|\phi_i\rangle_{\text{em}} = +\frac{ie\omega}{c}\mathbf{A}_0 \cdot \langle\phi_f|\mathbf{x}|\phi_i\rangle \qquad ;\text{ emission} \tag{6.168}$$

In the dipole approximation, the rate for absorption or emission of radiation in Eq. (6.155) depends on the *transition dipole matrix element*.

We discuss some applications of this analysis.

6.2.3.6 *Cross Sections*

Photoionization: Consider photoionization of an atom, where the final electron is in the continuum. Work in the dipole approximation. Assume an initially bound electron with energy $\varepsilon_i^0 = -|\varepsilon_i^0|$ and approximate the final electron by a free-particle state $\phi_f(\mathbf{x}) = e^{i\mathbf{k}_f \cdot \mathbf{x}}/\sqrt{\Omega}$ with energy $E_f = \hbar^2 k_f^2/2m$ (see Fig. 6.23).

---

[30]Here repeated Latin indices are summed from 1 to 3.

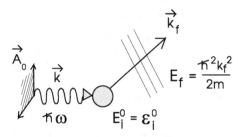

Fig. 6.23   Configuration for photoionization. The initial atomic electron is bound with an energy $\varepsilon_i^0 = -|\varepsilon_i^0|$ and the final electron is approximated by a free-particle state $\phi_f(\mathbf{x}) = e^{i\mathbf{k}_f \cdot \mathbf{x}}/\sqrt{\Omega}$ with energy $E_f = \hbar^2 k_f^2/2m$.

The transition rate in the Golden Rule, with one particle in the continuum in the final state, is calculated just as before

$$\mathcal{R}_{fi} = \int R_{fi}\, dn_f = \int R_{fi} \left(\frac{\partial n_f}{\partial E_f} \right) dE_f$$

$$= \frac{2\pi}{\hbar} \left[|\langle \phi_f | H_1 | \phi_i \rangle_{\text{abs}}|^2 \left(\frac{\partial n_f}{\partial E_f} \right) \right]_{E_f = \varepsilon_i^0 + \hbar\omega} \qquad (6.169)$$

- The density of final states is again obtained as

$$dn_f = \frac{\Omega d^3 k_f}{(2\pi)^3}$$

$$\frac{\partial n_f}{\partial E_f} = \frac{\Omega k_f^2}{(2\pi)^3} \left(\frac{m}{\hbar^2 k_f} \right) d\Omega_f \qquad (6.170)$$

- The incident photon flux is given in Eq. (6.148)

$$I_{\text{inc}} = \frac{\overline{\langle S_\gamma \rangle}}{\hbar\omega} = \frac{k}{2\pi\hbar} |\mathbf{A}_0|^2 \qquad (6.171)$$

- For generality, write the transverse vector potential $\mathbf{A}_0$ in Fig. (6.19) as

$$\mathbf{A}_0 = A_0 \mathbf{e}_s \qquad ; \; s = 1, 2 \qquad (6.172)$$

where $(\mathbf{e}_1, \mathbf{e}_2)$ form a complete set of orthonormal unit vectors transverse to $\mathbf{k}$. This corresponds to the two possible transverse polarizations of the radiation.

A combination of the above gives the photon cross section in the dipole approximation, for a given polarization s, as

$$\left(\frac{d\sigma}{d\Omega}\right)_{\text{ion}} = \frac{2\pi}{\hbar} \left|\frac{ew A_0}{c}\right|^2 |\langle\phi_f|\mathbf{x}|\phi_i\rangle \cdot \mathbf{e}_s|^2 \frac{\Omega k_f^2}{(2\pi)^3} \left(\frac{m}{\hbar^2 k_f}\right) \frac{1}{k|A_0|^2/2\pi\hbar}$$

$$(6.173)$$

A cancellation and combination of terms then gives the photoionization cross section[31]

$$\left(\frac{d\sigma}{d\Omega}\right)_{\text{ion}} = \frac{\alpha}{2\pi} \left(\frac{mc}{\hbar}\right) k k_f \left|\sqrt{\Omega} \langle\phi_f|\mathbf{x}|\phi_i\rangle \cdot \mathbf{e}_s\right|^2 \quad ; \text{ dipole approximation}$$

$$; \text{ given } s \qquad (6.174)$$

Note the following:

- Here $\alpha = e^2/\hbar c$ is the fine-structure constant;
- The final electron wave function is proportional to $1/\sqrt{\Omega}$, so this cross section is indeed independent of the quantization volume Ω;
- The strength of the field $|A_0|^2$ has also cancelled from this expression.

Photoabsorption to Discrete State: Consider photoabsorption to a discrete state, as illustrated in Fig. 6.21. The photon cross section is given by Eqs. (6.156)

$$\sigma_{fi}(\omega) = \frac{R_{fi}}{I_{\text{inc}}}$$

$$= \frac{(2\pi)^2}{k|\mathbf{A}_0|^2} |\langle\phi_f|H_1|\phi_i\rangle_{\text{abs}}|^2 \delta(E_f - E_i - \hbar\omega) \qquad (6.175)$$

For a given (E_f, E_i), this implies a sharp peak in the absorption cross section at $\hbar\omega_0 = E_f - E_i$. In fact, one observes a peak spread out over the *natural line width* for this transition, as sketched as a function of $\hbar\omega$ in Fig. 5.41.[32] What one *can* calculate with the current analysis is the *area under this curve*

$$\hbar \int d\omega\, \sigma_{fi}(\omega) = \frac{(2\pi)^2}{k|\mathbf{A}_0|^2} |\langle\phi_f|H_1|\phi_i\rangle_{\text{abs}}|^2 \qquad (6.176)$$

With the dipole approximation for the transition matrix element in

---

[31]A similar calculation of $\gamma + {}^2_1\text{H} \rightarrow n + p$ is covered by Prob. 7.5 in [Walecka (2004)].

[32]We later go through the Wigner-Weisskopf theory of the line width.

Eqs. (6.168), and the expression for $\mathbf{A}_0$ in Eq. (6.172), this becomes

$$\int_{\text{over peak}} d\omega\, \sigma_{fi}(\omega) = (2\pi)^2 \alpha\, \omega_0 \left| \mathbf{e}_s \cdot \langle \phi_f | \mathbf{x} | \phi_i \rangle \right|^2 \qquad (6.177)$$

; dipole approximation

With unpolarized incident radiation, one must take the *statistical average over the initial state*,[33] which here implies $(1/2)\sum_{s=1}^{2}$

$$\int_{\text{over peak}} d\omega\, \overline{\sigma_{fi}(\omega)} = \frac{1}{2}\sum_{s=1}^{2}(2\pi)^2 \alpha\omega_0 \left| \mathbf{e}_s \cdot \langle \phi_f | \mathbf{x} | \phi_i \rangle \right|^2 \qquad (6.178)$$

; unpolarized radiation

6.3   Higher-Order Perturbation Theory

So far, we have been working with transition probabilities and rates calculated to lowest order in the perturbation H_1. The goal here is to develop an expression for the transition probability that holds to all orders in H_1.

Let us return to the *exact coupled integral Eqs. (6.19)* for the coefficients $c_n^{(i)}(t)$ in Eq. (6.18)

$$c_n^{(i)}(t) - c_n^{(i)}(t_0) = -\frac{i}{\hbar}\sum_m \int_{t_0}^{t} dt\, \langle n|H_1|m\rangle c_m^{(i)}(t) e^{i(E_n - E_m)t/\hbar} \qquad (6.179)$$

; coupled integral eqns

Define the operator

$$\hat{H}_I(t) \equiv e^{i\hat{H}_0 t/\hbar}\, \hat{H}_1(t)\, e^{-i\hat{H}_0 t/\hbar} \qquad \text{; interaction picture}$$
$$(6.180)$$

This is referred to as the *interaction picture*.[34] Here the perturbation $\hat{H}_1$ may, or may not, have an explicit time dependence. Matrix elements of this operator between eigenstates of $\hat{H}_0$ then take the following form

$$\langle n|\hat{H}_I(t)|m\rangle = \langle n|\hat{H}_1(t)|m\rangle e^{i(E_n - E_m)t/\hbar} \qquad (6.181)$$

The coupled Eqs. (6.179) can therefore be written as

$$c_n^{(i)}(t) - c_n^{(i)}(t_0) = -\frac{i}{\hbar}\sum_m \int_{t_0}^{t} dt\, \langle n|\hat{H}_I(t)|m\rangle c_m^{(i)}(t) \qquad (6.182)$$

---

[33]Compare Eq. (7.164).

[34]We now return to abstract Hilbert space and our caret notation. See also Prob. 6.12.

Suppose the system is again prepared in a specific state $|\phi_i\rangle$ at time t_0, so that the initial coefficients take the form

$$c_n^{(i)}(t_0) = \delta_{ni} \qquad ; \text{ prepared initial state} \qquad (6.183)$$

Equations (6.182) can then be written as

$$c_n^{(i)}(t) = \delta_{ni} - \frac{i}{\hbar} \sum_m \int_{t_0}^t dt_1 \, \langle n|\hat{H}_I(t_1)|m\rangle c_m^{(i)}(t_1) \qquad (6.184)$$

Note that the time t appears as the upper limit of integration on the r.h.s., and in the integral, the coefficients $c_m^{(i)}(t_1)$ are required *at a time previous to t*.

The coupled integral Eqs. (6.184) can now be *iterated* to obtain a power series in $\hat{H}_1$

$$c_n^{(i)}(t) = \delta_{ni} - \frac{i}{\hbar} \int_{t_0}^t dt_1 \sum_m \langle n|\hat{H}_I(t_1)|m\rangle \delta_{mi} + \qquad (6.185)$$

$$\left(-\frac{i}{\hbar}\right)^2 \int_{t_0}^t dt_1 \sum_m \langle n|\hat{H}_I(t_1)|m\rangle \int_{t_0}^{t_1} dt_2 \sum_p \langle m|\hat{H}_I(t_2)|p\rangle \delta_{pi} + \cdots$$

Completeness can then be invoked on the intermediate-state sums

$$\sum_m |m\rangle\langle m| = \hat{1} \qquad ; \text{ completeness} \qquad (6.186)$$

This gives

$$c_n^{(i)}(t) = \delta_{ni} - \frac{i}{\hbar} \int_{t_0}^t dt_1 \, \langle n|\hat{H}_I(t_1)|i\rangle + \qquad (6.187)$$

$$\left(-\frac{i}{\hbar}\right)^2 \int_{t_0}^t dt_1 \int_{t_0}^{t_1} dt_2 \, \langle n|\hat{H}_I(t_1)\hat{H}_I(t_2)|i\rangle + \cdots$$

The generalization of this analysis to all orders results in

$$c_n^{(i)}(t) = \langle \phi_n|\hat{\mathcal{U}}(t,t_0)|\phi_i\rangle \qquad (6.188)$$

where the *time-development operator* $\hat{\mathcal{U}}(t,t_0)$ is given by

$$\hat{\mathcal{U}}(t,t_0) = \sum_{n=0}^{\infty} \left(-\frac{i}{\hbar}\right)^n \int_{t_0}^t dt_1 \int_{t_0}^{t_1} dt_2 \cdots \int_{t_0}^{t_{n-1}} dt_n \times$$
$$\hat{H}_I(t_1)\hat{H}_I(t_2)\cdots\hat{H}_I(t_n) \qquad (6.189)$$

Several comments:

- By convention, the $n = 0$ term is 1;
- The time-development operator $\hat{\mathcal{U}}(t, t_0)$ has now been constructed in the abstract Hilbert space. Matrix elements of this operator can be explicitly written out in any representation;
- The product of interaction-picture operators in Eq. (6.189) occurs in *time-ordered form*, with the latest time furthest to the left. In general, these operators do not commute at different times;
- This is an *exact result* that holds to all orders in $\hat{H}_1$;
- The depletion of the initial state is now fully taken into account.

From the general principles of quantum mechanics, the probability of making a transition from the initial state $|\phi_i\rangle$ to any other state $|\phi_f\rangle$ is now given by

$$P_{fi}(t, t_0) = \left| c_f^{(i)}(t) \right|^2 = \left| \langle \phi_f | \hat{\mathcal{U}}(t, t_0) | \phi_i \rangle \right|^2 \qquad ; \text{ exact result} \qquad (6.190)$$

The construction of the scattering operator $\hat{S} = \hat{\mathcal{U}}(\infty, -\infty)$ and corresponding S-matrix elements, an exact expression for the transition rate, and the relation to time-independent scattering theory, are all analyzed in detail in chapter 4 of Vol. II. The reader is referred to this material for further study.

Chapter 7

Electromagnetic Radiation and Quantum Electrodynamics

So far, we have talked about the quantum mechanics of a non-relativistic particle: free, in a central potential, in external (specified) fields, in a two-body system, *etc.* Now we want to discuss the quantum mechanics of the electromagnetic field itself, and its interaction with matter. We are immediately faced with two fundamental issues:

- The quanta, photons, are *massless*, and thus the problem is inherently relativistic;
- Photons can be *created and destroyed*. The emission process in Fig. 6.21, for example, actually occurs for an isolated system through the creation of a photon. We must be able to describe such processes within the framework of quantum mechanics.

7.1 Maxwell's Equations

A relativistic theory of the classical electromagnetic field is provided by *Maxwell's equations* in free space with sources.[1] In c.g.s. units they read

$$
\begin{aligned}
\boldsymbol{\nabla} \cdot \mathbf{B} &= 0 & ; \text{Maxwell's equations} \\
\boldsymbol{\nabla} \times \boldsymbol{\mathcal{E}} &= -\frac{1}{c}\frac{\partial \mathbf{B}}{\partial t} & ; \text{c.g.s. units} \\
\boldsymbol{\nabla} \cdot \boldsymbol{\mathcal{E}} &= 4\pi\varrho & \\
\boldsymbol{\nabla} \times \mathbf{B} &= \frac{1}{c}\frac{\partial \boldsymbol{\mathcal{E}}}{\partial t} + 4\pi\mathbf{j} & (7.1)
\end{aligned}
$$

---

[1]See Vol. I; for a discussion of units, see appendix K of that work.

Here $(\varrho, \mathbf{j})$ are the electromagnetic charge and current densities, satisfying *current conservation*[2]

$$\boldsymbol{\nabla} \cdot \mathbf{j} + \frac{1}{c}\frac{\partial \varrho}{\partial t} = 0 \qquad ; \text{ current conservation} \qquad (7.2)$$

In these units, the fine-structure constant is

$$\frac{e^2}{\hbar c} = \alpha = \frac{1}{137.04} \qquad ; \text{ fine-structure constant} \qquad (7.3)$$

The electric and magnetic fields $(\boldsymbol{\mathcal{E}}, \mathbf{B})$ are related to the electromagnetic *potentials* $(\mathbf{A}, \Phi)$ by

$$\mathbf{B} = \boldsymbol{\nabla} \times \mathbf{A} \qquad ; \text{ potentials}$$
$$\boldsymbol{\mathcal{E}} = -\boldsymbol{\nabla}\Phi - \frac{1}{c}\frac{\partial \mathbf{A}}{\partial t} \qquad (7.4)$$

The first two of Maxwell's equations, which are really constraint equations on the fields, are now satisfied identically. It is the last two of those equations that relate the fields to the sources. Substitution of the second of Eqs. (7.4) into the third of Maxwell's equations gives

$$\nabla^2 \Phi + \frac{1}{c}\frac{\partial}{\partial t}(\boldsymbol{\nabla} \cdot \mathbf{A}) = -4\pi\varrho \qquad (7.5)$$

Substitution of the first of Eqs. (7.4) into the last of Maxwell's equations yields

$$\boldsymbol{\nabla} \times (\boldsymbol{\nabla} \times \mathbf{A}) = \boldsymbol{\nabla}(\boldsymbol{\nabla} \cdot \mathbf{A}) - \nabla^2 \mathbf{A}$$
$$= -\frac{1}{c}\frac{\partial}{\partial t}(\boldsymbol{\nabla}\Phi) - \frac{1}{c^2}\frac{\partial^2}{\partial t^2}\mathbf{A} + 4\pi\mathbf{j} \qquad (7.6)$$

This is rearranged to read

$$\left(\nabla^2 - \frac{1}{c^2}\frac{\partial^2}{\partial t^2}\right)\mathbf{A} - \boldsymbol{\nabla}\left(\boldsymbol{\nabla} \cdot \mathbf{A} + \frac{1}{c}\frac{\partial \Phi}{\partial t}\right) = -4\pi\mathbf{j} \qquad (7.7)$$

The pair of Eqs. (7.5) and (7.7) now relate the potentials to their sources.

7.2 Gauges

Due to the gauge invariance of the physical fields, one has the freedom of choosing various *gauges*. The two most widely used are the following.

---

[2]Here ϱ is in e.s.u., and $\mathbf{j}$ is in e.m.u., where 1 e.m.u.=1 e.s.u./c. Consistent with our usage, we shall continue to refer to the current density $\mathbf{j}$ as the *current*.

7.2.1 Lorentz Gauge

The *Lorentz gauge* is defined by

$$\nabla \cdot \mathbf{A} + \frac{1}{c}\frac{\partial \Phi}{\partial t} = 0 \qquad ; \text{ Lorentz gauge} \qquad (7.8)$$

Equations (7.5) and (7.7) then take the form

$$\Box \Phi = -4\pi\varrho \qquad ; \Box \equiv \nabla^2 - \frac{1}{c^2}\frac{\partial^2}{\partial t^2}$$
$$\Box \mathbf{A} \equiv -4\pi\mathbf{j} \qquad (7.9)$$

This gauge has various advantages and disadvantages:

- The principal *advantage* is that the theory is now *explicitly* covariant. Define

$$j_\mu \equiv (\mathbf{j}, i\varrho) \qquad ; A_\mu \equiv (\mathbf{A}, i\Phi) \qquad ; x_\mu \equiv (\mathbf{x}, ict) \quad (7.10)$$

Then Eqs. (7.8)–(7.9) take the form

$$\Box A_\mu = -4\pi j_\mu \qquad ; \text{ Lorentz gauge}$$
$$\frac{\partial A_\mu}{\partial x_\mu} = 0 \qquad (7.11)$$

Here the summation convention employed is that repeated Greek indices are summed from 1 to 4;[3]

- The main *disadvantage* of the Lorentz gauge, is that when quantized, the theory contains *four* types of photons, two of which are unphysical.[4]

7.2.2 Coulomb Gauge

The *Coulomb gauge* is defined by

$$\nabla \cdot \mathbf{A} = 0 \qquad ; \text{ Coulomb gauge} \qquad (7.12)$$

In this gauge, Eq. (7.5) reads

$$\nabla^2 \Phi = -4\pi\varrho \qquad ; \text{ Poisson's eqn} \qquad (7.13)$$

---

[3]See Vol. I.

[4]Various contortions must be gone through to get rid of the unphysical photons, including the use of a Hilbert space with indefinite metric.

This is *Poisson's equation*. Just as in electrostatics, this equation is immediately solved to yield the potential $\Phi(\mathbf{x}, t)$ everywhere, at any instant in time, in terms of $\varrho(\mathbf{x}, t)$.

In the Coulomb gauge, Eq. (7.7) reads

$$\Box \mathbf{A} = -4\pi \left(\mathbf{j} - \frac{1}{4\pi c} \boldsymbol{\nabla} \frac{\partial \Phi}{\partial t} \right)$$

$$\equiv -4\pi \mathbf{j}_T \qquad \qquad \text{; transverse current} \qquad (7.14)$$

where the second line defines the *transverse current*. To establish the consistency with Eq. (7.12), just compute

$$\boldsymbol{\nabla} \cdot \mathbf{j}_T = \boldsymbol{\nabla} \cdot \mathbf{j} - \frac{1}{4\pi c} \frac{\partial}{\partial t} \nabla^2 \Phi$$

$$= \boldsymbol{\nabla} \cdot \mathbf{j} + \frac{1}{c} \frac{\partial \varrho}{\partial t} = 0 \qquad (7.15)$$

Here the last result follows from electromagnetic current conservation. Hence the divergence of both sides of Eq. (7.14) indeed vanishes.

The Coulomb gauge also has advantages and disadvantages:

- The principal *disadvantage* is that the theory is now not *explicitly* covariant;[5]
- The main *advantage* of the Coulomb gauge, and it is a major one, is that when quantized, the theory contains just *two* physical, transverse photons. One can then always work in a Hilbert space with *physical states*.

7.3   Free Radiation Field

Consider the free radiation field, with no sources anywhere inside a quantization volume $\Omega = L^3$ (see Fig. 6.10).

7.3.1   *Coulomb Gauge*

Work in the Coulomb gauge, where

$$\boldsymbol{\nabla} \cdot \mathbf{A} = 0 \qquad \text{; Coulomb gauge}$$

$$\Phi = 0 \qquad \text{; no sources} \qquad (7.16)$$

---

[5]The emphasis here is on *explicitly*, since when formulated in terms of the Dirac equation, the underlying theory is indeed covariant. Covariance is eventually recovered in the S-matrix (see Vol. II.)

The second line provides a solution to Eq. (7.13) in the source-free region. From Eq. (7.14), the vector potential then satisfies the free *wave equation*

$$\Box \mathbf{A} = 0 \qquad \text{; wave eqn} \qquad (7.17)$$

It follows from Eqs. (7.4) that in the Coulomb gauge the fields $(\mathcal{E}, \mathbf{B})$ are obtained as

$$\mathbf{B} = \boldsymbol{\nabla} \times \mathbf{A}$$
$$\mathcal{E} = -\frac{1}{c}\frac{\partial \mathbf{A}}{\partial t} \qquad\qquad (7.18)$$

In the volume Ω, the fields are therefore transverse and satisfy the wave equation

$$\boldsymbol{\nabla} \cdot \mathcal{E} = \boldsymbol{\nabla} \cdot \mathbf{B} = 0 \qquad \text{; transverse}$$
$$\Box \mathcal{E} = \Box \mathbf{B} = 0 \qquad \text{; wave eqn} \qquad (7.19)$$

7.3.2 General Expansion of $\mathbf{A}(\mathbf{x}, t)$

We develop a general expression for the vector potential $\mathbf{A}(\mathbf{x}, t)$ in the Coulomb gauge for the free radiation field inside the quantization volume $\Omega = L^3$.

(1) For each wavevector $\mathbf{k}$, introduce a complete set of orthonormal unit vectors $\mathbf{e}_{\mathbf{k},s}$ transverse to $\mathbf{k}$ (see Fig. 7.1)

$$\mathbf{e}_{\mathbf{k},s} \cdot \mathbf{e}_{\mathbf{k},s'} = \delta_{s,s'} \qquad ; (s, s') = 1, 2$$
$$\mathbf{e}_{\mathbf{k},s} \cdot \mathbf{k} = 0 \qquad\qquad (7.20)$$

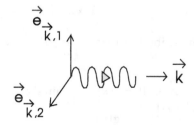

Fig. 7.1   Set of orthonormal unit vectors $\mathbf{e}_{\mathbf{k},s}$ transverse to $\mathbf{k}$.

A wave solution to Eqs. (7.16)–(7.17) can then be written

$$\mathbf{A}_{\mathbf{k},s}(\mathbf{x},t) = \mathbf{e}_{\mathbf{k},s}\frac{1}{\sqrt{\Omega}}e^{i(\mathbf{k}\cdot\mathbf{x}-\omega_k t)} \qquad ; \ \omega_k = |\mathbf{k}|c \qquad (7.21)$$

where $\omega_k \equiv |\mathbf{k}|c$;

(2) Apply *periodic boundary conditions* in the big box

$$k_i = \frac{2\pi n_i}{L} \qquad ; \ n_i = 0, \pm 1, \pm 2, \cdots, \pm\infty$$
$$; \ i = 1, 2, 3 \qquad (7.22)$$

The wave solutions in Eq. (7.21) then provide a complete, orthonormal set of vector basis functions for Eqs. (7.16)–(7.17) at each instant in time;

(3) Now use *superposition* to construct the general solution to these equations

$$\mathbf{A}(\mathbf{x},t) = \frac{1}{\sqrt{\Omega}}\sum_{\mathbf{k}}\sum_{s=1}^{2}\left[a(\mathbf{k},s)\mathbf{e}_{\mathbf{k},s}\,e^{i(\mathbf{k}\cdot\mathbf{x}-\omega_k t)} + a(\mathbf{k},s)^{\star}\mathbf{e}_{\mathbf{k},s}\,e^{-i(\mathbf{k}\cdot\mathbf{x}-\omega_k t)}\right]$$
$$; \ \text{general solution} \qquad (7.23)$$

Here $a(\mathbf{k},s)$ is an arbitrary complex amplitude. This expression has the following properties:

- It is *real*;
- It satisfies the *wave equation*

$$\Box\mathbf{A}(\mathbf{x},t) = 0 \qquad (7.24)$$

- It has a *vanishing divergence*

$$\boldsymbol{\nabla}\cdot\mathbf{A}(\mathbf{x},t) = 0 \qquad (7.25)$$

- It has enough flexibility to match an arbitrary set of *initial conditions*[6]

$$\mathbf{A}(\mathbf{x},0) = \mathbf{f}(\mathbf{x}) \qquad ; \ \text{initial conditions}$$
$$\left.\frac{\partial\mathbf{A}(\mathbf{x},t)}{\partial t}\right|_{t=0} = \mathbf{g}(\mathbf{x}) \qquad (7.26)$$

---

[6]See Prob. 7.1.

7.3.3 Energy in Field

The total energy in the electromagnetic radiation field is given in c.g.s. units by[7]

$$E_{\text{rad}} = \frac{1}{8\pi} \int_\Omega d^3x \, \left(\boldsymbol{\mathcal{E}}^2 + \mathbf{B}^2\right) \qquad ; \text{ energy in field} \qquad (7.27)$$

With the substitution of Eqs. (7.18) this becomes

$$E_{\text{rad}} = \frac{1}{8\pi} \int_\Omega d^3x \, \left[\frac{1}{c^2}\left(\frac{\partial \mathbf{A}}{\partial t}\right)^2 + (\boldsymbol{\nabla} \times \mathbf{A})^2\right] \qquad (7.28)$$

Now substitute the general expression for the vector potential in Eq. (7.23) into this expression

$$E_{\text{rad}} = \frac{1}{8\pi\Omega} \sum_{\mathbf{k}} \sum_{\mathbf{k'}} \sum_{s} \sum_{s'} \int_\Omega d^3x \times$$

$$\left[-\frac{1}{c^2}\omega_k \omega_{k'} \, \mathbf{e}_{\mathbf{k},s} \cdot \mathbf{e}_{\mathbf{k'},s'} - (\mathbf{k} \times \mathbf{e}_{\mathbf{k},s}) \cdot (\mathbf{k'} \times \mathbf{e}_{\mathbf{k'},s'})\right] \times$$

$$\left[a(\mathbf{k}, s) \, e^{i(\mathbf{k}\cdot\mathbf{x}-\omega_k t)} - a(\mathbf{k}, s)^\star \, e^{-i(\mathbf{k}\cdot\mathbf{x}-\omega_k t)}\right] \times$$

$$\left[a(\mathbf{k'}, s') \, e^{i(\mathbf{k'}\cdot\mathbf{x}-\omega_{k'} t)} - a(\mathbf{k'}, s')^\star \, e^{-i(\mathbf{k'}\cdot\mathbf{x}-\omega_{k'} t)}\right] \qquad (7.29)$$

This expression can be analyzed in the following manner:

(1) Use the orthonormality of the plane waves to evaluate the spatial integrals

$$\frac{1}{\Omega} \int_\Omega d^3x \, e^{i(\mathbf{k}-\mathbf{k'})\cdot\mathbf{x}} = \delta_{\mathbf{k},\mathbf{k'}} \qquad ; \text{ orthonormal} \qquad (7.30)$$

(2) There are then two types of terms in the sums in Eq. (7.29):

- The cross terms where $\mathbf{k'} = \mathbf{k}$. The factor in front for these is

$$\frac{\omega_k^2}{c^2} \mathbf{e}_{\mathbf{k},s} \cdot \mathbf{e}_{\mathbf{k},s'} + (\mathbf{k} \times \mathbf{e}_{\mathbf{k},s}) \cdot (\mathbf{k} \times \mathbf{e}_{\mathbf{k},s'}) = \frac{2\omega_k^2}{c^2}\delta_{s,s'} \qquad (7.31)$$

- The direct terms where $\mathbf{k'} = -\mathbf{k}$. Here the factor in front vanishes

$$-\frac{\omega_k^2}{c^2} \mathbf{e}_{\mathbf{k},s} \cdot \mathbf{e}_{-\mathbf{k},s'} + (\mathbf{k} \times \mathbf{e}_{\mathbf{k},s}) \cdot (\mathbf{k} \times \mathbf{e}_{-\mathbf{k},s'}) = \left(-\frac{\omega_k^2}{c^2} + \mathbf{k}^2\right) \mathbf{e}_{\mathbf{k},s} \cdot \mathbf{e}_{-\mathbf{k},s'}$$

$$= 0 \qquad (7.32)$$

---

[7]See Vol. II. (See also Prob. 7.3.)

(3) With the use of these two results, Eq. (7.29) reduces to

$$E_{\text{rad}} = \frac{1}{8\pi} \sum_{k} \sum_{s} \frac{2\omega_k^2}{c^2} \left[a(k, s)a(k, s)^\star + a(k, s)^\star a(k, s) \right] \quad (7.33)$$

Now define[8]

$$a(k, s) \equiv \left(\frac{2\pi \hbar c^2}{\omega_k} \right)^{1/2} a_{ks} \quad (7.34)$$

This yields

$$E_{\text{rad}} = \frac{1}{2} \sum_{k} \sum_{s=1}^{2} \hbar \omega_k \left(a_{ks} a_{ks}^\star + a_{ks}^\star a_{ks} \right) \quad ; \text{ normal modes} \quad (7.35)$$

The energy now receives additive contributions from a set of uncoupled simple harmonic oscillators.[9]

These are the *normal modes* of the radiation field in the box.

(4) In *summary*, substitution of Eq. (7.34) into (7.23) expresses the vector potential for the free radiation field in the Coulomb gauge in the quantization volume Ω, and the corresponding field energy, as

$$\mathbf{A}(\mathbf{x}, t) = \frac{1}{\sqrt{\Omega}} \sum_{k} \sum_{s=1}^{2} \left(\frac{2\pi \hbar c^2}{\omega_k} \right)^{1/2} \mathbf{e}_{ks} \left(a_{ks}\, e^{i\mathbf{k}\cdot\mathbf{x} - i\omega_k t} + a_{ks}^\star\, e^{-i\mathbf{k}\cdot\mathbf{x} + i\omega_k t} \right)$$

$$E_{\text{rad}} = \frac{1}{2} \sum_{k} \sum_{s=1}^{2} \hbar \omega_k \left(a_{ks} a_{ks}^\star + a_{ks}^\star a_{ks} \right) \quad ; \text{ radiation field} \quad (7.36)$$

7.3.4   Quantization

The problem has now been reduced to normal modes, and we know how to quantize the simple harmonic oscillator:

- Let the normal-mode amplitudes become *operators in an abstract Hilbert space*

$$a_{ks} \to \hat{a}_{ks} \quad ; \; a_{ks}^\star \to \hat{a}_{ks}^\dagger \quad ; \text{ operators} \quad (7.37)$$

- Impose *canonical commutation relations*

$$[\hat{a}_{ks}, \hat{a}_{k's'}^\dagger] = \delta_{k,k'} \delta_{s,s'} \quad ; \text{ C.C.R.} \quad (7.38)$$

---

[8] We henceforth use commas in subscripts only when necessary for clarity.
[9] Compare Eqs. (2.123).

Then, as shown previously, the properties of the operators follow entirely from the commutation relations.[10] In particular, the *number operator* $\hat{N}_{\mathbf{k}s}$ has a spectrum of the positive integers and zero

$$\hat{N}_{\mathbf{k}s} = \hat{a}^{\dagger}_{\mathbf{k}s}\hat{a}_{\mathbf{k}s} \qquad ; \text{number operator}$$
$$\hat{N}_{\mathbf{k}s}|n_{\mathbf{k}s}\rangle = n_{\mathbf{k}s}|n_{\mathbf{k}s}\rangle \qquad ; n_{\mathbf{k}s} = 0, 1, 2, \cdots, \infty \qquad (7.39)$$

- The energy becomes the *hamiltonian* for the free radiation field

$$E_{\text{rad}} \rightarrow \hat{H}_{\text{rad}} = \sum_{\mathbf{k},s} \hbar\omega_k \frac{1}{2}\left(\hat{a}_{\mathbf{k}s}\hat{a}^{\dagger}_{\mathbf{k}s} + \hat{a}^{\dagger}_{\mathbf{k}s}\hat{a}_{\mathbf{k}s}\right)$$

$$\hat{H}_{\text{rad}} = \sum_{\mathbf{k},s} \hbar\omega_k \left(\hat{N}_{\mathbf{k}s} + \frac{1}{2}\right) \qquad ; \text{hamiltonian} \qquad (7.40)$$

One then has the following results:

- $\hat{H}_{\text{rad}}$ is an *operator* in the abstract Hilbert space, and the dynamics is now governed by this hamiltonian;
- The basis vectors in the abstract Hilbert space are the *direct product* of the basis vectors for each mode

$$|n_1, n_2, \cdots, n_\infty\rangle = |n_1\rangle|n_2\rangle \cdots |n_\infty\rangle \qquad ; \text{basis vectors} \qquad (7.41)$$

Here we have simply ordered the modes $\{\mathbf{k}s\}$. The abstract Hilbert space is often referred to as the *occupation number space*;

- The eigenvalues of $\hat{H}_{\text{rad}}$ are

$$E_{n_1 n_2 \cdots n_\infty} = \sum_{\mathbf{k},s} \hbar\omega_k \left(n_{\mathbf{k}s} + \frac{1}{2}\right) \qquad ; \text{eigenvalues} \qquad (7.42)$$

The energy now comes in discrete quantum units called *photons*, with each photon having energy $\hbar\omega_k$ and (see later) momentum $\hbar\mathbf{k}$;

- There is a *zero-point* energy in the radiation field of

$$E_{\text{zero-pt}} = \frac{1}{2}\sum_{\mathbf{k},s} \hbar\omega_k \qquad ; \text{zero-point energy} \qquad (7.43)$$

This is the energy of the *vacuum*, and all energies are measured *relative to it*.[11]

---

[10]See section 3.8.3.
[11]It happens to be infinite!

7.3.4.1 *Quantum Mechanics*

The Schrödinger equation in the abstract occupation-number space is

$$i\hbar\frac{\partial}{\partial t}|\Psi(t)\rangle = \hat{H}_{\text{rad}}|\Psi(t)\rangle \qquad ; \text{Schrödinger eqn} \qquad (7.44)$$

The vector potential now becomes an *operator in this space that creates and destroys photons*. In the Schrödinger picture, where the operators are independent of time and all the time dependence is put into the state vector, the field operator $\hat{\mathbf{A}}(\mathbf{x})$ is defined through the $t = 0$ form of Eq. (7.36)

$$\hat{\mathbf{A}}(\mathbf{x}) \equiv \frac{1}{\sqrt{\Omega}}\sum_{\mathbf{k}}\sum_{s=1}^{2}\left(\frac{2\pi\hbar c^2}{\omega_k}\right)^{1/2}\mathbf{e}_{\mathbf{k}s}\left(\hat{a}_{\mathbf{k}s}\,e^{i\mathbf{k}\cdot\mathbf{x}} + \hat{a}_{\mathbf{k}s}^{\dagger}\,e^{-i\mathbf{k}\cdot\mathbf{x}}\right) \qquad (7.45)$$

$$; \text{Schrödinger picture}$$

From the commutation relations, the action of the creation and destruction operators on the states is given by

$$\hat{a}_{\mathbf{k}s}|n_{\mathbf{k}s}\rangle = \sqrt{n_{\mathbf{k}s}}\,|n_{\mathbf{k}s} - 1\rangle \qquad ; \text{destroys photon}$$

$$\hat{a}_{\mathbf{k}s}^{\dagger}|n_{\mathbf{k}s}\rangle = \sqrt{n_{\mathbf{k}s} + 1}\,|n_{\mathbf{k}s} + 1\rangle \qquad ; \text{creates photon} \qquad (7.46)$$

The classical field is now identified with the *expectation value*

$$\mathbf{A}(\mathbf{x}, t) = \langle\Psi(t)|\hat{\mathbf{A}}(\mathbf{x})|\Psi(t)\rangle \qquad ; \text{classical field} \qquad (7.47)$$

The time dependence of the expectation value in the Schrödinger picture arises from the time dependence of the state vector $|\Psi(t)\rangle$, and as in appendix B, this is governed by *Ehrenfest's theorem* in Eq. (3.171).

7.3.4.2 *Ehrenfest's Theorem*

In the Schrödinger picture, the time derivative of the operator $\hat{\mathbf{A}}(\mathbf{x})$ is given by Eq. (3.171) as[12]

$$\frac{d}{dt}\hat{\mathbf{A}}(\mathbf{x}) \equiv \dot{\hat{\mathbf{A}}}(\mathbf{x}) = \frac{i}{\hbar}[\hat{H}_{\text{rad}}, \hat{\mathbf{A}}(\mathbf{x})] \qquad (7.48)$$

The commutation relations in Eqs. (3.148) give

$$[\hat{H}_{\text{rad}}, \hat{a}_{\mathbf{k}s}] = -\hbar\omega_k\hat{a}_{\mathbf{k}s}$$

$$[\hat{H}_{\text{rad}}, \hat{a}_{\mathbf{k}s}^{\dagger}] = \hbar\omega_k\hat{a}_{\mathbf{k}s}^{\dagger} \qquad (7.49)$$

---

[12]One has to think through the notation. This is the time derivative of the operator $\hat{\mathbf{A}}(\mathbf{x})$, which has no explicit time dependence, at a given point in space. We shall call this $\dot{\hat{\mathbf{A}}}(\mathbf{x})$. For a function $\mathbf{A}(\mathbf{x}, t)$, this would be labeled as the partial derivative $\partial/\partial t$.

Thus

$$\hat{\mathbf{A}}(\mathbf{x}) = \frac{-i}{\sqrt{\Omega}} \sum_{\mathbf{k}} \sum_{s=1}^{2} \left(\frac{2\pi\hbar c^2}{\omega_k} \right)^{1/2} \omega_k \mathbf{e}_{\mathbf{k}s} \left(\hat{a}_{\mathbf{k}s} \, e^{i\mathbf{k}\cdot\mathbf{x}} - \hat{a}_{\mathbf{k}s}^{\dagger} \, e^{-i\mathbf{k}\cdot\mathbf{x}} \right) \quad (7.50)$$

Although the time dependence of the expectation value in Eq. (7.47) arises through the time dependence of the state vectors, $\hat{\mathbf{A}}(\mathbf{x})$ is exactly what one would get for the operator constructed from $[\partial\mathbf{A}(\mathbf{x},t)/\partial t]_{t=0}$ taken on the classical field in Eq. (7.36).

7.3.4.3 *Justification*

As justification for this quantization scheme, we show that Ehrenfest's theorem for quantum mechanics in the Schrödinger picture gives Maxwell's equations for the classical field.[13] We do this in steps:

(1) It is evident that the operator in Eq. (7.45) is transverse[14]

$$\nabla \cdot \hat{\mathbf{A}}(\mathbf{x}) = 0 \quad (7.51)$$

(2) Compute the *second* time derivative of the operator $\hat{\mathbf{A}}(\mathbf{x})$

$$\ddot{\hat{\mathbf{A}}}(\mathbf{x}) = \frac{i}{\hbar}[\hat{H}_{\text{rad}}, \dot{\hat{\mathbf{A}}}(\mathbf{x})] = \left(\frac{i}{\hbar} \right)^2 [\hat{H}_{\text{rad}}, [\hat{H}_{\text{rad}}, \hat{\mathbf{A}}(\mathbf{x})]] \quad (7.52)$$

With the use of Eqs. (7.49) this becomes

$$\ddot{\hat{\mathbf{A}}}(\mathbf{x}) = \frac{(-i)^2}{\sqrt{\Omega}} \sum_{\mathbf{k}} \sum_{s=1}^{2} \left(\frac{2\pi\hbar c^2}{\omega_k} \right)^{1/2} \omega_k^2 \, \mathbf{e}_{\mathbf{k}s} \left(\hat{a}_{\mathbf{k}s} \, e^{i\mathbf{k}\cdot\mathbf{x}} + \hat{a}_{\mathbf{k}s}^{\dagger} \, e^{-i\mathbf{k}\cdot\mathbf{x}} \right)$$
$$= c^2 \nabla^2 \hat{\mathbf{A}}(\mathbf{x}) \qquad\qquad ; \; \omega_k^2 = \mathbf{k}^2 c^2 \quad (7.53)$$

In the second line, the factor of $(-i\omega_k)^2$ in the sum has been converted to $c^2\nabla^2$ on the exponentials and then removed from the sum. Hence, as an *operator*, $\hat{\mathbf{A}}(\mathbf{x})$ satisfies the wave equation

$$\nabla^2 \hat{\mathbf{A}}(\mathbf{x}) - \frac{1}{c^2}\ddot{\hat{\mathbf{A}}}(\mathbf{x}) = 0 \qquad ; \text{ wave equation} \quad (7.54)$$

Since the operator in the Schrödinger picture satisfies the wave equation, so does the expectation value in Eq. (7.47);

(3) Since $\Phi = 0$, the above results imply that the *classical field in Eq. (7.47) satisfies Maxwell's equations in the Coulomb gauge.*

---

[13]Ultimately, the commutation relations are a *postulate* of quantum mechanics.

[14]Note that this condition does not involve the time and can be built *into the operator*.

7.3.5   *Momentum in Field*

Let $\mathcal{P}(\mathbf{x}, t)$ be the momentum density for the radiation field in classical electromagnetism. Then the *momentum flux* is given by

$$\mathcal{P}(\mathbf{x}, t)c = \text{momentum flux} = \frac{1}{c}\mathbf{S}_\gamma = \frac{1}{4\pi}\boldsymbol{\mathcal{E}} \times \mathbf{B} \qquad (7.55)$$

Here the *energy flux* is given by the Poynting vector $\mathbf{S}_\gamma$. Thus the total momentum in the field is given by[15]

$$\mathbf{P} = \frac{1}{4\pi c}\int_\Omega d^3x\, \boldsymbol{\mathcal{E}} \times \mathbf{B} \qquad ; \text{momentum in field} \qquad (7.56)$$

The substitution of Eqs. (7.18) then gives

$$\mathbf{P} = -\frac{1}{4\pi c^2}\int_\Omega d^3x\, \left(\frac{\partial\mathbf{A}}{\partial t}\right) \times (\boldsymbol{\nabla} \times \mathbf{A}) \qquad (7.57)$$

Now introduce the expansion of $\mathbf{A}(\mathbf{x}, t)$ in Eq. (7.36). One then has a calculation analogous to that carried out in Eqs. (7.28)–(7.35). The only differences are the following:

- The vector product that appears for $\mathbf{k}' = \mathbf{k}$ can be reduced as follows

$$\mathbf{e}_{\mathbf{k}s} \times (\mathbf{k} \times \mathbf{e}_{\mathbf{k}s'}) = \mathbf{k}\,(\mathbf{e}_{\mathbf{k}s} \cdot \mathbf{e}_{\mathbf{k}s'}) = \mathbf{k}\,\delta_{s,s'} \qquad (7.58)$$

- The antisymmetry of the summand under $(\mathbf{k} \rightleftharpoons -\mathbf{k},\ s \rightleftharpoons s')$ implies that the contribution from those terms with $\mathbf{k}' = -\mathbf{k}$ vanishes

$$\sum_{\mathbf{k}}\sum_{s,s'}\mathbf{k}\,(\mathbf{e}_{\mathbf{k}s} \cdot \mathbf{e}_{-\mathbf{k},s'})\,\left(a_{\mathbf{k}s}a_{-\mathbf{k},s'}\,e^{-2i\omega_k t} + a_{\mathbf{k}s}^\star a_{-\mathbf{k},s'}^\star\,e^{2i\omega_k t}\right) = 0$$
$$(7.59)$$

It follows that in the normal modes (see Prob. 7.2)

$$\mathbf{P} = \frac{1}{2}\sum_{\mathbf{k}}\sum_{s=1}^{2}\hbar\mathbf{k}\,(a_{\mathbf{k}s}a_{\mathbf{k}s}^\star + a_{\mathbf{k}s}^\star a_{\mathbf{k}s}) \qquad ; \text{normal modes} \qquad (7.60)$$

The quantization in Eqs. (7.37)–(7.39) then gives[16]

$$\mathbf{P} = \sum_{\mathbf{k}}\sum_{s}\hbar\mathbf{k}\,\hat{N}_{\mathbf{k}s} \qquad (7.61)$$

---

[15]Notice that as in Eq. (7.27), this expression is *gauge invariant*.
[16]Symmetry arguments again imply $\sum_{\mathbf{k}}\hbar\mathbf{k} = 0$; the vacuum has no net momentum.

Hence each photon in the mode $\{\mathbf{k}s\}$ carries momentum $\hbar\mathbf{k}$, as claimed previously.

7.3.6 Commutation Relations

For the subsequent discussion, we will need the canonical equal-time commutation relations for the vector field. From Eqs. (7.45) and (7.50) one has for the commutator of $\hat{\mathbf{A}}(\mathbf{x})$ and $\hat{\mathbf{A}}(\mathbf{x}')$

$$[\hat{A}_i(\mathbf{x}), \hat{A}_j(\mathbf{x}')] = i\left(\frac{2\pi\hbar c^2}{\Omega}\right)\sum_{k,s}(\mathbf{e}_{\mathbf{k}s})_i(\mathbf{e}_{\mathbf{k}s})_j\left[e^{i\mathbf{k}\cdot(\mathbf{x}-\mathbf{x}')} + e^{-i\mathbf{k}\cdot(\mathbf{x}-\mathbf{x}')}\right]$$

$$(7.62)$$

The sum over s goes over the two transverse unit vectors in Fig. 7.1. If the longitudinal unit vector $\mathbf{e}_{\mathbf{k}0} \equiv \mathbf{k}/|\mathbf{k}|$ were also included in the sum, then the complete orthonormal set of unit vectors would yield the *unit dyadic*

$$\sum_{s=0}^{2}(\mathbf{e}_{\mathbf{k}s})_i(\mathbf{e}_{\mathbf{k}s})_j = \delta_{ij} \qquad \text{; unit dyadic} \qquad (7.63)$$

Hence the sum over transverse unit vectors is given by

$$\sum_{s=1}^{2}(\mathbf{e}_{\mathbf{k}s})_i(\mathbf{e}_{\mathbf{k}s})_j = \delta_{ij} - (\mathbf{e}_{\mathbf{k}0})_i(\mathbf{e}_{\mathbf{k}0})_j \qquad ;\ \mathbf{e}_{\mathbf{k}0} \equiv \frac{\mathbf{k}}{|\mathbf{k}|}$$

$$= \delta_{ij} - \frac{k_i k_j}{\mathbf{k}^2} \qquad (7.64)$$

One can then change the summation variable $\mathbf{k} \to -\mathbf{k}$ in the second term in Eq. (7.62) to give

$$[\hat{A}_i(\mathbf{x}), \hat{A}_j(\mathbf{x}')] = -\frac{4\pi\hbar c^2}{i}\frac{1}{\Omega}\sum_{\mathbf{k}}\left(\delta_{ij} - \frac{k_i k_j}{\mathbf{k}^2}\right)e^{i\mathbf{k}\cdot(\mathbf{x}-\mathbf{x}')}$$

$$\equiv -\frac{4\pi\hbar c^2}{i}\delta_{ij}^{T}(\mathbf{x} - \mathbf{x}') \qquad (7.65)$$

Here the second line defines the *transverse delta-function*.[17] It is readily shown that at equal times, the fields themselves commute (see Prob. 7.4)

$$[\hat{A}_i(\mathbf{x}), \hat{A}_j(\mathbf{x}')] = 0 \qquad (7.66)$$

---

[17]Note that if the $k_i k_j/\mathbf{k}^2$ term were absent, then from the completeness relation for Fourier series $\delta_{ij}^{T}(\mathbf{x} - \mathbf{x}') = \delta_{ij}\delta^{(3)}(\mathbf{x} - \mathbf{x}')$. See also Prob. 7.6.

7.4   Quantum Electrodynamics

Consider now the full problem of a non-relativistic particle of mass m_0 and charge $e = -|e|$ at position $\mathbf{x}_e$ moving in the nuclear Coulomb potential, with a possible additional potential $V(\mathbf{x}_e)$, and interacting with the quantized radiation field. The radiation field now arises both from sources external to the quantization volume Ω in Fig. 6.10, and from the motion of the particle itself. This is a *coupled problem*, which combines the two we have already discussed: the motion of a charged particle in a given electromagnetic field, analyzed in detail in appendix B, and the free quantized radiation field, discussed previously in this chapter. With the caveats that the particle motion is non-relativistic, and there are as yet no antiparticles, this defines the problem of *quantum electrodynamics (QED)*.

7.4.1   *Formulation of Problem*

In abstract Hilbert space, the hamiltonian for this coupled problem is

$$\hat{H} = \frac{1}{2m_0} \left[\hat{\mathbf{p}}_e - \frac{e}{c}\hat{\mathbf{A}}(\hat{\mathbf{x}}_e) \right] \cdot \left[\hat{\mathbf{p}}_e - \frac{e}{c}\hat{\mathbf{A}}(\hat{\mathbf{x}}_e) \right] + e\Phi(\hat{\mathbf{x}}_e) + V(\hat{\mathbf{x}}_e) + \hat{H}_{\text{rad}}$$

$$\equiv \hat{H}_{\text{part}} + \hat{H}_{\text{rad}} \tag{7.67}$$

The field operator at the position of the particle $\hat{\mathbf{A}}(\hat{\mathbf{x}}_e)$ is given by Eq. (7.45) as

$$\hat{\mathbf{A}}(\hat{\mathbf{x}}_e) \equiv \frac{1}{\sqrt{\Omega}} \sum_{\mathbf{k}} \sum_{s=1}^{2} \left(\frac{2\pi\hbar c^2}{\omega_k} \right)^{1/2} \mathbf{e}_{\mathbf{k}s} \left(\hat{a}_{\mathbf{k}s}\, e^{i\mathbf{k}\cdot\hat{\mathbf{x}}_e} + \hat{a}_{\mathbf{k}s}^{\dagger}\, e^{-i\mathbf{k}\cdot\hat{\mathbf{x}}_e} \right) \tag{7.68}$$

The additional hamiltonian of the radiation field is that of Eq. (7.40)

$$\hat{H}_{\text{rad}} = \sum_{\mathbf{k}} \sum_{s=1}^{2} \hbar\omega_k \left(\hat{a}_{\mathbf{k}s}^{\dagger}\hat{a}_{\mathbf{k}s} + \frac{1}{2} \right) \tag{7.69}$$

Several comments:

- The canonical commutation relations are

$$[(\hat{p}_e)_i, (\hat{x}_e)_j] = \delta_{ij} \qquad\qquad \text{; C.C.R.}$$
$$[\hat{a}_{\mathbf{k}s}, \hat{a}_{\mathbf{k}'s'}^{\dagger}] = \delta_{\mathbf{k},\mathbf{k}'}\delta_{s,s'} \tag{7.70}$$

These particle and field operators *commute with each other;*

- The basis vectors in the abstract Hilbert space can be taken to be

$$|\phi_m; n_1 n_2 \cdots n_\infty\rangle = |\phi_m\rangle|n_1\rangle|n_2\rangle \cdots |n_\infty\rangle \qquad ; \text{ basis vectors}$$

$$\hat{H}_{\text{part}}^0|\phi_m\rangle = \varepsilon_m|\phi_m\rangle \tag{7.71}$$

- In the *coordinate representation* for the particle

$$\langle \mathbf{x}_e|\phi_m; n_1 n_2 \cdots n_\infty\rangle = \langle \mathbf{x}_e|\phi_m\rangle |n_1\rangle|n_2\rangle \cdots |n_\infty\rangle \qquad ; \text{ coordinate rep}$$

$$= \phi_m(\mathbf{x}_e)\,|n_1\rangle|n_2\rangle \cdots |n_\infty\rangle$$

$$\langle \mathbf{x}_e|\hat{H}(\hat{\mathbf{x}}_e,\,\hat{\mathbf{p}}_e)|\mathbf{x}_e'\rangle = \hat{H}\left(\mathbf{x}_e,\,\frac{\hbar}{i}\boldsymbol{\nabla}_e\right)\delta^{(3)}(\mathbf{x}_e - \mathbf{x}_e') \tag{7.72}$$

For convenience, we henceforth proceed directly to the coordinate representation for the particle, where one has wave functions, and the particle's position and momentum are $\{\mathbf{x}_e,\,\mathbf{p}_e\} = \{\mathbf{x}_e,\,(\hbar/i)\boldsymbol{\nabla}_e\}$. We continue to describe the radiation field in abstract Hilbert space;

- The hamiltonian in the coordinate representation for the particle is then (see Fig. 7.2)

$$\hat{H} = \frac{1}{2m_0}\left[\mathbf{p}_e - \frac{e}{c}\hat{\mathbf{A}}(\mathbf{x}_e)\right] \cdot \left[\mathbf{p}_e - \frac{e}{c}\hat{\mathbf{A}}(\mathbf{x}_e)\right] + e\Phi(\mathbf{x}_e) + V(\mathbf{x}_e) + \hat{H}_{\text{rad}}$$

$$; \; \mathbf{p}_e = \frac{\hbar}{i}\boldsymbol{\nabla}_e \qquad ; \text{ coordinate rep} \tag{7.73}$$

7.4.2 *Schrödinger Picture*

In the Schrödinger picture:

- All the operators in Eq. (7.73) are *independent of time*;

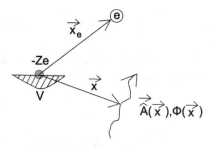

Fig. 7.2 A non-relativistic particle with charge $e = -|e|$ at the position $\mathbf{x}_e$ moving in the nuclear Coulomb potential, with an additional potential V, interacting with the quantized radiation field. The vector $\mathbf{x}$ locates the field position.

- The state vector is time-dependent and satisfies the *Schrödinger equation*

$$i\hbar \frac{\partial}{\partial t}|\Psi(t)\rangle = \hat{H}|\Psi(t)\rangle \qquad ; \text{S-eqn} \qquad (7.74)$$

- The observed particle position and classical field are now the *expectation values*

$$\langle \mathbf{x}_e \rangle = \langle \Psi(t)|\mathbf{x}_e|\Psi(t)\rangle \equiv \mathbf{x}_e(t)$$
$$\langle \hat{\mathbf{A}}(\mathbf{x}) \rangle = \langle \Psi(t)|\hat{\mathbf{A}}(\mathbf{x})|\Psi(t)\rangle \equiv \mathbf{A}(\mathbf{x}, t) \qquad (7.75)$$

- The time derivatives of the expectation values are related to the time-derivative of the operators through *Ehrenfest's theorem* in Eqs. (3.171).

7.4.3   *Justification*

Although again actually a postulate, the quantization scheme in Eqs. (7.70) will be justified by using Ehrenfest's theorem to show that

(1) The particle's motion is governed by Newton's second law with the Lorentz force equation;
(2) The field obeys Maxwell's equations with sources.

With the classical limit of a well-localized particle wave packet, one then reproduces classical electrodynamics.

7.4.3.1   *Lorentz Force*

This is the same calculation that is carried out in detail in appendix B. Ehrenfest's theorem with the hamiltonian in Eq. (7.73) gives

$$\frac{d\mathbf{x}_e}{dt} \equiv \hat{\dot{\mathbf{x}}}_e = \frac{i}{\hbar}[\hat{H}, \mathbf{x}_e] = \frac{1}{m_0}\left[\mathbf{p}_e - \frac{e}{c}\hat{\mathbf{A}}(\mathbf{x}_e)\right] \qquad ; \text{kinetic velocity} \qquad (7.76)$$

This relation defines the kinetic velocity operator $\hat{\dot{\mathbf{x}}}_e$ in the Schrödinger picture in the coordinate representation. Compute the second time derivative

$$\hat{\ddot{\mathbf{x}}}_e = \frac{i}{\hbar}[\hat{H}, \hat{\dot{\mathbf{x}}}_e] = \frac{i}{m_0\hbar}[\hat{H}_{\text{part}}, \mathbf{p}_e] -$$
$$\frac{e}{m_0 c}\left(\frac{i}{\hbar}[\hat{H}_{\text{part}}, \hat{\mathbf{A}}(\mathbf{x}_e)] + \frac{i}{\hbar}[\hat{H}_{\text{rad}}, \hat{\mathbf{A}}(\mathbf{x}_e)]\right) \qquad (7.77)$$

The first two terms on the r.h.s. are calculated exactly as in appendix B. The last term on the r.h.s. is identified from Eq. (7.48) as $\hat{\mathbf{A}}(\mathbf{x}_e)$. Now in appendix B, the time derivative of a specified *external* field at the particle location $\mathbf{x}_e$ is evaluated as

$$\frac{d}{dt}\mathbf{A}(\mathbf{x}_e,t) = \frac{i}{\hbar}[\hat{H}_{\text{part}}, \mathbf{A}(\mathbf{x}_e,t)] + \frac{\partial \mathbf{A}(\mathbf{x}_e,t)}{\partial t} \qquad ; \text{external field} \quad (7.78)$$

In the present case, all the dynamics is contained in the hamiltonian, and instead of $\partial \mathbf{A}(\mathbf{x}_e,t)/\partial t$, one has $\dot{\hat{\mathbf{A}}}(\mathbf{x}_e)$. The calculation then exactly parallels that in appendix B to give the following acceleration operator in the Schrödinger picture in the coordinate representation[18]

$$m_0\ddot{\hat{\mathbf{x}}}_e = -\boldsymbol{\nabla}_e V + e\left\{\hat{\boldsymbol{\mathcal{E}}}(\mathbf{x}_e) + \frac{1}{c}\left[\dot{\hat{\mathbf{x}}}_e \times \hat{\mathbf{B}}(\mathbf{x}_e)\right]_{\text{sym}}\right\} \qquad ; \text{Newton's second law}$$
$$; \text{Lorentz force} \quad (7.79)$$

This is Newton's second law with the Lorentz force.

7.4.3.2 *Maxwell's Equations*

We work in the Coulomb gauge, and show that Maxwell's equations are obtained as follows:

(1) It is evident from Eq. (7.45) that the field operator $\hat{\mathbf{A}}(\mathbf{x})$ is transverse

$$\boldsymbol{\nabla} \cdot \hat{\mathbf{A}}(\mathbf{x}) = 0 \qquad (7.80)$$

(2) The solution to Poisson's Eq. (7.13) for the Coulomb potential at the field point $\mathbf{x}$ at a given instant in time is

$$\Phi(\mathbf{x}) = -\frac{Ze}{|\mathbf{x}|} + \frac{e}{|\mathbf{x} - \mathbf{x}_e|} \qquad ; e = -|e| \qquad (7.81)$$

The first (static) term comes from the nucleus, while the second term arises from the particle with charge $e = -|e|$ located at the point $\mathbf{x}_e$ (see Fig. 7.2). This $\Phi(\mathbf{x})$ satisfies

$$\nabla^2\Phi = 4\pi\left[Ze\delta^{(3)}(\mathbf{x}) - e\delta^{(3)}(\mathbf{x} - \mathbf{x}_e)\right]$$
$$= -4\pi[\varrho_N(\mathbf{x}) + \varrho_e(\mathbf{x})] \qquad (7.82)$$

Here we identify the charge densities of both the nucleus and charged particle

$$\varrho_N(\mathbf{x}) = -Ze\delta^{(3)}(\mathbf{x}) \qquad ; \text{nuclear charge density}$$
$$\varrho_e(\mathbf{x}) = e\delta^{(3)}(\mathbf{x} - \mathbf{x}_e) \qquad ; \text{particle charge density} \qquad (7.83)$$

---

[18]Compare Eq. (B.34).

This is the correct *charge density operator* for the charged particle, since its expectation value in the state $|\phi_m\rangle$ in the coordinate representation yields

$$\langle\phi_m|\varrho_e(\mathbf{x})|\phi_m\rangle = e\int d^3x_e\ \phi_m^\star(\mathbf{x}_e)\delta^{(3)}(\mathbf{x} - \mathbf{x}_e)\phi_m(\mathbf{x}_e)$$

$$= e\,|\phi_m(\mathbf{x})|^2 \qquad\qquad \text{; charge density} \quad (7.84)$$

This is indeed the charge density at the point $\mathbf{x}$ arising from the particle.

(3) It remains to show that *at the field point* $\mathbf{x}$, the last of Maxwell's equation is satisfied

$$\nabla^2\hat{\mathbf{A}}(\mathbf{x}) - \frac{1}{c^2}\ddot{\hat{\mathbf{A}}}(\mathbf{x}) = -4\pi\,\hat{\mathbf{j}}_T(\mathbf{x}) \qquad (7.85)$$

The transverse current is given classically by Eqs. (7.14) as

$$\mathbf{j}_T = \mathbf{j} - \frac{1}{4\pi c}\boldsymbol{\nabla}\left(\frac{\partial\Phi}{\partial t}\right) \qquad (7.86)$$

In the Schrödinger picture, in the coordinate representation, the time derivative of Φ at a given field position $\mathbf{x}$ is given by Ehrenfest's theorem

$$\frac{d}{dt}\Phi(\mathbf{x}) \equiv \dot{\hat{\Phi}}(\mathbf{x}) = \frac{i}{\hbar}[\hat{H}, \Phi(\mathbf{x})] \qquad (7.87)$$

This is evaluated from Eqs. (7.73) and (7.81) as

$$\dot{\hat{\Phi}}(\mathbf{x}) = \frac{i}{\hbar}\left[\hat{H}_{\text{part}}, \frac{e}{|\mathbf{x} - \mathbf{x}_e|}\right]$$

$$= e\left\{\left(\frac{\partial}{\partial x_e}\right)_j\frac{1}{|\mathbf{x} - \mathbf{x}_e|}, \left(\dot{\hat{x}}_e\right)_j\right\}_{\text{sym}} \qquad (7.88)$$

It follows that[19]

$$\nabla_i\dot{\hat{\Phi}}(\mathbf{x}) = -e\left\{\frac{\partial}{\partial x_i}\frac{\partial}{\partial x_j}\frac{1}{|\mathbf{x} - \mathbf{x}_e|}, \left(\dot{\hat{x}}_e\right)_j\right\}_{\text{sym}} \qquad (7.89)$$

Now insert the Fourier series expansion of the Coulomb potential[20]

$$\frac{1}{|\mathbf{x} - \mathbf{x}_e|} = \frac{4\pi}{\Omega}\sum_{\mathbf{k}}\frac{1}{\mathbf{k}^2}\,e^{i\mathbf{k}\cdot(\mathbf{x}-\mathbf{x}_e)} \qquad (7.90)$$

---

[19]Note $\boldsymbol{\nabla}_e f(\mathbf{x} - \mathbf{x}_e) = -\boldsymbol{\nabla}f(\mathbf{x} - \mathbf{x}_e)$; repeated Latin indices are here summed from 1 to 3.

[20]We have done the corresponding Fourier transform.

This gives

$$-\frac{1}{4\pi c}\nabla_i\hat{\tilde{\Phi}}(\mathbf{x}) = -\frac{e}{c}\left\{\frac{1}{\Omega}\sum_{\mathbf{k}}\frac{k_ik_j}{\mathbf{k}^2}e^{i\mathbf{k}\cdot(\mathbf{x}-\mathbf{x}_e)}, \left(\hat{\dot{x}}_e\right)_j\right\}_{\text{sym}} \tag{7.91}$$

We are now in a position to compute the second time derivative of the vector potential field operator at the field point $\mathbf{x}$ (see Fig. 7.2)

$$\hat{\ddot{\mathbf{A}}}(\mathbf{x}) = \frac{i}{\hbar}[\hat{H}, \hat{\dot{\mathbf{A}}}(\mathbf{x})]$$
$$= \frac{i}{\hbar}[\hat{H}_{\text{part}}, \hat{\dot{\mathbf{A}}}(\mathbf{x})] + \frac{i}{\hbar}[\hat{H}_{\text{rad}}, \hat{\dot{\mathbf{A}}}(\mathbf{x})] \tag{7.92}$$

The two terms on the r.h.s. are evaluated as follows:

- From (7.52)–(7.53) the second term is

$$\frac{i}{\hbar}[\hat{H}_{\text{rad}}, \hat{\dot{\mathbf{A}}}(\mathbf{x})] = c^2\nabla^2\hat{\mathbf{A}}(\mathbf{x}) \tag{7.93}$$

- The first term is evaluated with the aid of Eq. (7.65) as

$$\frac{i}{\hbar}[\hat{H}_{\text{part}}, \hat{\dot{A}}_i(\mathbf{x})] = \frac{i}{\hbar}\left(\frac{-e}{c}\right)\left(\frac{-4\pi\hbar c^2}{i}\right)\times$$
$$\left\{\left(\hat{\dot{x}}_e\right)_j, \frac{1}{\Omega}\sum_{\mathbf{k}}\left(\delta_{ij}-\frac{k_ik_j}{\mathbf{k}^2}\right)e^{i\mathbf{k}\cdot(\mathbf{x}-\mathbf{x}_e)}\right\}_{\text{sym}} \tag{7.94}$$

Make use of Eq. (7.91) to re-write this as

$$\frac{i}{\hbar}[\hat{H}_{\text{part}}, \hat{\dot{A}}_i(\mathbf{x})] = 4\pi c^2\left[\frac{e}{c}\left\{\left(\hat{\dot{x}}_e\right)_i, \delta^{(3)}(\mathbf{x}-\mathbf{x}_e)\right\}_{\text{sym}} - \frac{1}{4\pi c}\nabla_i\hat{\tilde{\Phi}}(\mathbf{x})\right] \tag{7.95}$$

The *electromagnetic current* is now identified as

$$\hat{\mathbf{j}}(\mathbf{x}) = \frac{e}{c}\left\{\hat{\dot{\mathbf{x}}}_e, \delta^{(3)}(\mathbf{x}-\mathbf{x}_e)\right\}_{\text{sym}} \qquad ;\text{ electromagnetic current} \tag{7.96}$$

A combination of Eqs. (7.92)–(7.96) then gives

$$\nabla^2\hat{\mathbf{A}}(\mathbf{x}) - \frac{1}{c^2}\hat{\ddot{\mathbf{A}}}(\mathbf{x}) = -4\pi\hat{\mathbf{j}}_T(\mathbf{x}) \qquad ;\text{ Maxwell's eqn} \tag{7.97}$$

which is the desired result.

7.4.3.3　*Current Conservation*

It remains to show that the electromagnetic current in Eq. (7.96) due to the charged particle motion is *conserved*. Compute the time derivative of the charge density $\varrho_e(\mathbf{x})$ in Eq. (7.83) with the aid of Ehrenfest's theorem

$$\dot{\hat{\varrho}}_e(\mathbf{x}) = \frac{i}{\hbar}[\hat{H}, \varrho_e(\mathbf{x})] = \frac{i}{\hbar}[\hat{H}_{\text{part}}, e\delta^{(3)}(\mathbf{x} - \mathbf{x}_e)] \qquad (7.98)$$

Now compute the required commutator

$$\dot{\hat{\varrho}}_e(\mathbf{x}) = \frac{e}{2}\left[\dot{\hat{\mathbf{x}}}_e \cdot \boldsymbol{\nabla}_e\delta^{(3)}(\mathbf{x} - \mathbf{x}_e) + \boldsymbol{\nabla}_e\delta^{(3)}(\mathbf{x} - \mathbf{x}_e) \cdot \dot{\hat{\mathbf{x}}}_e\right]$$

$$= -\frac{e}{2}\left[\dot{\hat{\mathbf{x}}}_e \cdot \boldsymbol{\nabla}\delta^{(3)}(\mathbf{x} - \mathbf{x}_e) + \boldsymbol{\nabla}\delta^{(3)}(\mathbf{x} - \mathbf{x}_e) \cdot \dot{\hat{\mathbf{x}}}_e\right]$$

$$= -e\,\boldsymbol{\nabla} \cdot \left\{\dot{\hat{\mathbf{x}}}_e, \delta^{(3)}(\mathbf{x} - \mathbf{x}_e)\right\}_{\text{sym}} \qquad (7.99)$$

This is the operator statement in the Schrödinger picture, in the coordinate representation, of *current conservation*

$$\frac{1}{c}\dot{\hat{\varrho}}_e(\mathbf{x}) + \boldsymbol{\nabla} \cdot \hat{\mathbf{j}}(\mathbf{x}) = 0 \qquad ;\text{ current conservation} \qquad (7.100)$$

In *summary*:

- Equation (7.73) presents the hamiltonian for the quantum electrodynamics of a non-relativistic particle;
- The dynamics is governed by the Schrödinger Eq. (7.74);
- The problem is here formulated in the Schrödinger picture, and in the coordinate representation for the particle; the radiation field is described in abstract Hilbert space;
- One derives Newton's second law and the Lorentz force equation for expectation values of the particle coordinate $\mathbf{x}_e$;
- Maxwell's equations with sources hold for expectation values of the field operators at all field points $\mathbf{x}$;
- The electomagnetic charge density and current operators for the particle are identified in Eqs. (7.83) and (7.96), and the current is conserved;
- The expectation value of the charge density operator at the point $\mathbf{x}$ is

$$\langle\phi_m; n_1 \cdots n_\infty|\hat{\rho}(\mathbf{x})|\phi_m; n_1 \cdots n_\infty\rangle = e\int d^3x_e\, \phi_m^\star(\mathbf{x}_e)\delta^{(3)}(\mathbf{x} - \mathbf{x}_e)\phi_m(\mathbf{x}_e)$$

$$= e|\phi_m(\mathbf{x})|^2 \qquad (7.101)$$

A similar result holds for the current.[21]

---

[21]See Prob. 7.5.

Recall that $\Phi(\mathbf{x})$ is the Coulomb potential at an arbitrary field point

$$\Phi(\mathbf{x}) = -\frac{Ze}{|\mathbf{x}|} + \frac{e}{|\mathbf{x} - \mathbf{x}_e|} \tag{7.102}$$

In Eq. (7.73) one requires this potential at the position of the charged particle $\Phi(\mathbf{x}_e)$, and the second term in Eq. (7.102) blows up as $\mathbf{x} \to \mathbf{x}_e$. This represents the *Coulomb self-energy* of the charge e, which is infinite for a point charge.[22] In any event, it is just a *constant*, call it $E_{\text{self-en}}$, and at this stage we are free to take $E_{\text{self-en}} + E_{\text{zero-pt}}$ to define our zero of energy. Our effective hamiltonian in abstract Hilbert space is then[23]

$$\hat{H} = \hat{H}_{\text{QED}} - E_{\text{self-en}} - E_{\text{zero-pt}} \tag{7.103}$$

$$= \frac{1}{2m_0}\left[\hat{\mathbf{p}}_e - \frac{e}{c}\hat{\mathbf{A}}(\hat{\mathbf{x}}_e)\right] \cdot \left[\hat{\mathbf{p}}_e - \frac{e}{c}\hat{\mathbf{A}}(\hat{\mathbf{x}}_e)\right] - \frac{Ze^2}{|\hat{\mathbf{x}}_e|} + V(\hat{\mathbf{x}}_e) + \sum_{\mathbf{k},s}\hbar\omega_k\hat{a}^\dagger_{\mathbf{k}s}\hat{a}_{\mathbf{k}s}$$

This is readily separated into an unperturbed part and a perturbation

$$\hat{H} = \hat{H}_0 + \hat{H}_1 \qquad\qquad ; \text{ effective hamiltonian for QED}$$

$$\hat{H}_0 = \frac{1}{2m_0}\hat{\mathbf{p}}_e^2 - \frac{Ze^2}{|\hat{\mathbf{x}}_e|} + V(\hat{\mathbf{x}}_e) + \sum_{\mathbf{k},s}\hbar\omega_k\hat{a}^\dagger_{\mathbf{k}s}\hat{a}_{\mathbf{k}s} \equiv \hat{H}_0^{\text{part}} + \hat{H}_0^{\text{rad}}$$

$$\hat{H}_1 = -\frac{e}{2m_0c}\left[\hat{\mathbf{p}}_e \cdot \hat{\mathbf{A}}(\hat{\mathbf{x}}_e) + \hat{\mathbf{A}}(\hat{\mathbf{x}}_e) \cdot \hat{\mathbf{p}}_e\right] + \frac{e^2}{2m_0c^2}\hat{\mathbf{A}}(\hat{\mathbf{x}}_e)^2 \tag{7.104}$$

Note the definition in the second line, and $\hat{\mathbf{A}}(\hat{\mathbf{x}}_e)$ is given by Eq. (7.68). Once the hamiltonian is in this form, rates and cross sections follow directly.

7.4.4 Photoemission

Consider the *spontaneous* emission of a photon by a system that makes a transition from a state $|\phi_i\rangle$ to a state $|\phi_f\rangle$ (see Fig. 7.3). The transition rate is then calculated through the series of steps detailed below.

7.4.4.1 Transition Rate

- Start with eigenstates of $\hat{H}_0$ in Eqs. (7.104)

$$\hat{H}_0|\Phi_m\rangle = E_m|\Phi_m\rangle \tag{7.105}$$

---

[22]With an extended charge distribution, this term would be $E_{\text{self-en}} = \int d^3x'\, \rho(x')/x'$.

[23]The Coulomb self-energy is actually one contribution to the *self-mass* of the particle, and a proper treatment of the full electromagnetic self-energy, and corresponding mass renormalization, requires the covariant version of QED (see Vol. II).

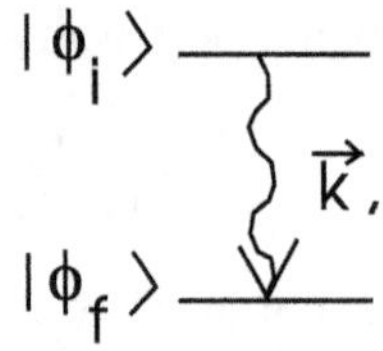

$$|\phi_i\rangle \underset{\displaystyle \Big\} \vec{K},s}{\rule{3cm}{0.4pt}}$$

$$|\phi_f\rangle \rule{3cm}{0.4pt}$$

Fig. 7.3　Photoemission where the system goes from $|\phi_i\rangle$ to $|\phi_f\rangle$.

The initial and final states in abstract Hilbert space are then

$$|\Phi_i\rangle = |\phi_i; 0, 0, \cdots, 0\rangle \qquad ; \text{ system in } |\phi_i\rangle; \text{ no photons}$$
$$|\Phi_f\rangle = |\phi_f; 0, \cdots, 1_{\mathbf{k}s}, \cdots, 0\rangle \qquad ; \text{ system in } |\phi_f\rangle; \text{ 1 photon } \{\mathbf{k}s\}$$

$$(7.106)$$

- To lowest order in e, the matrix element of the perturbation in Eqs. (7.104) is then given by

$$\langle \Phi_f | \hat{H}_1 | \Phi_i \rangle = -\frac{e}{2m_0 c} \langle \phi_f; 1_{\mathbf{k}s} | \hat{\mathbf{p}}_e \cdot \hat{\mathbf{A}}(\hat{\mathbf{x}}_e) + \hat{\mathbf{A}}(\hat{\mathbf{x}}_e) \cdot \hat{\mathbf{p}}_e | \phi_i; 0 \rangle \quad (7.107)$$

- Take the required matrix elements of the photon field

$$\langle 1_{\mathbf{k}s} | \hat{a}^\dagger_{\mathbf{k}'s'} | 0 \rangle = \delta_{\mathbf{k},\mathbf{k}'} \delta_{s,s'}$$
$$\langle 1_{\mathbf{k}s} | \hat{a}_{\mathbf{k}'s'} | 0 \rangle = 0 \qquad\qquad (7.108)$$

This picks out one term in the sums in Eq. (7.68); all other photon modes give $\langle 0|0 \rangle = 1$. Hence

$$\langle \Phi_f | \hat{H}_1 | \Phi_i \rangle = -\frac{e}{2m_0 c} \left(\frac{2\pi \hbar c^2}{\Omega \omega_k} \right)^{1/2} \mathbf{e}_{\mathbf{k}s} \cdot \langle \phi_f | \hat{\mathbf{p}}_e \, e^{-i\mathbf{k}\cdot\hat{\mathbf{x}}_e} + e^{-i\mathbf{k}\cdot\hat{\mathbf{x}}_e} \, \hat{\mathbf{p}}_e | \phi_i \rangle$$

$$(7.109)$$

- Now go to the coordinate representation for the particle, utilizing the eigenstates of position $\hat{\mathbf{x}}_e|\boldsymbol{\xi}\rangle = \boldsymbol{\xi}|\boldsymbol{\xi}\rangle$ satisfying

$$\int d^3\xi \, |\boldsymbol{\xi}\rangle\langle\boldsymbol{\xi}| = \hat{1} \qquad\qquad (7.110)$$

The matrix element then takes the form

$$\frac{1}{2m_0} \langle \phi_f | \hat{\mathbf{p}}_e \, e^{-i\mathbf{k}\cdot\hat{\mathbf{x}}_e} + e^{-i\mathbf{k}\cdot\hat{\mathbf{x}}_e} \, \hat{\mathbf{p}}_e | \phi_i \rangle = \qquad\qquad (7.111)$$

$$\frac{1}{2m_0} \int d^3\xi \, \phi_f^\star(\boldsymbol{\xi}) \left[\frac{\hbar}{i} \boldsymbol{\nabla}_\xi \, e^{-i\mathbf{k}\cdot\boldsymbol{\xi}} + e^{-i\mathbf{k}\cdot\boldsymbol{\xi}} \, \frac{\hbar}{i} \boldsymbol{\nabla}_\xi \right] \phi_i(\boldsymbol{\xi})$$

A partial integration gives

$$\frac{1}{2m_0}\langle\phi_f|\hat{\mathbf{p}}_e\, e^{-i\mathbf{k}\cdot\hat{\mathbf{x}}_e} + e^{-i\mathbf{k}\cdot\hat{\mathbf{x}}_e}\,\hat{\mathbf{p}}_e|\phi_i\rangle = \tag{7.112}$$

$$\frac{\hbar}{2im_0}\int d^3\xi\, e^{-i\mathbf{k}\cdot\boldsymbol{\xi}}\left\{\phi_f^\star(\boldsymbol{\xi})\boldsymbol{\nabla}_\xi\,\phi_i(\boldsymbol{\xi}) - [\boldsymbol{\nabla}_\xi\,\phi_f(\boldsymbol{\xi})]^\star\,\phi_i(\boldsymbol{\xi})\right\}$$

This result can be written as the Fourier transform of the *transition probability current* $\mathbf{S}_{fi}(\mathbf{x})$

$$\mathbf{S}_{fi}(\mathbf{x}) = \frac{\hbar}{2im_0}\left\{\phi_f^\star(\mathbf{x})\boldsymbol{\nabla}\phi_i(\mathbf{x}) - [\boldsymbol{\nabla}\phi_f(\mathbf{x})]^\star\,\phi_i(\mathbf{x})\right\}$$

$$\text{; transition probability current} \tag{7.113}$$

Thus

$$\frac{1}{2m_0}\langle\phi_f|\hat{\mathbf{p}}_e\, e^{-i\mathbf{k}\cdot\hat{\mathbf{x}}_e} + e^{-i\mathbf{k}\cdot\hat{\mathbf{x}}_e}\,\hat{\mathbf{p}}_e|\phi_i\rangle = \int d^3x\, e^{-i\mathbf{k}\cdot\mathbf{x}}\,\mathbf{S}_{fi}(\mathbf{x}) \tag{7.114}$$

It follows that the matrix element in Eq. (7.109) becomes[24]

$$\langle\Phi_f|\hat{H}_1|\Phi_i\rangle = -\left(\frac{2\pi\hbar c^2}{\Omega\omega_k}\right)^{1/2}\mathbf{e}_{\mathbf{k}s}\cdot\int d^3x\, e^{-i\mathbf{k}\cdot\mathbf{x}}\,\mathbf{j}_{fi}(\mathbf{x})$$

$$\mathbf{j}_{fi}(\mathbf{x}) = \frac{e}{c}\mathbf{S}_{fi}(\mathbf{x}) \qquad\text{; transition current} \tag{7.115}$$

Here $\mathbf{j}_{fi}(\mathbf{x})$ is the *transition electromagnetic current.*

- To *summarize* the manipulations on the current, the abstract form of the electromagnetic current operator is to order e

$$\hat{\mathbf{j}}(\mathbf{x}) = \frac{e}{2m_0c}\{\hat{\mathbf{p}}_e,\,\delta^{(3)}(\mathbf{x}-\hat{\mathbf{x}}_e)\}_{\text{sym}} \quad\text{; abstract form} \tag{7.116}$$

The required transition matrix element of the current is

$$\mathbf{j}_{fi}(\mathbf{x}) = \langle\phi_f|\hat{\mathbf{j}}(\mathbf{x})|\phi_i\rangle \qquad\text{; transition current} \tag{7.117}$$

The coordinate representation introduces eigenstates of position

$$\hat{\mathbf{x}}_e|\boldsymbol{\xi}\rangle = \boldsymbol{\xi}|\boldsymbol{\xi}\rangle \qquad\text{; coordinate rep} \tag{7.118}$$

In the coordinate representation, the transition matrix element of the current becomes

$$\mathbf{j}_{fi}(\mathbf{x}) = \frac{e}{2m_0c}\int d^3\xi\, \phi_f^\star(\boldsymbol{\xi})\left[\frac{\hbar}{i}\boldsymbol{\nabla}_\xi\delta^{(3)}(\mathbf{x}-\boldsymbol{\xi}) + \delta^{(3)}(\mathbf{x}-\boldsymbol{\xi})\frac{\hbar}{i}\boldsymbol{\nabla}_\xi\right]\phi_i(\boldsymbol{\xi})$$

$$\tag{7.119}$$

---

[24]Compare Prob. 7.5(b).

A partial integration then gives the expression in Eqs. (7.113) and (7.115)

$$\mathbf{j}_{fi}(\mathbf{x}) = \frac{e\hbar}{2im_0c}\left\{\phi_f^\star(\mathbf{x})\boldsymbol{\nabla}\phi_i(\mathbf{x}) - [\boldsymbol{\nabla}\phi_f(\mathbf{x})]^\star\,\phi_i(\mathbf{x})\right\}$$

$$\text{; transition current} \qquad (7.120)$$

- The *transition rate* for the process in Fig. 7.3 is given by Dirac perturbation theory as

$$R_{fi} = \frac{2\pi}{\hbar}|\langle\Phi_f|\hat{H}_1|\Phi_i\rangle|^2\delta(E_f - E_i) \qquad (7.121)$$

Here the total initial and final energies are

$$E_i = \varepsilon_i$$
$$E_f = \varepsilon_f + \hbar\omega_k = \varepsilon_f + \hbar kc \qquad (7.122)$$

where the system energies are the eigenvalues of

$$\hat{H}_0^{\text{part}}|\phi_m\rangle = \varepsilon_m|\phi_m\rangle \qquad (7.123)$$

The emitted photon is in the *continuum* in the final state. Since we impose the periodic boundary conditions of Eqs. (7.22) in the quantization volume of Fig. 6.10, the number of modes in the volume element d^3k about $\mathbf{k}$ is the familiar expression

$$dn_f = \frac{\Omega d^3k}{(2\pi)^3} \qquad \text{; \# of modes in volume element } d^3k \qquad (7.124)$$

The transition rate into the detector then follows as previously

$$\mathcal{R}_{fi} = \int R_{fi}dn_f = \int R_{fi}\left(\frac{\partial n_f}{\partial E_f}\right)dE_f \qquad (7.125)$$

This yields the Golden Rule

$$\mathcal{R}_{fi} = \frac{2\pi}{\hbar}|\langle\Phi_f|\hat{H}_1|\Phi_i\rangle|^2\left(\frac{\partial n_f}{\partial E_f}\right)_{\varepsilon_f+\hbar\omega_k=\varepsilon_i} \qquad (7.126)$$

Now use

$$\frac{\partial E_f}{\partial k} = \hbar c$$
$$\frac{\partial n_f}{\partial E_f} = \frac{\Omega k^2 d\Omega_k}{(2\pi)^3}\frac{1}{\hbar c} \qquad (7.127)$$

This gives the transition rate as

$$\mathcal{R}_{fi} = \frac{2\pi}{\hbar} \frac{\Omega k^2 d\Omega_k}{(2\pi)^3 \hbar c} \left(\frac{2\pi \hbar c^2}{\Omega \omega_k} \right) \left| \mathbf{e}_{\mathbf{k}s} \cdot \int d^3x \, e^{-i\mathbf{k}\cdot\mathbf{x}} \langle \phi_f | \hat{\mathbf{j}}(\mathbf{x}) | \phi_i \rangle \right|^2 \quad (7.128)$$

A cancellation of factors gives

$$\mathcal{R}_{fi} = \frac{1}{2\pi} \frac{\omega_k}{\hbar c} \left| \mathbf{e}_{\mathbf{k}s} \cdot \int d^3x \, e^{-i\mathbf{k}\cdot\mathbf{x}} \langle \phi_f | \hat{\mathbf{j}}(\mathbf{x}) | \phi_i \rangle \right|^2 d\Omega_k \quad ; \text{ transition rate}$$

$$= \frac{\alpha}{2\pi} \frac{\omega_k}{c^2} \left| \mathbf{e}_{\mathbf{k}s} \cdot \int d^3x \, e^{-i\mathbf{k}\cdot\mathbf{x}} \langle \phi_f | \hat{\mathbf{S}}(\mathbf{x}) | \phi_i \rangle \right|^2 d\Omega_k \quad (7.129)$$

This is a *general result* for the transition rate to lowest order in the fine-structure constant $\alpha = e^2/\hbar c$. It holds for any finite quantum system.[25] The situation is sketched in Fig. 7.4.

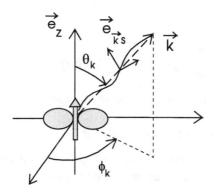

Fig. 7.4 Configuration for the photoemission rate in Eq. (7.129). Here $\mathbf{e}_z$ denotes the axis for spatial quantization of the z-component of the angular momentum of the system.

- For the total transition rate, it is necessary to sum over everything that gets into the detector. Hence, if the polarization of the final photon is

---

[25] Although derived here for non-relativistic motion, the result is more general (see Probs. 12.3–12.4 in Vol. I).

not detected, one must *sum over final photon polarizations*. Use[26]

$$\sum_{s=1}^{2} |\mathbf{e}_{\mathbf{k}s} \cdot \mathbf{v}|^2 = \sum_{s=1}^{2} (\mathbf{e}_{\mathbf{k}s})_i (\mathbf{e}_{\mathbf{k}s})_j v_i v_j^{\star} \tag{7.130}$$

$$= [\delta_{ij} - (\mathbf{e}_{\mathbf{k}0})_i (\mathbf{e}_{\mathbf{k}0})_j] v_i v_j^{\star}$$

$$= \mathbf{v} \cdot \mathbf{v}^{\star} - (\mathbf{v} \cdot \mathbf{e}_{\mathbf{k}0})(\mathbf{v}^{\star} \cdot \mathbf{e}_{\mathbf{k}0}) \qquad ; \ \mathbf{e}_{\mathbf{k}0} = \frac{\mathbf{k}}{|\mathbf{k}|}$$

This expression can also be written

$$\sum_{s=1}^{2} |\mathbf{e}_{\mathbf{k}s} \cdot \mathbf{v}|^2 = |\mathbf{v} - \mathbf{e}_{\mathbf{k}0}(\mathbf{e}_{\mathbf{k}0} \cdot \mathbf{v})|^2 \equiv |\mathbf{v}_{\perp}|^2 \tag{7.131}$$

where this relation defines the transverse vector $\mathbf{v}_{\perp}$. Thus when summed over final photon polarizations, Eq. (7.129) becomes

$$\sum_{s=1}^{2} \mathcal{R}_{fi} = \frac{\alpha}{2\pi} \frac{\omega_k}{c^2} \left| \int d^3x \, e^{-i\mathbf{k}\cdot\mathbf{x}} \langle \phi_f | \hat{\mathbf{S}}(\mathbf{x}) | \phi_i \rangle_{\perp} \right|^2 d\Omega_k \tag{7.132}$$

The radiated *power* is given by

$$d\mathcal{P} = \hbar\omega_k \sum_{s} \mathcal{R}_{fi} \tag{7.133}$$

$$\frac{d\mathcal{P}}{d\Omega_k} = \frac{e^2}{2\pi} \frac{\omega_k^2}{c^3} \left| \int d^3x \, e^{-i\mathbf{k}\cdot\mathbf{x}} \langle \phi_f | \hat{\mathbf{S}}(\mathbf{x}) | \phi_i \rangle_{\perp} \right|^2 \qquad ; \ \text{radiated power}$$

Here $\alpha = e^2/\hbar c$ has been used in the second line.

7.4.4.2 *Comparison with Classical Result*

It is instructive to compare the above expression with the *classical result* for the power radiated from an accelerated charge distribution. Assume

- The radiation is coming from a finite system;
- The system is oscillating with a given frequency ω;
- The electromagnetic current has the form

$$\mathbf{j}(\mathbf{x}, t) = \frac{e}{c} \mathcal{S}_C(\mathbf{x}, t)$$

$$\mathcal{S}_C(\mathbf{x}, t) = \mathcal{S}(\mathbf{x}) e^{-i\omega t} + \mathcal{S}^{\star}(\mathbf{x}) e^{i\omega t}$$

$$= 2\text{Re}\, \mathcal{S}(\mathbf{x}) e^{-i\omega t} \tag{7.134}$$

---

[26]See Eq. (7.64). Once again, we employ the summation convention that repeated Latin indices are summed from 1 to 3.

Notice the factor of two in this expression;

- The classical radiated power is then given by[27]

$$\frac{d\mathcal{P}}{d\Omega_k} = \frac{e^2}{8\pi} \frac{\omega^2}{c^3} 4 \left| \int d^3x \, e^{-i\mathbf{k}\cdot\mathbf{x}} \, \mathcal{S}(\mathbf{x})_\perp \right|^2 \qquad ; \text{classical result} \qquad (7.135)$$

Thus

- This result is now in *exactly the same form* as that in Eqs. (7.133);
- To make the results coincide, one only has to replace the classical current $\mathcal{S}(\mathbf{x})$ by the transition probability current $\mathbf{S}_{fi}(\mathbf{x})$

$$\mathcal{S}(\mathbf{x}) \rightarrow \langle \phi_f | \hat{\mathbf{S}}(\mathbf{x}) | \phi_i \rangle \qquad ; \text{in Q.M.} \qquad (7.136)$$

- From the analysis in Eqs. (6.152)–(6.155), the term $\mathcal{S}(\mathbf{x})e^{-i\omega t} \rightarrow \mathbf{S}_{if}(\mathbf{x})e^{-i\omega t}$ corresponds to *absorption*, and its complex conjugate $\mathbf{S}_{fi}(\mathbf{x})e^{i\omega t}$ to *emission*, in quantum mechanics.
- The calculation in Eq. (7.133) has been done with the quantized field, and all factors have been correctly determined.

7.4.4.3 *Dipole Approximation*

If $ka \ll 1$, where a is the size of the system built into the wave functions, then in Eq. (7.129) one can replace[28]

$$e^{-i\mathbf{k}\cdot\mathbf{x}} \approx 1 \qquad ; \text{dipole approximation}$$
$$; ka \ll 1 \qquad (7.137)$$

Then, just as before and as verified in Prob. 7.9,

$$\int d^3x \, \langle \phi_f | \hat{\mathbf{S}}(\mathbf{x}) | \phi_i \rangle = \frac{1}{2m_0} \int d^3x \, \langle \phi_f | \hat{\mathbf{p}}_e \delta^{(3)}(\mathbf{x} - \hat{\mathbf{x}}_e) + \delta^{(3)}(\mathbf{x} - \hat{\mathbf{x}}_e)\hat{\mathbf{p}}_e | \phi_i \rangle$$
$$= \frac{1}{m_0} \langle \phi_f | \hat{\mathbf{p}}_e | \phi_i \rangle \qquad (7.138)$$

From Eqs. (7.104) and (7.123), this can be written

$$\frac{1}{m_0} \langle \phi_f | \hat{\mathbf{p}}_e | \phi_i \rangle = \frac{i}{\hbar} \langle \phi_f | [\hat{H}_0^{\text{part}}, \hat{\mathbf{x}}_e] | \phi_i \rangle = \frac{i(\varepsilon_f - \varepsilon_i)}{\hbar} \langle \phi_f | \hat{\mathbf{x}}_e | \phi_i \rangle$$
$$= -i\omega_k \langle \phi_f | \hat{\mathbf{x}}_e | \phi_i \rangle \qquad (7.139)$$

---

[27] See [Jackson (1998)] Secs. 9.1–9.2; also [Schiff (1968)] pp. 407–409.

[28] The dipole approximation is most relevant to atomic and molecular physics, where the dominant transitions are indeed dipole. For gamma transitions in nuclei, a full multipole analysis of the expression in Eq. (7.129) is required (see [Blatt and Weisskopf (1952); Walecka (2004)]).

Hence in the dipole approximation, the transition rate becomes

$$\mathcal{R}_{fi} = \frac{\alpha}{2\pi} \frac{\omega_k^3}{c^2} |\mathbf{e}_{\mathbf{k}s} \cdot \langle \phi_f | \hat{\mathbf{x}}_e | \phi_i \rangle|^2 \, d\Omega_k \qquad ; \text{dipole approx} \quad (7.140)$$

This is the abstract form, which in the coordinate representation becomes

$$\mathcal{R}_{fi} = \frac{\alpha}{2\pi} \frac{\omega_k^3}{c^2} \left| \mathbf{e}_{\mathbf{k}s} \cdot \int d^3x \, \phi_f^\star(\mathbf{x}) \, \mathbf{x} \, \phi_i(\mathbf{x}) \right|^2 \, d\Omega_k \quad ; \text{coordinate rep} \quad (7.141)$$

This expression is reduced as follows:

- For no measurement of the final photon polarization, the sum over photon polarizations is carried out using Eq. (7.130);
- If one does not observe the angle between the quantization axis $\mathbf{e}_z$ and $\mathbf{k}$ but only the *total* decay rate, then it is necessary to do the $\int d\Omega_k$

$$\int d\Omega_k = 4\pi$$

$$\int d\Omega_k (\mathbf{e}_{\mathbf{k}0})_i (\mathbf{e}_{\mathbf{k}0})_j = \int d\Omega_k \frac{k_i k_j}{\mathbf{k}^2} = \frac{4\pi}{3} \delta_{ij} \qquad (7.142)$$

Hence from Eq. (7.130)

$$\int d\Omega_k \sum_{s=1}^{2} |\mathbf{e}_{\mathbf{k}s} \cdot \mathbf{v}|^2 = 4\pi |\mathbf{v}|^2 \left(1 - \frac{1}{3} \right) = \frac{8\pi}{3} |\mathbf{v}|^2 \qquad (7.143)$$

The total decay rate in the dipole approximation is then given by

$$\int \sum_s \mathcal{R}_{fi} = \frac{4\alpha}{3} \frac{\omega_k^3}{c^2} \left| \int d^3x \, \phi_f^\star(\mathbf{x}) \, \mathbf{x} \, \phi_i(\mathbf{x}) \right|^2 \qquad ; \text{dipole approximation}$$

$$(7.144)$$

- Introduce the *spherical components* of a vector

$$v_{\pm 1} \equiv \mp \frac{1}{\sqrt{2}} (v_x \pm i v_y) \qquad ; \text{spherical components}$$

$$v_0 \equiv v_z \qquad (7.145)$$

Then

$$\mathbf{v} \cdot \mathbf{v}^\star = v_x v_x^\star + v_y v_y^\star + v_z v_z^\star$$

$$= \frac{1}{2}(v_x + iv_y)(v_x^\star - iv_y^\star) + \frac{1}{2}(v_x - iv_y)(v_x^\star + iv_y^\star) + v_z v_z^\star$$

$$= v_{+1} v_{+1}^\star + v_{-1} v_{-1}^\star + v_0 v_0^\star \qquad (7.146)$$

Hence

$$|\mathbf{v}|^2 = \sum_m v_m v_m^\star = \sum_m |v_m|^2 \qquad ; \; m = 0, \pm 1 \qquad (7.147)$$

Now express the spherical components of the dipole vector $\mathbf{x}$ in terms of the spherical coordinates for the system in Fig. 7.5.

$$x_{\pm 1} = \mp \frac{1}{\sqrt{2}}(x \pm iy) = \mp \frac{1}{\sqrt{2}} r \sin\theta(\cos\phi \pm i\sin\phi)$$

$$x_0 = z = r\cos\theta \qquad (7.148)$$

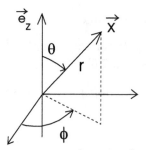

Fig. 7.5 Spherical coordinates for system.

Recall the spherical harmonics[29]

$$Y_{1,\pm 1}(\theta,\phi) = \mp \left(\frac{3}{8\pi}\right)^{1/2} \sin\theta \, e^{\pm i\phi}$$

$$Y_{1,0}(\theta,\phi) = \left(\frac{3}{4\pi}\right)^{1/2} \cos\theta \qquad (7.149)$$

Hence the spherical components of $\mathbf{x}$ can be expressed in terms of spherical harmonics as

$$x_m = \left(\frac{4\pi}{3}\right)^{1/2} r \, Y_{1,m}(\theta,\phi) \qquad (7.150)$$

The wave functions for a particle in a central field have the form

$$\phi_{nlm}(\mathbf{x}) = R_{nl}(r)Y_{lm}(\Omega_x) \qquad ; \; \text{central field} \qquad (7.151)$$

---

[29]See Prob. 2.14.

Thus the transition rate in Eq. (7.144) can be written in spherical coordinates as

$$\int \sum_s \mathcal{R}_{fi} = \frac{16\pi\alpha}{9} \frac{\omega_k^3}{c^2} \left| \int_0^\infty r^2 dr\, R_{n_f l_f}(r) r R_{n_i l_i}(r) \right|^2 \sum_m \times$$

$$\left| \int d\Omega_x\, Y^\star_{l_f, m_f}(\Omega_x) Y_{1,m}(\Omega_x) Y_{l_i, m_i}(\Omega_x) \right|^2 \qquad (7.152)$$

Notice that the radial integral *factors*. Given any two solutions in a central field in Eq. (7.151), one can compute the transition rate for photomission in the dipole approximation from this result.

7.4.4.4 *Average Over Initial States*

What about the *initial states* of the target? Suppose there is a g-fold degeneracy, such that (see Fig. 7.6)

$$\hat{H}_0^{\text{part}} |\phi_{i,\nu}\rangle = \varepsilon_i |\phi_{i,\nu}\rangle \qquad ; \nu = 1, 2, \cdots, g \qquad (7.153)$$

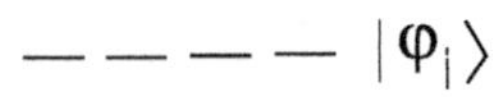

$$- - - - \; |\varphi_i\rangle$$

Fig. 7.6   Degeneracy in the initial state. Here $g = 4$.

The state vector for the initial system can then be written as

$$|\psi_i\rangle = \sum_{\nu=1}^g c_\nu |\phi_{i,\nu}\rangle \qquad ; \text{depends on preparation}$$

$$\sum_\nu |c_\nu|^2 = 1 \qquad (7.154)$$

The coefficients c_ν depend on just *how the initial state is prepared*.

The required square of the transition matrix element now takes the form

$$|\langle \phi_f | \hat{H}_1 | \psi_i \rangle|^2 = \sum_\nu \sum_{\nu'} c_\nu c_{\nu'}^\star \langle \phi_f | \hat{H}_1 | \phi_{i,\nu} \rangle \langle \phi_f | \hat{H}_1 | \phi_{i,\nu'} \rangle^\star \qquad (7.155)$$

The combination $c_\nu c_{\nu'}^\star$ is referred to as the *density matrix* for the initial state

$$\rho_{\nu\nu'} \equiv c_\nu c_{\nu'}^\star \qquad ; \text{density matrix} \qquad (7.156)$$

Since the initial state is normalized, one has[30]

$$\sum_\nu \rho_{\nu\nu} = \sum_\nu |c_\nu|^2 = 1 \qquad (7.157)$$

We observe that

- The density matrix depends on just how the initial state is prepared;
- Introduce the modulus and phase of the coefficients c_ν

$$c_\nu = \rho_\nu e^{i\phi_\nu}$$
$$\rho_{\nu\nu'} = \rho_\nu \rho_{\nu'} e^{i(\phi_\nu - \phi_{\nu'})} \qquad (7.158)$$

Then with a *random ensemble*, which is randomly prepared, the ensemble average of the density matrix will be

$$\left\langle \rho_\nu \rho_{\nu'} e^{i(\phi_\nu - \phi_{\nu'})} \right\rangle_{\text{ensemble}} = \delta_{\nu,\nu'} \langle \rho_\nu^2 \rangle_{\text{ensemble}} \qquad ; \text{ ensemble average}$$
$$(7.159)$$

This is the *random-phase approximation*. It assumes that $e^{i(\phi_\nu - \phi_{\nu'})}$ will average to zero over the ensemble if $\nu \neq \nu'$;

- With no other information, the degenerate states are then given *equal a priori probability*

$$\langle \rho_{\nu\nu'} \rangle_{\text{random ensemble}} = \frac{1}{g} \delta_{\nu,\nu'} \qquad ; \text{ equal a priori probability}$$
$$(7.160)$$

- Equation (7.155) now becomes

$$\left\langle |\langle \phi_f | \hat{H}_1 | \psi_i \rangle|^2 \right\rangle_{\text{random ensemble}} = \frac{1}{g} \sum_\nu |\langle \phi_f | \hat{H}_1 | \phi_{i,\nu} \rangle|^2 \qquad (7.161)$$

- This implies that in Eq. (7.152) one should take

$$\frac{1}{g} \sum_{m_i} \sum_{m_f} = \frac{1}{2l_i + 1} \sum_{m_i} \sum_{m_f} \qquad (7.162)$$

Hence the total transition rate for photoemission ω_{fi}, in the dipole

---

[30]In matrix notation, $\sum_\nu \rho_{\nu\nu} = \text{Trace}\,\underline{\rho}$.

approximation, becomes

$$\omega_{fi} = \frac{16\pi\alpha}{9}\frac{\omega_k^3}{c^2}\left|\int_0^\infty r^2 dr\, R_{n_f l_f}(r) r R_{n_i l_i}(r)\right|^2 \frac{1}{2l_i+1}\sum_{m_i}\sum_{m_f}\sum_m \times$$

$$\left|\int d\Omega_x\, Y_{l_f,m_f}^\star(\Omega_x) Y_{1,m}(\Omega_x) Y_{l_i,m_i}(\Omega_x)\right|^2 \quad ;\text{ total transition rate}$$

$$\text{; dipole approx}\quad (7.163)$$

On the basis of the above arguments, we come up with the *important general rule*. In calculating rates and cross sections, one must:

<u>Average</u> over initial states, if randomly prepared

<u>Sum</u> over final states, that get into detector

$$(7.164)$$

7.4.4.5 *Example: 2p → 1s Transition in H-Like Atom*

As an application of Eq. (7.163), consider the $2p \to 1s$ transition in a hyrdogen-like atom (see Fig. 7.7).

Fig. 7.7 $2p \to 1s$ transition in H-like atom.

(1) The radial wave functions are given in Prob. 2.22

$$R_{1s}(r) = \frac{1}{\sqrt{2}}\left(\frac{2Z}{a_0}\right)^{3/2} e^{-Zr/a_0} \qquad ; a_0 = \frac{1}{\alpha}\frac{\hbar}{m_0 c}$$

$$R_{2p}(r) = \frac{1}{\sqrt{3}}\left(\frac{Z}{2a_0}\right)^{3/2}\frac{Zr}{a_0}e^{-Zr/2a_0} \qquad (7.165)$$

The required radial integral is then evaluated as[31]

$$\int_0^\infty r^2 dr \, R_{2p}(r) r R_{1s}(r) = \frac{1}{\sqrt{6}} \left(\frac{Z}{a_0}\right)^4 \int_0^\infty r^4 e^{-3Zr/2a_0} \, dr$$

$$= \frac{1}{\sqrt{6}} \left(\frac{a_0}{Z}\right) \left(\frac{2}{3}\right)^5 \int_0^\infty t^4 e^{-t} \, dt$$

$$= \frac{1}{\sqrt{6}} \frac{2^8}{3^4} \frac{a_0}{Z} \tag{7.166}$$

(2) The angular integral is easy in this case

$$\int d\Omega_x \, Y_{00}^\star(\Omega_x) Y_{1,m}(\Omega_x) Y_{l_i,m_i}(\Omega_x) = \frac{(-1)^m}{\sqrt{4\pi}} \int d\Omega_x Y_{1,-m}^\star(\Omega_x) Y_{l_i,m_i}(\Omega_x)$$

$$= \frac{(-1)^m}{\sqrt{4\pi}} \delta_{m_i,-m} \, \delta_{l_i,1} \tag{7.167}$$

Note the angular momentum selection rule that only a p-state with $l_i = 1$ can make a transition to the $1s$ ground state;

(3) Since $l_i = 1$ and $l_f = 0$, one has[32]

$$\frac{1}{2l_i + 1} \sum_{m_i} \sum_{m_f} \sum_m \delta_{m_i,-m} = \frac{1}{3} 3 = 1 \tag{7.168}$$

(4) The energy levels of the H-like atom, and photon energy, are given by

$$\varepsilon_n = -\frac{1}{2} \frac{(Z\alpha)^2 m_0 c^2}{n^2}$$

$$\hbar\omega_k = \varepsilon_{2p} - \varepsilon_{1s} = \frac{3}{8}(Z\alpha)^2 m_0 c^2 \quad ; \text{ photon energy} \tag{7.169}$$

A combination of these results evaluates the total decay rate for the $2p \to 1s$ transition in H-like atoms from Eq. (7.163) as

$$\omega(2p \to 1s) = \frac{16\pi\alpha}{9c^2} \left[\frac{3^3}{2^9}(Z\alpha)^6 \left(\frac{m_0 c^2}{\hbar}\right)^3\right] \frac{1}{4\pi} \left[\frac{1}{6} \frac{2^{16}}{3^8} \frac{1}{(Z\alpha)^2} \left(\frac{\hbar}{m_0 c}\right)^2\right]$$

$$\tag{7.170}$$

This simplifies to

$$\omega(2p \to 1s) = \left(\frac{2}{3}\right)^8 \alpha(Z\alpha)^4 \frac{m_0 c^2}{\hbar} \quad ; \text{ decay rate} \tag{7.171}$$

---

[31] Use $\int_0^\infty t^n e^{-t} dt = n!$.

[32] Note $[\delta_{m_i,-m}]^2 = \delta_{m_i,-m}$.

7.4.4.6   *Lifetimes*

Suppose one has N systems in the state $|\phi_i\rangle$. The previous analysis calculated the *decay rate* ω_{fi}, that is, the probability per unit time that a single system will decay. The number of decays of the N systems in the time interval dt is therefore

$$dN = -N\omega_{fi}\, dt \qquad ;\text{ number of decays} \qquad (7.172)$$

This differential equation can be integrated to give

$$N = N_0 e^{-\omega_{fi}t} \equiv N_0 e^{-t/\tau} \qquad (7.173)$$

This relation defines the *mean life* τ

$$\tau = \frac{1}{\omega_{fi}} \qquad ;\text{ mean life} \qquad (7.174)$$

Suppose there are *many final states* for the decay out of the state $|\phi_i\rangle$ (see Fig. 7.8).

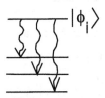

Fig. 7.8   Photoemission where there are many final states for decay out of $|\phi_i\rangle$.

Then

$$dN = -\sum_f N\omega_{fi}\, dt = -N\left(\sum_f \omega_{fi}\right) dt \qquad (7.175)$$

A repetition of the above leads to

$$N = N_0 e^{-\left(\sum_f \omega_{fi}\right)t} \equiv N_0 e^{-t/\tau}$$
$$\tau = \frac{1}{\sum_f \omega_{fi}} \qquad ;\text{ mean life} \qquad (7.176)$$

Recall that for an electron

$$\frac{\hbar}{m_e c} = 3.862 \times 10^{-11} \, \text{cm}$$

$$c = 2.998 \times 10^{10} \, \frac{\text{cm}}{\text{sec}}$$

$$\alpha^{-1} = 137.0 \tag{7.177}$$

The units of ω_{fi} are sec^{-1}, and the value of the transition rate and mean life for the $2p \rightarrow 1s$ transition in hydrogen calculated from Eq. (7.171) are shown in the second column of Table 7.1.

Table 7.1 $2p \rightarrow 1s$ decay rates and mean lives.

	Hydrogen	μ-Mesic Atom
$\omega(2p \rightarrow 1s)$	6.28×10^8 sec^{-1}	$1.30 \times 10^{11} Z^4$ sec^{-1}
$\tau(2p \rightarrow 1s)$	1.59×10^{-9} sec	$7.71 \times 10^{-12} Z^{-4}$ sec

The corresponding values for the $2p \rightarrow 1s$ transition in a μ-mesic atom are shown in the third column of Table 7.1.[33] Note the competing effects:

- The H-atom is bigger, and has a larger *transition dipole moment*;[34]
- There is a larger *phase space* for the μ-mesic decay, and the decay rate goes as ω^3.

At least for this author, it is truly remarkable that starting from scratch, one is able to calculate the lifetime of the $2p$-state in hydrogen.

7.4.4.7   Dipole Sum Rule

Recall the expression for the decay rate for photoemission in the dipole approximation in Eq. (7.141), which depends on the transition dipole matrix elements $\langle \phi_f | \mathbf{x} | \phi_i \rangle$

$$\mathcal{R}_{fi} = \frac{\alpha}{2\pi} \frac{\omega_k^3}{c^2} \left| \mathbf{e}_{\mathbf{k}s} \cdot \langle \phi_f | \mathbf{x} | \phi_i \rangle \right|^2 d\Omega_k \qquad ; \text{dipole approx} \quad (7.178)$$

Here we work in the coordinate representation, and simply refer to the position and momentum of the system as $(\mathbf{x}, \mathbf{p})$. The canonical commutation relations are

$$[p_i, \, x_j] = \frac{\hbar}{i} \delta_{ij} \qquad ; \text{C.C.R.} \qquad (7.179)$$

---

[33] Use $m_\mu / m_e = 206.7$.
[34] Compare Fig. 4.11.

The unperturbed system hamiltonian is

$$H_0 \equiv H_0^{\text{part}} = \frac{\mathbf{p}^2}{2m_0} - \frac{Ze^2}{|\mathbf{x}|} + V(|\mathbf{x}|) \tag{7.180}$$

Consider, as before,

$$\frac{i}{\hbar}[H_0, x_i] = \frac{1}{m_0} p_i \tag{7.181}$$

Now take one more commutator

$$\left(\frac{i}{\hbar}\right)^2 [[H_0, x_i], x_j] = \frac{1}{m_0} \delta_{ij} \tag{7.182}$$

This is *just a number!* It follows that

$$[[H_0, x_j], x_j] = -\frac{3\hbar^2}{m_0} \tag{7.183}$$

Now take the matrix element of this expression in the ground state $|\phi_i\rangle$, and insert a complete set $\sum_n |\phi_n\rangle\langle\phi_n|$ of eigenstates of H_0[35]

$$\sum_n \left[\langle\phi_i|[H_0, x_j]|\phi_n\rangle\langle\phi_n|x_j|\phi_i\rangle - \langle\phi_i|x_j|\phi_n\rangle\langle\phi_n|[H_0, x_j]|\phi_i\rangle\right] = -\frac{3\hbar^2}{m_0}$$
$$\tag{7.184}$$

Let H_0 act on the eigenstates, using

$$H_0|\phi_i\rangle = \varepsilon_i|\phi_i\rangle \qquad ; H_0|\phi_n\rangle = \varepsilon_n|\phi_n\rangle \tag{7.185}$$

The two terms in Eq. (7.184) then *add*, and the result is

$$\sum_n (\varepsilon_n - \varepsilon_i) \, |\langle\phi_n|\mathbf{x}|\phi_i\rangle|^2 = \frac{3\hbar^2}{2m_0} \qquad ; \text{dipole sum rule} \tag{7.186}$$

A few comments:

- This is the Thomas-Reiche-Kuhn (TRK) *dipole sum rule;*
- The sum over the dipole matrix elements is *energy-weighted* with a factor $(\varepsilon_n - \varepsilon_i)$, and since $|\phi_i\rangle$ is the ground state, the contributions on the l.h.s. are positive definite;
- This sum rule holds for any central potential $V(|\mathbf{x}|)$, no matter how complicated, and therefore it holds no matter how complicated the wave functions;

---

[35]Although we intend to evaluate the dipole matrix element in Eq. (7.186) in the coordinate representation, the derivation is presented in abstract form.

- The sum rule tells us about the summed transition dipole strength to the ground state, for example, the summed strength for all $np \rightarrow 1s$ transitions in hydrogen.[36]

7.4.5 Planck Distribution Law

The celebrated Planck distribution law, an originally phenomenological result that formed a starting point for quantum mechanics,[37] can now be obtained by a detailed-balance argument on photon emission and absorption rates. The argument goes as follows:

(1) Suppose one has *many independent systems* in states $|\phi_1\rangle$ and $|\phi_2\rangle$ in a cavity, as well as many photons in the mode $\{\mathbf{k}s\}$, where $|\mathbf{k}|$ is determined by energy conservation. Suppose, further, that the whole ensemble is in *thermal equilibrium* at a temperature T, so that its properties do not change with time (see Fig. 7.9).

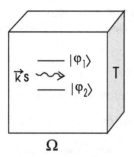

Fig. 7.9 Thermodynamic equilibrium with many independent systems in states $|\phi_1\rangle$ and $|\phi_2\rangle$ in a cavity of volume Ω and temperature T, as well as many photons in the mode $\{\mathbf{k}s\}$ where $|\mathbf{k}|$ is determined by energy conservation.

At equilibrium, let there be

$$N_1 \text{ systems in state } |\phi_1\rangle$$
$$N_2 \text{ systems in state } |\phi_2\rangle$$
$$n_{\mathbf{k}s} \text{ photons in mode } \{\mathbf{k}s\} \tag{7.187}$$

(2) Now focus on one of the independent systems. A system in the state

---

[36]Compare Prob. 7.10.
[37]See Vol. I.

$|\phi_1\rangle$ can *emit* a photon and go to the state $|\phi_2\rangle$. In this case, the initial and final states of the system and photons in the abstract Hilbert space are

$$|\Phi_i\rangle = |\phi_1; \cdots, n_{\mathbf{ks}}, \cdots\rangle$$
$$|\Phi_f\rangle = |\phi_2; \cdots, n_{\mathbf{ks}} + 1, \cdots\rangle \qquad (7.188)$$

From the commutation relations, we know that the matrix element of the photon creation operator is

$$\langle n_{\mathbf{ks}} + 1|\hat{a}_{\mathbf{ks}}^\dagger|n_{\mathbf{ks}}\rangle = \sqrt{n_{\mathbf{ks}} + 1} \qquad (7.189)$$

The *rate* for this process is given by Eqs. (7.115) and (7.121) as[38]

$$R_{1\to 2} = \frac{2\pi}{\hbar}\left(\frac{2\pi\hbar c^2}{\omega_k\Omega}\right)\left(\frac{e}{c}\right)^2 (n_{\mathbf{ks}} + 1) \times \qquad (7.190)$$

$$\left|\mathbf{e}_{\mathbf{ks}} \cdot \int d^3x\, e^{-i\mathbf{k}\cdot\mathbf{x}}\langle\phi_2|\hat{\mathbf{S}}(\mathbf{x})|\phi_1\rangle\right|^2 \delta(\varepsilon_1 - \varepsilon_2 - \hbar\omega)$$

(3) Similarly, a system in the state $|\phi_2\rangle$ can *absorb* a photon and go to the state $|\phi_1\rangle$. In this case, the initial and final states in Eqs. (7.188) become

$$|\Phi_i\rangle = |\phi_2; \cdots, n_{\mathbf{ks}}, \cdots\rangle$$
$$|\Phi_f\rangle = |\phi_1; \cdots, n_{\mathbf{ks}} - 1, \cdots\rangle \qquad (7.191)$$

Again, we know that the matrix element of the destruction operator is

$$\langle n_{\mathbf{ks}} - 1|\hat{a}_{\mathbf{ks}}|n_{\mathbf{ks}}\rangle = \sqrt{n_{\mathbf{ks}}} \qquad (7.192)$$

The rate for this process is

$$R_{2\to 1} = \frac{2\pi}{\hbar}\left(\frac{2\pi\hbar c^2}{\omega_k\Omega}\right)\left(\frac{e}{c}\right)^2 (n_{\mathbf{ks}}) \times \qquad (7.193)$$

$$\left|\mathbf{e}_{\mathbf{ks}} \cdot \int d^3x\, e^{i\mathbf{k}\cdot\mathbf{x}}\langle\phi_1|\hat{\mathbf{S}}(\mathbf{x})|\phi_2\rangle\right|^2 \delta(\varepsilon_2 + \hbar\omega - \varepsilon_1)$$

(4) The condition for thermodynamic equilibrium can be obtained by *detailed balance*. The requirement that the rates in the two directions be balanced for the ensemble is

$$N_1 R_{1\to 2} = N_2 R_{2\to 1} \qquad ; \text{ detailed balance} \qquad (7.194)$$

---

[38]In the factor $(n_{\mathbf{ks}} + 1)$, the first term gives rise to the *stimulated*, and the second the *spontaneous*, photoemission. Stimulated emission is particularly relevant for a laser.

(5) The Boltzmann distribution can now be invoked *for the systems*

$$\frac{N_1}{N_2} = \frac{e^{-\varepsilon_1/k_B T}}{e^{-\varepsilon_2/k_B T}} = e^{-(\varepsilon_1 - \varepsilon_2)/k_B T} = e^{-\hbar\omega/k_B T}$$

; Boltzmann for systems   (7.195)

(6) A combination of the above gives

$$\frac{R_{2\to 1}}{R_{1\to 2}} = \frac{n_{ks}}{n_{ks} + 1} = \frac{N_1}{N_2} = e^{-\hbar\omega/k_B T} \qquad (7.196)$$

Hence

$$n_{ks} = \frac{1}{e^{\hbar\omega/k_B T} - 1} \qquad \text{; Planck distribution} \qquad (7.197)$$

This is indeed the Planck distribution for the photons.

7.4.6 *Photoabsorption*

Consider the *absorption* of a photon with $\{ks\}$ by a system that makes a transition from a state $|\phi_i\rangle$ to a state $|\phi_f\rangle$ (see Fig. 7.10).

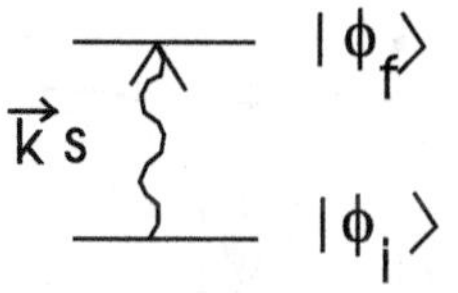

Fig. 7.10   Photoabsorption where the system goes from $|\phi_i\rangle$ to $|\phi_f\rangle$.

The initial and final states in the abstract Hilbert space are

$$|\Phi_i\rangle = |\phi_i; 0, \cdots, 1_{ks}, \cdots, 0\rangle \qquad \text{; system in } |\phi_i\rangle; \ 1 \text{ photon } \{ks\}$$
$$|\Phi_f\rangle = |\phi_f; 0, 0, \cdots, 0\rangle \qquad \text{; system in } |\phi_f\rangle; \text{ no photons}$$

(7.198)

The incident flux is calculated from one photon in the volume Ω

$$I_{\text{inc}} = \rho v = \frac{1}{\Omega} c \qquad \text{; incident flux} \qquad (7.199)$$

The transition rate is given in Eq. (7.193), and the *photoabsorption cross section* is then given by

$$\sigma_{fi}(\omega_k) = \frac{R_{fi}}{I_{\text{inc}}}$$

$$= \frac{2\pi}{\hbar} \left(\frac{2\pi\hbar c^2}{\omega_k \Omega}\right) \left(\frac{e}{c}\right)^2 \left|\mathbf{e}_{\mathbf{k}s} \cdot \int d^3x\, e^{i\mathbf{k}\cdot\mathbf{x}} \langle \phi_f | \hat{\mathbf{S}}(\mathbf{x}) | \phi_i \rangle\right|^2 \times$$

$$\delta(\varepsilon_i + \hbar\omega_k - \varepsilon_f)\frac{\Omega}{c} \tag{7.200}$$

This yields[39]

$$\int_{\text{abs line}} d\omega_k\, \sigma_{fi}(\omega_k) = \frac{(2\pi)^2 \alpha}{\omega_k^0} \left|\mathbf{e}_{\mathbf{k}s} \cdot \int d^3x\, e^{i\mathbf{k}\cdot\mathbf{x}} \langle \phi_f | \hat{\mathbf{S}}(\mathbf{x}) | \phi_i \rangle\right|^2 \tag{7.201}$$

This is *exactly the same answer* as was obtained for a classical electromagnetic field in Eqs. (6.153) and (6.176).

7.4.7 *Wigner-Weisskopf Theory of Line Width*

The transition rates recalled in Eqs. (7.190) and (7.193) have Dirac delta-functions in the energy. This implies an infinitely sharp photon line. In fact, the photon energy is spread out by the natural line width, as indicated in Fig. 7.11.

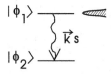

Fig. 7.11 Photoemission with natural line width.

We go through the theory of the line width due to [Wigner and Weisskopf (1930)]:

- These authors deal with time-dependent perturbation theory and explicitly include the *depletion of the initial state*;
- The physics is the following: the initial state has a finite lifetime, which implies that its energy cannot be specified precisely. This implies a spread in photon energies;[40]

---

[39]Here $\hbar\omega_k^0 \equiv \varepsilon_f - \varepsilon_i$.

[40]Compare Prob. 7.12.

- Their calculation, which includes radiation reaction on the initial state, makes this quantitative by computing the *shape of the spectral line.*

7.4.7.1 *Schrödinger Equation*

The Schrödinger equation based on the hamiltonian in Eqs. (7.104) is

$$i\hbar\frac{\partial}{\partial t}|\Psi(t)\rangle = \hat{H}|\Psi(t)\rangle \qquad ; \text{S-eqn} \qquad (7.202)$$

Expand the solution in the complete set of eigenstates of $\hat{H}_0$

$$|\Psi(t)\rangle = \sum_n c_n(t)e^{-iE_n t/\hbar}|\Phi_n\rangle$$

$$\hat{H}_0|\Phi_n\rangle = E_n|\Phi_n\rangle \qquad (7.203)$$

Substitute this expansion into the Schrödinger equation

$$\sum_n \left(i\hbar\dot{c}_n + E_n c_n\right)e^{-iE_n t/\hbar}|\Phi_n\rangle = \sum_n c_n e^{-iE_n t/\hbar}(E_n + \hat{H}_1)|\Phi_n\rangle \qquad (7.204)$$

Now use the orthonormality of the states to project out $\dot{c}_n = dc_n(t)/dt$

$$i\hbar\frac{dc_n(t)}{dt} = \sum_m \langle\Phi_n|\hat{H}_1|\Phi_m\rangle e^{i(E_n - E_m)t/\hbar}\, c_m(t) \qquad (7.205)$$

- Previously, we used Dirac perturbation theory with $c_m(t) \approx c_m(0) = \delta_{mi}$ to solve these equations;
- Now an alternative procedure is employed, and these equations are *solved in a truncated basis*

7.4.7.2 *Formulation in Truncated Basis*

The problem is defined through the following steps:

(1) Truncate the basis in the abstract Hilbert space to the following states

$$|\phi_1; 0, 0, \cdots, 0\rangle \equiv |\phi_1\rangle \qquad ; \text{system in excited state}$$

$$|\phi_2; 0, \cdots, 1_{\mathbf{k}s}, \cdots, 0\rangle \equiv |\phi_2; \mathbf{k}s\rangle \qquad ; \text{system in ground state,}$$
$$\text{one photon in } any \ \{\mathbf{k}s\} \qquad (7.206)$$

(2) Keep $\hat{H}_1$ to order e, which includes terms that both emit and absorb photons, and define

$$(\varepsilon_1 - \varepsilon_2) \equiv \hbar\omega_{12} \quad ; \ kc \equiv \omega \qquad (7.207)$$

Equations (7.205) then become

$$i\hbar\frac{dc_{2,\mathbf{ks}}(t)}{dt} = \langle\phi_2;\mathbf{ks}|\hat{H}_1|\phi_1\rangle e^{-i(\omega_{12}-\omega)t}\,c_1(t) \qquad ; \text{ all } \{\mathbf{ks}\}$$

$$i\hbar\frac{dc_1(t)}{dt} = \sum_{\mathbf{ks}}\langle\phi_1|\hat{H}_1|\phi_2;\mathbf{ks}\rangle e^{i(\omega_{12}-\omega)t}\,c_{2,\mathbf{ks}}(t) \qquad (7.208)$$

Since the first equation holds for all $\{\mathbf{ks}\}$, this is an infinite set of coupled first-order differential equations in the time;

(3) If the system is prepared in the state $|\phi_1\rangle$ at time $t = 0$, then one has the *initial conditions*

$$c_1(0) = 1 \qquad ; \text{ initial conditions}$$

$$c_{2,\mathbf{ks}}(0) = 0 \qquad (7.209)$$

7.4.7.3   Solution

We *guess* the solution

$$c_1(t) = e^{-\gamma t/2} \qquad ; \text{ guess solution} \qquad (7.210)$$

Now plug this in, and show that it satisfies the equations.

(1) Substitute Eq. (7.210) into the first of Eqs. (7.208), integrate on the time, and use the second initial condition in Eqs. (7.209)

$$i\hbar c_{2,\mathbf{ks}}(t) = -\langle\phi_2;\mathbf{ks}|\hat{H}_1|\phi_1\rangle\left[\frac{e^{-[\gamma/2+i(\omega_{12}-\omega)]t} - 1}{\gamma/2 + i(\omega_{12} - \omega)}\right] \qquad (7.211)$$

(2) Substitute this result into the second of Eqs. (7.208)

$$i\hbar\left(-\frac{\gamma}{2}\right)e^{-\gamma t/2} = \frac{1}{i\hbar}\sum_{\mathbf{ks}}\frac{|\langle\phi_1|\hat{H}_1|\phi_2;\mathbf{ks}\rangle|^2}{\gamma/2 + i(\omega_{12} - \omega)}\left[e^{-i[(\omega-\omega_{12})+i\gamma/2]t} - 1\right]e^{-\gamma t/2}$$

$$(7.212)$$

Note the role of the terms has been interchanged in the last two factors. Cancel the factors of $e^{-\gamma t/2}$, and define the deviation of the photon angular frequency ω from the difference of eigenvalues $\omega_{12} = (\varepsilon_1 - \varepsilon_2)/\hbar$ as[41]

$$\nu \equiv \omega - \omega_{12} \qquad ; \text{ photon angular frequency} \qquad (7.213)$$

---

[41]Note that, internal to this calculation, ν is the deviation of photon *angular* frequencies.

This gives

$$i\hbar\left(-\frac{\gamma}{2}\right) = \frac{1}{i\hbar}\sum_{ks}|\langle\phi_1|\hat{H}_1|\phi_2;ks\rangle|^2\left[\frac{e^{-i(\nu+i\gamma/2)t}-1}{-i(\nu+i\gamma/2)}\right] \quad (7.214)$$

- For this to be a solution, the r.h.s. must be *independent of time*;
- Make the replacement

$$\sum_k \rightarrow \frac{\Omega}{(2\pi)^3}\int\cdots\int d^3k$$

$$= \int\cdots\int dn_f = \int\cdots\int\left(\frac{\partial n_f}{\partial\nu}\right)d\nu \quad (7.215)$$

Thus Eq. (7.214) becomes

$$\gamma = \int d\nu\left[\frac{2}{\hbar^2}\int\sum_s|\langle\phi_1|\hat{H}_1|\phi_2;ks\rangle|^2\left(\frac{\partial n_f}{\partial\nu}\right)\right]\left[\frac{e^{-i(\nu+i\gamma/2)t}-1}{-i(\nu+i\gamma/2)}\right]$$
$$(7.216)$$

Here the integral inside the square brackets goes over $d\Omega_k$.[42]

(3) Now observe the following:

- If $\gamma \to 0$ and $t \to \infty$, the last factor in the integrand in Eq. (7.216) is *strongly peaked about* $\nu = 0$;
- If the first factor in the integral is a slowly varying function of ν, it can then be removed from the integral;
- The essence of the calculation will not be lost if a *further* approximation is made to truncate the sum over photon states to $\int_{-\nu_0}^{\nu_0}d\nu$.

Equation (7.216) then becomes

$$\gamma = \left[\frac{2}{\hbar^2}\int\sum_s|\langle\phi_1|\hat{H}_1|\phi_2;ks\rangle|^2\left(\frac{\partial n_f}{\partial\nu}\right)\right]_{\nu=0}I$$

$$I \equiv \int_{-\nu_0}^{\nu_0}d\nu\left[\frac{e^{-i(\nu+i\gamma/2)t}-1}{-i(\nu+i\gamma/2)}\right] \quad (7.217)$$

Call $z \equiv \nu + i\gamma/2$, and consider I as an integral in the complex z-plane along the contour C_1 running from $-\nu_0 + i\gamma/2$ to $\nu_0 + i\gamma/2$ (see Fig. 7.12)

$$I = \int_{C_1}dz\left(\frac{e^{-izt}-1}{-iz}\right) \quad ; z \equiv \nu + \frac{i\gamma}{2} \quad (7.218)$$

---

[42]Recall Eq. (7.127) and Eqs. (7.142).

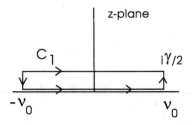

Fig. 7.12   Contour integral for I. The contour C_1 runs from $-\nu_0 + i\gamma/2$ to $\nu_0 + i\gamma/2$.

This integral can be analyzed as follows:

- The integrand is analytic inside the closed contour in Fig. 7.12;[43]
- Thus the contour C_1 can be shifted down as indicated;
- If $\gamma/\nu_0 \ll 1$, the contributions from the vertical segments on the two ends can be neglected. This yields an integral along the x-axis

$$
\begin{aligned}
I &= \int_{-\nu_0}^{\nu_0} dx \left(\frac{e^{-ixt} - 1}{-ix} \right) && ; \, x \equiv \nu \\
&= \int_{-\nu_0}^{\nu_0} dx \left[\frac{\cos xt - 1}{-ix} + \frac{\sin xt}{x} \right] \\
&= \int_{-\nu_0 t}^{\nu_0 t} du \left[\frac{\cos u - 1}{-iu} + \frac{\sin u}{u} \right] && ; \, u \equiv \nu t \qquad (7.219)
\end{aligned}
$$

The first integral in the last line disappears since the integrand is odd, and as $\nu_0 t \to \infty$, the second integral is just π

$$
I = \int_{-\nu_0 t}^{\nu_0 t} du \, \frac{\sin u}{u} = \pi \qquad ; \, \nu_0 t \to \infty \qquad (7.220)
$$

- The integrand in this last expression is strongly peaked in the frequency range $\nu \ll 1/t$ (see Fig. 7.13), and one has to *wait long enough* so that $\nu_0 \gg 1/t$ to recover the result in Eq. (7.220). Equation (7.217) then gives

$$
\gamma = \frac{2\pi}{\hbar^2} \left[\int \sum_s |\langle \phi_1 | \hat{H}_1 | \phi_2 ; \mathbf{k}s \rangle|^2 \left(\frac{\partial n_f}{\partial \nu} \right) \right]_{\varepsilon_1 = \varepsilon_2 + \hbar\omega} \qquad (7.221)
$$

This is just our *previous result for the inverse lifetime*,[44] the lifetime

---

[43]There is no singularity at $z = 0$. We take $\nu_0 \sim \omega_{12}$ and assume $\gamma \ll \omega_{12}$.

[44]Compare Eqs. (7.126) and (7.174). Note $\partial n_f / \partial \hbar\nu = \partial n_f / \partial \hbar\omega = \partial n_f / \partial E_f$.

identification here following from

$$|c_1(t)|^2 = e^{-\gamma t} \qquad \text{; probability in state } |\phi_1\rangle \qquad (7.222)$$

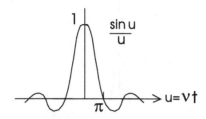

Fig. 7.13 Sketch of the integrand in Eq. (7.220).

(4) It follows from Eq. (7.211) that if $\gamma t \gg 1$, then

$$|c_{2,\mathbf{k}s}(t)|^2 = \frac{1}{\hbar^2}|\langle\phi_2; \mathbf{k}s|\hat{H}_1|\phi_1\rangle|^2 \frac{1}{\nu^2 + \gamma^2/4} \qquad ; \gamma t \gg 1 \qquad (7.223)$$

Again replace

$$\sum_{\mathbf{k}s} \rightarrow \int \cdots \int \sum_s \left(\frac{\partial n_f}{\partial \nu}\right) d\nu \qquad (7.224)$$

Then

$$\sum_{\mathbf{k}s} |c_{2,\mathbf{k}s}(t)|^2 = \int d\nu \left[\int \sum_s \frac{1}{\hbar^2}|\langle\phi_2; \mathbf{k}s|\hat{H}_1|\phi_1\rangle|^2 \left(\frac{\partial n_f}{\partial \nu}\right)\right] \frac{1}{\nu^2 + \gamma^2/4}$$

$$\equiv \int d\nu\, N(\nu) \qquad \text{; probability in state } |\phi_2\rangle,$$

$$\text{photon present} \qquad (7.225)$$

- This is the probability of finding the system in the state $|\phi_2\rangle$, and a photon present, after a time $\gamma t \gg 1$;
- Since $\nu = \omega - \omega_{12}$, this provides the *distribution of angular frequency* of the radiation (recall $\omega = kc$);
- With the assumption that γ is small, and the factor in square brackets in the first of Eqs. (7.225) is a slowly varying function of ν near $\nu = 0$, this result can be re-written

$$N(\nu) = \frac{\gamma}{2\pi} \frac{1}{\nu^2 + \gamma^2/4} \qquad \text{; line shape} \qquad (7.226)$$

where γ is identified from Eq. (7.221). This expression provides the *shape of the spectral line* (see Fig. 7.14);[45]

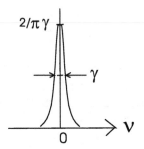

Fig. 7.14   Shape of the spectral line $N(\nu)$ in Eq. (7.226). Here $\nu = \omega - \omega_{12}$ is the angular frequency of the radiation. (Compare Fig. 5.41.)

• Note

$$\int_{-\infty}^{\infty} N(\nu)d\nu = \frac{\gamma}{2\pi}\frac{2}{\gamma}\int_{-\infty}^{\infty}\frac{dx}{1+x^2} \qquad ; \; x \equiv \frac{2\nu}{\gamma}$$
$$= 1 \qquad\qquad\qquad (7.227)$$

This represents *conservation of probability*. After a long time, all of the initial state $|\phi_1\rangle$ must end up in the final state $|\phi_2\rangle$ with a photon present;[46]

• It is readily established that as the width $\gamma \to 0$, the function $N(\nu)$ produces a Dirac delta-function (see Prob. 7.14)

$$\mathrm{Lim}_{\gamma\to 0}\, N(\nu) = \delta(\nu) \qquad ; \; \gamma \to 0 \qquad (7.228)$$

This provides the basis for what we did in integrating the photoabsorption cross section for the two levels over the photon frequency.

---

[45]Since $\gamma^2 \propto e^4$, one has a term $\sim \alpha^2$ in the *denominator* of the expression for the line shape. This can only be obtained by summing an infinite number of terms in perturbation theory. This was accomplished here by solving the time-dependent Schrödinger equation in a truncated basis.

[46]Note that the results in Eqs. (7.221) and (7.226) are *independent of* ν_0, as long as $\gamma/\nu_0 \ll 1$. See also Prob. 7.13.

PART 3

Relativistic Quantum Field Theory

Chapter 8

Discrete Symmetries

Chapter 6 of Vol. II contains a discussion of the role of continuous symmetries in a relativistic quantum field theory based on a local, Lorentz-invariant lagrangian density. In this chapter, that discussion is extended to the discrete symmetries of parity P, charge conjugation, or more properly particle-antiparticle conjugation C, and time reversal T. All of these are individually symmetries of the observed strong and electromagnetic interactions. In 1956, based on some puzzling experimental results, [Lee and Yang (1956)] suggested that *parity* is not conserved in the *weak interactions*. This proposal was subsequently brilliantly confirmed by an experiment on the β-decay of ^{60}Co [Wu, Ambler, Hayward, Hoppes, and Hudson (1957)].

The parity-violation result set off a flurry of work on the role of discrete symmetries in quantum field theory. It was immediately realized that there existed a previously-somewhat-academic theorem [Schwinger (1951); Lüders (1954)] that with a local, Lorentz-invariant lagrangian density, the *combined* operation CPT would always remain a good symmetry. Lee and Yang (among others) showed that it is a consequence of this theorem that particles and antiparticles must have identical masses and lifetimes.

In 1964, a violation of CP-invariance was observed in the neutral kaon system [Christenson, Cronin, Fitch, and Turlay (1964)]. Many subsequent experiments have similarly observed CP-violation [Particle Data Group (2012)]. CP-violation is incorporated theoretically in the standard model of the strong and electroweak interactions through an element in the quark mixing matrix.[1] The role of CP-violation in the ultra-weak interactions in the neutrino sector is the subject of active ongoing investigations.

Direct measurement of T-violation is difficult, and there is continuing experimental effort in this regard. The most celebrated experiments here

---

[1]See, for example, [Walecka (2004)].

335

are the attempts to measure the electric dipole moments of the electron and neutron, an electric dipole moment being forbidden by T-invariance.[2]

The latest experimental results in all these areas can always be found in the most recent version of [Particle Data Group (2012)].

The goal of this chapter is to provide an introduction to the theoretical description of discrete symmetries in quantum field theory. We start with the simplest case of a spin-zero field. The specific spin-zero fields we have in mind here are the *pions* $(\pi^0, \pi^{\pm})$.

8.1 Spin-Zero Field

A spin-zero field is expanded in terms of creation and destruction operators in the interaction picture according to[3]

$$\hat{\phi}(x) = \sum_{\mathbf{k}} \left(\frac{\hbar}{2\omega_k \Omega} \right)^{1/2} \left(c_{\mathbf{k}} e^{ik \cdot x} + c_{\mathbf{k}}^{\dagger} e^{-ik \cdot x} \right) \qquad ; \text{ neutral}$$

$$= \sum_{\mathbf{k}} \left(\frac{\hbar}{2\omega_k \Omega} \right)^{1/2} \left(a_{\mathbf{k}} e^{ik \cdot x} + b_{\mathbf{k}}^{\dagger} e^{-ik \cdot x} \right) \qquad ; \text{ charged} \qquad (8.1)$$

Here $k \cdot x = \mathbf{k} \cdot \mathbf{x} - \omega_k t$.

8.1.1 *Parity*

We will define an operation on the states. If one knows what an operator does to all the states, then the operator is well-defined. All the operations are defined in the abstract Hilbert space; they work there.

We want the operation to preserve the norm of the states. This implies that the defined operator should be *unitary*

$$\hat{P}^{\dagger} = \hat{P}^{-1} \qquad ; \text{ unitary} \qquad (8.2)$$

For parity, we seek an operator that *reflects the space coordinates*. Consider the neutral one-particle state

$$|\mathbf{k}\rangle = c_{\mathbf{k}}^{\dagger}|0\rangle \qquad ; \text{ one-particle state} \qquad (8.3)$$

---

[2]For a proof of this fact, see appendix E of [Walecka (2001)]; see also Prob. 9.5.

[3]The interaction picture is also the Heisenberg picture for the free field. We now suppress the hats on the creation and destruction operators since their nature is evident.

Define an operator that reverses all momenta and multiplies by a phase η_p

$$\hat{P}|\mathbf{k}\rangle \equiv \eta_p|-\mathbf{k}\rangle \qquad (8.4)$$

Two applications of the operator should restore the original state

$$\hat{P}^2|\mathbf{k}\rangle = |\mathbf{k}\rangle$$
$$\implies \quad \hat{P}^2 = 1 \qquad (8.5)$$

The phase must therefore satisfy

$$\eta_p^2 = 1$$
$$\implies \quad \eta_p = \pm 1 \qquad (8.6)$$

Now write Eq. (8.4) as

$$\hat{P}c_{\mathbf{k}}^{\dagger}|0\rangle = \hat{P}c_{\mathbf{k}}^{\dagger}\hat{P}^{-1}\hat{P}|0\rangle = \hat{P}c_{\mathbf{k}}^{\dagger}\hat{P}^{-1}|0\rangle \qquad (8.7)$$

Here it has been observed that the vacuum should be invariant under the parity transformation

$$\hat{P}|0\rangle = |0\rangle \qquad \text{; vacuum invariant} \qquad (8.8)$$

In *summary*, we take the definition of the parity operator in this case to be

$$\hat{P}c_{\mathbf{k}}^{\dagger}\hat{P}^{-1} = \eta_p c_{-\mathbf{k}}^{\dagger} \qquad ; \ \hat{P}c_{\mathbf{k}}\hat{P}^{-1} = \eta_p c_{-\mathbf{k}}$$
$$\hat{P}|0\rangle = |0\rangle$$
$$\hat{P}^{\dagger} = \hat{P}^{-1} \qquad (8.9)$$

Several comments:

- We now know what the operator $\hat{P}$ does to any state formed from the $(c_{\mathbf{k}}^{\dagger}, c_{\mathbf{k}})$, and this defines the parity operator for a neutral spin-zero field;
- Exactly analogous relations hold for a charged spin-zero field

$$\hat{P}a_{\mathbf{k}}^{\dagger}\hat{P}^{-1} = \eta_p a_{-\mathbf{k}}^{\dagger} \qquad ; \ \hat{P}a_{\mathbf{k}}\hat{P}^{-1} = \eta_p a_{-\mathbf{k}}$$
$$\hat{P}b_{\mathbf{k}}^{\dagger}\hat{P}^{-1} = \eta_p b_{-\mathbf{k}}^{\dagger} \qquad ; \ \hat{P}b_{\mathbf{k}}\hat{P}^{-1} = \eta_p b_{-\mathbf{k}} \qquad (8.10)$$

- From the second of Eqs. (8.5), and the last of Eqs. (8.9), one has

$$\hat{P} = \hat{P}^{-1} = \hat{P}^{\dagger} \qquad \text{; hermitian} \qquad (8.11)$$

Hence the parity operator is also *hermitian*;

- In the *rest frame* of the particle

$$\hat{P}c_0^\dagger|0\rangle = \eta_p c_0^\dagger|0\rangle \qquad ; \text{eigenstate} \qquad (8.12)$$

Hence a particle at rest is in an *eigentate* of parity, and $\eta_p = \pm 1$ is the *intrinsic parity* of the particle. For the pseudoscalar pion

$$\eta_{p_\pi} = -1 \qquad ; \text{pion} \qquad (8.13)$$

- One can now compute the behavior of the field under the parity transformation

$$\hat{P}\hat{\phi}(\mathbf{x},t)\hat{P}^{-1} = \sum_k \left(\frac{\hbar}{2\omega_k\Omega}\right)^{1/2}\left[\left(\hat{P}c_\mathbf{k}\hat{P}^{-1}\right)e^{ik\cdot x} + \left(\hat{P}c_\mathbf{k}^\dagger\hat{P}^{-1}\right)e^{-ik\cdot x}\right]$$

$$= \sum_k \left(\frac{\hbar}{2\omega_k\Omega}\right)^{1/2}\eta_p\left(c_{-\mathbf{k}}e^{ik\cdot x} + c_{-\mathbf{k}}^\dagger e^{-ik\cdot x}\right) \qquad (8.14)$$

Now change dummy summation variables from $\mathbf{k} \to -\mathbf{k}$, and use $e^{ik\cdot\mathbf{x}} = e^{i(-\mathbf{k})\cdot(-\mathbf{x})}$

$$\hat{P}\hat{\phi}(\mathbf{x},t)\hat{P}^{-1} = \eta_p\hat{\phi}(-\mathbf{x},t) \qquad ; \text{parity} \qquad (8.15)$$

The parity operator *reflects the spatial location of the field*, exactly as we want it to!

- One could just as well have expanded the field in terms of the complete set of spherical single-particle wave functions $\phi_{nlm}(\mathbf{x})$

$$\hat{\phi}(\mathbf{x},t) = \sum_{nlm}\left(\frac{\hbar}{2\omega_{nl}}\right)^{1/2}\left[c_{nlm}\phi_{nlm}(\mathbf{x})e^{-i\omega_{nl}t} + c_{nlm}^\dagger\phi_{nlm}^\dagger(\mathbf{x})e^{i\omega_{nl}t}\right]$$

$$(8.16)$$

In this case, in order to have the correct transformation property of the field in Eq. (8.15), one would write

$$\hat{P}c_{nlm}\hat{P}^{-1} = \eta_p(-1)^l c_{nlm}$$
$$\hat{P}c_{nlm}^\dagger\hat{P}^{-1} = \eta_p(-1)^l c_{nlm}^\dagger \qquad (8.17)$$

and use the reflection property of the wave functions

$$(-1)^l\phi_{nlm}(\mathbf{x}) = \phi_{nlm}(-\mathbf{x}) \qquad (8.18)$$

Note that in this case, $c_{nlm}^\dagger|0\rangle$ is an *eigenstate* of parity

$$\hat{P}c_{nlm}^\dagger|0\rangle = \eta_p(-1)^l c_{nlm}^\dagger|0\rangle \qquad ; \text{eigenstate} \qquad (8.19)$$

- Consider next a *two-particle* state formed from a (π^+, π^-) particle-antiparticle pair in the center-of-momentum (C-M) frame. The most general state vector with angular momentum L is (see Prob. 8.1)

$$|ELM\rangle = \int d^3k \, a_E(k) Y_{LM}(\Omega_k) \, a_{\mathbf{k}}^\dagger b_{-\mathbf{k}}^\dagger |0\rangle \qquad (8.20)$$

Let the parity operator act on this state

$$\hat{P}|ELM\rangle = \int k^2 dk \, a_E(k) \int d\Omega_k Y_{LM}(\Omega_k) \, \hat{P} a_{\mathbf{k}}^\dagger b_{-\mathbf{k}}^\dagger |0\rangle$$

$$= \int k^2 dk \, a_E(k) \int d\Omega_k Y_{LM}(\Omega_k) \, \eta_p^2 a_{-\mathbf{k}}^\dagger b_{\mathbf{k}}^\dagger |0\rangle \qquad (8.21)$$

Now change variables $\mathbf{k} \to -\mathbf{k}$, and use $Y_{LM}(-\mathbf{k}/k) = (-1)^L Y_{LM}(\mathbf{k}/k)$

$$\hat{P}|ELM\rangle = (-1)^L |ELM\rangle \qquad \text{; eigenstate} \qquad (8.22)$$

This is again an eigenstate of parity, with eigenvalue $(-1)^L$.

8.1.2 Charge Conjugation

The operator $\hat{C}$ should change particles to antiparticles, so we want

$$\hat{C} a_{\mathbf{k}}^\dagger |0\rangle = \eta_c b_{\mathbf{k}}^\dagger |0\rangle$$
$$\hat{C} b_{\mathbf{k}}^\dagger |0\rangle = \eta_c^\star a_{\mathbf{k}}^\dagger |0\rangle \qquad ; |\eta_c|^2 = 1 \qquad (8.23)$$

The choice of phase is to ensure that the field has a definite transformation property under $\hat{C}$; here $|\eta_c|^2 = 1$. The operator should again be unitary, so the analog of Eqs. (8.9) for charge conjugation is[4]

$$\hat{C} a_{\mathbf{k}}^\dagger \hat{C}^{-1} = \eta_c b_{\mathbf{k}}^\dagger \qquad ; \hat{C} a_{\mathbf{k}} \hat{C}^{-1} = \eta_c^\star b_{\mathbf{k}}$$
$$\hat{C} b_{\mathbf{k}}^\dagger \hat{C}^{-1} = \eta_c^\star a_{\mathbf{k}}^\dagger \qquad ; \hat{C} b_{\mathbf{k}} \hat{C}^{-1} = \eta_c a_{\mathbf{k}}$$
$$\hat{C}|0\rangle = |0\rangle$$
$$\hat{C}^\dagger = \hat{C}^{-1} \qquad (8.24)$$

These relations define the charge-conjugation operator $\hat{C}$ for the charged spin-zero field. The implied transformation property of the field is imme-

---

[4]If the spin-zero particle and antiparticle differ only in their electromagnetic charge, as is the case for the pions, then charge conjugation *is* particle-antiparticle conjugation.

diately calculated from the second of Eqs. (8.1) as

$$\hat{C}\hat{\phi}(x)\hat{C}^{-1} = \sum_k \left(\frac{\hbar}{2\omega_k\Omega}\right)^{1/2} \left[\left(\hat{C}a_k\hat{C}^{-1}\right)e^{ik\cdot x} + \left(\hat{C}b_k^\dagger\hat{C}^{-1}\right)e^{-ik\cdot x}\right]$$

$$= \sum_k \left(\frac{\hbar}{2\omega_k\Omega}\right)^{1/2} \eta_c^\star \left(b_k e^{ik\cdot x} + a_k^\dagger e^{-ik\cdot x}\right) \qquad (8.25)$$

Hence

$$\hat{C}\hat{\phi}(x)\hat{C}^{-1} = \eta_c^\star\hat{\phi}^\star(x) \qquad ; \text{ charge-conjugation} \qquad (8.26)$$

Charge conjugation takes the field into its complex conjugate.[5]
In the case of a neutral field

$$\hat{C}c_k^\dagger|0\rangle = \eta_c c_k^\dagger|0\rangle \qquad (8.27)$$

In this case, since $\hat{C}^2 = 1$,

$$\eta_c = \pm 1 \qquad (8.28)$$

For the neutral field, with no other quantum numbers, charge-conjugation by itself does not mean anything. It is only when we look at all particles with strong and electromagnetic interactions that we can ask whether $\eta_c = \pm 1$. It turns out that

$$\eta_{c_\pi} = +1 \qquad ; \text{ pions}$$
$$\eta_{c_\gamma} = -1 \qquad ; \text{ photons} \qquad (8.29)$$

Call the neutral field $\hat{\phi}_3(x)$. Then for the neutral field

$$\hat{C}\hat{\phi}_3(x)\hat{C}^{-1} = \eta_{c_3}\hat{\phi}_3(x) \quad ; \eta_{c_3} = \pm 1 \qquad (8.30)$$

Consider the behavior of the (π^+, π^-) state in Eq. (8.20) under $\hat{C}$

$$\hat{C}|ELM\rangle = \int k^2 dk\, a_E(k) \int d\Omega_k Y_{LM}(\Omega_k)\, \hat{C}a_k^\dagger b_{-k}^\dagger|0\rangle$$

$$= \int k^2 dk\, a_E(k) \int d\Omega_k Y_{LM}(\Omega_k)\, |\eta_c|^2 b_k^\dagger a_{-k}^\dagger|0\rangle \qquad (8.31)$$

Use the fact that $[a^\dagger, b^\dagger] = 0$, and again change $k \to -k$. It follows that

$$\hat{C}|ELM\rangle = (-1)^L|ELM\rangle \qquad (8.32)$$

---

[5]We write it this way to emphasize the classical correspondence; in quantum field theory, this is really the hermitian adjoint $\hat{\phi}^\star \equiv \hat{\phi}^\dagger$.

This particle-antiparticle state is *also* an eigenstate of $\hat{C}$, with the same eigenvalue as $\hat{P}$.

So far, we have just defined transformations. We can always do this. Whether or not they are *useful*, depends on the form of the interaction.[6]

8.1.3 *Time Reversal*

Time reversal should relate the scattering process in Fig. 8.1(a) to the time-reversed scattering process in Fig. 8.1(b).

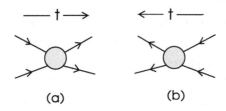

(a)   (b)

Fig. 8.1 The goal is to relate the scattering process in (a) to the time-reversed process in (b).

We make several observations:

- The time-reversal operator should somewhere switch around the initial and final states;
- Recall the definition of a unitary operator

$$\langle b|\hat{T}^{-1}|a\rangle = \langle b|\hat{T}^{\dagger}|a\rangle \qquad \text{; unitary}$$
$$= \langle Tb|a\rangle = \langle a|\hat{T}|b\rangle^{\star} \qquad (8.33)$$

- In quantum mechanics, one has the additional freedom of working with *anti-unitary* operators, while still preserving the norm of the states. An anti-unitary operator is defined by

$$\langle b|\hat{T}^{-1}|a\rangle = \langle b|\hat{T}^{\dagger}|a\rangle^{\star} \qquad \text{; anti-unitary}$$
$$= \langle Tb|a\rangle^{\star} = \langle a|\hat{T}|b\rangle \qquad (8.34)$$

---

[6]For example, suppose $\hat{V}^{\dagger} = \hat{V}^{-1}$ is a unitary operator that commutes with the interaction hamiltonian $\hat{V}\hat{H}'\hat{V}^{-1} = \hat{H}'$, and therefore also with the scattering operator $\hat{V}\hat{S}\hat{V}^{-1} = \hat{S}$. Matrix elements of $\hat{S}$ are then related as follows

$$\langle f|\hat{S}|i\rangle = \langle f|\hat{V}^{-1}\hat{V}\hat{S}\hat{V}^{-1}\hat{V}|i\rangle = \langle Vf|\hat{V}\hat{S}\hat{V}^{-1}|Vi\rangle = \langle Vf|\hat{S}|Vi\rangle$$

If the states $|i\rangle$ and $|f\rangle$ are also eigenstates of $\hat{V}$, this leads to *selection rules* on $\langle f|\hat{S}|i\rangle$.

It is the complex conjugation in the first relation that makes the operator anti-unitary. The last two equalities follow from the definition of the adjoint;

- To be consistent with this definition, the operator $\hat{T}$ can no longer commute with the complex numbers in the theory, but must take their complex conjugate. Thus it must satisfy

$$\hat{T}i\hat{T}^{-1} = -i \tag{8.35}$$

To see this, let c be a complex number, and use the rules of quantum mechanics, together with the above,

$$
\begin{aligned}
\langle b|\hat{T}^{-1}c^*|a\rangle &= \langle b|\hat{T}^\dagger c^*|a\rangle^\star \qquad\qquad\text{; anti-unitary} \\
&= \langle Tb|c^*|a\rangle^\star = c\langle Tb|a\rangle^\star = c\langle b|\hat{T}^\dagger|a\rangle^\star \\
&= c\langle b|\hat{T}^{-1}|a\rangle = \langle b|c\hat{T}^{-1}|a\rangle
\end{aligned}
\tag{8.36}
$$

Hence

$$\hat{T}^{-1}c^* = c\hat{T}^{-1} \tag{8.37}$$

This is just Eq. (8.35);

- The time-reversal operator should reverse the direction of the momentum of a state

$$\hat{T}a_{\mathbf{k}}^\dagger|0\rangle = \eta_t a_{-\mathbf{k}}^\dagger|0\rangle \tag{8.38}$$

Since $\hat{T}^2 = 1$, a second application of $\hat{T}$ will just restore the state

$$\hat{T}^2 a_{\mathbf{k}}^\dagger|0\rangle = |\eta_t|^2 a_{\mathbf{k}}^\dagger|0\rangle = a_{\mathbf{k}}^\dagger|0\rangle \tag{8.39}$$

This implies

$$|\eta_t|^2 = 1 \tag{8.40}$$

The time-reversal operator $\hat{T}$ is thus an anti-unitary operator satisfying

$$
\begin{aligned}
\langle b|\hat{T}^{-1}|a\rangle &= \langle b|\hat{T}^\dagger|a\rangle^\star = \langle Tb|a\rangle^\star = \langle a|\hat{T}|b\rangle \\
\hat{T}i\hat{T}^{-1} &= -i \\
\hat{T}|0\rangle &= |0\rangle
\end{aligned}
\tag{8.41}
$$

Acting on the creation and destruction operators, $\hat{T}$ is defined to give

$$
\begin{aligned}
\hat{T}a_{\mathbf{k}}^\dagger\hat{T}^{-1} &= \eta_t a_{-\mathbf{k}}^\dagger \qquad && ; \ \hat{T}a_{\mathbf{k}}\hat{T}^{-1} = \eta_t^\star a_{-\mathbf{k}} \\
\hat{T}b_{\mathbf{k}}^\dagger\hat{T}^{-1} &= \eta_t^\star b_{-\mathbf{k}}^\dagger \qquad && ; \ \hat{T}b_{\mathbf{k}}\hat{T}^{-1} = \eta_t b_{-\mathbf{k}}
\end{aligned}
\tag{8.42}
$$

The phases are again chosen to give the field a nice transformation property[7]

$$\hat{T}\hat{\phi}(x)\hat{T}^{-1} = \sum_{\mathbf{k}} \left(\frac{\hbar}{2\omega_k\Omega}\right)^{1/2} \hat{T}\left(a_{\mathbf{k}}e^{ik\cdot x} + b_{\mathbf{k}}^{\dagger}e^{-ik\cdot x}\right)\hat{T}^{-1}$$

$$= \sum_{\mathbf{k}} \left(\frac{\hbar}{2\omega_k\Omega}\right)^{1/2} \eta_t^{\star}\left(a_{-\mathbf{k}}e^{-ik\cdot x} + b_{-\mathbf{k}}^{\dagger}e^{ik\cdot x}\right) \qquad (8.43)$$

A change of summation variable $\mathbf{k} \to -\mathbf{k}$ then yields

$$\hat{T}\hat{\phi}(\mathbf{x},t)\hat{T}^{-1} = \eta_t^{\star}\hat{\phi}(\mathbf{x},-t) \qquad ; \text{ time-reversal} \qquad (8.44)$$

The time-reversal operator evaluates the field at the reversed time, which is again just what we want.

8.1.4 Transformation Properties of the E-M Current

The electromagnetic current operator for a charged spin-zero field is given to lowest order in e by

$$\hat{\mathbf{j}}(\mathbf{x},t) = \frac{ec}{i\hbar}\left(\hat{\phi}^{\star}\boldsymbol{\nabla}\hat{\phi} - \hat{\phi}\boldsymbol{\nabla}\hat{\phi}^{\star}\right)$$

$$\hat{\rho}(\mathbf{x},t) = \frac{ie}{\hbar}\left(\hat{\phi}^{\star}\hat{\pi}^{\star} - \hat{\phi}\hat{\pi}\right) \qquad (8.45)$$

Here $\hat{\pi}$ is the canonical momentum, which in the interaction picture is[8]

$$\hat{\pi}(\mathbf{x},t) = \sum_{\mathbf{k}} \left(\frac{\hbar\omega_k}{2\Omega}\right)^{1/2} i\left(a_{\mathbf{k}}^{\dagger}e^{-ik\cdot x} - b_{\mathbf{k}}e^{ik\cdot x}\right) \qquad (8.46)$$

It follows from Eqs. (8.10), (8.24), and (8.42) that

$$\hat{P}\hat{\pi}(\mathbf{x},t)\hat{P}^{-1} = \eta_p\hat{\pi}(-\mathbf{x},t)$$

$$\hat{C}\hat{\pi}(\mathbf{x},t)\hat{C}^{-1} = \eta_c\hat{\pi}^{\star}(\mathbf{x},t)$$

$$\hat{T}\hat{\pi}(\mathbf{x},t)\hat{T}^{-1} = -\eta_t\hat{\pi}(\mathbf{x},-t) \qquad (8.47)$$

One can then just read off the transformation properties of the electromagnetic current

$$\hat{P}\hat{\mathbf{j}}(\mathbf{x},t)\hat{P}^{-1} = -\hat{\mathbf{j}}(-\mathbf{x},t) \qquad ; \; \hat{P}\hat{\rho}(\mathbf{x},t)\hat{P}^{-1} = \hat{\rho}(-\mathbf{x},t)$$

$$\hat{C}\hat{\mathbf{j}}(\mathbf{x},t)\hat{C}^{-1} = -\hat{\mathbf{j}}(\mathbf{x},t) \qquad ; \; \hat{C}\hat{\rho}(\mathbf{x},t)\hat{C}^{-1} = -\hat{\rho}(\mathbf{x},t)$$

$$\hat{T}\hat{\mathbf{j}}(\mathbf{x},t)\hat{T}^{-1} = -\hat{\mathbf{j}}(\mathbf{x},-t) \qquad ; \; \hat{T}\hat{\rho}(\mathbf{x},t)\hat{T}^{-1} = \hat{\rho}(\mathbf{x},-t) \qquad (8.48)$$

[7]Note the complex conjugation of the plane waves!
[8]Note that classically, $\pi = \partial\phi^{\star}/\partial t$, and $\pi^{\star} = \partial\phi/\partial t$. Here $e = e_p$.

This is precisely what we want the transformations $\hat{P}$, $\hat{C}$, and $\hat{T}$ to do for us, based on our classical concept of the behavior of the electromagnetic current under spatial inversion, charge conjugation, and time reversal!

8.2   Spin-1/2 Particles and Fields

We extend the discussion of discrete symmetries to spin-1/2 particles and fields through the use of the Dirac equation, as examined in Vol. I. The Dirac field in the interaction picture is given by

$$\hat{\psi}(x) = \frac{1}{\sqrt{\Omega}} \sum_{k,\lambda} \left[a_{\mathbf{k}\lambda} u(\mathbf{k}\lambda) e^{ik\cdot x} + b^{\dagger}_{\mathbf{k}\lambda} v(-\mathbf{k}\lambda) e^{-ik\cdot x} \right] \qquad (8.49)$$

where $k \cdot x = \mathbf{k} \cdot \mathbf{x} - \omega_k t$. The positive and negative energy spinors satisfy the Dirac equation

$$(i\not{k} + M)u(\mathbf{k}) = \bar{u}(\mathbf{k})(i\not{k} + M) = 0$$
$$(i\not{k} - M)v(-\mathbf{k}) = \bar{v}(-\mathbf{k})(i\not{k} - M) = 0 \qquad (8.50)$$

The sum in Eq. (8.49) goes over the *helicities* $\lambda = (\uparrow, \downarrow)$, and the helicities are defined *with respect to the momentum variables*[9]

$$\boldsymbol{\sigma} \cdot \left(\frac{\mathbf{k}}{k} \right) u(\mathbf{k}\uparrow) = +u(\mathbf{k}\uparrow) \quad ; \quad \boldsymbol{\sigma} \cdot \left(-\frac{\mathbf{k}}{k} \right) v(-\mathbf{k}\uparrow) = +v(-\mathbf{k}\uparrow)$$

$$\boldsymbol{\sigma} \cdot \left(\frac{\mathbf{k}}{k} \right) u(\mathbf{k}\downarrow) = -u(\mathbf{k}\downarrow) \quad ; \quad \boldsymbol{\sigma} \cdot \left(-\frac{\mathbf{k}}{k} \right) v(-\mathbf{k}\downarrow) = -v(-\mathbf{k}\downarrow) \qquad (8.51)$$

Recall from our discussion of hole theory that the negative energy solution $v(-\mathbf{k}\lambda)$ is the wave function for an antiparticle with momentum $\hbar\mathbf{k}$ and helicity λ with respect to $\mathbf{k}$.

We use the standard representation of the Dirac matrices

$$\gamma_4 = \beta \qquad\qquad ; \ \boldsymbol{\gamma} = i\boldsymbol{\alpha}\beta$$

$$\beta = \begin{pmatrix} 1 & 0 \\ 0 & -1 \end{pmatrix} \qquad ; \ \boldsymbol{\alpha} = \begin{pmatrix} 0 & \boldsymbol{\sigma} \\ \boldsymbol{\sigma} & 0 \end{pmatrix} \qquad (8.52)$$

---

[9]The numerical values $\lambda = \pm 1$ are the eigenvalues in Eqs. (8.51).

Explicitly[10]

$$\gamma_1 = \begin{bmatrix} & & & -i \\ & & -i & \\ & i & & \\ i & & & \end{bmatrix} \quad \gamma_2 = \begin{bmatrix} & & & -1 \\ & & 1 & \\ & 1 & & \\ -1 & & & \end{bmatrix} \quad \gamma_3 = \begin{bmatrix} & & -i & \\ & & & i \\ i & & & \\ & -i & & \end{bmatrix} \quad \gamma_4 = \begin{bmatrix} 1 & & & \\ & 1 & & \\ & & -1 & \\ & & & -1 \end{bmatrix}$$

$$(8.53)$$

The Dirac spinors can then be written as shown in Table 8.1.[11]

Table 8.1 Dirac spinors. The quantities (a, b, r, s) are defined in Eqs. (8.54).

$\sqrt{2}\,u(\mathbf{k}\uparrow)$	$\sqrt{2}\,u(\mathbf{k}\downarrow)$	$\sqrt{2}\,v(-\mathbf{k}\downarrow)$	$\sqrt{2}\,v(-\mathbf{k}\uparrow)$
$a(r+s)$	$-b^\star(r+s)$	$a(r-s)$	$-b^\star(r-s)$
$b(r+s)$	$a^\star(r+s)$	$b(r-s)$	$a^\star(r-s)$
$a(r-s)$	$b^\star(r-s)$	$a(r+s)$	$b^\star(r+s)$
$b(r-s)$	$-a^\star(r-s)$	$b(r+s)$	$-a^\star(r+s)$

The quantities (a, b, r, s) appearing in Table 8.1 are defined by

$$a \equiv \cos\frac{\theta_k}{2}\, e^{-i\phi_k/2} \quad ; \quad b \equiv \sin\frac{\theta_k}{2}\, e^{i\phi_k/2}$$

$$r \equiv \cos\frac{\chi}{2} \quad ; \quad s \equiv \sin\frac{\chi}{2} \quad ; \quad \cot\chi \equiv \frac{pc}{m_0 c^2} \quad (8.54)$$

8.2.1 *Parity*

The parity operation should reverse the momentum, while leaving the spin unchanged. Thus it also *reverses the helicity* (see Fig. 8.2).

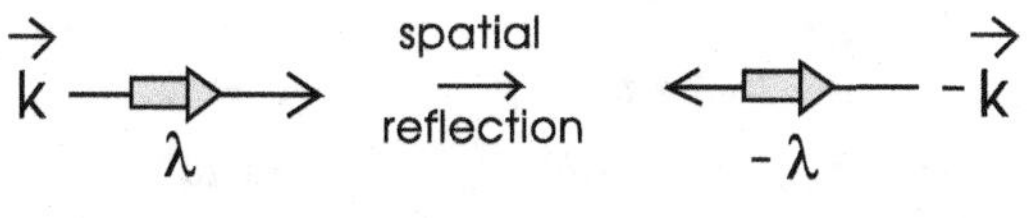

Fig. 8.2 Spatial reflection of a single-particle state with spin 1/2.

Thus

$$\hat{P}a^\dagger_{\mathbf{k}\lambda}|0\rangle = \eta_p^\star a^\dagger_{-\mathbf{k},-\lambda}|0\rangle$$

$$\hat{P}b^\dagger_{\mathbf{k}\lambda}|0\rangle = \eta_p b^\dagger_{-\mathbf{k},-\lambda}|0\rangle \quad (8.55)$$

---

[10]Recall the Clifford algebra $\gamma_\mu\gamma_\nu + \gamma_\nu\gamma_\mu = 2\delta_{\mu\nu}$.

[11]See Prob. 8.2.

The parity operator for spin-1/2 fields is therefore defined through the following relations

$$\hat{P}a^\dagger_{\mathbf{k}\lambda}\hat{P}^{-1} = \eta_p^\star a^\dagger_{-\mathbf{k},-\lambda} \qquad ; \hat{P}b^\dagger_{\mathbf{k}\lambda}\hat{P}^{-1} = \eta_p b^\dagger_{-\mathbf{k},-\lambda}$$

$$\hat{P}|0\rangle = |0\rangle$$

$$\hat{P}^\dagger = \hat{P}^{-1} \tag{8.56}$$

What about the phase η_p, with $|\eta_p|^2 = 1$, introduced here so that the field has a well-defined transformation property? Recall that the spin-1/2 state vectors are *double-valued*; that is, if $\mathbf{k}$ lies along the z-axis, then $\exp{(-2\pi i \hat{J}_z)}|\mathbf{k}\lambda\rangle = -|\mathbf{k}\lambda\rangle$. Therefore all we might expect to ask for spin-1/2 is

$$\hat{P}^4 = 1$$

$$\implies \qquad \eta_p^4 = 1 \qquad ; \eta_p = \pm 1, \pm i \tag{8.57}$$

The field now transforms according to

$$\hat{P}\hat{\psi}(x)\hat{P}^{-1} = \frac{\eta_p}{\sqrt{\Omega}} \sum_{\mathbf{k},\lambda} \left[a_{-\mathbf{k},-\lambda}u(\mathbf{k}\lambda)e^{ik\cdot x} + b^\dagger_{-\mathbf{k},-\lambda}v(-\mathbf{k}\lambda)e^{-ik\cdot x} \right] \tag{8.58}$$

We observe the following:

- The fields are originally summed over all $\mathbf{k}$ such that $(0 \leq \theta_k \leq \pi, 0 \leq \phi_k \leq 2\pi)$; however, the wave functions are defined and single-valued over the *extended* range $(0 \leq \theta_k \leq \pi, -\pi \leq \phi_k \leq 3\pi)$;
- We now change the dummy summation variable $\mathbf{k} \to -\mathbf{k}$ in Eq. (8.58), but we do it in such a way that all phases are completely defined

$$\theta_k \to \pi - \theta_k \qquad ; \phi_k \to \phi_k + \pi$$

$$\implies \qquad a \to -ib^\star \qquad ; b \to ia^\star \tag{8.59}$$

Equation (8.58) then becomes

$$\hat{P}\hat{\psi}(x)\hat{P}^{-1} = \frac{\eta_p}{\sqrt{\Omega}} \sum_{\mathbf{k},\lambda} \left[a_{\mathbf{k},-\lambda}u(-\mathbf{k}\lambda)e^{i\mathbf{k}\cdot(-\mathbf{x})-i\omega_k t} + \right.$$

$$\left. b^\dagger_{\mathbf{k},-\lambda}v(\mathbf{k}\lambda)e^{-i\mathbf{k}\cdot(-\mathbf{x})+i\omega_k t} \right] \tag{8.60}$$

From Eqs. (8.59) and Table 8.1, the transformed Dirac spinors $u(-\mathbf{k}\lambda)$ and $v(\mathbf{k}\lambda)$ are now given by the expressions in Table 8.2.

Table 8.2 Spatially-reflected spinors.

$\sqrt{2}\,u(-\mathbf{k}\uparrow)$	$\sqrt{2}\,u(-\mathbf{k}\downarrow)$	$\sqrt{2}\,v(\mathbf{k}\downarrow)$	$\sqrt{2}\,v(\mathbf{k}\uparrow)$
$-ib^\star(r+s)$	$ia(r+s)$	$-ib^\star(r-s)$	$ia(r-s)$
$ia^\star(r+s)$	$ib(r+s)$	$ia^\star(r-s)$	$ib(r-s)$
$-ib^\star(r-s)$	$-ia(r-s)$	$-ib^\star(r+s)$	$-ia(r+s)$
$ia^\star(r-s)$	$-ib(r-s)$	$ia^\star(r+s)$	$-ib(r+s)$

- The goal now is to relate these reflected spinors back to the original spinors in Eq. (8.49). This is readily done with the aid of Eqs. (8.53)

$$u(-\mathbf{k}\uparrow) = i\gamma_4 u(\mathbf{k}\downarrow) \qquad ; \ v(\mathbf{k}\downarrow) = i\gamma_4 v(-\mathbf{k}\uparrow)$$
$$u(-\mathbf{k}\downarrow) = i\gamma_4 u(\mathbf{k}\uparrow) \qquad ; \ v(\mathbf{k}\uparrow) = i\gamma_4 v(-\mathbf{k}\downarrow) \qquad (8.61)$$

Hence

$$\hat{P}\hat{\psi}(\mathbf{x},t)\hat{P}^{-1} = i\eta_p \gamma_4 \hat{\psi}(-\mathbf{x},t)$$
$$\hat{P}\hat{\bar{\psi}}(\mathbf{x},t)\hat{P}^{-1} = -i\eta_p^\star \hat{\bar{\psi}}(-\mathbf{x},t)\gamma_4 \qquad (8.62)$$

Notice the γ_4 appearing in the parity operation for the Dirac field;
- Note that one cannot just keep substituting $\mathbf{k} \rightleftharpoons -\mathbf{k}$ willy-nilly, since the wave functions are double-valued. Care must be taken in defining the domain of the transformations, as done above. In particular, for the operators in Eq. (8.56), $\mathbf{k} \to -\mathbf{k}$ implies $(\theta_k \to \pi - \theta_k,\ \phi_k \to \pi - \phi_k)$.

8.2.2 Charge Conjugation

The charge-conjugation operator is again defined to take particles to antiparticles

$$\hat{C}a_{\mathbf{k}\lambda}\hat{C}^{-1} = \eta_c^\star b_{\mathbf{k}\lambda} \qquad ; \ \hat{C}b_{\mathbf{k}\lambda}\hat{C}^{-1} = \eta_c a_{\mathbf{k}\lambda}$$
$$\hat{C}|0\rangle = |0\rangle$$
$$\hat{C}^\dagger = \hat{C}^{-1} \qquad (8.63)$$

where $|\eta_c|^2 = 1$. The field then transforms according to

$$\hat{C}\hat{\psi}(x)\hat{C}^{-1} = \frac{\eta_c^\star}{\sqrt{\Omega}} \sum_{\mathbf{k},\lambda} \left[b_{\mathbf{k}\lambda} u(\mathbf{k}\lambda)e^{ik\cdot x} + a_{\mathbf{k}\lambda}^\dagger v(-\mathbf{k}\lambda)e^{-ik\cdot x} \right] \qquad (8.64)$$

The goal now is to relate $u(\mathbf{k}\lambda)$ to $v^\star(-\mathbf{k}\lambda)$ and $v(-\mathbf{k}\lambda)$ to $u^\star(\mathbf{k}\lambda)$ so we get the complex conjugate (hermitian adjoint) of the field back again. With the use of γ_2 from Eqs. (8.53), the Dirac spinors in Table 8.1, and a little matrix algebra, one establishes the relations[12]

$$u(\mathbf{k}\lambda)_\alpha = [v(-\mathbf{k}\lambda)^\dagger \gamma_2]_\alpha = [\bar{v}(-\mathbf{k}\lambda)\gamma_4\gamma_2]_\alpha$$
$$v(-\mathbf{k}\lambda)_\alpha = [u(\mathbf{k}\lambda)^\dagger \gamma_2]_\alpha = [\bar{u}(\mathbf{k}\lambda)\gamma_4\gamma_2]_\alpha \tag{8.65}$$

Here the matrix index α has been made explicit. Hence

$$\hat{C}\hat{\psi}_\alpha(x)\hat{C}^{-1} = \eta_c^\star[\hat{\bar{\psi}}(x)\gamma_4\gamma_2]_\alpha$$
$$\hat{C}\hat{\bar{\psi}}_\alpha(x)\hat{C}^{-1} = \eta_c[\gamma_4\gamma_2\hat{\psi}(x)]_\alpha \tag{8.66}$$

The second relation follows from the first by using the fact that $\gamma_4 = \gamma_4^T$, and then[13]

$$[\gamma_2\hat{\psi}]_\rho[\gamma_4]_{\rho\alpha} = [\gamma_4^T]_{\alpha\rho}[\gamma_2\hat{\psi}]_\rho = [\gamma_4\gamma_2\hat{\psi}]_\alpha \tag{8.67}$$

The unitary, 4×4, charge-conjugation matrix S_c is now defined by

$$S_c \equiv \gamma_2\gamma_4 \qquad\qquad ; \text{ charge-conjugation matrix}$$
$$S_c^\dagger = \gamma_4\gamma_2 = S_c^{-1} \tag{8.68}$$

It has the property that

$$S_c\gamma_\mu S_c^{-1} = \gamma_2\gamma_4\gamma_\mu\gamma_4\gamma_2 = -\gamma_\mu \qquad ; \mu = 2, 4$$
$$= +\gamma_\mu \qquad ; \mu = 1, 3 \tag{8.69}$$

Hence, from Eqs. (8.53),

$$S_c\gamma_\mu S_c^{-1} = -\gamma_\mu^T \tag{8.70}$$

8.2.3   *Time Reversal*

The time-reversal operation should reverse the momentum and also the spin. Thus it *leaves the helicity unchanged* (see Fig. 8.2).

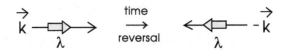

Fig. 8.3   Time reversal of a single-particle state with spin $1/2$.

---

[12]See Prob. 8.3.

[13]Repeated Greek indices are again summed from one to four.

The operator $\hat{T}$ is again defined to be anti-unitary, and to leave the vacuum invariant

$$\langle b|\hat{T}^{-1}|a\rangle = \langle b|\hat{T}^\dagger|a\rangle^\star = \langle Tb|a\rangle^\star = \langle a|\hat{T}|b\rangle \quad ; \text{ anti-unitary}$$

$$\hat{T}i\hat{T}^{-1} = -i$$

$$\hat{T}|0\rangle = |0\rangle \tag{8.71}$$

Its effect on the creation and destruction operators is defined as follows

$$\hat{T}a_{\mathbf{k}\lambda}\hat{T}^{-1} = \eta_t(\delta_{\lambda\downarrow} - \delta_{\lambda\uparrow})a_{-\mathbf{k}\lambda} \quad ; \hat{T}a_{\mathbf{k}\lambda}^\dagger\hat{T}^{-1} = \eta_t^\star(\delta_{\lambda\downarrow} - \delta_{\lambda\uparrow})a_{-\mathbf{k}\lambda}^\dagger$$

$$\hat{T}b_{\mathbf{k}\lambda}\hat{T}^{-1} = \eta_t^\star(\delta_{\lambda\uparrow} - \delta_{\lambda\downarrow})b_{-\mathbf{k}\lambda} \quad ; \hat{T}b_{\mathbf{k}\lambda}^\dagger\hat{T}^{-1} = \eta_t(\delta_{\lambda\uparrow} - \delta_{\lambda\downarrow})b_{-\mathbf{k}\lambda}^\dagger \tag{8.72}$$

Once again, the phases, with $|\eta_t|^2 = 1$, are chosen so that the *field* has a nice transformation property.

Under time-reversal, the Dirac field then transforms according to

$$\hat{T}\hat{\psi}(x)\hat{T}^{-1} = \frac{\eta_t}{\sqrt{\Omega}}\sum_{\mathbf{k}}\left[u(\mathbf{k}\downarrow)^\star a_{-\mathbf{k}\downarrow}e^{-ik\cdot x} - u(\mathbf{k}\uparrow)^\star a_{-\mathbf{k}\uparrow}e^{-ik\cdot x} + \right.$$

$$\left. v(-\mathbf{k}\uparrow)^\star b_{-\mathbf{k}\uparrow}^\dagger e^{ik\cdot x} - v(-\mathbf{k}\downarrow)^\star b_{-\mathbf{k}\downarrow}^\dagger e^{ik\cdot x}\right] \tag{8.73}$$

We now change summation variable $\mathbf{k} \to -\mathbf{k}$ by taking $(\theta_k \to \pi - \theta_k,\ \phi_k \to \phi_k + \pi)$ as in Eqs. (8.59). This gives the time-reversed spinors in Table 8.3.

Table 8.3   Time-reversed spinors.

$\sqrt{2}\,u(-\mathbf{k}\uparrow)^\star$	$\sqrt{2}\,u(-\mathbf{k}\downarrow)^\star$	$\sqrt{2}\,v(\mathbf{k}\downarrow)^\star$	$\sqrt{2}\,v(\mathbf{k}\uparrow)^\star$
$ib(r+s)$	$-ia^\star(r+s)$	$ib(r-s)$	$-ia^\star(r-s)$
$-ia(r+s)$	$-ib^\star(r+s)$	$-ia(r-s)$	$-ib^\star(r-s)$
$ib(r-s)$	$ia^\star(r-s)$	$ib(r+s)$	$ia^\star(r+s)$
$-ia(r-s)$	$ib^\star(r-s)$	$-ia(r+s)$	$ib^\star(r+s)$

It follows from Eqs. (8.53) that

$$\gamma_1\gamma_3 = \begin{bmatrix} & -1 & & \\ 1 & & & \\ & & -1 & \\ & & & 1 \end{bmatrix} \tag{8.74}$$

A little matrix algebra, and reference to Table 8.1, then establishes

$$u(-\mathbf{k}\uparrow)^\star = -i\gamma_1\gamma_3 u(\mathbf{k}\uparrow) \quad ; v(\mathbf{k}\downarrow)^\star = -i\gamma_1\gamma_3 v(-\mathbf{k}\downarrow)$$

$$u(-\mathbf{k}\downarrow)^\star = +i\gamma_1\gamma_3 u(\mathbf{k}\downarrow) \quad ; v(\mathbf{k}\uparrow)^\star = +i\gamma_1\gamma_3 v(-\mathbf{k}\uparrow) \tag{8.75}$$

Therefore

$$\hat{T}\hat{\psi}(\mathbf{x},t)\hat{T}^{-1} = i\eta_t \gamma_1 \gamma_3 \hat{\psi}(\mathbf{x},-t)$$

$$\hat{T}\hat{\bar{\psi}}(\mathbf{x},t)\hat{T}^{-1} = -i\eta_t^\star \hat{\bar{\psi}}(\mathbf{x},-t)\gamma_3 \gamma_1 \tag{8.76}$$

The second relation is arrived at in exactly the same fashion as the first. We now define a unitary, 4×4, time-reversal matrix S_t by

$$S_t \equiv \gamma_1 \gamma_3 \qquad\qquad ; \text{ time-reversal matrix}$$

$$S_t^\dagger = \gamma_3 \gamma_1 = S_t^{-1} \tag{8.77}$$

It has the property that

$$S_t \gamma_\mu S_t^{-1} = \gamma_1 \gamma_3 \gamma_\mu \gamma_3 \gamma_1 = -\gamma_\mu \qquad ; \mu = 1,3$$

$$= +\gamma_\mu \qquad ; \mu = 2,4 \tag{8.78}$$

Hence, from Eqs. (8.53)

$$S_t \gamma_\mu S_t^{-1} = \gamma_\mu^\star \tag{8.79}$$

8.2.4 Transformation Properties of the E-M Current

The Dirac electromagnetic current is given in the interaction picture by

$$\hat{j}_\mu(x) = ie\left[\hat{\bar{\psi}}(x)\gamma_\mu \hat{\psi}(x) - \langle 0|\hat{\bar{\psi}}(x)\gamma_\mu \hat{\psi}(x)|0\rangle \right] \tag{8.80}$$

The subtraction of the vacuum expectation value serves to normal-order the current.[14]

Consider first the *parity* transformation on the current. It follows from Eqs. (8.62) that

$$\hat{P}ie\hat{\bar{\psi}}(x)\gamma_\mu \hat{\psi}(x)\hat{P}^{-1} = ie\hat{\bar{\psi}}(-\mathbf{x},t)\gamma_4 \gamma_\mu \gamma_4 \hat{\psi}(-\mathbf{x},t) \tag{8.81}$$

Furthermore

$$\langle 0|\hat{\bar{\psi}}(x)\gamma_\mu \hat{\psi}(x)|0\rangle = \langle 0|\hat{P}\hat{\bar{\psi}}(x)\gamma_\mu \hat{\psi}(x)\hat{P}^{-1}|0\rangle \tag{8.82}$$

Hence, from Eq. (8.81),[15]

$$\hat{P}\hat{\mathbf{j}}(\mathbf{x},t)\hat{P}^{-1} = -\hat{\mathbf{j}}(-\mathbf{x},t)$$

$$\hat{P}\hat{\rho}(\mathbf{x},t)\hat{P}^{-1} = +\hat{\rho}(-\mathbf{x},t) \tag{8.83}$$

---

[14]In Vol. II, this is written as $ie:\hat{\bar{\psi}}(x)\gamma_\mu \hat{\psi}(x):$.

[15]Use $\gamma_4 \boldsymbol{\gamma} \gamma_4 = -\boldsymbol{\gamma}$, and $\gamma_4 \gamma_4 \gamma_4 = \gamma_4$.

The behavior under *charge conjugation* follows from Eqs. (8.66)

$$\hat{C}ie\bar{\psi}(x)\gamma_\mu\hat{\psi}(x)\hat{C}^{-1} = ie[\gamma_4\gamma_2\psi(x)]_\alpha[\gamma_\mu]_{\alpha\beta}[\bar{\psi}(x)\gamma_4\gamma_2]_\beta$$
$$= -ie\bar{\psi}(x)\gamma_4\gamma_2[\gamma_\mu]^T\gamma_4\gamma_2\psi(x) + \text{c-number} \qquad (8.84)$$

Now note the following:

- We have used the fact that fermion fields *anti-commute*;
- The c-number term is cancelled by an identical term in

$$\langle 0|\bar{\psi}(x)\gamma_\mu\hat{\psi}(x)|0\rangle = \langle 0|\hat{C}\bar{\psi}(x)\gamma_\mu\hat{\psi}(x)\hat{C}^{-1}|0\rangle \qquad (8.85)$$

- Equation (8.70) can then be employed to give

$$\gamma_4\gamma_2[\gamma_\mu]^T\gamma_4\gamma_2 = -S_c^{-1}\gamma_\mu^T S_c = \gamma_\mu \qquad (8.86)$$

It follows that[16]

$$\hat{C}j_\mu(x)\hat{C}^{-1} = -j_\mu(x) \qquad (8.87)$$

For *time-reversal*, Eqs. (8.76) are employed to give

$$\hat{T}ie\bar{\psi}(x)\gamma_\mu\hat{\psi}(x)\hat{T}^{-1} = -ie\bar{\psi}(\mathbf{x}, -t)\gamma_3\gamma_1\gamma_\mu^\star\gamma_1\gamma_3\psi(\mathbf{x}, -t) \qquad (8.88)$$

We now observe:

- The second of Eq. (8.71) implies the indicated complex conjugates;
- There is also the factor of i to be complex conjugated in $j_\mu = (\mathbf{j}, i\rho)$;
- From Eq. (8.79)

$$\gamma_3\gamma_1[\gamma_\mu]^\star\gamma_1\gamma_3 = S_t^{-1}\gamma_\mu^\star S_t = \gamma_\mu \qquad (8.89)$$

- The current is hermitian, and therefore

$$\langle 0|\hat{\mathbf{j}}(x)|0\rangle = \langle 0|\hat{\mathbf{j}}(x)|0\rangle^\star$$
$$= \langle 0|\hat{T}^{-1}\hat{T}\hat{\mathbf{j}}(x)\hat{T}^{-1}\hat{T}|0\rangle^\star = \langle 0|\hat{T}\hat{\mathbf{j}}(x)\hat{T}^{-1}|0\rangle \qquad (8.90)$$

The last equality holds since $\hat{T}$ is anti-unitary. A similar relation holds for $\langle 0|\hat{\rho}(x)|0\rangle$.

---

[16]Heisenberg originally suggested that one get rid of the vacuum infinities by taking

$$\hat{j}_\mu^c \equiv (ie/2)[\bar{\psi}(x)\gamma_\mu\hat{\psi}(x) - \hat{C}^{-1}\bar{\psi}(x)\gamma_\mu\hat{\psi}(x)\hat{C}]$$

Since $\hat{C}^2 = 1$ and $\hat{C} = \hat{C}^{-1}$, it follows that $\hat{C}\hat{j}_\mu^c\hat{C}^{-1} = -\hat{j}_\mu^c$, so the current is now *explicitly* odd under $\hat{C}$. This definition has the immediate consequence that $\langle 0|\hat{j}_\mu^c|0\rangle = 0$, and the result is the same as normal ordering.

It follows that

$$\hat{T}\hat{j}(\mathbf{x},t)\hat{T}^{-1} = -\hat{j}(\mathbf{x},-t)$$
$$\hat{T}\hat{\rho}(\mathbf{x},t)\hat{T}^{-1} = +\hat{\rho}(\mathbf{x},-t) \tag{8.91}$$

Thus all three operations $(\hat{P},\hat{C},\hat{T})$ give exactly the same results on the Dirac electromagnetic current as our previous results in Eqs. (8.48).

8.2.5   Two-Particle Fermion–Antifermion States

We construct the two-particle fermion-antifermion state in the non-relativistic limit, starting from the Dirac spinors in Table 8.1.

(1) The non-relativistic limit is obtained from $\cot\chi = pc/m_0c^2 \to 0$, or $\chi = \pi/2$. Thus

$$\chi = \frac{\pi}{2} \qquad\qquad ; \text{non-relativistic limit}$$
$$r = s = \sin\frac{\pi}{4} = \cos\frac{\pi}{4} = \frac{1}{\sqrt{2}} \tag{8.92}$$

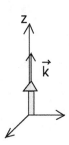

Fig. 8.4   Non-relativistic Dirac spinors for spin ($\uparrow\downarrow$) along the z-axis.

(2) Furthermore, the spinors for spin $\uparrow$ or spin $\downarrow$ along the z-axis (Fig. 8.4) are then obtained by setting[17]

$$a = \cos\frac{\theta_k}{2}\, e^{-i\phi_k/2} = 1$$
$$b = \sin\frac{\theta_k}{2}\, e^{i\phi_k/2} = 0 \tag{8.93}$$

The results of inserting these values are shown in Table 8.4.

---

[17]Start with $\phi_k = 0$, and then let $\theta_k \to 0$.

Table 8.4 Non-relativistic Dirac spinors for spin
($\uparrow\downarrow$) along the z-axis (see Fig. 8.4). The indicated
two-component spinors will be labeled as $\chi_m^{(\pm)}$,
where the superscript is the charge and subscript
the z-component of the spin (see last row).

$u(\mathbf{k}\uparrow)$	$u(\mathbf{k}\downarrow)$	$v(-\mathbf{k}\downarrow)$	$v(-\mathbf{k}\uparrow)$
$\begin{pmatrix}1\\0\end{pmatrix}$	$\begin{pmatrix}0\\1\end{pmatrix}$	0	0
		0	0
0	0	$\begin{pmatrix}1\\0\end{pmatrix}$	$\begin{pmatrix}0\\-1\end{pmatrix}$
0	0		
$e^-\uparrow$	$e^-\downarrow$	$e^+\downarrow$	$e^+\uparrow$

The wave functions are *independent of* $\mathbf{k}$ *in this limit, and the spin and orbital motion are decoupled.*[18]

Call the indicated two-component spinors in Table 8.4 $\chi_m^{(\pm)}$, where the superscript denotes the charge and the subscript denotes the z-component of the spin, and consider the expansion of the field in this non-relativistic basis

$$\hat{\psi}(x) \doteq \frac{1}{\sqrt{\Omega}} \sum_{\mathbf{k},m} \left[a_{\mathbf{k}m} \chi_m^{(-)} e^{ik\cdot x} + b_{\mathbf{k}m}^\dagger \chi_m^{(+)} e^{-ik\cdot x} \right] \qquad (8.94)$$

The transformation properties of the full field under $\hat{P}$ and $\hat{C}$ are given in Eqs. (8.62) and (8.66), while the relevant gamma matrices are given in Eqs. (8.53)

$$\hat{P}\hat{\psi}(\mathbf{x},t)\hat{P}^{-1} = i\eta_p\gamma_4\hat{\psi}(-\mathbf{x},t) \qquad ; \quad \hat{C}\hat{\psi}_\alpha(\mathbf{x},t)\hat{C}^{-1} = \eta_c^\star[\hat{\psi}^\dagger(\mathbf{x},t)\gamma_2]_\alpha$$

$$\gamma_4 = \begin{bmatrix} 1 & & & \\ & 1 & & \\ & & -1 & \\ & & & -1 \end{bmatrix} \qquad ; \quad \gamma_2 = \begin{bmatrix} & & & -1 \\ & & 1 & \\ & 1 & & \\ -1 & & & \end{bmatrix} \qquad (8.95)$$

One can then just read off the appropriate transformation properties of the creation and destruction operators appearing in Eq. (8.94)

$$\hat{P}a_{\mathbf{k}m}\hat{P}^{-1} = i\eta_p a_{-\mathbf{k}m} \qquad ; \quad \hat{C}a_{\mathbf{k}m}\hat{C}^{-1} = \eta_c^\star b_{\mathbf{k}m}$$

$$\hat{P}b_{\mathbf{k}m}^\dagger\hat{P}^{-1} = -i\eta_p b_{-\mathbf{k}m}^\dagger \qquad ; \quad \hat{C}b_{\mathbf{k}m}^\dagger\hat{C}^{-1} = \eta_c^\star a_{\mathbf{k}m}^\dagger \qquad (8.96)$$

---

[18]The analysis is readily generalized using relativistic helicity states (see [Jacob and Wick (1959)]).

We are now in a position to construct the two-particle fermion-antifermion state in the center-of-momentum system

$$|ELSJM\rangle = \sum_{m_a,m_b} \int d^3k \, a_E(k)\phi_{m_a m_b}\,(\mathbf{e_k}LSJM)\,a^\dagger_{\mathbf{k}m_a}b^\dagger_{-\mathbf{k}m_b}|0\rangle \quad (8.97)$$

Here the unit vector $\mathbf{e_k} \equiv \mathbf{k}/k$ indicates the direction, and

$$\phi_{m_a m_b} = \sum_{M_L,M_S} Y_{LM_L}(\Omega_k)\langle s_a m_a s_b m_b|s_a s_b SM_S\rangle\langle LM_L SM_S|LSJM_J\rangle \tag{8.98}$$

The behavior of this state under $\hat{P}$ and $\hat{C}$ can now just be read off from Eqs. (8.96)–(8.98)

$$\hat{P}|ELSJM\rangle = (-i)^2|\eta_p|^2(-1)^L|ELSJM\rangle = (-1)^{L+1}|ELSJM\rangle$$
$$\hat{C}|ELSJM\rangle = (-1)(-1)^L(-1)^{1/2+1/2-S}|ELSJM\rangle = (-1)^{L-S}|ELSJM\rangle \tag{8.99}$$

Some comments:

- Note the extra factor of (-1) in the parity relation; fermions and antifermions have *opposite intrinsic parity*;
- In the charge-conjugation relation, we have used the fact that the fermion operators anti-commute, $\{a^\dagger, b^\dagger\} = 0$;
- We have also used the following relation to change the order of coupling in the Clebsch-Gordan coefficient

$$\langle \tfrac{1}{2}m_a \tfrac{1}{2}m_b|\tfrac{1}{2}\tfrac{1}{2}SM_S\rangle = (-1)^{1/2+1/2-S}\langle \tfrac{1}{2}m_b \tfrac{1}{2}m_a|\tfrac{1}{2}\tfrac{1}{2}SM_S\rangle \quad (8.100)$$

The two-particle fermion-antifermion state in Eq. (8.97) is thus an eigenstate of $\hat{P}$ with eigenvalue $(-1)^{L+1}$, and of $\hat{C}$ with eigenvalue $(-1)^{L-S}$.

8.3   Spin-1 Field (Massless Photons)

We confine our discussion of spin-1 fields to the case of massless photons. The vector potential in the Coulomb gauge for the radiation field in the interaction picture is given in Prob. 7.8 as[19]

$$\hat{\mathbf{A}}(x) = \sum_{\mathbf{k}} \sum_{s=1}^{2} \left(\frac{2\pi\hbar c^2}{\omega_k} \right)^{1/2} \left(a_{\mathbf{k}s}\mathbf{e}_{\mathbf{k}s}e^{ik\cdot x} + a_{\mathbf{k}s}^{\dagger}\mathbf{e}_{\mathbf{k}s}e^{-ik\cdot x} \right) \quad (8.101)$$

The sum goes over the transverse unit vectors illustrated in Fig. 8.5.

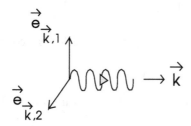

Fig. 8.5 Set of orthonormal unit vectors $\mathbf{e}_{\mathbf{k}s}$, with $s = 1, 2$, transverse to $\mathbf{k}$. We will later also refer to $\mathbf{e}_{\mathbf{k}0} \equiv \mathbf{k}/k$.

Now define

$$\mathbf{e}_{\mathbf{k},\pm 1} \equiv \mp \frac{1}{\sqrt{2}}(\mathbf{e}_{\mathbf{k}1} \pm i\mathbf{e}_{\mathbf{k}2}) \quad ; \quad \mathbf{e}_{\mathbf{k},\pm 1}^{\dagger} = \mp \frac{1}{\sqrt{2}}(\mathbf{e}_{\mathbf{k}1} \mp i\mathbf{e}_{\mathbf{k}2})$$

$$a_{\mathbf{k},\pm 1}^{\dagger} \equiv \mp \frac{1}{\sqrt{2}}(a_{\mathbf{k}1}^{\dagger} \pm ia_{\mathbf{k}2}^{\dagger}) \quad ; \quad a_{\mathbf{k},\pm 1} = \mp \frac{1}{\sqrt{2}}(a_{\mathbf{k}1} \mp ia_{\mathbf{k}2}) \quad (8.102)$$

It is readily verified that this transformation is canonical

$$[a_{\mathbf{k}\lambda}, a_{\mathbf{k}\lambda'}^{\dagger}] = \delta_{\lambda\lambda'} \quad ; \quad (\lambda, \lambda') = \pm 1 \quad (8.103)$$

and that the new polarization vectors satisfy

$$\mathbf{e}_{\mathbf{k}\lambda} \cdot \mathbf{e}_{\mathbf{k}\lambda'}^{\dagger} = \delta_{\lambda\lambda'} \quad ; \quad \mathbf{e}_{\mathbf{k}\lambda}^{\dagger} = (-1)^{\lambda}\mathbf{e}_{\mathbf{k},-\lambda} \quad (8.104)$$

Furthermore

$$a_{\mathbf{k}1}\mathbf{e}_{\mathbf{k}1} + a_{\mathbf{k}2}\mathbf{e}_{\mathbf{k}2} = a_{\mathbf{k},+1}\mathbf{e}_{\mathbf{k},+1} + a_{\mathbf{k},-1}\mathbf{e}_{\mathbf{k},-1} \quad (8.105)$$

The photon field in Eq. (8.101) can thus be written as

$$\hat{\mathbf{A}}(x) = \sum_{\mathbf{k}} \sum_{\lambda=\pm 1} \left(\frac{2\pi\hbar c^2}{\omega_k} \right)^{1/2} \left(a_{\mathbf{k}\lambda}\mathbf{e}_{\mathbf{k}\lambda}e^{ik\cdot x} + a_{\mathbf{k}\lambda}^{\dagger}\mathbf{e}_{\mathbf{k}\lambda}^{\dagger}e^{-ik\cdot x} \right) \quad (8.106)$$

---

[19]Recall the interaction picture is identical to the Heisenberg picture for the free field. Here $k \cdot x = \mathbf{k} \cdot \mathbf{x} - \omega_{\mathbf{k}}t$.

We *claim* that $\lambda = \pm 1$ is now the *helicity* of the photon, that is, its angular momentum along the direction of motion. Thus $a_{\mathbf{k}\lambda}^{\dagger}$ creates a photon with momentum $\hbar\mathbf{k}$ and helicity λ with respect to $\mathbf{k}$.

The *proof* of this fact follows from the verification in chapter 3 of Vol. II that the operator $e^{-i\boldsymbol{\omega}\cdot\hat{\mathbf{J}}}$ rotates the state vector by an angle ω about the $\boldsymbol{\omega}/\omega$ axis. Let $\mathbf{k}$ lie along the z-axis as in Fig. 8.6. Now let $e^{-i\phi\hat{J}_z}$ act on the state $a_{\mathbf{k},\pm 1}^{\dagger}|0\rangle = \mp(1/\sqrt{2})\,(|\mathbf{k}1\rangle \pm i|\mathbf{k}2\rangle)$

$$e^{-i\phi\hat{J}_z}\, a_{\mathbf{k},\pm 1}^{\dagger}|0\rangle = \mp\frac{1}{\sqrt{2}}\left[\cos\phi|\mathbf{k}1\rangle + \sin\phi|\mathbf{k}2\rangle \pm i\cos\phi|\mathbf{k}2\rangle \mp i\sin\phi|\mathbf{k}1\rangle\right]$$

$$= e^{-i\phi(\pm 1)}\, a_{\mathbf{k},\pm 1}^{\dagger}|0\rangle \qquad (8.107)$$

Thus $\hat{J}_z$ has eigenvalues ± 1 for the configuration in Fig. 8.6.[20]

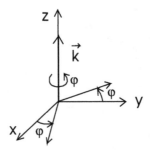

Fig. 8.6   Rotation by ϕ around the $\mathbf{k}$-axis. Here the 1-axis is denoted by x, and the 2-axis by y.

8.3.1   *Two-Photon States in the C-M System*

Consider the two-photon states in the center-of-momentum (C-M) frame (see Fig. 8.7). There are four states we can make $|\mathbf{k}; \lambda_1\lambda_2\rangle$ (see Fig. 8.7).

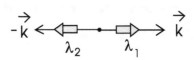

Fig. 8.7   Two-photon states in the center-of-momentum system.

---

[20]That λ is the helicity is also established by examining the angular momentum carried off in an arbitary radiative transition in appendix A of Vol. II.

Write these as

$$\frac{1}{\sqrt{2}}\left(|+1,+1\rangle \pm |-1,-1\rangle\right) = \frac{1}{\sqrt{2}}\left(a^\dagger_{\mathbf{k},+1}a^\dagger_{-\mathbf{k},+1} \pm a^\dagger_{\mathbf{k},-1}a^\dagger_{-\mathbf{k},-1}\right)|0\rangle$$

$$|+1,-1\rangle = a^\dagger_{\mathbf{k},+1}a^\dagger_{-\mathbf{k},-1}|0\rangle$$

$$|-1,+1\rangle = a^\dagger_{\mathbf{k},-1}a^\dagger_{-\mathbf{k},+1}|0\rangle \tag{8.108}$$

The projection of the angular momentum along the **k**-axis for each of these states is indicated in Table 8.5. The parity of each state (calculated below) is also indicated.

Table 8.5 Properties of the two-photon state in the C-M system.

State	J_k	P
$(|+1,+1\rangle + |-1,-1\rangle)/\sqrt{2}$	0	+
$(|+1,+1\rangle - |-1,-1\rangle)/\sqrt{2}$	0	−
$|+1,-1\rangle$	+2	+
$|-1,+1\rangle$	−2	+

8.3.2 Symmetry Properties of the Field

The behavior of $\hat{\mathbf{A}}(\mathbf{x},t)$ under the various transformations is not arbitrary since one has a *classical correspondence*. The electromagnetic potentials $(\mathbf{A},\Phi)$ are given in the Coulomb gauge by

$$\Box\mathbf{A} = -4\pi\mathbf{j} + \boldsymbol{\nabla}\frac{1}{c}\frac{\partial\Phi}{\partial t}$$

$$\nabla^2\Phi = -4\pi\varrho \tag{8.109}$$

Thus $\mathbf{A}(\mathbf{x},t)$ must behave just like a classical current $\mathbf{j}(\mathbf{x},t)$ under the transformations if they are to be *useful* and preserve the equations of motion. This definition of phases is used as the starting point, and all other phases are referred to it.

8.3.2.1 Parity

The *parity* operation on the photon field is defined by

$$\hat{P}a^\dagger_{\mathbf{k}\lambda}\hat{P}^{-1} = \eta_p a^\dagger_{-\mathbf{k},-\lambda}$$

$$\hat{P}|0\rangle = |0\rangle$$

$$\hat{P}^{-1} = \hat{P}^\dagger \tag{8.110}$$

Furthermore, since $\hat{P}^2 = 1$, one has $\eta_p = \pm 1$. The field then transforms according to

$$\hat{P}\hat{\mathbf{A}}(x)\hat{P}^{-1} = \sum_{\mathbf{k}} \sum_{\lambda=\pm 1} \left(\frac{2\pi\hbar c^2}{\omega_k}\right)^{1/2} \eta_p \left(a_{-\mathbf{k},-\lambda}\mathbf{e}_{\mathbf{k}\lambda}e^{ik\cdot x} + a^{\dagger}_{-\mathbf{k},-\lambda}\mathbf{e}^{\dagger}_{\mathbf{k}\lambda}e^{-ik\cdot x}\right)$$

(8.111)

A change of dummy summation variables gives

$$\hat{P}\hat{\mathbf{A}}(x)\hat{P}^{-1} = \sum_{\mathbf{k}} \sum_{\lambda=\pm 1} \left(\frac{2\pi\hbar c^2}{\omega_k}\right)^{1/2} \eta_p \left(a_{\mathbf{k},\lambda}\mathbf{e}_{-\mathbf{k},-\lambda}\, e^{i\mathbf{k}\cdot(-\mathbf{x})-i\omega_k t}+\right.$$
$$\left. a^{\dagger}_{\mathbf{k},\lambda}\mathbf{e}^{\dagger}_{-\mathbf{k},-\lambda}\, e^{-i\mathbf{k}\cdot(-\mathbf{x})+i\omega_k t}\right)$$

(8.112)

We cannot go any further until we specify how the polarization vectors are related for $\pm \mathbf{k}$. So far, for every $\mathbf{k}$, we have simply picked some orthogonal basis. We must now define the *relative phases* of the states in order to discuss these discrete symmetries.

Pick 1/2 of the $\mathbf{k}$,[21] and define (recall Fig. 8.5)

$$\mathbf{e}_{\mathbf{k},\pm 1} \equiv \mp \frac{1}{\sqrt{2}} \left(\mathbf{e}_{\mathbf{k}1} \pm i\mathbf{e}_{\mathbf{k}2}\right)$$

(8.113)

Now define the polarization vectors for $-\mathbf{k}$ through a rotation of π about the 2-axis (see Fig. 8.8).

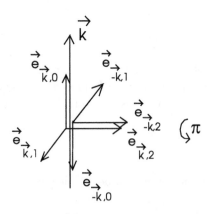

Fig. 8.8   Polarization vectors for $-\mathbf{k}$ defined through a rotation by π about the 2-axis.

---

[21]Say those with $k_z > 0$.

It follows that

$$\mathbf{e}_{-\mathbf{k},\pm 1} \equiv \mp\frac{1}{\sqrt{2}}\left(\mathbf{e}_{-\mathbf{k}1} \pm i\mathbf{e}_{-\mathbf{k}2}\right) = \mp\frac{1}{\sqrt{2}}\left(-\mathbf{e}_{\mathbf{k}1} \pm i\mathbf{e}_{\mathbf{k}2}\right) = \pm\frac{1}{\sqrt{2}}\left(\mathbf{e}_{\mathbf{k}1} \mp i\mathbf{e}_{\mathbf{k}2}\right)$$
$$= \mathbf{e}_{\mathbf{k},\mp 1} \tag{8.114}$$

Thus, with this phase convention,

$$\mathbf{e}_{\mathbf{k}\lambda} = \mathbf{e}_{-\mathbf{k},-\lambda} \tag{8.115}$$

Substitution of this result into Eq. (8.112) leads to

$$\hat{P}\hat{\mathbf{A}}(\mathbf{x},t)\hat{P}^{-1} = \eta_p\hat{\mathbf{A}}(-\mathbf{x},t) \tag{8.116}$$

From the equations of motion in Eqs. (8.109), one must take $\eta_p = -1$ so that $\hat{\mathbf{A}}$ behaves as a true vector under spatial reflection. Therefore

$$\hat{P}\hat{\mathbf{A}}(\mathbf{x},t)\hat{P}^{-1} = -\hat{\mathbf{A}}(-\mathbf{x},t)$$
$$\eta_{p_\gamma} = -1 \tag{8.117}$$

One can now just read off the parity eigenvalues of the two-photon states shown in Table 8.5.

8.3.2.2 *Charge Conjugation*

Again, $\mathbf{A}$ is a neutral field, and by itself, $\hat{C}$ does not mean anything for the free field. We *can* ask what happens to the $\mathbf{A}$ in Eqs. (8.109) when all particles are taken into antiparticles. It must behave as $\mathbf{j}$, and thus

$$\hat{C}\hat{\mathbf{A}}(x)\hat{C}^{-1} = -\hat{\mathbf{A}}(x) \tag{8.118}$$

In the interaction picture, this transformation now leaves the interaction of the photon field with the current *invariant*

$$\mathcal{H}_1(x) = -\hat{\mathbf{j}}(x) \cdot \hat{\mathbf{A}}(x) \qquad ; \text{ invariant} \tag{8.119}$$

The charge-conjugation operator for the photon field is thus defined by

$$\hat{C}a_{\mathbf{k}\lambda}^\dagger\hat{C}^{-1} = \eta_c a_{\mathbf{k}\lambda}^\dagger$$
$$\hat{C}|0\rangle = |0\rangle$$
$$\hat{C}^{-1} = \hat{C}^\dagger \tag{8.120}$$

To reproduce Eq. (8.118), one must take

$$\eta_{c_\gamma} = -1 \tag{8.121}$$

It is an immediate consequence of these relations that the n-photon state is an *eigenstate* of charge conjugation, with eigenvalues $(-1)^{n}$ [22]

$$\hat{C}|n\gamma\rangle = (-1)^{n}|n\gamma\rangle \qquad ; \text{ eigenstate} \qquad (8.122)$$

8.3.2.3 Time Reversal

Classical correspondence dictates that

$$\hat{T}\hat{\mathbf{A}}(\mathbf{x},t)\hat{T}^{-1} = -\hat{\mathbf{A}}(\mathbf{x},-t) \qquad (8.123)$$

The time-reversal operator for photons is thus defined to be anti-unitary

$$\langle b|\hat{T}^{-1}|a\rangle = \langle b|\hat{T}^{\dagger}|a\rangle^{\star} = \langle Tb|a\rangle^{\star} = \langle a|\hat{T}|b\rangle$$
$$\hat{T}i\hat{T}^{-1} = -i \qquad (8.124)$$

and its effect on the creation and destruction operators is given by

$$\hat{T}a^{\dagger}_{\mathbf{k}\lambda}\hat{T}^{-1} = \eta_t a^{\dagger}_{-\mathbf{k}\lambda} \qquad ; \hat{T}a_{\mathbf{k}\lambda}\hat{T}^{-1} = \eta_t a_{-\mathbf{k}\lambda}$$
$$\hat{T}|0\rangle = |0\rangle \qquad (8.125)$$

The field then transforms according to

$$\hat{T}\hat{\mathbf{A}}(x)\hat{T}^{-1} = \sum_{\mathbf{k}}\sum_{\lambda=\pm 1}\left(\frac{2\pi\hbar c^2}{\omega_k}\right)^{1/2}\eta_t\left(a_{-\mathbf{k},\lambda}\mathbf{e}^{\dagger}_{\mathbf{k}\lambda}e^{-ik\cdot x} + a^{\dagger}_{-\mathbf{k},\lambda}\mathbf{e}_{\mathbf{k}\lambda}e^{ik\cdot x}\right) \qquad (8.126)$$

A change of dummy summation variables then gives

$$\hat{T}\hat{\mathbf{A}}(x)\hat{T}^{-1} = \sum_{\mathbf{k}}\sum_{\lambda=\pm 1}\left(\frac{2\pi\hbar c^2}{\omega_k}\right)^{1/2}\eta_t\Big[a_{\mathbf{k},\lambda}\mathbf{e}^{\dagger}_{-\mathbf{k},\lambda}\,e^{i\mathbf{k}\cdot\mathbf{x}-i\omega_k(-t)}+$$
$$a^{\dagger}_{\mathbf{k},\lambda}\mathbf{e}_{-\mathbf{k},\lambda}\,e^{-i\mathbf{k}\cdot\mathbf{x}+i\omega_k(-t)}\Big] \qquad (8.127)$$

Now use from Eqs. (8.104) and (8.115)

$$\mathbf{e}^{\dagger}_{-\mathbf{k}\lambda} = \mathbf{e}^{\dagger}_{\mathbf{k},-\lambda} = (-1)^{\lambda}\mathbf{e}_{\mathbf{k}\lambda} = -\mathbf{e}_{\mathbf{k}\lambda} \qquad ; \lambda = \pm 1 \qquad (8.128)$$

Hence

$$\hat{T}\hat{\mathbf{A}}(\mathbf{x},t)\hat{T}^{-1} = -\eta_t\hat{\mathbf{A}}(\mathbf{x},-t) \qquad (8.129)$$

---

[22] Use $\hat{C}a^{\dagger}_{\mathbf{k}\lambda}|0\rangle = -a^{\dagger}_{\mathbf{k}\lambda}|0\rangle$, etc.

Here, to reproduce Eq. (8.123), one must take

$$\eta_{t_\gamma} = +1 \qquad (8.130)$$

8.4 Selection Rules

It will be assumed that the interaction hamiltonian density in the interaction picture is a scalar under the discrete symmetries of $\hat{C}$, $\hat{P}$ and $\hat{T}$

$$\hat{C}\hat{\mathcal{H}}_1(x)\hat{C}^{-1} = \hat{\mathcal{H}}_1(x)$$
$$\hat{P}\hat{\mathcal{H}}_1(\mathbf{x},t)\hat{P}^{-1} = \hat{\mathcal{H}}_1(-\mathbf{x},t)$$
$$\hat{T}\hat{\mathcal{H}}_1(\mathbf{x},t)\hat{T}^{-1} = \hat{\mathcal{H}}_1(\mathbf{x},-t) \qquad (8.131)$$

- We have shown that this is true for the electromagnetic interaction;
- This is built into current theories of the strong interactions;
- It is *not* true for the weak interactions.

The scattering operator is given in terms of $\hat{\mathcal{H}}_1(x)$ by[23]

$$\hat{S} = \sum_{n=0}^{\infty} \frac{1}{n!} \left(\frac{-i}{\hbar c} \right)^n \int \cdots \int d^4x_1 \cdots d^4x_n P[\hat{\mathcal{H}}(x_1)\hat{\mathcal{H}}(x_2)\cdots\hat{\mathcal{H}}(x_n)] \qquad (8.132)$$

Here $d^4x = d^3x\,cdt$ and $P[\cdots]$ indicates the P-product of the operators. The relations in Eqs. (8.131) then imply the following conditions on the scattering operator

$$\hat{C}\hat{S}\hat{C}^{-1} = \hat{S}$$
$$\hat{P}\hat{S}\hat{P}^{-1} = \hat{S}$$
$$\hat{T}\hat{S}\hat{T}^{-1} = \hat{S}^\dagger \qquad (8.133)$$

We give a few examples of the applications of these results.

8.4.1 *Some Examples*

(1) $\pi^0 \to n\gamma$: Consider the decay of the π^0 into photons. In the rest-frame of the π^0, the S-matrix can be written

$$\langle n\gamma|\hat{S}|\pi^0\rangle = \langle n\gamma|\hat{C}^{-1}\hat{C}\hat{S}\hat{C}^{-1}\hat{C}|\pi^0\rangle$$
$$= (-1)^n \eta_{c_\pi} \langle n\gamma|\hat{C}\hat{S}\hat{C}^{-1}|\pi^0\rangle \qquad (8.134)$$

---

[23]See chapter 4 in Vol. II; see also Prob. 8.6.

Thus

$$[1 - (-1)^n \eta_{c_\pi}] \langle n\gamma | \hat{S} | \pi^0 \rangle = 0 \qquad (8.135)$$

The S-matrix element therefore vanishes unless

$$(-1)^n \eta_{c_\pi} = 1 \qquad (8.136)$$

The π^0 is observed to decay into 2 photons, hence $\eta_{c_\pi} = +1$. It is then forbidden by charge conjugation from decaying into an *odd* number of photons.

(2) $(\pi^+ \pi^-) \to n\gamma$; We have shown that for the two-particle $(\pi^+ \pi^-)$ states in the C-M system

$$\hat{C} | ELM \rangle = (-1)^L | ELM \rangle$$
$$\hat{P} | ELM \rangle = (-1)^L | ELM \rangle \qquad (8.137)$$

A repetition of the argument in Eqs. (8.134)–(8.135) gives

$$[1 - (-1)^{n+L}] \langle n\gamma | \hat{S} | ELM \rangle = 0 \qquad (8.138)$$

Hence

$$(-1)^{n+L} = 1 \qquad (8.139)$$

States of even L can only decay into an even number of photons, and states of odd l into an odd number of photons.

(3) $(e^+ e^-) \to n\gamma$: Consider positronium in the C-M system. The properties of the two-particle fermion-antifermion states have been shown to be

$$\hat{P} | ELSJM \rangle = (-1)^{L+1} | ELSJM \rangle$$
$$\hat{C} | ELSJM \rangle = (-1)^{L-S} | ELSJM \rangle \qquad (8.140)$$

A repetition of the above arguments then leads to the results summarized in Table 8.6.

Table 8.6   Decay of positronium.

State	C	$n\gamma$	P	$\mathbf{E}_1 \cdot \mathbf{E}_2$
$^3S_1$	-1	3 (odd)	-1	
$^1S_0$	$+1$	2 (even)	-1	$\perp$

Several comments:

- The eigenvalues of the $^1S_0$ and $^3S_1$ states under $\hat{C}$ and $\hat{P}$ are indicated;
- As above, the state with $C = +1$ can only decay to an even number of photons, and the state with $C = -1$ into an odd number;
- It is the $^1S_0$ state that can decay into two photons, and hence has the shortest lifetime;
- Since there is no angular momentum in the $^1S_0$ state, and since the parity is odd, it can only decay into the second two-photon state in Table 8.5;
- The second two-photon state in Table 8.5 can be re-written in terms of plane polarizations as

$$\frac{1}{\sqrt{2}} \left(a^\dagger_{\mathbf{k},+1} a^\dagger_{-\mathbf{k},+1} - a^\dagger_{\mathbf{k},-1} a^\dagger_{-\mathbf{k},-1} \right) |0\rangle = \frac{i}{\sqrt{2}} \left(a^\dagger_{\mathbf{k}1} a^\dagger_{-\mathbf{k}2} + a^\dagger_{\mathbf{k}2} a^\dagger_{-\mathbf{k}1} \right) |0\rangle$$

(8.141)

It is evident that the planes of polarization of the **E**-vectors of the two photons are *perpendicular* in this state;

- Since the π^0 has all the same quantum numbers as the $^1S_0$ state of positronium, an identical analysis applies to it![24]

(4) *An application of T-invariance*: Consider a scattering S-matrix element $\langle f|\hat{S}|i\rangle$. Write

$$\begin{aligned}
\langle f|\hat{S}|i\rangle &= \langle f|\hat{T}^{-1}\hat{T}\hat{S}\hat{T}^{-1}\hat{T}|i\rangle = \langle f|\hat{T}^\dagger\hat{T}\hat{S}\hat{T}^{-1}\hat{T}|i\rangle^\star \\
&= \langle f_T|\hat{T}\hat{S}\hat{T}^{-1}|i_T\rangle^\star = \langle f_T|\hat{S}^\dagger|i_T\rangle^\star \\
&= \langle i_T|\hat{S}|f_T\rangle
\end{aligned}$$

(8.142)

where the second equality follows from the anti-unitarity of $\hat{T}$, and

$$|i_T\rangle = \hat{T}|i\rangle \qquad ; \quad |f_T\rangle = \hat{T}|f\rangle$$

(8.143)

Thus time reversal relates the S-matrix element for a scattering process to that for the time-reversed states going in the opposite direction, exactly as in our original motivation in Fig. 8.1.[25]

---

[24]In fact, the π^0 can be modeled as the $^1S_0$ state of a quark–antiquark pair.
[25]Compare Prob. 8.7.

8.5   Lorentz Transformations

Continuous Lorentz transformations in quantum field theory are discussed in some detail in appendix E of Vol. II. It is not our intention to repeat that material here; however, it is important to summarize enough of that material so that our presentation of the CPT theorem in the next section makes some sense. We start by summarizing Lorentz transformations of the spin-zero field.

8.5.1   *Spin-Zero Field*

First, define new creation and destruction operators by

$$\tilde{c}^\dagger_\mathbf{k} \equiv \sqrt{2k_0\Omega}\, c^\dagger_\mathbf{k} \qquad ; \; k_\mu = (\mathbf{k}, ik_0) \tag{8.144}$$

Here the four-vector $k_\mu = (\mathbf{k}, ik_0)$, with $k_0 = \sqrt{\mathbf{k}^2 + \mu^2}$, μ is the inverse Compton wavelength, and Ω is the quantization volume. The scalar field in the interaction picture can then be re-written as[26]

$$\hat{\phi}(x) = \left(\frac{\hbar}{c}\right)^{1/2} \frac{1}{(2\pi)^3} \int d^4k\, \delta(k^2 + \mu^2)\theta(k_0)\left(\tilde{c}_\mathbf{k} e^{ik\cdot x} + \tilde{c}^\dagger_\mathbf{k} e^{-ik\cdot x}\right) \tag{8.145}$$

The unitary Lorentz transformation operator $\hat{U}(L)$ is then defined by the relations

$$\hat{U}(L)\tilde{c}^\dagger_\mathbf{k}\hat{U}(L)^{-1} = \tilde{c}^\dagger_{\mathbf{k}'} \qquad ; \; k'_\mu = a_{\mu\nu}(-v)k_\nu$$
$$\hat{U}(L)|0\rangle = |0\rangle$$
$$\hat{U}(L)^\dagger = \hat{U}(L)^{-1} \tag{8.146}$$

Here $a_{\mu\nu}(-v)$ is the Lorentz transformation matrix to a new frame moving with velocity $-v$ relative to the first.[27]

Under this Lorentz transformation, the scalar field transforms as

$$\hat{U}(L)\hat{\phi}(x)\hat{U}(L)^{-1} = \hat{\phi}(x') \qquad ; \; x'_\mu = a_{\mu\nu}(-v)x_\nu \tag{8.147}$$

The scalar field simply gets evaluated at the Lorentz-transformed position.

---

[26] Here $k\cdot x = \mathbf{k}\cdot\mathbf{x} - ik_0 ct$, $k^2 = \mathbf{k}^2 - k_0^2$, $d^4k = d^3k\, dk_0$, and $\theta(k_0)$ is the step-function.

[27] For simplicity in these arguments, let $\mathbf{v}$ lie along the z-axis as in Vol. II. This Lorentz transformation is defined to be an *active* transformation ("boost"), hence the $-v$.

8.5.2 Spin-1/2 Field

For the Dirac field, we first define new spinors with invariant norms

$$\mathcal{U}(\mathbf{k}\lambda) \equiv \left(\frac{k_0}{M}\right)^{1/2} u(\mathbf{k}\lambda) \quad ; \bar{\mathcal{U}}\mathcal{U} = 1$$

$$\mathcal{V}(-\mathbf{k}\lambda) \equiv \left(\frac{k_0}{M}\right)^{1/2} v(-\mathbf{k}\lambda) \quad ; \bar{\mathcal{V}}\mathcal{V} = -1 \qquad (8.148)$$

Here $k_0 = \sqrt{\mathbf{k}^2 + M^2}$. New creation and destruction operators are then defined as in Eq. (8.144)

$$\tilde{a}_{\mathbf{k}\lambda}^{\dagger} \equiv \sqrt{2k_0\Omega}\, a_{\mathbf{k}\lambda}^{\dagger} \qquad ; \tilde{b}_{\mathbf{k}\lambda}^{\dagger} \equiv \sqrt{2k_0\Omega}\, b_{\mathbf{k}\lambda}^{\dagger} \qquad (8.149)$$

The Dirac field is re-written in terms of these quantities as

$$\hat{\psi}(x) = \frac{\sqrt{2M}}{(2\pi)^3} \int d^4k\, \delta(k^2 + M^2)\theta(k_0) \times$$

$$\sum_{\lambda} \left(\tilde{a}_{\mathbf{k}\lambda} \mathcal{U}(\mathbf{k}\lambda)e^{ik\cdot x} + \tilde{b}_{\mathbf{k}\lambda}^{\dagger} \mathcal{V}(-\mathbf{k}\lambda)e^{-ik\cdot x}\right) \qquad (8.150)$$

The unitary Lorentz transformation operator $\hat{U}(L)$ for the Dirac field is then defined by

$$\hat{U}(L)\tilde{a}_{\mathbf{k}\lambda}^{\dagger}\hat{U}(L)^{-1} = \sum_{\lambda'} \alpha_{\lambda\lambda'}\, \tilde{a}_{\mathbf{k}'\lambda'}^{\dagger} \quad ; \hat{U}(L)\tilde{b}_{\mathbf{k}\lambda}^{\dagger}\hat{U}(L)^{-1} = \sum_{\lambda'} \alpha_{\lambda\lambda'}^{\star}\, \tilde{b}_{\mathbf{k}'\lambda'}^{\dagger}$$

$$\hat{U}(L)|0\rangle = |0\rangle$$

$$\hat{U}(L)^{\dagger} = \hat{U}(L)^{-1} \qquad (8.151)$$

Here $\alpha_{\lambda\lambda'}$ is a unitary 2×2 matrix defined through the relation[28]

$$S^{-1}\mathcal{U}(\mathbf{k}\lambda) = \sum_{\lambda'} \alpha_{\lambda\lambda'}\, \mathcal{U}(\mathbf{k}'\lambda') \qquad (8.152)$$

The 4×4 matrix S characterizing the Lorentz transformation is given by

$$S(\bar{\Omega}) = \exp\left(\frac{i}{4}\bar{\Omega}\,\alpha_{\mu\nu}\sigma_{\mu\nu}\right) \quad ; \sigma_{\mu\nu} = \frac{1}{2i}[\gamma_\mu, \gamma_\nu] \qquad (8.153)$$

Here $i\bar{\Omega}$ is the angle of rotation in the plane characterized by the four-dimensional tensor $\alpha_{\mu\nu}$.[29] This matrix has the following

---

[28]See Prob. 8.4.

[29]For Lorentz transformations in the z-direction, $\alpha_{\mu\nu} = (n_1)_\mu(n_2)_\nu - (n_2)_\mu(n_1)_\nu$, with $n_1 = (0,0,1,0)$ and $n_2 = (0,0,0,i)$, and $\tan i\bar{\Omega} = iv/c$. Compare Prob. E.4 in Vol. II.

important properties

$$\gamma_4 S^\dagger \gamma_4 = S^{-1}$$
$$S\gamma_\mu S^{-1} = a_{\mu\nu}(-v)\gamma_\nu \qquad (8.154)$$

The behavior of the Dirac field in Eq. (8.150) under a Lorentz transformation is then given by

$$\hat{U}(L)\hat{\psi}(x)\hat{U}(L)^{-1} = S\hat{\psi}(x') \qquad ; x'_\mu = a_{\mu\nu}(-v)x_\nu \quad (8.155)$$

As an example, it follows that the behavior of the Dirac electromagnetic current in Eq. (8.80) under a Lorentz transformation is now properly given by

$$\hat{U}(L)\hat{j}_\mu(x)\hat{U}(L)^{-1} = a_{\nu\mu}(-v)\hat{j}_\nu(x') \qquad (8.156)$$

where $a_{\nu\mu}(-v)$ is the transpose of the Lorentz transformation matrix.[30]

8.5.3 *Some Invariant Interactions*

A local lagrangian density $\hat{\mathcal{L}}_I(x)$ in the interaction picture is defined to be Lorentz-invariant if it is a *scalar* under Lorentz transformations

$$\hat{U}(L)\hat{\mathcal{L}}_I(x)\hat{U}(L)^{-1} = \hat{\mathcal{L}}_I(x') \qquad ; \text{Lorentz-invariant} \quad (8.157)$$

We give two examples:

(1) Let $\hat{\psi}$ be a Dirac field, $\hat{\phi}$ a pseudoscalar field, and consider

$$\hat{\mathcal{L}}_I(x) = ig\hat{\bar{\psi}}(x)\gamma_5\hat{\psi}(x)\hat{\phi}(x) \qquad (8.158)$$

Under a Lorentz transformation, this transforms as

$$\begin{aligned}
\hat{U}(L)\hat{\mathcal{L}}_I(x)\hat{U}(L)^{-1} &= ig\hat{\psi}^\dagger(x')S^\dagger\gamma_4\gamma_5 S\hat{\psi}(x')\hat{\phi}(x') \\
&= ig\hat{\bar{\psi}}(x')S^{-1}\gamma_5 S\hat{\psi}(x')\hat{\phi}(x') \\
&= ig\hat{\bar{\psi}}(x')\gamma_5\hat{\psi}(x')\hat{\phi}(x') \qquad (8.159)
\end{aligned}$$

The last relation follows since $[\gamma_5, S] = 0$.

Thus Eq. (8.157) is satisfied for the interaction in Eq. (8.158);

---

[30]This corresponds to the classical relation $j_\mu(x') = a_{\mu\nu}(-v)j_\nu(x)$. The second of Eqs. (8.154) differs from the result in Vol. II through the sign of v.

(2) Let $\hat{\psi}$ be a Dirac field, $\hat{\phi}$ be a scalar field, and consider

$$\hat{\mathcal{L}}_I(x) = ig\hat{\bar{\psi}}(x)\gamma_\mu\hat{\psi}(x)\frac{\partial\hat{\phi}(x)}{\partial x_\mu} \tag{8.160}$$

Under a Lorentz transformation this becomes

$$\hat{U}(L)\hat{\mathcal{L}}_I(x)\hat{U}(L)^{-1} = ig\hat{\bar{\psi}}(x')S^{-1}\gamma_\mu S\hat{\psi}(x')\frac{\partial\hat{\phi}(x')}{\partial x_\mu} \tag{8.161}$$

Now use $x'_\mu = a_{\mu\nu}x_\nu$, and $a_{\mu\nu}a_{\mu\lambda} = \delta_{\nu\lambda}$, to compute

$$\frac{\partial}{\partial x'_\mu} = a_{\mu\nu}\frac{\partial}{\partial x_\nu} \qquad ; \qquad \frac{\partial}{\partial x_\mu} = a_{\nu\mu}\frac{\partial}{\partial x'_\nu}$$
$$S^{-1}a_{\nu\mu}\gamma_\mu S = \gamma_\nu \tag{8.162}$$

Hence

$$\hat{U}(L)\hat{\mathcal{L}}_I(x)\hat{U}(L)^{-1} = ig\hat{\bar{\psi}}(x')\gamma_\nu\hat{\psi}(x')\frac{\partial\hat{\phi}(x')}{\partial x'_\nu} \tag{8.163}$$

Thus Eq. (8.157) is again satisfied for the interaction in Eq. (8.160).

The reader now has enough tools to easily build Lorentz-invariant lagrangian densities.

8.6 The *CPT* Theorem

The *CPT* theorem due to [Schwinger (1951); Lüders (1954)] states the following: Given

- A local quantum field theory with an hermitian interaction;
- Invariance under proper Lorentz transformations.

Then the intrinsic phases can always be chosen so that the *theory is invariant under the combined operations CPT, taken in any order.*
    We will not attempt a general proof here,[31] but will be content to illustrate the ideas with two specific interactions: one a fermion-boson model,

---

[31]See [Lüders (1957)].

with potential applicability to the strong interactions, and another a contact four-fermion model, applicable to the weak interactions[32]

$$\mathcal{L}_S(x) = \bar{\psi}_1(a + ib\gamma_5)\psi_2\phi + \bar{\psi}_1 i(c + d\gamma_5)\gamma_\mu\psi_2\frac{\partial\phi}{\partial x_\mu} + \text{h.a.}$$

$$\mathcal{L}_W(x) = \left[\bar{\psi}_1(a + ib\gamma_5)\psi_2\right]\left[\bar{\psi}_3(c + id\gamma_5)\psi_4\right] +$$
$$\left[\bar{\psi}_1 i(e + f\gamma_5)\gamma_\mu\psi_2\right]\left[\bar{\psi}_3 i(g + h\gamma_5)\gamma_\mu\psi_4\right] +$$
$$k\left[\bar{\psi}_1\sigma_{\mu\nu}\psi_2\right]\left[\bar{\psi}_3\sigma_{\mu\nu}\psi_4\right] + \text{h.a.} \qquad (8.164)$$

Here ϕ is a spin-zero field, $(\psi_1, \cdots, \psi_4)$ are independent Dirac fields, $(a, b, \cdots, k)$ are c-number coupling constants, and in both cases the lagrangian density is a sum of the indicated interaction and its hermitian adjoint (h.a.).

A few comments:

- These lagrangian densities are evidently scalars under proper Lorentz transformations;
- The interactions are hermitian by construction;
- There is no difficulty showing that the *free* lagrangians are invariant under the combined operation TCP

$$TCP\mathcal{L}_0(x)P^{-1}C^{-1}T^{-1} = \mathcal{L}_0(x') \qquad ; x'_\mu = -x_\mu \qquad (8.165)$$

- The electromagnetic interaction (not illustrated here) has been shown to be *separately* invariant under P, C, and T;
- No attempt has been made to impose *separate* P, C, or T symmetry on the interaction densities in Eqs. (8.164).

Let us now make use of our previous results to compute the effect of the *combined* operation TCP on the interactions in Eqs. (8.164).

For the spin-zero field

$$TCP\phi(x)P^{-1}C^{-1}T^{-1} = \eta^*_{p_\phi}\eta_{c_\phi}\eta_{t_\phi}\phi^*(x') \qquad ; x'_\mu = -x_\mu$$

$$TCP\frac{\partial\phi(x)}{\partial x_\mu}P^{-1}C^{-1}T^{-1} = \varepsilon\,\eta^*_{p_\phi}\eta_{c_\phi}\eta_{t_\phi}\frac{\partial\phi^*(x')}{\partial x'_\mu} \qquad (8.166)$$

Here the sign factor ε in the last expression is

$$\varepsilon = -1 \qquad ; \mu = 1, 2, 3$$
$$= +1 \qquad ; \mu = 4 \qquad (8.167)$$

---

[32]At this point the reader should be sufficiently sophisticated that for ease of writing, we can proceed to suppress the hats over *all* the operators in the abstract Hilbert space.

Let O^i be one of the sixteen hermitian Dirac matrices in Table 8.7.

Table 8.7   Dirac matrices O^i and sign factor ε^i in Eq. (8.169).

O^i	1	γ_5	γ_μ	$i\gamma_5\gamma_\mu$	$\sigma_{\mu\nu}$
ε^i	$+1$	-1	$+1\ \mu=1,2,3$ $-1\ \mu=4$	$-1\ \mu=1,2,3$ $+1\ \mu=4$	$+1\ \mu,\nu=1,2,3$ $-1\ \mu\ \text{or}\ \nu=4$

Then for the Dirac fields

$$TCP\bar\psi_1(x)O^i\psi_2(x)P^{-1}C^{-1}T^{-1} = \eta_{p_1}\eta_{p_2}^\star\eta_{c_1}^\star\eta_{c_2}\eta_{t_1}\eta_{t_2}^\star \times$$
$$\bar\psi_2(x')\gamma_3\gamma_1\gamma_4\gamma_2(\gamma_4 O^i\gamma_4)^\dagger\gamma_2\gamma_4\gamma_1\gamma_3\psi_1(x')$$
$$;\ x'_\mu = -x_\mu \qquad (8.168)$$

Here we have used the following:

- The fermion fields anticommute;
- $\gamma_4\gamma_2 = -\gamma_2\gamma_4$ in the charge conjugation;
- $(\gamma_4\gamma_2)^\star = \gamma_4\gamma_2$ in the time reversal;
- For a matrix, $(M^T)^\star = M^\dagger$;
- Note that $(\gamma_4 O^i\gamma_4)^\dagger = \gamma_4 O^i\gamma_4$ since the O^i are hermitian.

For the Dirac matrix product in Eq. (8.168) we thus arrive at

$$\bar\psi_2(x')\gamma_3\gamma_1\gamma_4\gamma_2(\gamma_4 O^i\gamma_4)^\dagger\gamma_2\gamma_4\gamma_1\gamma_3\psi_1(x') = \bar\psi_2(x')\gamma_5\gamma_4 O^i\gamma_4\gamma_5\psi_1(x')$$
$$= \varepsilon^i\bar\psi_2(x')O^i\psi_1(x') \qquad (8.169)$$

where the appropriate sign factors ε^i are listed in Table 8.7.[33] A combination of these expressions gives

$$TCP\bar\psi_1(x)O^i\psi_2(x)P^{-1}C^{-1}T^{-1} = \varepsilon^i\,\eta_{p_1}\eta_{p_2}^\star\eta_{c_1}^\star\eta_{c_2}\eta_{t_1}\eta_{t_2}^\star\bar\psi_2(x')O^i\psi_1(x') \qquad (8.170)$$

Note that, just as in Eq. (8.167), the sign factors ε^i reflect only the Lorentz transformation properties of these Dirac bilinear covariants!

Let us now *choose* to impose the following conditions on the intrinsic phases[34]

- For theory-S:

$$\eta_{p_\phi}^\star\eta_{c_\phi}\eta_{t_\phi}\eta_{p_1}\eta_{p_2}^\star\eta_{c_1}^\star\eta_{c_2}\eta_{t_1}\eta_{t_2}^\star = +1 \qquad (8.171)$$

---

[33]There is no sum on i in Eq. (8.169). Recall $\gamma_5 = \gamma_1\gamma_2\gamma_3\gamma_4$, and $\gamma_5\gamma_\mu + \gamma_\mu\gamma_5 = 0$.
[34]Remember, we are free to define the transformation any way we want; the only question is whether it is *useful*.

- For theory-W:

$$(\eta_{p_1}\eta_{p_2}^\star\eta_{p_3}\eta_{p_4}^\star)(\eta_{c_1}^\star\eta_{c_2}\eta_{c_3}^\star\eta_{c_4})(\eta_{t_1}\eta_{t_2}^\star\eta_{t_3}\eta_{t_4}^\star) = +1 \qquad (8.172)$$

The interactions in Eqs. (8.164) then transform under TCP as follows:

- For $\mathcal{L}_S(x)$:

$$TCP\mathcal{L}_S(x)P^{-1}C^{-1}T^{-1} = \bar{\psi}_2(a^\star + ib^\star\gamma_5)\psi_1\phi^\star(x') +$$
$$\bar{\psi}_2 i(c^\star + d^\star\gamma_5)\gamma_\mu\psi_1\frac{\partial\phi^\star(x')}{\partial x'_\mu} + \text{h. a.}$$
$$= \mathcal{L}_S(x') \qquad\qquad ; \ x'_\mu = -x_\mu \qquad (8.173)$$

- For $\mathcal{L}_W(x)$:

$$TCP\mathcal{L}_W(x)P^{-1}C^{-1}T^{-1} = \left[\bar{\psi}_2(a^\star + ib^\star\gamma_5)\psi_1\right]\left[\bar{\psi}_4(c^\star + id^\star\gamma_5)\psi_3\right] +$$
$$\left[\bar{\psi}_2 i(e^\star + f^\star\gamma_5)\gamma_\mu\psi_1\right]\left[\bar{\psi}_4 i(g^\star + h^\star\gamma_5)\gamma_\mu\psi_3\right] +$$
$$k^\star\left[\bar{\psi}_2\sigma_{\mu\nu}\psi_1\right]\left[\bar{\psi}_4\sigma_{\mu\nu}\psi_3\right] + \text{h. a.}$$
$$= \mathcal{L}_W(x') \qquad\qquad ; \ x'_\mu = -x_\mu \qquad (8.174)$$

The TCP transformation simply interchanges the role of the interaction and its hermitian adjoint (h.a.), and hence in both cases

$$TCP\mathcal{L}(x)P^{-1}C^{-1}T^{-1} = \mathcal{L}(x') \qquad\qquad ; \ x'_\mu = -x_\mu \qquad (8.175)$$

and the theorem is established.

One can now go to the scattering operator S, using[35]

$$TCP\mathcal{H}_I(x)P^{-1}C^{-1}T^{-1} = \mathcal{H}_I(x') \qquad ; \ x'_\mu = -x_\mu \qquad (8.176)$$

It follows as in Eq. (8.133) that

$$TCPSP^{-1}C^{-1}T^{-1} = S^\dagger \qquad (8.177)$$

We now know how to exploit this important relation.

In *summary*, experiment indicates that[36]

- The strong interactions have P, C, T separately as symmetries;
- The electromagnetic interactions also have P, C, T separately as symmetries;
- The weak interactions violate P and CP symmetry;
- TCP is an exact symmetry of nature.

---

[35]One must still go from $\mathcal{L} \to \mathcal{H}$ (see Prob. 8.9).
[36]See [Particle Data Group (2012)].

8.6.1 *Two Consequences*

We proceed to derive two consequences of the CPT theorem due to Lee and Yang (and others).

Theorem I: Particles and Antiparticles Have the Same Mass.
To prove this result, go to the rest frame of particle a where[37]

$$H|M_a\rangle = M_a|M_a\rangle$$
$$H = \int d^3x\, \mathcal{H}(\mathbf{x}, 0) \tag{8.178}$$

TCP symmetry implies

$$TCPHP^{-1}C^{-1}T^{-1} = H$$
$$\text{or;} \qquad (TCP)H = H(TCP) \tag{8.179}$$

Now act on Eq. (8.178) with (TCP)

$$(TCP)H|M_a\rangle = H(TCP)|M_a\rangle = \eta_t\eta_c^\star\eta_p H|M_{\bar{a}}\rangle$$
$$= M_a(TCP)|M_a\rangle = \eta_t\eta_c^\star\eta_p M_a|M_{\bar{a}}\rangle \tag{8.180}$$

Here $|M_{\bar{a}}\rangle$ is the corresponding antiparticle state. Hence

$$H|M_{\bar{a}}\rangle = M_a|M_{\bar{a}}\rangle \tag{8.181}$$

and the result is established.

Theorem II: Particles and Antiparticles Have the Same Lifetime.
Consider the decay processes

$$a \to b + c$$
$$\bar{a} \to \bar{b} + \bar{c} \tag{8.182}$$

We assume these decays are governed by the weak interaction H_W, which does not separately conserve P, C, or T, and we work to lowest order in H_W.

The decay rate for the first process in Eqs. (8.182) is given by

$$\omega_a = \frac{2\pi}{\hbar} \sum_n |\langle n|H_W|M_a\rangle|^2 \delta(E_n - M_a) \tag{8.183}$$

---

[37] Here $M_a \equiv m_a c^2$.

The combined operation TCP *is* a good symmetry, and it follows that

$$|\langle n|H_W|M_a\rangle|^2 = |\langle n|P^{-1}C^{-1}T^{-1}TCPH_WP^{-1}C^{-1}T^{-1}TCP|M_a\rangle|^2$$
$$= |\langle M_{\bar{a}}|H_W|\bar{n}_{t,p}\rangle|^2 = |\langle \bar{n}_{t,p}|H_W|M_{\bar{a}}\rangle|^2 \qquad (8.184)$$

The last equality holds since H_W is hermitian.

Now use $\sum_{\bar{n}} = \sum_{\bar{n}_{t,p}}$, which just involves a change of dummy variables (spins, momenta), and both go over a complete set of final states for a given final energy $E_{\bar{n}} = E_n$ equal to the initial energy $M_{\bar{a}} = M_a$.[38] Hence Eq. (8.183) can be re-written

$$\omega_a = \frac{2\pi}{\hbar} \sum_{\bar{n}_{t,p}} |\langle \bar{n}_{t,p}|H_W|M_{\bar{a}}\rangle|^2 \delta(E_{\bar{n}_{t,p}} - M_{\bar{a}})$$
$$= \frac{2\pi}{\hbar} \sum_{\bar{n}} |\langle \bar{n}|H_W|M_{\bar{a}}\rangle|^2 \delta(E_{\bar{n}} - M_{\bar{a}})$$
$$= \omega_{\bar{a}} \qquad (8.185)$$

This establishes the second theorem.[39]

---

[38] This is true even if particles (b, c) interact strongly.
[39] At least to leading order in H_W.

Chapter 9

Heisenberg Picture

This chapter is concerned with some general results obtained using only the previously-discussed symmetry properties, without actually solving the theory. We start by recalling the transformations between one picture and another. The analysis will be carried out for *finite times*, and it is the interaction picture, with adiabatic damping, that is used to go off to $t = \pm\infty$.[1]

9.1 Change of Picture

Start with the Schrödinger picture.

9.1.1 *Schrödinger Picture*

In the Schrödinger picture, the state vector depends on time, and the time development is given by the Schrödinger equation[2]

$$i\hbar\frac{\partial}{\partial t}|\Psi_S(t)\rangle = H|\Psi_S(t)\rangle \qquad ; \text{S-eqn} \qquad (9.1)$$

The operators are independent of time. This is quantum mechanics as we have developed it.

9.1.2 *Interaction Picture*

The interaction picture is useful for scattering analysis. For example, it is in the interaction picture that the Feynman rules are developed with

---

[1]This chapter assumes familiarity with chapters 4,6,7 and appendices E,F in Vol. II.

[2]Again, we are now familiar enough with the material that for simplicity of presentation, the carets can be suppressed on the operators in abstract Hilbert space.

373

the aid of Wick's theorem.[3] The interaction picture is obtained from the Schrödinger picture through a unitary transformation at the time t

$$|\Psi_I(t)\rangle = e^{iH_0t/\hbar}|\Psi_S(t)\rangle$$
$$H_I(t) = e^{iH_0t/\hbar}H_1e^{-iH_0t/\hbar} \qquad (9.2)$$

The Schrödinger equation in the interaction picture becomes

$$i\hbar\frac{\partial}{\partial t}|\Psi_I(t)\rangle = H_I(t)|\Psi_I(t)\rangle \qquad ; \text{S-eqn} \qquad (9.3)$$

The state vectors and operators in the Schrödinger and interaction pictures coincide at the time $t = 0$

$$|\Psi_I(0)\rangle = |\Psi_S(0)\rangle$$
$$O_I(0) = O_S \qquad (9.4)$$

The state vector in the interaction picture propagates in time according to

$$|\Psi_I(t)\rangle = U(t,t_0)|\Psi_I(t_0)\rangle \qquad (9.5)$$

For finite times, and with a time-independent H, the time-development operator $U(t,t_0)$ is given by

$$U(t,t_0) = e^{iH_0t/\hbar}e^{-iH(t-t_0)/\hbar}e^{-iH_0t_0/\hbar} \qquad ; \text{finite times} \qquad (9.6)$$

An alternate expression for this quantity is [compare Eq. (6.189)]

$$U(t,t_0) = \sum_{n=0}^{\infty}\frac{1}{n!}\left(\frac{-i}{\hbar c}\right)^n\int_{t_0}^{t}\cdots\int_{t_0}^{t}d^4x_1\cdots d^4x_n P\left[\mathcal{H}_I(x_1)\cdots\mathcal{H}_I(x_n)\right] \quad (9.7)$$

Here

- The four-dimensional volume element is $d^4x = cd^3xdt$;
- The symbol $P[\cdots]$ indicates a *time-ordered* product of the operators;[4]
- $\mathcal{H}_I(x)$ is the hamiltonian *density* in the interaction picture;
- With adiabatic damping, one replaces

$$\mathcal{H}_I(x) \to e^{-\epsilon|t|}\mathcal{H}_I(x) \qquad ; \epsilon \to 0 \qquad (9.8)$$

The theory is then defined in the limit $\epsilon \to 0$;

---

[3]See also [Fetter and Walecka (2003)].
[4]More generally, the P-product includes a factor of (-1) raised to the number of interchanges of fermion operators required to achieve the time-ordering—see Vol. II.

- "Finite time" in the sense of Eq. (9.6) implies that

$$|t| \ll 1/\epsilon \qquad ; \text{finite time} \qquad (9.9)$$

It will henceforth be assumed that under this condition, H is independent of time.

9.1.3 Heisenberg Picture

The Heisenberg picture is also obtained from the Schrödinger picture through a unitary transformation, but in this case utilizing the full hamiltonian at a finite time

$$|\Psi_H\rangle = e^{iHt/\hbar}|\Psi_S(t)\rangle$$
$$O_H(t) = e^{iHt/\hbar}O_S e^{-iHt/\hbar} \qquad (9.10)$$

The state vector in the Heisenberg picture is independent of time

$$i\hbar\frac{\partial}{\partial t}|\Psi_H\rangle = 0 \qquad ; \text{S-eqn} \qquad (9.11)$$

In the Heisenberg picture, all the time dependence goes into the operators.

The relation between the Heisenberg and interaction pictures follows directly from the above

$$|\Psi_H\rangle = e^{iHt/\hbar}e^{-iH_0t/\hbar}|\Psi_I(t)\rangle$$
$$O_H(t) = e^{iHt/\hbar}e^{-iH_0t/\hbar}O_I(t)e^{iH_0t/\hbar}e^{-iHt/\hbar} \qquad (9.12)$$

Note that *all three pictures coincide* at time $t = 0$

$$|\Psi_H\rangle = |\Psi_S(0)\rangle = |\Psi_I(0)\rangle$$
$$O_S = O_I(0) = O_H(0) \qquad (9.13)$$

Note also from Eqs. (9.6) and (9.12) that

$$O_H(t) = U(0,t)O_I(t)U(t,0) \qquad (9.14)$$

The fact that we know how the interaction state vector propagates in time allows us to express the Heisenberg state vector as

$$|\Psi_H\rangle = |\Psi_I(0)\rangle = U(0,t_0)|\Psi_I(t_0)\rangle \qquad (9.15)$$

One can now let the time $t_0 \to \pm\infty$. Then, with adiabatic switching of the interaction H_1, the state in the interaction picture goes to an eigenstate of

H_0 with the *correct mass*

$$|\Psi_I(t_0)\rangle \rightarrow |\psi\rangle \qquad ; t_0 \rightarrow \pm\infty$$
$$H_0|\psi\rangle = E_0|\psi\rangle \qquad\qquad\qquad (9.16)$$

It is convenient to denote the Heisenberg state by simply underlining the symbol

$$|\Psi_H\rangle \equiv |\underline{\Psi}\rangle \qquad ; \text{notation} \qquad (9.17)$$

We then define the following Heisenberg states

$$|\underline{\Psi}_i^{(+)}\rangle \equiv U(0,-\infty)|\psi_i\rangle \qquad ; \text{"incoming" states}$$
$$|\underline{\Psi}_f^{(-)}\rangle \equiv U(0,+\infty)|\psi_f\rangle \qquad ; \text{"outgoing" states} \qquad (9.18)$$

Some comments:

- The incoming states start out as $|\psi_i\rangle$ at $t = -\infty$ in the interaction picture. They then propagate up to time $t = 0$ where they produce the Heisenberg state $|\underline{\Psi}_i^{(+)}\rangle$;
- The outgoing states propagate back from $|\psi_f\rangle$ at $t = +\infty$ in the interaction picture to a time $t = 0$ where they coincide with the Heisenberg state $|\underline{\Psi}_f^{(-)}\rangle$;
- In the absence of bound states, *both the incoming and outgoing states form a complete orthonormal set.*

Recall the properties of $U(t,t_0)$, which follow directly from Eq. (9.6)[5]

$$U(t,t_0)^\dagger = U(t_0,t) = U(t,t_0)^{-1} \qquad ; \text{unitary}$$
$$U(t,t_1)U(t_1,t_0) = U(t,t_0) \qquad ; \text{group property} \qquad (9.19)$$

We proceed to prove two theorems which relate matrix elements in the Heisenberg and interaction pictures. The theorems are invaluable, since they provide the key to analyzing Heisenberg matrix elements in terms of Feynman diagrams.

---

[5]They also hold for the series in Eq. (9.7), even when the adiabatic damping factor is included—see Vol. II.

9.1.3.1 *Two Theorems*

Theorem Ia: The first theorem states that

$$\langle \underline{\Psi}_f^{(-)}|O_H(t)|\underline{\Psi}_i^{(+)}\rangle = \langle \psi_f| \sum_{\nu=0}^{\infty} \frac{1}{\nu!} \left(\frac{-i}{\hbar c}\right)^{\nu} \int_{-\infty}^{\infty} \cdots \int_{-\infty}^{\infty} d^4x_1 \cdots d^4x_\nu \times$$
$$P\left[\mathcal{H}_I(x_1)\cdots\mathcal{H}_I(x_\nu)O_I(t)\right]|\psi_i\rangle \qquad (9.20)$$

The proof goes as follows. Use Eqs. (9.14), (9.18), and (9.19) to write the l.h.s. as

$$\begin{aligned}\text{l.h.s.} &= \langle \psi_f|U(\infty,0)U(0,t)O_I(t)U(t,0)U(0,-\infty)|\psi_i\rangle \\ &= \langle \psi_f|U(\infty,t)O_I(t)U(t,-\infty)|\psi_i\rangle\end{aligned} \qquad (9.21)$$

Now write the operator in the last matrix element as

$$U(\infty,t)O_I(t)U(t,-\infty) = \qquad (9.22)$$
$$\sum_{n=0}^{\infty} \frac{1}{n!} \left(\frac{-i}{\hbar c}\right)^{n} \int_{t}^{\infty} \cdots \int_{t}^{\infty} d^4x_1 \cdots d^4x_n P\left[\mathcal{H}_I(x_1)\cdots\mathcal{H}_I(x_n)\right]O_I(t) \times$$
$$\sum_{m=0}^{\infty} \frac{1}{m!} \left(\frac{-i}{\hbar c}\right)^{m} \int_{-\infty}^{t} \cdots \int_{-\infty}^{t} d^4x_1 \cdots d^4x_m P\left[\mathcal{H}_I(x_1)\cdots\mathcal{H}_I(x_m)\right]$$

The r.h.s. of this expression can, in turn, be written as

$$\text{r.h.s.} = \sum_{\nu=0}^{\infty} \frac{1}{\nu!} \left(\frac{-i}{\hbar c}\right)^{\nu} \int_{-\infty}^{\infty} \cdots \int_{-\infty}^{\infty} d^4x_1 \cdots d^4x_\nu P\left[\mathcal{H}_I(x_1)\cdots\mathcal{H}_I(x_\nu)O_I(t)\right]$$
$$(9.23)$$

To see this:

(1) For each value of ν, split the multiple integral into contributions from regions where
   - n of the integration variables $(x_i) = (\mathbf{x}_i, t_i)$ have $t_i > t$
   - m of the integration variables $(x_i) = (\mathbf{x}_i, t_i)$ have $t_i < t$

   There are $\nu!/n!m!$ ways to form this partition, and the contributions are shown to be identical by a change of dummy variables;
(2) The factor $(-i/\hbar c)^{\nu} = (-i/\hbar c)^{n+m}$;
(3) Now sum over all values of (m,n) with $m + n = \nu$. This gives all possible regions for the integrand for a given ν;

(4) Finally, sum over ν to obtain all terms in the series, but this is then the same as summing over all (m, n) independently

$$\sum_{\nu} \sum_{m+n=\nu} = \sum_{m} \sum_{n} \qquad (9.24)$$

This establishes Theorem Ia.

Theorem Ib: The second theorem states that

$$\langle \underline{\Psi}_f^{(-)} | P[O_H(t_x)O_H(t_y)] | \underline{\Psi}_i^{(+)} \rangle = \langle \psi_f | \sum_{\nu=0}^{\infty} \frac{1}{\nu!} \left(\frac{-i}{\hbar c} \right)^{\nu} \times$$

$$\int_{-\infty}^{\infty} \cdots \int_{-\infty}^{\infty} d^4x_1 \cdots d^4x_{\nu} \, P\left[\mathcal{H}_I(x_1) \cdots \mathcal{H}_I(x_{\nu}) O_I(t_x) O_I(t_y) \right] | \psi_i \rangle$$

$$(9.25)$$

Start with $t_x > t_y$. Use Eqs. (9.14), (9.18), and (9.19) to write the l.h.s. as

$$\begin{aligned}
\text{l.h.s.} &= \langle \psi_f | U(\infty, 0) U(0, t_x) O_I(t_x) U(t_x, 0) \times \\
&\qquad U(0, t_y) O_I(t_y) U(t_y, 0) U(0, -\infty) | \psi_i \rangle \\
&= \langle \psi_f | U(\infty, t_x) O_I(t_x) U(t_x, t_y) O_I(t_y) U(t_y - \infty) | \psi_i \rangle \quad (9.26)
\end{aligned}$$

Now repeat the previous arguments using three time intervals, rather than two. The result clearly holds for both time orderings of (t_x, t_y), and hence for the P-product. This proves Theorem Ib.[6]

9.1.3.2 *Special Cases*

We discuss two special cases of the above results:

(1) *Vacuum State.* There is only *one vacuum state* of the interacting system, assumed here to be non-degenerate. Consider

$$\begin{aligned}
|\underline{0}^{(+)}\rangle &= U(0, -\infty)|0\rangle \\
|\underline{0}^{(-)}\rangle &= U(0, +\infty)|0\rangle
\end{aligned} \qquad (9.27)$$

These two states are orthogonal to all the other $|\underline{\Psi}^{(\pm)}\rangle$ states by energy-momentum considerations.[7] Compute their inner product

$$\begin{aligned}
\langle \underline{0}^{(-)} | \underline{0}^{(+)} \rangle &= \langle 0 | U(\infty, 0) U(0, -\infty) | 0 \rangle \\
&= \langle 0 | U(\infty, -\infty) | 0 \rangle \\
&= \langle 0 | S | 0 \rangle \qquad \text{; scattering operator } S \quad (9.28)
\end{aligned}$$

---

[6] Compare Prob. 9.2.
[7] See the following, and Prob. 9.3.

The third line identifies the vacuum expectation value of the *scattering operator S*. It was shown in Vol. II that this quantity is a pure phase

$$\langle 0|S|0\rangle = e^{i\Phi} \tag{9.29}$$

Thus the two states in Eqs. (9.27) *differ by a phase*.
It is convenient to take this phase out and define two sets of states[8]

$$|\Psi_i^{(+)}\rangle \equiv e^{-i\Phi/2}\, U(0,-\infty)|\psi_i\rangle$$
$$|\Psi_f^{(-)}\rangle \equiv e^{i\Phi/2}\, U(0,+\infty)|\psi_f\rangle \tag{9.30}$$

Each set consists of orthonormal states that are complete in the absence of bound states. The sets differ only in their *boundary conditions*. With the phase removed, one can now *drop the* ($\pm$) *labels on the vacuum*

$$|\underline{0}^{(\pm)}\rangle = |\underline{0}\rangle \qquad ; \text{ vacuum} \tag{9.31}$$

(2) *Stable Single-Particle States.* Consider the single-particle states

$$|\underline{p}^{(+)}\rangle = e^{-i\Phi/2}\, U(0,-\infty)|p\rangle$$
$$|\underline{p}^{(-)}\rangle = e^{i\Phi/2}\, U(0,+\infty)|p\rangle \tag{9.32}$$

There is only one such state if the particle is stable.[9] Compute the inner product of the two states in Eqs. (9.32)

$$\langle \underline{p}^{(-)}|\underline{p}^{(+)}\rangle = e^{-i\Phi}\langle p|S|p\rangle$$
$$= e^{-i\Phi}\cdot 1 \cdot \langle 0|S|0\rangle = 1 \tag{9.33}$$

Here the factorization of the vacuum-vacuum diagrams has been used in the second line. Thus one can also *drop the* ($\pm$) *labels on the stable single-particle states*

$$|\underline{p}^{(\pm)}\rangle = |\underline{p}\rangle \qquad ; \text{ stable single-particle states} \tag{9.34}$$

The labels ($\pm$) are therefore only required on the two (or more)-particle states, where they correspond to different boundary conditions.

---

[8]The states are identified by the tilde under the state label.
[9]Again, by energy-momentum considerations (see Prob. 9.3).

9.2   Properties of Heisenberg Operators and States

Consider first the previously-discussed discrete transformations of parity, charge conjugation, and time reversal (P, C, T).

9.2.1   *Discrete Transformations*

For illustration, consider the *scalar (spin-zero) field.* Assume the parity operator P commutes with the hamiltonian $H = H_0 + H_1$, as relevant to the strong and electromagnetic interactions

$$PH_0P^{-1} = H_0$$
$$PH_1P^{-1} = H_1 \qquad \text{; same for } C, T \qquad (9.35)$$

and similarly for C, T. Then, when applied to the field operator in the Heisenberg picture $\underline{\phi}(x_\mu)$, one has

$$P\underline{\phi}(x_\mu)P^{-1} = Pe^{iHt/\hbar}e^{-iH_0t/\hbar}\phi_I(x_\mu)e^{iH_0t/\hbar}e^{-iHt/\hbar}P^{-1}$$
$$= e^{iHt/\hbar}e^{-iH_0t/\hbar}P\phi_I(x_\mu)P^{-1}e^{iH_0t/\hbar}e^{-iHt/\hbar} \qquad (9.36)$$

Hence, with the aid of our previous results for the transformation properties of the field in the interaction picture

$$P\underline{\phi}(x_\mu)P^{-1} = \eta_p\,\underline{\phi}(x'_\mu) \qquad \text{; } x'_\mu = (-\mathbf{x}, ict) \qquad (9.37)$$

In a similar manner, one has the results that

$$P\underline{\phi}(x_\mu)P^{-1} = \eta_p\,\underline{\phi}(x'_\mu) \qquad \text{; } x'_\mu = (-\mathbf{x}, ict)$$
$$C\underline{\phi}(x_\mu)C^{-1} = \eta_c^\star\,\underline{\phi}^\star(x_\mu)$$
$$T\underline{\phi}(x_\mu)T^{-1} = \eta_t^\star\,\underline{\phi}(x'_\mu) \qquad \text{; } x'_\mu = (\mathbf{x}, -ict) \qquad (9.38)$$

Here the last relation follows because the time-reversal operator T includes the complex conjugation of any c-numbers.

When applied to the *states*, one finds for P, C

$$Pe^{-i\Phi/2}U(0, -\infty)|\psi_i\rangle = e^{-i\Phi/2}U(0, -\infty)P|\psi_i\rangle$$
$$Ce^{-i\Phi/2}U(0, -\infty)|\psi_i\rangle = e^{-i\Phi/2}U(0, -\infty)C|\psi_i\rangle \qquad (9.39)$$

Thus

$$P|\Psi_i^{(\pm)}\rangle = |\Psi_{i_P}^{(\pm)}\rangle \qquad \text{; } P|\psi_i\rangle \equiv |\psi_{i_P}\rangle$$
$$C|\Psi_i^{(\pm)}\rangle = |\Psi_{i_C}^{(\pm)}\rangle \qquad \text{; } C|\psi_i\rangle \equiv |\psi_{i_C}\rangle \qquad (9.40)$$

For T, however, one has[10]

$$Te^{-i\Phi/2}U(0,-\infty)|\psi_i\rangle = e^{i\Phi/2}U(+\infty,0)^\dagger T|\psi_i\rangle$$
$$= e^{i\Phi/2}U(0,+\infty)T|\psi_i\rangle \qquad (9.41)$$

Hence

$$T|\underline{\Psi}_i^{(\pm)}\rangle = |\underline{\Psi}_{i_T}^{(\mp)}\rangle \qquad ; T|\psi_i\rangle \equiv |\psi_{i_T}\rangle \qquad (9.42)$$

Notice the reversal of the boundary conditions in this last expression.

9.2.2 Spectrum

We claim the Heisenberg states are eigenstates of the total four-momentum operator $P_\mu = (\mathbf{P}, iH/c)$

$$P_\mu|\underline{p}_n^{(\pm)}\rangle = p_{n,\mu}|\underline{p}_n^{(\pm)}\rangle \qquad ; \text{ eigenstates of } P_\mu \qquad (9.43)$$

The proof for $|\underline{p}_n^{(+)}\rangle$, for example, goes as follows:

(1) For the three-momentum $\mathbf{P}$ this is obvious, since it commutes with the hamiltonian

$$\mathbf{P}e^{-i\Phi/2}U(0,-\infty) = e^{-i\Phi/2}U(0,-\infty)\mathbf{P} \qquad (9.44)$$

(2) For the total hamiltonian H, consider[11]

$$e^{iH_0\tau/\hbar}|\underline{p}_n^{(+)}\rangle = e^{-i\Phi/2}e^{iH_0\tau/\hbar}U(0,-\infty)e^{-iH_0\tau/\hbar}e^{iH_0\tau/\hbar}|p_n\rangle \qquad (9.45)$$

The last factor is

$$e^{iH_0\tau/\hbar}|p_n\rangle = e^{iE_n\tau/\hbar}|p_n\rangle \qquad (9.46)$$

Now use the following expression from Eq. (9.6) for the combination

$$e^{iH_0\tau/\hbar}U(t,t_0)e^{-iH_0\tau/\hbar} = U(t+\tau,t_0+\tau) \qquad (9.47)$$

Thus

$$e^{iH_0\tau/\hbar}U(0,-\infty)e^{-iH_0\tau/\hbar} = U(\tau,-\infty) = U(\tau,0)U(0,-\infty) \qquad (9.48)$$

---

[10]For finite times, $TU(0,-\infty)T^{-1} = U(+\infty,0)^\dagger = U(0,+\infty)$ follows immediately from Eq. (9.6). For the expression in Eq. (9.7), see Prob. 9.1.

[11]In Vol. II, it is shown that the states $|\underline{\Psi}^{(\pm)}\rangle$ are eigenstates of H by doing the time integrals and going over to a time-independent analysis; here we give an alternate proof. (See also Prob. 9.4.)

Let $\tau \to 0$, and use

$$U(\tau, 0) = 1 - \frac{i}{\hbar} \int_0^\tau H_I(t) dt + O(\tau^2)$$

$$= 1 - \frac{i}{\hbar}\tau H_1 \qquad\qquad ; \tau \to 0 \quad (9.49)$$

Hence

$$e^{iH_0\tau/\hbar} U(0, -\infty) e^{-iH_0\tau/\hbar} = \left(1 - \frac{i}{\hbar}\tau H_1\right) U(0, -\infty) \quad ; \tau \to 0 \quad (9.50)$$

Now collect terms of order τ in Eq. (9.45)

$$\tau \frac{i}{\hbar}(H_0 + H_1)e^{-i\Phi/2}U(0, -\infty)|p_n\rangle = \tau \frac{i}{\hbar}E_n e^{-i\Phi/2}U(0, -\infty)|p_n\rangle \quad (9.51)$$

Thus

$$H|\underset{\sim}{p}_n^{(+)}\rangle = E_n|\underset{\sim}{p}_n^{(+)}\rangle \qquad\qquad (9.52)$$

This completes the verification of Eq. (9.43).[12]

9.2.3 *Continuous Transformations*

We consider continuous transformations of a scalar field in the Heisenberg picture, and start with translations.

9.2.3.1 *Translations*

Continuous transformations of the Heisenberg field in time and space are achieved through

$$\underline{\phi}(x_\mu) = e^{iHt/\hbar}\phi(\mathbf{x})e^{-iHt/\hbar}$$

$$= e^{iHt/\hbar}e^{-i\mathbf{P}\cdot\mathbf{x}/\hbar}\phi(0)e^{i\mathbf{P}\cdot\mathbf{x}/\hbar}e^{-iHt/\hbar} \qquad (9.53)$$

Here we have used the fact that the three-momentum $\mathbf{P}$ is the generator of spatial translations. Equation (9.53) can be re-written as[13]

$$\underline{\phi}(x_\mu) = e^{-iP_\mu x_\mu/\hbar}\phi(0)e^{iP_\mu x_\mu/\hbar} \qquad ; \text{ translation by } x_\mu = (\mathbf{x}, ict)$$

$$P_\mu = \left(\mathbf{P}, \frac{i}{c}H\right) \qquad\qquad ; \text{ four-momentum} \qquad (9.54)$$

---

[12]The proof for $|\underset{\sim}{p}_n^{(-)}\rangle$ follows in the same fashion [see Prob. 9.4(b)].

[13]We again invoke the summation convention that repeated Greek indices are summed from 1 to 4. It is again assumed here that $[H, \mathbf{P}] = 0$.

The generator of translations in space-time is the four-momentum operator.

9.2.3.2 *Lorentz Transformations*

We assume a set of operators $(P_\mu, M_{\mu\nu})$ for the interacting field theory that satisfy the commutation relations for the Poincaré, or inhomogeneous Lorentz group.[14] Consider the behavior of the Heisenberg field under a homogeneous Lorentz transformation

$$U(L)\underline{\phi}(x_\mu)U(L)^{-1} = U(L)e^{-iP_\mu x_\mu/\hbar}\phi(0)e^{iP_\mu x_\mu/\hbar}U(L)^{-1} \tag{9.55}$$

$$= U(L)e^{-iP_\mu x_\mu/\hbar}U(L)^{-1}\phi(0)U(L)e^{iP_\mu x_\mu/\hbar}U(L)^{-1}$$

The second line follows from the behavior of a scalar field at the origin under such a transformation

$$U(L)\phi(0)U(L)^{-1} = \phi(0) \tag{9.56}$$

Now use the transformation law for a vector operator, *derived entirely from the commutation relations*

$$U(L)P_\mu U(L)^{-1} = a_{\mu\nu}(+v)P_\nu \qquad ; \text{ operator transformation (Boost)}$$

if ; $\qquad a_{\mu\nu}(-v)x_\nu = x'_\mu \qquad ; \text{ coordinate transformation}$

$$\tag{9.57}$$

These relations have the following implications:

(1) From the properties of the Lorentz transformation matrix $a_{\mu\nu}(v)$

$$x_\mu a_{\mu\nu}(+v) = x_\mu[a^T(+v)]_{\nu\mu} = x_\mu[a^{-1}(+v)]_{\nu\mu} = a(-v)_{\nu\mu}x_\mu \tag{9.58}$$

Hence from Eqs. (9.57)–(9.58), the term $P_\mu x_\mu$ in the exponents in Eq. (9.55) takes the form

$$x_\mu[a_{\mu\nu}(+v)P_\nu] = x'_\nu P_\nu$$

$$x'_\nu = a_{\nu\mu}(-v)x_\mu \tag{9.59}$$

(2) Similarly

$$U(L)P_\mu U(L)^{-1} = a_{\mu\nu}(+v)P_\nu = a_{\nu\mu}(-v)P_\nu$$

$$U(L)P_\mu = a_{\nu\mu}(-v)P_\nu U(L)$$

$$U(L)[a_{\lambda\mu}(-v)P_\mu] = P_\lambda U(L) \tag{9.60}$$

---

[14]All these operators are constructed from the stress tensor $T_{\mu\nu}$, and, for example, $H \neq H_0$.

where the last line follows from summing the second with $a_{\lambda\mu}(-v)$. This operator relation is re-written as

$$P_\mu U(L) = U(L)[a_{\mu\nu}(-v)P_\nu] \tag{9.61}$$

It then follows from Eqs. (9.55) and (9.59) that under a Lorentz transformation, the Heisenberg field transforms according to

$$U(L)\underline{\phi}(x_\mu)U(L)^{-1} = \underline{\phi}(x'_\mu) \tag{9.62}$$

This is simply the scalar field evaluated at the Lorentz-transformed point. Furthermore, it follows from Eq. (9.61) that the Lorentz-transformed state vectors satisfy

$$P_\mu U(L)|\underline{p}_n^{(\pm)}\rangle = [a_{\mu\nu}(-v)p_{n,\nu}]U(L)|\underline{p}_n^{(\pm)}\rangle \tag{9.63}$$

Hence $U(L)$ indeed boosts the four-momentum of the state[15]

$$U(L)|\underline{p}_n^{(\pm)}) = |\underline{p}_n'^{(\pm)}) \qquad ; p'_{n,\mu} = a_{\mu\nu}(-v)p_{n,\nu} \tag{9.64}$$

9.2.3.3 *Isotopic Spin*

Consider isospin, or any other continuous symmetry operator that commutes with H_0 and H_1. The finite transformations then move directly onto the interaction-picture states and fields, so that, for example

$$e^{-i\boldsymbol{\omega}\cdot\mathbf{T}}e^{-i\Phi/2}U(0,-\infty)|\psi_i\rangle = e^{-i\Phi/2}U(0,-\infty)e^{-i\boldsymbol{\omega}\cdot\mathbf{T}}|\psi_i\rangle$$
$$e^{-i\boldsymbol{\omega}\cdot\mathbf{T}}\underline{\phi}_i e^{i\boldsymbol{\omega}\cdot\mathbf{T}} = a_{ij}(\boldsymbol{\omega})\underline{\phi}_j \tag{9.65}$$

All the previous symmetry arguments then apply.

9.2.4 *Notation*

We *summarize* the results so far, and confirm the notation.

$$|\underline{\Psi}^{(\pm)}\rangle \qquad\qquad\qquad ; \text{Heisenberg state}$$
$$|\underline{\Psi}^{(\pm)}\rangle \equiv e^{\mp i\Phi/2}|\underline{\Psi}^{(\pm)}\rangle \qquad ; \text{H-state with vac-vac phase extracted}$$
$$|\underline{\Psi}^{(\pm)}) \equiv \prod_i \sqrt{2k_{0i}\Omega}\,|\underline{\Psi}^{(\pm)}\rangle \quad ; \text{H-state with invariant norm} \tag{9.66}$$

---

[15]Here we use the states with invariant norm defined in Eqs. (9.66), and denoted by the round bracket on the state vector [recall Eqs. (8.144) and (8.149)].

Note the *exception for the vacuum*

$$|\underline{0}\,) \equiv |\underline{0}\,\rangle \qquad ; \text{vacuum} \qquad (9.67)$$

9.3 Boson Propagator

Recall that the lowest-order Feynman propagator for the scalar field with inverse Compton wavelength μ is defined by the following matrix element in the interaction picture[16]

$$\frac{\hbar}{ic}\Delta_F(x-y) = \langle 0|P[\phi_I(x)\phi_I(y)]|0\rangle$$
$$= \frac{\hbar}{ic}\frac{1}{(2\pi)^4}\int \frac{d^4q}{q^2+\mu^2-i\eta}\,e^{iq\cdot(x-y)} \qquad (9.68)$$

Thus

$$\Delta_F(q^2) = \frac{1}{q^2+\mu^2-i\eta} \qquad (9.69)$$

Consider a theory with an interaction hamiltonian density

$$\mathcal{H}_1(x) = -g\bar{\psi}(x)\psi(x)\phi(x) \qquad (9.70)$$

where ψ is a fermion field. The definition of the *exact* scalar propagator in terms of Feynman diagrams for this theory is illustrated in Fig. 9.1.

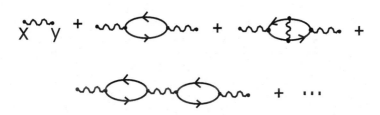

Fig. 9.1   Graphical expansion of the exact scalar propagator $(\hbar/ic)\Delta'_F(x,y)$ in a theory with $\mathcal{H}_1 = -g\bar{\psi}\psi\phi$. The Feynman rules for this theory are given in chapter 7 of Vol. II.

---

[16]Here $q \cdot (x-y) = \mathbf{q} \cdot (\mathbf{x}-\mathbf{y}) - q_0 c(t_x - t_y)$.

The corresponding analytic expression is

$$\frac{\hbar}{ic}\Delta'_F(x,y) = \langle 0| \sum_{\nu=0}^{\infty} \frac{1}{\nu!} \left(\frac{-i}{\hbar c}\right)^{\nu} \int_{-\infty}^{\infty} \cdots \int_{-\infty}^{\infty} d^4x_1 \cdots d^4x_{\nu} \times \qquad (9.71)$$

$$P\left[\mathcal{H}_I(x_1)\cdots\mathcal{H}_I(x_{\nu})\phi_I(x)\phi_I(y)\right]|0\rangle \,/\, \mathcal{D}$$

$$\mathcal{D} = \langle 0| \sum_{\nu=0}^{\infty} \frac{1}{\nu!} \left(\frac{-i}{\hbar c}\right)^{\nu} \int_{-\infty}^{\infty} \cdots \int_{-\infty}^{\infty} d^4x_1 \cdots d^4x_{\nu} \, P\left[\mathcal{H}_I(x_1)\cdots\mathcal{H}_I(x_{\nu})\right]|0\rangle$$

The denominator $\mathcal{D} = \langle 0|S|0\rangle$ serves to cancel the disconnected diagrams. With the aid of Wick's theorem, it is straightforward to see that the expression in Eq. (9.71) gives rise to the connected series of Feynman diagrams shown in Fig. 9.1.

Now observe that:

- The denominator is just the vacuum-vacuum phase $\mathcal{D} = e^{i\Phi}$;
- From our previous Theorem Ib, the expression in Eq. (9.71) is the Heisenberg matrix element

$$\frac{\hbar}{ic}\Delta'_F(x,y) = \langle \underline{0}\,|P[\underline{\phi}(x),\underline{\phi}(y)]|\underline{0}\,\rangle \qquad (9.72)$$

It is extremely convenient to have a closed expression for quantities such as this, and one can now make use of all the properties of the Heisenberg fields and states to derive a general Lehmann representation for this propagator.[17]

---

[17] See, for example, [Fetter and Walecka (2003)]; for the relativistic version, see [Bjorken and Drell (1965)].

Chapter 10

Feynman Rules for QCD

Quantum chromodynamics (QCD) is the theory of the strong interactions binding quarks into the existing hadrons, baryons and mesons. The goal of this final chapter is to show how one obtains the Feynman rules for QCD. Lest the reader underestimate the challenge here, he or she is referred to [Feynman (1963)]. We start with a summary of the theory of QCD, the basic elements of which were covered in Vol. I.[1]

10.1 Quantum Chromodynamics (QCD)

Quantum chromodynamics (QCD) is a Yang-Mills non-abelian gauge theory theory built on an $SU(3)_C$ color symmetry. The fundamental spin-1/2 fermion fields are those of the *quarks*:

- Quarks come in many *flavors* (u, d, s, c, b, t);
- Each flavor of quark also comes in three *colors* (R, G, B)

10.1.1 *Quark Fields*

The quark fields thus take the form[2]

$$\psi = \begin{pmatrix} u \\ d \\ s \\ c \end{pmatrix} \rightarrow \begin{pmatrix} u_R & u_G & u_B \\ d_R & d_G & d_B \\ s_R & s_G & s_B \\ c_R & c_G & c_B \end{pmatrix} \equiv (\psi_R, \psi_G, \psi_B) \equiv \psi_i \quad ; \ i = R, G, B \quad (10.1)$$

---

[1]This chapter assumes familiarity with chapters 5–8, 10 and appendices D, H of Vol. II.
[2]Here we specialize to the first four flavors of quarks and the low-lying hadrons, the (b, t) quarks having a much larger mass. The observed hadrons are color singlets.

Let us re-write this as a column matrix denoting a three-component color field

$$\underline{\psi} = \begin{pmatrix} \psi_R \\ \psi_G \\ \psi_B \end{pmatrix} \qquad ; \text{ color field} \qquad (10.2)$$

This is a very compact notation:[3]

- Each color field has many flavors;
- Each flavor is a four-component Dirac field.

The simplest *Dirac lagrangian density* for free, massless, quark fields is

$$\mathcal{L} = -\bar{\underline{\psi}}\gamma_\mu \frac{\partial}{\partial x_\mu}\underline{\psi}$$
$$= -\left(\bar{\psi}_R \gamma_\mu \frac{\partial}{\partial x_\mu}\psi_R + \bar{\psi}_G \gamma_\mu \frac{\partial}{\partial x_\mu}\psi_G + \bar{\psi}_B \gamma_\mu \frac{\partial}{\partial x_\mu}\psi_B \right) \qquad (10.3)$$

There is an additive contribution for each *color* of quark. Each of these terms gets an additive contribution from each *flavor*

$$\bar{\psi}_R \gamma_\mu \frac{\partial}{\partial x_\mu}\psi_R = \bar{u}_R \gamma_\mu \frac{\partial}{\partial x_\mu}u_R + \bar{d}_R \gamma_\mu \frac{\partial}{\partial x_\mu}d_R + \bar{s}_R \gamma_\mu \frac{\partial}{\partial x_\mu}s_R + \bar{c}_R \gamma_\mu \frac{\partial}{\partial x_\mu}c_R$$
$$(10.4)$$

Each of these, in turn, is a Dirac matrix product.

10.1.2   Yang-Mills Theory Based on Color

The lagrangian density in Eq. (10.3) has a global SU(3) invariance with respect to color:

- Introduce the eight hermitian, traceless, 3×3 Gell-Mann matrices $\underline{\lambda}^a$, the analogs of the Pauli matrices for SU(2), satisfying the Lie algebra for SU(3)

$$\left[\frac{1}{2}\underline{\lambda}^a, \frac{1}{2}\underline{\lambda}^b \right] = if^{abc}\frac{1}{2}\underline{\lambda}^c \qquad ; \text{ Lie algebra for SU(3)}$$
$$(a,b,c) = 1,2,\cdots,8 \qquad (10.5)$$

---

[3]An underlined symbol now represents a matrix in the color space; the absence of underlining implies a suppressed unit matrix with respect to color (it may still be a matrix with respect to flavor and in Dirac space). Alternatively, one could simply carry along all the subscripts, one each for color, flavor, and Dirac component (see Prob. 10.4).

The f^{abc} are the *structure constants*, they are completely antisymmetric in the indices (abc), and here we adopt a *new* summation convention that repeated Latin color indices are now summed from 1 to 8;[4]

- Introduce the eight hermitian generators G^a for SU(3) transformations of the quark field $\underline{\psi}$, the analogs of the isospin operators T_i for SU(2), satisfying

$$[G^a, \underline{\psi}] = -\frac{1}{2}\lambda^a \underline{\psi} \qquad ; \text{generators} \qquad (10.6)$$

- Let θ^a with $a = 1, 2, \cdots, 8$ be a set of real constants. Then the operator

$$R(\omega) = e^{iG^a \theta^a} \qquad ; \omega \equiv (\theta^1, \cdots, \theta^8) \qquad (10.7)$$

produces the finite transformation

$$R(\omega)\underline{\psi}R(\omega)^{-1} = \left[e^{-i\theta^a \underline{\lambda}^a/2}\right]\underline{\psi} \equiv \underline{U}(\theta)\underline{\psi} \qquad (10.8)$$

Here the 3×3 matrices $\underline{U}(\theta)$ are unitary and unimodular

$$\underline{U}(\theta)^\dagger = \underline{U}(\theta)^{-1}$$
$$\det \underline{U}(\theta) = 1 \qquad (10.9)$$

- The quark field $\underline{\psi}$ forms a basis for the fundamental representation of SU(3) based on color. Call this symmetry SU(3)$_C$;
- It is evident that this global symmetry SU(3)$_C$ leaves the lagrangian density in Eq. (10.3) *invariant*.

10.1.2.1 *Local Gauge Invariance*

We now convert the global SU(3)$_C$ invariance of the lagrangian density in Eq. (10.3) into a *local* gauge invariance, where the transformation parameters $\theta^a(x)$ in Eq. (10.7) depend on the position in space-time. The procedure for doing this due to [Yang and Mills (1954)] is detailed in Vol. II:

(1) Introduce vector fields $A_\mu^a(x)$, *one for each generator*, with $a = 1, 2, \cdots 8$

$$A_\mu^a(x) \qquad ; a = 1, 2, \cdots, 8$$
$$\qquad ; \text{gluon fields} \qquad (10.10)$$

These are the known as the *gluon* fields;

---

[4]See Prob. 10.1; the matrices $\underline{\lambda}^a$ and structure constants f^{abc} are given there.

(2) Use the *covariant derivative* in the lagrangian density

$$\frac{\partial}{\partial x_\mu}\underline{\psi} \to \left[\frac{\partial}{\partial x_\mu} - \frac{ig}{2}\underline{\lambda}^a A_\mu^a(x)\right]\underline{\psi} \qquad ; \text{ covariant derivative} \quad (10.11)$$

Here g is a coupling constant.

(3) Include a *kinetic energy* term for the gluons

$$\mathcal{L}_{\text{K.E.}} = -\frac{1}{4}\mathcal{F}_{\mu\nu}^a\mathcal{F}_{\mu\nu}^a$$

$$\mathcal{F}_{\mu\nu}^a \equiv \frac{\partial}{\partial x_\mu}A_\nu^a - \frac{\partial}{\partial x_\nu}A_\mu^a + gf^{abc}A_\mu^b A_\nu^c \qquad ; \text{ field tensor} \quad (10.12)$$

Note the gluon *field tensor* $\mathcal{F}_{\mu\nu}^a$ is antisymmetric in its Lorentz indices

$$\mathcal{F}_{\mu\nu}^a = -\mathcal{F}_{\nu\mu}^a \qquad\qquad (10.13)$$

(4) The *lagrangian density of QCD*, with massless quarks, then takes the form

$$\mathcal{L}_{\text{QCD}} = -\underline{\bar{\psi}}\gamma_\mu\left[\frac{\partial}{\partial x_\mu} - \frac{ig}{2}\underline{\lambda}^a A_\mu^a(x)\right]\underline{\psi} - \frac{1}{4}\mathcal{F}_{\mu\nu}^a\mathcal{F}_{\mu\nu}^a \qquad (10.14)$$

(5) This lagrangian density is invariant under *local gauge transformations*. For *infinitesimal* $\theta^a \to 0$, with $a = 1, 2, \cdots, 8$, these transformations take the form[5]

$$\delta\underline{\psi} = -\frac{i}{2}\theta^a\underline{\lambda}^a\underline{\psi} \qquad\qquad ; \theta^a \to 0$$

$$\delta A_\mu^a = -\frac{1}{g}\frac{\partial\theta^a}{\partial x_\mu} + f^{abc}\theta^b A_\mu^c$$

$$\delta\mathcal{F}_{\mu\nu}^a = f^{abc}\theta^b\mathcal{F}_{\mu\nu}^c \qquad\qquad (10.15)$$

This local $SU(3)_C$ color symmetry is *exact*;

(6) A mass term for the quarks can be included which preserves exact local $SU(3)_C$ gauge invariance[6]

$$\delta\mathcal{L}_{\text{mass}} = -\underline{\bar{\psi}}M\underline{\psi} \qquad ; M = \begin{pmatrix} M_u & & & \\ & M_d & & \\ & & M_s & \\ & & & M_c \end{pmatrix} \qquad (10.16)$$

---

[5] Here, once again, f^{abc} are the *structure constants*.

[6] This mass term actually arises from spontaneous symmetry breaking in the electroweak sector [see, for example, [Walecka (2004)]].

Here M is the unit matrix with respect to color. In more detail

$$\delta\mathcal{L}_{\text{mass}} = - \left[\bar{\psi}_R M \psi_R + \bar{\psi}_G M \psi_G + \bar{\psi}_B M \psi_B\right] \qquad (10.17)$$

Different colored quarks all have the same mass, while it may be anything with respect to *flavor*.

(7) A mass term for the gluons *destroys* the local $SU(3)_C$ gauge invariance

$$\delta\mathcal{L}_{\text{mass}} = -\frac{1}{2}m_A^2 A_\mu^a A_\mu^a \qquad ; \underline{\text{not}} \text{ gauge invariant} \qquad (10.18)$$

As a consequence, the gluons must be *massless*

$$\implies \quad m_A^2 = 0 \qquad ; \text{ gluons massless} \qquad (10.19)$$

Some comments:

- Local $SU(3)_C$ color symmetry is *exact* in QCD;[7]
- The strong interactions *confine* the quarks;[8]
- The observed hadrons are *color singlets*, belonging to the [1] representation of $SU(3)_C$.

10.1.2.2 *Electromagnetic Current Operator*

The electromagnetic current operator in this theory is given by

$$J_\mu^\gamma = ie_p \bar{\underline{\psi}} \gamma_\mu Q \underline{\psi} \qquad ; Q = \begin{pmatrix} 2/3 & & & \\ & -1/3 & & \\ & & -1/3 & \\ & & & 2/3 \end{pmatrix} \qquad (10.20)$$

Several comments:

- Q is again the unit matrix with respect to color, so in more detail

$$J_\mu^\gamma = ie_p \left[\bar{\psi}_R \gamma_\mu Q \psi_R + \bar{\psi}_G \gamma_\mu Q \psi_G + \bar{\psi}_B \gamma_\mu Q \psi_B\right] \qquad (10.21)$$

- The different colored quarks all have the same electromagnetic interaction; in otherwords, the electromagnetic interaction is *colorblind*;
- The electromagnetic interaction couples only to the quarks; the gluons are absolutely *neutral*;

---

[7]Quark states form a basis for the [3], or fundamental, representation of $SU(3)_C$, while gluons belong to the [8].

[8]Lattice gauge theory calculations for QCD produce a confining potential [see again, for example, [Walecka (2004)]].

- The different flavored quarks have *third-integral* charges. Since the observed hadrons are colored singlets, their charges are integral.

10.1.3 *Lagrangian Density of QCD*

Let us investigate the lagrangian density of QCD in Eq. (10.14) in more detail. Make an expansion in powers of g, the one coupling constant in the theory,

$$\mathcal{L} = \mathcal{L}_0 + \mathcal{L}_1 + \mathcal{L}_2 \qquad \text{; powers of g} \qquad (10.22)$$

Then

$$\mathcal{L}_0 = -\bar{\psi}\gamma_\mu \frac{\partial}{\partial x_\mu}\psi - \frac{1}{4}F^a_{\mu\nu}F^a_{\mu\nu}$$

$$\mathcal{L}_1 = \frac{ig}{2}\bar{\psi}\gamma_\mu \lambda^a \psi A^a_\mu - \frac{g}{2}f^{abc}F^a_{\mu\nu}A^b_\mu A^c_\nu$$

$$\mathcal{L}_2 = -\frac{g^2}{4}f^{abc}f^{ade}A^b_\mu A^c_\nu A^d_\mu A^e_\nu \qquad (10.23)$$

Here $F_{\mu\nu}$ is the familiar field tensor

$$F^a_{\mu\nu} \equiv \frac{\partial}{\partial x_\mu}A^a_\nu - \frac{\partial}{\partial x_\nu}A^a_\mu \qquad (10.24)$$

Some comments:

- $\mathcal{L}_0$ is the sum of lagrangians for free, massless quarks and gluons;
- $\mathcal{L}_1$ contains a Yukawa coupling of the quarks to the gluons [see Fig. 10.1(a)]; the structure of this coupling was analyzed in Vol. I;[9]
- The coupling strength of the quarks to gluons is independent of the *flavor* of the quark;
- $\mathcal{L}_1$ contains a specific *cubic gluon self-coupling* [Fig. 10.1(b)];
- $\mathcal{L}_2$ contains a specific *quartic gluon self-coupling* [Fig. 10.1(c)].[10]
- In the gluon sector alone, QCD presents a non-linear, interacting field theory.

---

[9] See also Prob. 10.2

[10] In quantum electrodynamics (QED), photons couple to each other only through (e^+e^-) pairs—there are *no* photon self-couplings.

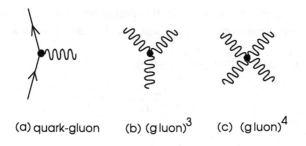

(a) quark-gluon   (b) (gluon)^3 (c) (gluon)^4

Fig. 10.1   Couplings in the lagrangian density of QCD.

10.1.3.1   *Equations of Motion*

The Euler-Lagrange equation for a generalized coordinate $q_i(x)$ is

$$\frac{\partial}{\partial x_\mu} \frac{\partial \mathcal{L}}{\partial(\partial q_i / \partial x_\mu)} = \frac{\partial \mathcal{L}}{\partial q_i} \qquad \text{; E-L eqn}$$

$$\text{; each } q_i \qquad (10.25)$$

This allows one to compute the following equations of motion for QCD from the lagrangian density $\mathcal{L}_{\text{QCD}}$ in Eq. (10.14):

- $\underline{\bar{\psi}}$ *Field*: It is evident that

$$\frac{\partial \mathcal{L}}{\partial(\partial \underline{\bar{\psi}} / \partial x_\mu)} = 0 \qquad (10.26)$$

Thus

$$\gamma_\mu \left[\frac{\partial}{\partial x_\mu} - \frac{ig}{2} \lambda^a A_\mu^a(x) \right] \underline{\psi} = 0 \qquad (10.27)$$

There are 48 components to the $\underline{\bar{\psi}}$ field obtained from Eq. (10.2), and Eq. (10.27) represents one equation for each component;

- $\underline{\psi}$ *Field*: One has

$$\frac{\partial \mathcal{L}}{\partial(\partial \underline{\psi} / \partial x_\mu)} = -\underline{\bar{\psi}} \gamma_\mu \qquad (10.28)$$

Therefore

$$\underline{\bar{\psi}} \gamma_\mu \left[\frac{\overleftarrow{\partial}}{\partial x_\mu} + \frac{ig}{2} \lambda^a A_\mu^a(x) \right] = 0 \qquad (10.29)$$

Here the arrow indicates that the derivative acts to the left. Once again, this represents 48 equations, one for each component of the $\underline{\psi}$ field;

- A_ν^a *Field*: With a little effort, one finds

$$\frac{\partial \mathcal{L}}{\partial(\partial A_\nu^a/\partial x_\mu)} = \mathcal{F}_{\nu\mu}^a \tag{10.30}$$

Hence

$$\frac{\partial}{\partial x_\mu} \mathcal{F}_{\nu\mu}^a = \frac{ig}{2} \underline{\bar\psi}\gamma_\nu\underline{\lambda}^a\underline{\psi} + gf^{abc}\mathcal{F}_{\nu\mu}^b A_\mu^c \tag{10.31}$$

This represents 32 equations, one for each (a, ν)

10.1.3.2   *Conserved Currents*

It follows from these equations of motion that the following *currents are conserved*:[11]

(1) *Baryon Current*

$$\frac{\partial}{\partial x_\mu} \left(i\underline{\bar\psi}\gamma_\mu\underline{\psi} \right) = 0 \tag{10.32}$$

(2) *Color Currents*

$$\frac{\partial}{\partial x_\mu} \left(\frac{i}{2}\underline{\bar\psi}\gamma_\mu\underline{\lambda}^a\underline{\psi} + f^{abc}\mathcal{F}_{\mu\nu}^b A_\nu^c \right) = 0 \tag{10.33}$$

There is one such current for each $a = 1, 2, \cdots, 8$;

(3) *Flavor Currents*

$$\frac{\partial}{\partial x_\mu} \left(i\underline{\bar\psi}\gamma_\mu\Sigma\,\underline{\psi} \right) = 0 \tag{10.34}$$

Here Σ is the unit matrix with respect to color, but *anything* with respect to flavor

$$\Sigma = \begin{pmatrix} \sigma_u & & & X \\ & \sigma_d & & \\ & & \sigma_s & \\ X & & & \sigma_c \end{pmatrix} \tag{10.35}$$

---

[11]See Prob. 10.3. The currents are unnormalized, and it is assumed here that Σ is real.

10.2   Feynman Rules for the S-matrix in QCD

The goal is to start from the lagrangian density in Eq. (10.23) and generate the Feynman rules for QCD. We make several comments:

- It is possible to generate the Feynman rules by using canonical quantization in the Coulomb gauge as done for QED in Vol. II:[12]
  - Only "physical" states of quarks and gluons are involved;
  - The S-matrix can again be rearranged into covariant, gauge-invariant form;
  - The non-linear, derivative couplings make this procedure prohibitively difficult;

- Here we give the set of *effective Feynman rules*, as derived below using path-integral techniques. The basic references are [Feynman (1963); Feynman and Hibbs (1965); Faddeev and Popov (1967); Abers and Lee (1973)];

- The path-integral approach has the great advantage that it formulates quantum mechanics in terms of the *classical lagrangian density*, and corresponding *classical action*, which can be maintained throughout in covariant, gauge-invariant form;

- The covariant vertices for the quark-gluon coupling, and cubic and quartic gluon self-couplings, can then be read off directly from the lagrangian density;

- There are additional loop graphs in the gluon propagator arising from Faddeev-Popov "ghosts";

- The rules are stated for the gluon propagator in the Landau gauge;

- The rules are stated for the S-matrix, *even though asymptotically-free states of quarks and gluons do not exist in QCD*;

- These Feynman rules can be used to calculate the Green's functions, whose use is frequently more appropriate for a confining theory.

The following are the effective Feynman rules for constructing a covariant, unitary S-matrix from the lagrangian density for QCD in Eqs. (10.23):

(1) Draw all topologically distinct connected diagrams;

(2) Include a factor of $(-i)$ for each order in perturbation theory;[13]

---

[12]For canonical quantization in the Coulomb gauge, see Probs. 10.5–10.7.

[13]The order is the number of vertices. We henceforth in chapter 10 simplify to units where $\hbar = c = 1$. All four-momenta are now inverse lengths.

(3) The *quark propagator* illustrated in Fig. 10.2(a) is given by

$$\frac{1}{(2\pi)^4 i}\delta_{ij}\delta_{lm}\frac{1}{i\not{p}}$$; quark propagator

                                                ; $(i,j) = R, G, B$; color
                                                ; $(l,m) = u, d, s, c$; flavor   (10.36)

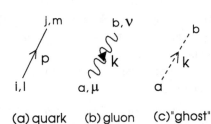

(a) quark   (b) gluon   (c) "ghost"

Fig. 10.2   Propagators. For the quark $(i,j) = R, G, B$ are color indices, and $(l,m) = u, d, s, c$ are flavor indices. For the gluon and ghost, $(a,b) = 1, 2, \cdots, 8$ are color indices.

Here, and in the following, the color and flavor indices are now made explicit;

(4) The *gluon propagator* illustrated in Fig. 10.2(b) is given in the Landau gauge by

$$\frac{1}{(2\pi)^4 i}\delta^{ab}\frac{1}{k^2}\left(\delta_{\mu\nu} - \frac{k_\mu k_\nu}{k^2}\right)$$; gluon propagator   ; Landau gauge

                                                ; $(a,b) = 1, 2, \cdots, 8$; color   (10.37)

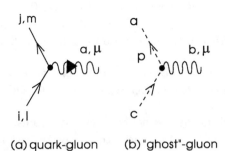

(a) quark-gluon   (b) "ghost"-gluon

Fig. 10.3   Gluon couplings. Here $(a, b, c) = 1, 2, \cdots, 8$ are color indices.

Gluons and ghosts are "dichromatic", and the color indices run over $(a, b) = 1, 2, \cdots, 8$.

(5) The *quark-gluon vertex* shown in Fig. 10.3(a) is given by[14]

$$-\frac{ig}{2}\lambda^a_{ji}\delta_{lm}\gamma_\mu \qquad ; \text{ quark-gluon vertex} \qquad (10.38)$$

(6) The *cubic gluon self-coupling vertex* shown in Fig. 10.4(a) is given by

$$igf^{abc}\left[(q - r)_\lambda\delta_{\mu\nu} + (p - q)_\nu\delta_{\lambda\mu} + (r - p)_\mu\delta_{\nu\lambda}\right]$$
$$; \text{ cubic gluon self-coupling} \qquad (10.39)$$

Here all four-momenta are *outgoing* from the vertex.

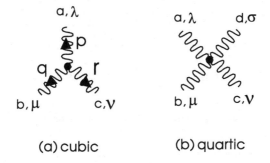

(a) cubic   (b) quartic

Fig. 10.4   Gluon self-couplings. In the cubic coupling, the four-momenta (p, q, r) are outgoing.

(7) The *quartic gluon self-coupling vertex* shown in Fig. 10.4(b) is given by

$$g^2\left[f^{abe}f^{cde}\left(\delta_{\lambda\nu}\delta_{\sigma\mu} - \delta_{\lambda\sigma}\delta_{\mu\nu}\right) + f^{ace}f^{bde}\left(\delta_{\lambda\mu}\delta_{\sigma\nu} - \delta_{\lambda\sigma}\delta_{\mu\nu}\right) + \right.$$
$$\left. f^{ade}f^{cbe}\left(\delta_{\lambda\nu}\delta_{\sigma\mu} - \delta_{\sigma\nu}\delta_{\lambda\mu}\right)\right]$$
$$; \text{ quartic gluon self-coupling} \qquad (10.40)$$

(8) The *ghost-gluon vertex* shown in Fig. 10.3(b) is given by

$$-igf^{abc}p_\mu \qquad ; \text{ ghost-gluon vertex} \qquad (10.41)$$

(9) The *ghost propagator* shown in Fig. 10.2(c) is given by

$$\frac{1}{(2\pi)^4i}\delta^{ab}\frac{1}{k^2} \qquad ; \text{ ghost propagator} \qquad (10.42)$$

---

[14]The additional factor of (-1) arises in going from an effective lagrangian interaction density to an effective hamiltonian interaction density, $\mathcal{H}' = -\mathcal{L}'$.

Note that *ghosts only contribute to internal loops in the gluon propagator;*

(10) Include a factor of (-1) for each closed fermion loop;

(11) Include a factor of (-1) for each closed ghost loop;[15]

(12) Include wave functions for the *external particles*, as illustrated in Fig. (10.5)

$$\frac{1}{\sqrt{\Omega}}\eta_i^{\text{color}}\zeta_l^{\text{flavor}}u(p) \qquad ; \text{ incoming quark}$$

$$\frac{1}{\sqrt{2\omega\Omega}}\varepsilon_\mu^a \qquad ; \text{ incoming gluon} \qquad ; \text{ etc.} \qquad (10.43)$$

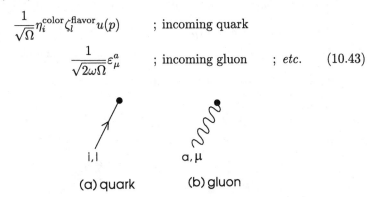

(a) quark   (b) gluon

Fig. 10.5   External lines. Illustrated here for an incoming quark and incoming gluon.

(13) Read along fermion lines;

(14) Read along ghost lines;

(15) Include a delta-function $(2\pi)^4\delta^{(4)}(\Delta p)$ for energy-momentum conservation at each vertex;

(16) Integrate over all internal four-momenta $\int d^4q$.

At very large four-momentum transfer q^2, the effective coupling constant in QCD behaves as

$$g(q^2) = \frac{g^2}{1 + (g^2/16\pi^2)(33/3 - 2N_f/3)\ln q^2/\lambda^2} \qquad ; \text{ asymptotic freedom}$$

$$(10.44)$$

where N_f is the number of quark flavors. This result, obtained by summing leading logarithms using renormalization-group techniques, is due to [Gross and Wilczek (1973); Politzer (1973)]. It is known as *asymptotic freedom.*[16] Equation (10.44) implies that at high q^2, the coupling constant becomes

---

[15]The ghosts obey Fermi statistics!

[16]In QED the corresponding result is $\alpha(q^2) = \alpha[1 - (\alpha/3\pi)\ln q^2/\lambda^2]^{-1}$; note the sign! (See Probs. 27.6–27.10 in [Walecka (2004)].)

small, and one can do *perturbation theory* in this regime. It is just here that having the Feynman rules for QCD in one's arsenal provides a formidable weapon. We proceed to the derivation of the Feynman rules.

10.3 Derivation of Feynman Rules

We start with a review of the derivation of the Feynman rules for quantum electrodynamics (QED) using path-integral methods and the Faddeev-Popov identity.[17]

10.3.1 *Review of Derivation for QED*

Consider the free electromagnetic field coupled to an external source current $J_\mu(x)$.[18]

10.3.1.1 *Generating Functional for Free Field*

The photon propagator is determined by taking variational derivatives with respect to the source current of the *generating functional*

$$\tilde{W}_0(J) = \int \mathcal{D}(A_\mu) \exp\left\{ i \int d^4x \left[\mathcal{L}_0 + J_\mu A_\mu \right] \right\} / [\cdots]_{J=0}$$
$$; \text{ generating functional}$$
$$\mathcal{L}_0 = -\frac{1}{4} F_{\mu\nu} F_{\mu\nu} \qquad ; \text{ free field} \qquad (10.45)$$

The path integral $\int \mathcal{D}(A_\mu)$ goes over all values of the field at a given point in space-time. Note that the generating functional is a *ratio*, where the denominator is obtained from the numerator by setting $J = 0$.

A difficulty is that the physical fields are unchanged under a gauge transformation

$$A_\mu \to A_\mu + \frac{\partial \Lambda}{\partial x_\mu} \qquad ; \text{ gauge transformation} \qquad (10.46)$$

Hence the path integral $\int \mathcal{D}(A_\mu)$ contains many equivalent contributions. The Faddeev-Popov identity [Faddeev and Popov (1967)] *factors* the path

---

[17]The reader is now assumed to have studied in detail the material in chapter 10 and appendix H of Vol. II. Recall that here the electromagnetic field is described in H-L units, with $\hbar = c = 1$. Note that in the present units, $\exp\{(i/\hbar c) \int d^4x \, \mathcal{L}\} \to \exp\{i \int d^4x \, \mathcal{L}\}$.

[18]It is assumed that the source current is conserved $\partial J_\mu/\partial x_\mu = 0$. Recall in Eq. (10.45) that $F_{\mu\nu} = \partial A_\nu/\partial x_\mu - \partial A_\mu/\partial x_\nu$.

integral into a part that is an *integral over all gauges* and a part that is *gauge invariant*

$$\int \mathcal{D}(A_\mu) = \int \mathcal{D}(\Lambda) \int \mathcal{D}(A_\mu) \left(\det \varepsilon^4 M_f \right) \delta_P \left[f(A_\mu) - \lambda(x) \right]$$

; Faddeev-Popov   (10.47)

Several comments:

- It is assumed here that the rest of the integrand in the path integral is gauge invariant;
- The condition

$$f(A_\mu) = \lambda(x) \qquad ; \text{ selects gauge} \qquad (10.48)$$

selects a particular gauge;
- The path integral is to be performed by dividing space-time into n cells of volume ε^4, and in the end letting $n \to \infty$ and $\varepsilon \to 0$ (see Fig. 10.6);

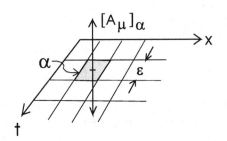

Fig. 10.6   Path integral in discretized space-time. Here the index α labels the cell.

- The quantity δ_P is defined as a product of delta-functions over all cells

$$\delta_P \left[f(A_\mu) - \lambda(x) \right] \equiv \delta(f_1 - \lambda_1) \cdots \delta(f_n - \lambda_n) \qquad (10.49)$$

Its role is to enforce the condition in Eq. (10.48) at all points in space-time;
- The *Faddeev-Popov determinant* $\left(\det \varepsilon^4 M_f \right)$ is defined through the linear response of the function $f(A_\mu)$ to an infinitesimal gauge transfor-

mation $\delta\Lambda(x)$

$$A_\mu \to A_\mu + \frac{\partial}{\partial x_\mu} \delta\Lambda(x)$$

$$\delta f(x) = \int d^4y \, M_f(x-y) \, \delta\Lambda(y)$$

$$\delta f_\alpha = \sum_\beta \left(\varepsilon^4 M_f\right)_{\alpha\beta} \delta\Lambda_\beta \qquad (10.50)$$

Here the third line is the discretized-version of the second, and the sum goes over all cells;

- The proof of Eq. (10.47) is quite simple, and illustrative. The path integral is a multiple integral. Carry out $\int \mathcal{D}(\Lambda)\delta_P \left[f(A_\mu) - \lambda(x)\right]$ first, and change variables in the path integral

$$\int \mathcal{D}(\Lambda)\delta_P \left[f(A_\mu) - \lambda(x)\right] = \int \cdots \int d\Lambda_1 \cdots d\Lambda_n \, \delta(f_1 - \lambda_1)\cdots\delta(f_n - \lambda_n)$$

$$= \int \cdots \int df_1 \cdots df_n \, \delta(f_1 - \lambda_1)\cdots\delta(f_n - \lambda_n)\frac{\partial(\Lambda_1 \cdots \Lambda_n)}{\partial(f_1 \cdots f_n)} \qquad (10.51)$$

The jacobian for the transformation is immediately evaluated from the last of Eqs. (10.50)

$$\frac{\partial(f_1 \cdots f_n)}{\partial(\Lambda_1 \cdots \Lambda_n)} = \left[\frac{\partial(\Lambda_1 \cdots \Lambda_n)}{\partial(f_1 \cdots f_n)}\right]^{-1} = \left(\det \varepsilon^4 M_f\right) \qquad (10.52)$$

Hence Eq. (10.47) reduces to the identity

$$\int \mathcal{D}(A_\mu) \equiv \int \mathcal{D}(A_\mu) \qquad (10.53)$$

As an *example*, assume a gauge-fixing term where

$$f(A_\mu) \equiv \frac{\partial A_\mu}{\partial x_\mu} \qquad ; \text{ example} \qquad (10.54)$$

It follows from the first of Eqs. (10.50) that

$$\delta f(x) = \Box \, \delta\Lambda(x) \qquad (10.55)$$

Thus, from the second of Eqs. (10.50)

$$M_f(x-y) = \Box_x \delta^{(4)}(x-y) \qquad (10.56)$$

This expression is

- Independent of $\Lambda(x)$;

- Independent of $\lambda(x)$.

We will henceforth assume that $M_f(x-y)$ has these properties.

One now finds the following (remarkable!) results when Eq. (10.47) is substituted into Eq. (10.45):

(1) The *integrand* for the path integral $\int \mathcal{D}(\Lambda)$ is *independent of* Λ.

  - Thus that integrand is *gauge-invariant*;
  - The integral over gauges $\int \mathcal{D}(\Lambda)$ then *factors, and cancels in the ratio*.

(2) The integrand is also *independent of* $\lambda(x)$.

  - Different choices of the gauge-fixing term yield the same result for the generating functional;
  - One can therefore take a path integral of an arbitrary function of $\lambda(x)$ in both the numerator and denominator, for example

$$\int \mathcal{D}(\lambda) \exp\left\{ -\frac{i}{2\xi} \int d^4x [\lambda(x)]^2 \right\} \qquad ;\text{ example} \qquad (10.57)$$

  where ξ is some constant. This expression again *factors, and cancels in the ratio*;

  - Instead of cancelling it as a factor, the product of delta-functions in Eq. (10.49) can be used to actually *perform the path integral* $\int \mathcal{D}(\lambda)$ in numerator and denominator

$$\int \mathcal{D}(\lambda) \exp\left\{ -\frac{i}{2\xi} \int d^4x [\lambda(x)]^2 \right\} \delta_P \left[f(A_\mu) - \lambda(x) \right] =$$
$$\exp\left\{ -\frac{i}{2\xi} \int d^4x \left[f(A_\mu) \right]^2 \right\} \qquad (10.58)$$

(3) The measure $(\varepsilon^4)^n$ also *factors, and cancels in the ratio*.

With these observations and manipulations, and with the use of Eq. (10.54), the generating functional in Eq. (10.45) takes the form

$$\tilde{W}_0(J) = \int \mathcal{D}(A_\mu) \, (\det M_f) \times \qquad (10.59)$$
$$\exp\left\{ i \int d^4x \left[\mathcal{L}_0 + J_\mu A_\mu - \frac{1}{2\xi} \left(\frac{\partial A_\mu}{\partial x_\mu} \right)^2 \right] \right\} / \left[\cdots \right]_{J=0}$$

(4) Now use a nice *trick*, and write the Faddeev-Popov determinant as a *path integral over Grassmann variables* $[\bar{c}(x), c(x)]$[19]

$$\int \mathcal{D}(\bar{c})\mathcal{D}(c) \exp\left\{i \int d^4x\, \bar{c}(x)\Box_x c(x)\right\} = \left(\frac{\varepsilon^8}{i}\right)^n \det M_f \quad (10.60)$$

The factor of $(\varepsilon^8/i)^n$ also cancels in the ratio, and with a partial integration in the exponent in Eq. (10.60), the generating functional in Eq. (10.59) becomes

$$\tilde{W}_0(J) = \int \mathcal{D}(A_\mu)\mathcal{D}(\bar{c})\mathcal{D}(c) \times \qquad\qquad (10.61)$$

$$\exp\left\{i \int d^4x \left[\mathcal{L}_0 + J_\mu A_\mu - \frac{1}{2\xi}\left(\frac{\partial A_\mu}{\partial x_\mu}\right)^2 - \frac{\partial \bar{c}}{\partial x_\mu}\frac{\partial c}{\partial x_\mu}\right]\right\} / [\cdots]_{J=0}$$

The original problem has been transformed into a new problem with an effective lagrangian containing a gauge-fixing term and additional "ghost" fields.

In QED, the ghost fields do not couple to anything, and their contribution therefore again *factors and cancels in the ratio.*[20] Thus Eq. (10.61) can equally well be written

$$\tilde{W}_0(J) = \int \mathcal{D}(A_\mu) \exp\left\{i \int d^4x \left[\mathcal{L}_0 + J_\mu A_\mu - \frac{1}{2\xi}\left(\frac{\partial A_\mu}{\partial x_\mu}\right)^2\right]\right\} / [\cdots]_{J=0}$$

$$(10.62)$$

10.3.1.2 *Photon Propagator*

The effective action is bilinear in A_μ, and the path integral in Eq. (10.62) can be evaluated with gaussian integration to give (see Vol. II)

$$\tilde{W}_0(J) = \exp\left\{\frac{i}{2}\int d^4x \int d^4y\, J_\mu(x) D_{\mu\nu}(x-y) J_\nu(y)\right\} \quad (10.63)$$

$$D_{\mu\nu}(x-y) = \int \frac{d^4k}{(2\pi)^4} \tilde{D}_{\mu\nu}(k) e^{ik\cdot(x-y)}$$

$$\tilde{D}_{\mu\nu}(k) = \frac{1}{k^2 - i\eta}\left[\delta_{\mu\nu} - (1-\xi)\frac{k_\mu k_\nu}{k^2}\right] \qquad \text{; photon propagator}$$

---

[19]Grassmann variables are anti-commuting c-numbers. Recall, here, Eq. (10.56).
[20]This will *not* be the case with QCD.

Two variational derivatives with respect to the source current $J_\mu(x)$ identify the *photon propagator*, as indicated. Various values of the parameter ξ now correspond to different gauges:

- $\xi = 1$ gives the *Feynman gauge* and *Feynman propagator*;
- $\xi = 0$ gives the *Landau gauge*;
- $\xi = \infty$ gives an undefined result; note that with this value, the gauge-fixing term in Eq. (10.62) disappears.

10.3.1.3   *Generating Functional for QED*

To obtain the theory of QED with interactions, one uses the full lagrangian density

$$\mathcal{L}_0 \to \mathcal{L} = -\frac{1}{4} F_{\mu\nu} F_{\mu\nu} - \bar{\psi} \left[\gamma_\mu \left(\frac{\partial}{\partial x_\mu} - ie A_\mu \right) + M \right] \psi \quad (10.64)$$

Now fermions $(\bar{\psi}, \psi)$, fermion sources $(\bar{\zeta}, \zeta)$, and ghost sources $(\bar{\eta}, \eta)$ must be included in the generating functional. The interaction $\mathcal{L}_1 = ie\bar{\psi}\gamma_\mu\psi A_\mu$ can be factored from the path integral in the generating functional by making use of variational derivatives with respect to the sources

$$e^{i \int d^4x \, \mathcal{L}_1} \to \exp \left\{ i \int d^4x \left[ie \left(-\frac{1}{i} \frac{\delta}{\delta \zeta(x)} \right) \gamma_\mu \left(\frac{1}{i} \frac{\delta}{\delta \bar{\zeta}(x)} \right) \left(\frac{1}{i} \frac{\delta}{\delta J_\mu(x)} \right) \right] \right\}$$

$$(10.65)$$

The remaining path integral is of the form $\tilde{W}_0 \times$ [factor independent of sources] where $\tilde{W}_0$ is calculated for free fields. The last factor cancels in the ratio, and the generating functional for QED thus takes the form

$$\tilde{W}_{\text{QED}}[J, \bar{\zeta}, \zeta, \bar{\eta}, \eta] = \exp \left\{ i \int d^4x \times \right. \hspace{3cm} (10.66)$$

$$\left. \left[ie \left(-\frac{1}{i} \frac{\delta}{\delta \zeta(x)} \right) \gamma_\mu \left(\frac{1}{i} \frac{\delta}{\delta \bar{\zeta}(x)} \right) \left(\frac{1}{i} \frac{\delta}{\delta J_\mu(x)} \right) \right] \right\} \tilde{W}_0 \, / \, [\cdots]_{\text{sources}=0}$$

where

$$\tilde{W}_0[J, \bar{\zeta}, \zeta, \bar{\eta}, \eta] = \exp \left\{ \frac{i}{2} \int d^4x \int d^4y \, J_\mu(x) D_{\mu\nu}(x-y) J_\nu(y) \right\} \times$$

$$\exp \left\{ -i \int d^4x \int d^4y \, \bar{\zeta}(x) S_F(x-y) \zeta(y) \right\} \times$$

$$\exp \left\{ i \int d^4x \int d^4y \, \bar{\eta}(x) \Delta_F(x-y) \eta(y) \right\} \quad (10.67)$$

The interaction does not involve $(\bar{\eta}, \eta)$, and thus the last term factors out in Eq. (10.66). The *ghost propagator*, if we wanted it, is calculated from this term as[21]

$$\left[\frac{1}{i} \frac{\delta}{\delta\bar{\eta}(x)} \tilde{W}_0 \frac{1}{i} \frac{\delta}{\delta\eta(y)} \right]_{\text{sources}=0} = \frac{1}{i}\Delta_F(x-y) = \langle 0 | P[c(x), \bar{c}(y)] | 0 \rangle_C$$

$$; \text{ ghost propagator} \qquad (10.68)$$

Several comments:

- Equations (10.66) and (10.67) now *generate the Feynman rules for QED*;[22]
- One can just read off the various components;
- These results are *covariant and gauge invariant*.

10.3.2   *Derivation for Quantum Chromodynamics (QCD)*

10.3.2.1   *Lagrangian Density*

We start from the lagrangian density in Eqs. (10.14) and (10.23). This is invariant under local $SU(3)_C$ transformations, whose infinitesimal form is that in Eqs. (10.15)

$$\delta\underline{\psi} = -\frac{i}{2}\theta^a\underline{\lambda}^a\underline{\psi} \qquad\qquad ; \theta^a \to 0$$

$$\delta A_\mu^a = -\frac{1}{g}\frac{\partial\theta^a}{\partial x_\mu} + f^{abc}\theta^b A_\mu^c \qquad\qquad (10.69)$$

It is the last term in the second line that is new for the gauge fields in a non-abelian gauge theory.

10.3.2.2   *Gauge Fields Alone*

To start, consider the theory with only gauge fields. We again work with the generating functional. The measure in the path integral over the gauge fields has one term for each type of gluon

$$\mathcal{D}(A_\mu) \equiv \prod_{a=1}^{8} \mathcal{D}(A_\mu^a) \qquad\qquad (10.70)$$

---

[21]Recall $\Delta_F(x-y) = (2\pi)^{-4} \int d^4k \, e^{ik\cdot(x-y)}(k^2 - i\eta)^{-1}$.
[22]See Prob. 10.8 for an exercise in producing the Feynman rules from the generating functional.

The Faddeev-Popov identity is again employed[23]

$$\int \mathcal{D}(A_\mu) = \int \mathcal{D}(\Lambda) \int \mathcal{D}(A_\mu)\, \delta_P\left[f(A_\mu) - \lambda(x)\right]\left(\det \varepsilon^4 M_f\right) \qquad (10.71)$$

As an example, consider again a gauge-fixing function of the form

$$f(A_\mu^a) \equiv f^a = \frac{\partial A_\mu^a}{\partial x_\mu} \qquad\qquad ; a = 1, 2, \cdots, 8 \qquad (10.72)$$

Note there is one such expression for each $a = 1, \cdots, 8$.[24] The Faddeev-Popov determinant $\left(\det \varepsilon^4 M_f\right)$ is obtained from the change in these functions under an infinitesimal gauge transformation. Write this infinitesimal gauge transformation as

$$\theta^a(x) = -g\, \delta\Lambda^a(x) \qquad (10.73)$$

Then from Eqs. (10.69) and (10.72)

$$\delta f^a(x) = \square\, \delta\Lambda^a(x) - g\frac{\partial}{\partial x_\mu}\left[f^{abc}\delta\Lambda^b(x)\, A_\mu^c(x)\right] \qquad (10.74)$$

As in Eqs. (10.50), we write

$$\delta f^a(x) = \int d^4y\, M_f^{ab}(x-y)\, \delta\Lambda^b(y)$$

$$M_f^{ab}(x-y) = \left[\delta^{ab}\square_x - gf^{abc}\frac{\overrightarrow{\partial}}{\partial x_\mu}A_\mu^c(x)\right]\delta^{(4)}(x-y) \qquad (10.75)$$

Also as in Eqs. (10.50), the discretized form of this relation is

$$\delta f_\alpha^a = \sum_\beta \left(\varepsilon^4 M_f^{ab}\right)_{\alpha\beta} \delta\Lambda_\beta^b \qquad (10.76)$$

The Faddeev-Popov determinant is again written as an integral over Grassmann variables

$$\int \mathcal{D}(\bar{c})\mathcal{D}(c)\exp\left\{i\int d^4x\, \bar{c}^a\left[\delta^{ab}\square_x - gf^{abc}\frac{\overrightarrow{\partial}}{\partial x_\mu}A_\mu^c\right]c^b\right\} = \det\left(\frac{\varepsilon^8}{i}M_f\right)$$

$$(10.77)$$

---

[23]The proof of this result must be extended to non-abelian symmetry groups [see [Faddeev and Popov (1967)]].

[24]Now, as in Eq. (10.70), $\delta_P\left[f(A_\mu) - \lambda(x)\right] \equiv \prod_{a=1}^{8} \delta_P\left[f(A_\mu^a) - \lambda^a(x)\right]$.

Partial integrations in the exponent allow this to be re-written as

$$\int \mathcal{D}(\bar{c})\mathcal{D}(c) \exp\left\{i \int d^4x \left[-\frac{\partial \bar{c}^a}{\partial x_\mu}\frac{\partial c^a}{\partial x_\mu} + gf^{abc}\bar{c}^a \frac{\overleftarrow{\partial}}{\partial x_\mu} A_\mu^c c^b\right]\right\} =$$

$$\det\left(\frac{\varepsilon^8}{i}M_f\right) \qquad (10.78)$$

10.3.2.3 *Generating Functional for QCD*

With the addition of quark fields $(\bar{\psi}, \psi)$, quark sources $(\bar{\zeta}, \zeta)$, and ghost sources $(\bar{\eta}^a, \eta^a)$, the generating functional for QCD takes the form

$$\tilde{W}_{\mathrm{QCD}}[J, \bar{\zeta}, \zeta, \bar{\eta}, \eta] = \int \mathcal{D}(A_\mu)\mathcal{D}(\bar{\psi})\mathcal{D}(\psi)\mathcal{D}(\bar{c})\mathcal{D}(c) \times$$

$$\exp\left\{i \int d^4x \left[\mathcal{L}_0 + \mathcal{L}_1 + \mathcal{L}_2 + J_\mu^a A_\mu^a + \bar{\zeta}\psi + \bar{\psi}\zeta + \bar{\eta}^a c^a + \bar{c}^a \eta^a\right.\right.$$

$$\left.\left. -\frac{1}{2\xi}\left(\frac{\partial A_\mu^a}{\partial x_\mu}\right)^2 + \left(-\frac{\partial \bar{c}^a}{\partial x_\mu}\frac{\partial c^a}{\partial x_\mu} + gf^{abc}\bar{c}^a \frac{\overleftarrow{\partial}}{\partial x_\mu} A_\mu^c c^b\right)\right]\right\} \qquad (10.79)$$

Several comments:

- The generating functional is obtained from an effective lagrangian density $\mathcal{L}_{\mathrm{eff}}$, containing quark, gluon, and ghost fields;[25]
- The term $\mathcal{L}_0 + \mathcal{L}_1 + \mathcal{L}_2 = \mathcal{L}$ is the classical lagrangian density of QCD; it produces the classical action;
- The terms $J_\mu^a A_\mu^a + \cdots + \bar{c}^a \eta^a$ represent the sources;
- The term $-(1/2\xi)(\partial A_\mu^a/\partial x_\mu)^2$ in $\mathcal{L}_{\mathrm{eff}}$ fixes the gauge;
- The contribution to $\mathcal{L}_{\mathrm{eff}}$ involving $(\bar{c}^a, c^a)$ is that of the ghost fields;
- Here, in contrast to QED, the ghosts *do* couple to the gauge fields through the final term in $\mathcal{L}_{\mathrm{eff}}$.

10.3.2.4 *Feynman Rules*

The Feynman rules for QCD follow directly from the generating functional in Eq. (10.79):

(1) The Feynman rules, expressed in terms of those for the S-matrix, are as previously stated in Sec. 10.2;

---

[25]The effective action is $S_{\mathrm{eff}} = \int d^4x\,\mathcal{L}_{\mathrm{eff}}$.

(2) The *quark-gluon* and *gluon self-coupling vertices* in Eqs. (10.38)–(10.40) can just be read off from $\mathcal{L}_1$ and $\mathcal{L}_2$.[26] The resulting interaction is covariant and gauge invariant;

(3) This analysis gives a more general *gluon propagator* [see Eq. (10.63)]; the Landau gauge is obtained with $\xi = 0$;

$$\frac{1}{(2\pi)^4 i} \delta^{ab} \frac{1}{k^2 - i\eta} \left(\delta_{\mu\nu} - \frac{k_\mu k_\nu}{k^2} \right) \qquad ; \text{ gluon propagator}$$

$$; \text{ Landau gauge} \qquad (10.80)$$

(4) The *ghost propagator* in Eq. (10.42) has already been obtained in Eq. (10.68)

$$\delta^{ab} \frac{1}{i} \Delta_F(x - y) = \delta^{ab} \frac{1}{i} \int \frac{d^4 k}{(2\pi)^4} \frac{1}{k^2 - i\eta} e^{ik \cdot (x-y)}$$

$$; \text{ ghost propagator} \qquad (10.81)$$

(5) The *ghost-gluon vertex* in Fig. 10.3(b) and Eq. (10.41) is obtained from the last term in Eq. (10.79) as follows

$$\left[\frac{\partial}{\partial x_\mu} \left(\text{create } e^{-ip_\mu x_\mu} \right) \right] (\text{effective } \mathcal{H})(\text{color factor}) = (-ip_\mu)(-1)(gf^{acb})$$

$$= -ig f^{abc} p_\mu$$

$$; \text{ ghost-gluon vertex} \qquad (10.82)$$

This is the result quoted in Eq. (10.41);

(6) The factor of (-1) for closed ghost loops arises since the ghost fields are Grassmann variables, and hence behave as *fermions.*

With this introduction to the Feynman rules for QCD, together with the introduction to lattice gauge theory in [Walecka (2004)], where QCD is solved numerically on a finite space-time lattice, the reader has at his or her disposal the essential tools of quantum chromodynamics, the remarkably successful theory of the strong interactions binding quarks and gluons into the observed hadrons.

---

[26]In order to evaluate *derivatives* at a vertex, a wave function $e^{-ip_\mu x_\mu}$ is employed for the creation of a quantum with four-momentum p_μ.

PART 4

Problems and Appendices

Chapter 11

Problems

2.1 Assume the potential $V(\mathbf{x})$ is real, and show that the continuity Eq. (2.6) then follows from the Schrödinger Eq. (2.8).

2.2 (a) Prove from the continuity equation that for a localized disturbance, the normalization integral is independent of time

$$\frac{d}{dt} \int \Psi^\star \Psi \, d^3x = \frac{d}{dt} \int \rho \, d^3x = 0 \qquad ; \text{ localized disturbance}$$

(b) Use the Schrödinger equation to show quite generally that this condition holds if H is hermitian.

2.3 The expectation value of a physical quantity represented by the operator O is obtained in general from the following matrix element

$$\langle O \rangle = \frac{\langle \Psi(t) | O | \Psi(t) \rangle}{\langle \Psi | \Psi \rangle} \equiv \frac{\int \Psi^\star(\mathbf{x}, t) O \Psi(\mathbf{x}, t) \, d^3x}{\int \Psi^\star \Psi \, d^3x}$$

(a) Show the expectation value is real if O is hermitian;
(b) Show that for normal modes the energy is $\langle H \rangle = E$;[1]
(c) What is the particle's momentum in the coordinate representation?

2.4 Start from the eigenvalue Eq. (2.17) and its complex conjugate. Show

$$\int [\psi_n^\star H \psi_n - (H\psi_n)^\star \psi_n] d^3x = (E_n - E_n^\star) \int |\psi_n|^2 d^3x$$

Now use the fact that H is hermitian to conclude the eigenvalues E_n must be *real*.

---

[1] *Hint*: Use the Schrödinger equation.

411

2.5 Start from the eigenvalue Eq. (2.17) for ψ_n, and the complex conjugate equation for ψ_m. Show

$$\int [\psi_m^\star H\psi_n - (H\psi_m)^\star \psi_n]d^3x = (E_n - E_m^\star) \int \psi_m^\star \psi_n \, d^3x$$

Use the fact that H is hermitian, and the result from Prob. 2.4, to conclude that the eigenfunctions corresponding to distinct eigenvalues are *orthogonal*.

2.6 Suppose the eigenfunctions (ψ_1, ψ_2) correspond to the same eigenvalue. Introduce the linear combinations

$$\psi_1' \equiv \psi_1 \qquad ; \psi_2' \equiv \psi_2 - \left[\frac{\int \psi_1^\star \psi_2 \, d^3y}{\int |\psi_1|^2 d^3y}\right] \psi_1$$

Show these degenerate wave functions are now *orthogonal*.[2]

2.7 If the eigenfunctions $\psi_n(\mathbf{x})$ are complete, an arbitrary acceptable function $\phi(\mathbf{x})$ can be expanded as

$$\phi(\mathbf{x}) = \sum_{n=0}^{\infty} a_n \psi_n(\mathbf{x})$$

$$a_n = \int \psi_n^\star(\mathbf{y})\phi(\mathbf{y}) \, d^3y$$

where the last relation follows from the orthonormality of the eigenfunctions. Now justify the statement of completeness in Eq. (2.21).

2.8 (a) Use Eq. (2.81) to generate $h_n(\xi)$ for $n = 1, 2, \ldots, 5$;

(b) Reproduce the results in (a) with Eq. (2.79), starting from $h_0(\xi) = 1$;

(c) Make a good plot of $(\hbar/m\omega_0)^{1/4}\psi_n$ and $(\hbar/m\omega_0)^{1/2}|\psi_n|^2$ as a function of ξ for the harmonic oscillator using these results.

2.9 (a) Write, or obtain, a program to numerically integrate the differential Eq. (2.42) given an initial (ψ, ψ');

(b) Start from $\xi \ll 0$ and the actual ground-state solution $\psi_0 \sim e^{-\xi^2/2}$. Chose an eigenvalue near, but not equal to, $\varepsilon_0 = 1$. What does your numerical solution look like for $\xi \gg 0$? Why? Show it is only for $\varepsilon_0 = 1$ that you can again match onto the decaying solution $\psi \sim e^{-\xi^2/2}$. Discuss;

(c) Repeat for the first excited state with $\psi_1 \sim \xi e^{-\xi^2/2}$ for $\xi \ll 0$.

2.10 Show the raising and lowering operators can be written

$$a^\dagger = \left(\frac{m\omega_0}{2\hbar}\right)^{1/2}\left(x - \frac{i}{m\omega_0}p\right) \qquad ; a = \left(\frac{m\omega_0}{2\hbar}\right)^{1/2}\left(x + \frac{i}{m\omega_0}p\right)$$

---

[2]This is the simplest example of the Schmidt orthogonalization procedure.

2.11 Show from the definition that the adjoint of a product is the product of the adjoints in the reverse order

$$(AB)^\dagger = B^\dagger A^\dagger$$

2.12 (a) Show a matrix element of the number operator satisfies

$$\langle \psi | a^\dagger a | \psi \rangle \equiv \int \psi^* a^\dagger a \psi \, dx = \langle a\psi | a\psi \rangle \geq 0$$

(b) Use the result in (a) to prove that the eigenvalues of the hamiltonian in Eq. (2.125) are positive definite.

2.13 (a) Use the commutation relation $[a, a^\dagger] = 1$, and the fact that $a\psi_0 = 0$, to show that

$$\langle \psi_0 | aa^\dagger | \psi_0 \rangle = 1$$

(b) Show through repeated use of the approach in (a) that

$$\langle \psi_0 | a^n (a^\dagger)^n | \psi_0 \rangle = n!$$

Hence establish the normalization of the eigenfunctions in Eq. (2.134).

2.14 Show the first few spherical harmonics are given by

$$Y_{00} = \frac{1}{\sqrt{4\pi}}$$

$$Y_{10} = \sqrt{\frac{3}{4\pi}} \cos\theta \qquad ; \; Y_{1,\pm 1} = \mp\sqrt{\frac{3}{8\pi}} \sin\theta \, e^{\pm i\phi}$$

$$Y_{20} = \sqrt{\frac{5}{16\pi}} (3\cos^2\theta - 1) \quad ; \; Y_{2,\pm 1} = \mp\sqrt{\frac{15}{8\pi}} \sin\theta \cos\theta \, e^{\pm i\phi}$$

$$; \; Y_{2,\pm 2} = \sqrt{\frac{15}{32\pi}} \sin^2\theta \, e^{\pm 2i\phi}$$

2.15 (a) Draw a more detailed picture, and provide a good geometric derivation of Eqs. (2.186) and (2.189);

(b) Derive Eqs. (2.187) and (2.190) algebraically through the transformation from cartesian to spherical coordinates

$$x = r\sin\theta\cos\phi \qquad ; \; y = r\sin\theta\sin\phi \qquad ; \; z = r\cos\theta$$

2.16 Show that all three components of the angular momentum in Eqs. (2.187) are hermitian operators.

2.17 Use the formulas for the spherical Bessel functions in [Schiff (1968)], or any other source, to verify the normalization constant in Eq. (2.222).

2.18 (a) Plot the radial probability densities for the states in Fig. 2.12;

(b) Use the results in Prob. 2.14 to make a polar plot of the probability densities for these states.

2.19 Two solutions $\{\psi_1, \psi_2\}$ to a second-order differential equation for $\psi(r)$ are said to form a *fundamental system* if their *wronksian* is non-zero

$$\psi_1 \psi_2' - \psi_2 \psi_1' \neq 0 \qquad ; \text{ wronskian}$$

Show this is just the condition that allows one to match the initial value and slope $\{\psi, \psi'\}$ at an arbitrary point with a linear combination

$$\psi(r) = \alpha \psi_1(r) + \beta \psi_2(r)$$

Hence conclude that any solution can be expanded in terms of a fundamental system.

2.20 (a) Discuss the graphical solution to the eigenvalue Eq. (2.229);

(b) Show the condition on the combination $V_0 a^2$ that there be just one bound state with $|E| \to 0$ is[3]

$$\frac{2m_0}{\hbar^2} V_0 a^2 = \frac{\pi^2}{4} \qquad ; \text{ zero-energy bound state}$$

(c) Sketch the corresponding wave function $u_0(r)$ in part (b).

2.21 [Schiff (1968)] gives the following expression for the associated Laguerre polynomials

$$L_\kappa^q(\rho) = \sum_{\nu=0}^{\kappa-q} \frac{(-1)^{\nu+q}(\kappa!)^2 \rho^\nu}{(\kappa-q-\nu)!(q+\nu)!\nu!}$$

Start from Eqs. (2.277) and (2.271) and derive this result.

2.22 Show the first three radial wave functions in the hydrogen atom

---

[3]In the two-body problem, one must use the *reduced mass* $\mu = m_1 m_2/(m_1 + m_2)$ in this expression (see appendix A).

are given by

$$R_{10}(r) = \left(\frac{Z}{a_0}\right)^{3/2} 2e^{-Zr/a_0}$$

$$R_{20}(r) = \left(\frac{Z}{2a_0}\right)^{3/2} \left(2 - \frac{Zr}{a_0}\right) e^{-Zr/2a_0}$$

$$R_{21}(r) = \left(\frac{Z}{2a_0}\right)^{3/2} \frac{Zr}{a_0\sqrt{3}} e^{-Zr/2a_0}$$

2.23 It is the s-states that get into the origin in the hydrogen atom. Start from Eq. (2.267) and compute the radial derivative $\partial\psi_{n00}(r)/\partial r|_{r=0}$. Make a three-dimensional sketch of $\psi_{n00}(r)$ for small r. Discuss.

2.24 The capture of negative muons on nuclei takes place between a muon in a $1s$ atomic orbit and a proton in the nucleus.

(a) Use the Bohr model to demonstrate that for light nuclei, the muon orbit lies well inside the atomic electron cloud and outside of the nucleus;

(b) The reaction proceeds through the contact weak interaction

$$\mu^- + p \to \nu_\mu + n$$

Show that the capture-rate is proportional to Z^4 for light nuclei.

2.25 Consider the radial part of the three-dimensional stationary-state Schrödinger equation $R(r) = u(r)/r$ with an attractive exponential potential[4]

$$\frac{d^2u}{dr^2} + \left[\frac{2m_0}{\hbar^2}[E - V(r)] - \frac{l(l+1)}{r^2}\right] u = 0$$

$$V(r) = -V_0 e^{-r/a}$$

(a) Assume s-waves ($l=0$). Change variables from r to $z \equiv e^{-r/2a}$ and show that Bessel's equation results;

(b) What boundary conditions are to be imposed on $u_0(r)$ as a function of z, and how can these be used to determine the energy levels?

(c) What is the lower limit on V_0 for which a bound state exists?

2.26[5] Consider the stationary-state Schrödinger equation in an isotropic

---

[4]See [Schiff (1968)], Prob. 4.8.

[5]This problem is longer, but the results are very important, and the steps all parallel those detailed in the text for the hydrogen atom.

three-dimensional harmonic-oscillator potential

$$V(r) = -V_0 + \frac{1}{2}k\mathbf{r}^2 = -V_0 + \frac{1}{2}kr^2$$

Introduce the length b through

$$\omega_0 = \left(\frac{k}{m_0}\right)^{1/2} \equiv \frac{\hbar}{m_0 b^2} \qquad ; q \equiv \frac{r}{b}$$

(a) Show the solutions can be written in the form

$$\psi_{nlm}(\mathbf{r}) = \frac{u_{nl}(r)}{r} Y_{lm}(\theta, \phi)$$

$$u_{nl}(r) = N_{nl} q^{l+1} e^{-q^2/2} \mathcal{L}_{n-1}^{l+1/2}(q^2)$$

where the *generalized Laguerre polynomials* are defined in terms of the confluent hypergeometric series by[6]

$$\mathcal{L}_p^a(z) \equiv \frac{[\Gamma(a+p+1)]^2}{p!\,\Gamma(a+1)} F(-p|a+1|z)$$

$$= \frac{\Gamma(a+p+1)}{\Gamma(p+1)} \frac{e^z}{z^a} \frac{d^p}{dz^p}\left(z^{a+p} e^{-z}\right)$$

(b) Show the eigenvalue spectrum is given by (see Fig. 11.1)

$$E_{nl} = -V_0 + \hbar\omega_0\left(N + \frac{3}{2}\right)$$

$$N \equiv 2(n-1) + l \qquad ; n = 1, 2, 3, \cdots, \infty$$
$$l = 0, 1, 2, \cdots, \infty$$

(c) Show the normalization constant is given by

$$N_{nl}^2 = \frac{2(n-1)!}{b[\Gamma(n+l+1/2)]^3}$$

These wave functions form the basis for many nuclear physics calculations.[7]

2.27 (a) The isotropic three-dimensional harmonic-oscillator potential in Prob. 2.26 can be re-written using $\mathbf{r}^2 = x^2 + y^2 + z^2$. Show the problem

---

[6]Note the *generalized* Laguerre polynomials reduce to the *associated* Laguerre polynomials in the case a = integer.

[7]See [Walecka (2004)].

separates into three one-dimensional oscillators with an eigenvalue spectrum

$$E_{n_x n_y n_z} = -V_0 + \hbar\omega_0 \left(N + \frac{3}{2}\right)$$

$$N \equiv n_x + n_y + n_z \qquad ; \; (n_x, n_y, n_z) = 0, 1, 2, \cdots, \infty$$

(b) Show the degeneracy of the levels with $N = 0, \cdots, 4$ is exactly that obtained in Prob. 2.26(b).

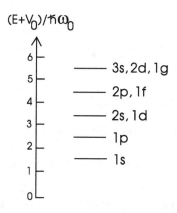

Fig. 11.1 First few levels in the spectrum of the three-dimensional oscillator. This is in the nuclear physics notation (n, l) where these quantum numbers are defined in Prob. 2.26(b). Note that n corresponds to the number of nodes in the radial wave function, excluding the origin and including the point at infinity. (Compare Fig. 2.11.)

2.28 Compute and plot the radial wave functions for the first four states in Fig. 11.1 as a function of r/b. Compare with the results in Fig. 2.12.

2.29 (a) In two dimensions, in polar coordinates, the laplacian is[8]

$$\nabla^2 = \frac{1}{r}\frac{\partial}{\partial r} r \frac{\partial}{\partial r} + \frac{1}{r^2}\frac{\partial^2}{\partial \phi^2}$$

Show that for a particle in a constant potential $-V_0$ in two dimensions, the separated solutions to the time-independent Schrödinger equation take the form

$$\psi(r, \phi) = R(r)e^{\pm im\phi} \qquad ; \; m = 0, 1, 2, \cdots$$

$$z^2 \frac{d^2 R}{dz^2} + z \frac{dR}{dz} + (z^2 - m^2)R = 0 \qquad ; \; z \equiv \kappa r$$

---

[8]See Vol. I.

where $\kappa^2 \equiv (2m_0/\hbar^2)(E + V_0)$;

(b) A fundamental system of solutions to the above radial equation is given by the cylindrical Bessel function $J_m(z)$ of Eq. (2.207) and the cylindrical Neumann function $N_m(z)$.[9] The Neumann function is singular at the origin and has a wronskian $W[J_m(z), N_m(z)] = 2/\pi z$. Construct an argument that eliminates the singular solution.

2.30 (a) Use the results in Prob. 2.29 to show that in two dimensions the eigenvalues and eigenfunctions for a particle in a circular box of radius a are given by

$$E_{nm} = -V_0 + \frac{\hbar^2}{2m_0 a^2} Z_{nm}^2 \qquad ; m = 0, 1, 2, \cdots, \infty$$

$$; n = 1, 2, \cdots, \infty$$

$$\psi_{n, \pm m} = N_{nm} J_m \left(Z_{nm} \frac{r}{a} \right) e^{\pm im\phi}$$

Here Z_{nm} is the nth zero of the mth cylindrical Bessel function, excluding the origin;

(b) Plot the first four eigenvalues and corresponding radial wave functions $J_m(Z_{nm}r/a)$.

2.31 Solve the eigenvalue Eqs. (2.336) numerically for a few values of λ, and plot the curves in the band structure and extended band structure of the Kronig-Penney model in Figs. 2.21 and 2.22.

2.32 As a preliminary to the next two problems, this problem reviews and extends the discussion in Vol. I of barrier penetration in one dimension. Assume a repulsive step potential of height V_0 and width l, with $E < V_0$ (see Fig. 11.2). The Schrödinger equation and solutions in the various regions are given in Eqs. (2.24)–(2.29). With the utilization of scattering states, the solutions in the three regions take the following form

$$\psi_I = e^{ikx} + re^{-ikx} \qquad ; \psi_{II} = ae^{-\kappa x} + be^{\kappa x} \qquad ; \psi_{III} = te^{ikx}$$

(a) Match the boundary conditions at both sides of the potential, and show that the transmission and reflection amplitudes satisfy the following relations

$$te^{ikl} = e^{-\kappa l} + re^{-\kappa l}e^{-i\xi} \qquad ; e^{i\xi} \equiv \frac{\kappa - ik}{\kappa + ik}$$

$$te^{ikl} = e^{\kappa l} + re^{\kappa l}e^{i\xi}$$

---

[9]See [Fetter and Walecka (2003a)].

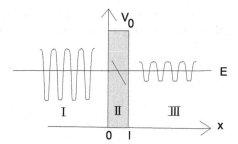

Fig. 11.2 Barrier penetration in one dimension, with $E < V_0$. There are both incident and reflected waves $e^{ikx} + re^{-ikx}$ in region I, and a transmitted wave te^{ikx} in region III.

(b) Show the transmission amplitude and transmission coefficient are

$$t = e^{-ikl}\left[\frac{e^{i\xi} - e^{-i\xi}}{e^{\kappa l}e^{i\xi} - e^{-\kappa l}e^{-i\xi}}\right]$$

$$T \equiv |t|^2 = \frac{(2k\kappa)^2}{(2k\kappa)^2 + (\kappa^2 + k^2)^2 \sinh^2 \kappa l}$$

(c) Show the reflection amplitude and reflection coefficient are

$$r = -\left[\frac{e^{\kappa l} - e^{-\kappa l}}{e^{\kappa l}e^{i\xi} - e^{-\kappa l}e^{-i\xi}}\right]$$

$$R \equiv |r|^2 = \frac{(\kappa^2 + k^2)^2 \sinh^2 \kappa l}{(2k\kappa)^2 + (\kappa^2 + k^2)^2 \sinh^2 \kappa l}$$

(d) Verify that $R + T = 1$, and interpret this result.

2.33 To obtain the boundary condition on the wave function at a delta-function potential, take the following limit of the results in Prob. 2.32

$$V_0 \to \infty \qquad ; l \to 0 \qquad ; \int V(x)dx = V_0 l = \text{constant}$$

(a) Show $t = 1 + r$;

(b) Hence conclude that the wave function must be continuous across the potential, $\psi_{III}(0) = \psi_I(0)$.

2.34 Use the results in Probs. 2.32–2.33 to show that $T \to 0$ as $\lambda \to \infty$ for the delta-function potential in Eq. (2.315), and hence demonstrate that there is no barrier penetration in the tight-binding limit of the Kronig-Penney model.

2.35 A particle of mass m is confined by rigid walls to the interior of a rectangular box with dimensions $0 \leq x \leq a$, $0 \leq y \leq a$, and $0 \leq z \leq l$ (see Fig. 11.3).

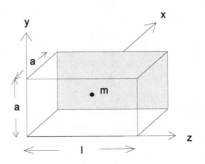

Fig. 11.3   Particle of mass m in a rectangular box with dimensions $0 \leq x \leq a$, $0 \leq y \leq a$, and $0 \leq z \leq l$.

(a) Construct the general solution to the time-dependent Schrödinger equation;

(b) Suppose the particle is injected into the box at time $t = 0$ in such a manner that its wave function at that time is

$$\Psi(\mathbf{x}, 0) = \left(\frac{8}{a^2 l}\right)^{1/2} \sin \frac{\pi x}{a} \sin \frac{\pi y}{a} \left[\sin \frac{\pi z}{l} + \sin \frac{2\pi z}{l}\right]$$

Plot the initial density distribution, and determine the wave function at all subsequent times;

(c) Compute the integrated probability flux through the plane $z = l/2$ as a function of time. Then use the continuity equation to compute the probability of finding the particle in the right-hand half of the box as a function of time. Plot the latter quantity. After what time (or times) will it be maximized?

3.1 Consider a two-dimensional problem with coefficient matrix

$$\begin{bmatrix} (H_{11} - E) & H_{12} \\ H_{12}^\star & (H_{22} - E) \end{bmatrix}$$

(a) Determine the eigenvalues $E^{(s)}$ and eigenvectors $\underline{c}^{(s)}$, for $s = 1, 2$;

(b) Set $H_{11} = H_{22}$ and let $H_{12} \to 0$ so that the eigenvalues become *degenerate*. Show that information is lost, and *any* new orthonormal pair of eigenvectors then provides a solution.

3.2 Take a matrix element of the completeness relation in Eq. (3.97) between the states $|\xi'\rangle$ and $|\xi''\rangle$, and show that one obtains the correct expression for $\langle\xi'|\xi''\rangle$.

3.3 Show that the Schrödinger equation in the momentum representation in Eq. (3.121) is obtained as the Fourier tranform of the Schrödinger equation in the coordinate representation in Eq. (3.113).

3.4 Use completeness to derive the following relations in the coordinate and momentum representations

$$\langle\boldsymbol{\xi}|\hat{\mathbf{p}}|\boldsymbol{\xi}'\rangle = \frac{\hbar}{i}\boldsymbol{\nabla}_\xi\langle\boldsymbol{\xi}|\boldsymbol{\xi}'\rangle \qquad ; \ \langle\mathbf{k}|\hat{\mathbf{x}}|\mathbf{k}'\rangle = i\hbar\boldsymbol{\nabla}_p\langle\mathbf{k}|\mathbf{k}'\rangle$$

3.5 (a) Consider the harmonic oscillator, and verify the example of Ehrenfest's theorem in Eqs. (3.163);[10]

$$\frac{d}{dt}\langle\Psi(t)|\hat{x}|\Psi(t)\rangle = \langle\Psi(t)|\frac{i}{\hbar}[\hat{H},\hat{x}]|\Psi(t)\rangle = \frac{1}{m}\langle\Psi(t)|\hat{p}|\Psi(t)\rangle$$

(b) Then show, in addition, that

$$\frac{d}{dt}\langle\Psi(t)|\hat{p}|\Psi(t)\rangle = \langle\Psi(t)|\frac{i}{\hbar}[\hat{H},\hat{p}]|\Psi(t)\rangle = -k\langle\Psi(t)|\hat{x}|\Psi(t)\rangle$$

(c) Use these results to verify Newton's second law for the classical limit of the motion of a particle in an oscillator potential.

3.6 (a) Consider the following operator $\hat{F}(\lambda) \equiv e^{i\lambda\hat{A}}\hat{B}e^{-i\lambda\hat{A}}$. Show

$$\frac{d^n\hat{F}(\lambda)}{d\lambda^n}\bigg|_{\lambda=0} = i^n[\hat{A},\cdots,[\hat{A},[\hat{A},\hat{B}]]\cdots] \qquad ; \ n \text{ terms}$$

(b) Make a Taylor series expansion of $\hat{F}(\lambda)$ in λ, and then set $\lambda=1$ to verify Eq. (3.122).

3.7 Consider the harmonic oscillator in the Heisenberg picture. Use Eqs. (3.148), (3.158), and (3.122) to show that

$$\hat{a}_H(t) = e^{i\hat{H}t/\hbar}\,\hat{a}\,e^{-i\hat{H}t/\hbar} = \hat{a}\,e^{-i\omega_0 t}$$
$$\hat{a}_H^\dagger(t) = e^{i\hat{H}t/\hbar}\,\hat{a}^\dagger e^{-i\hat{H}t/\hbar} = \hat{a}^\dagger e^{i\omega_0 t}$$

3.8 (a) Start from Eqs. (3.161), use the results in Prob. 3.7, and show by explicit differentiation with respect to time that for the harmonic oscillator

---

[10] Compare the discussion in Eqs. (3.169)–(3.171).

in the Heisenberg picture

$$m\frac{d\hat{x}_H(t)}{dt} = \hat{p}_H(t) \qquad ; \qquad \frac{d\hat{p}_H(t)}{dt} = -k\,\hat{x}_H(t)$$

Compare with the results in Prob. 3.5;

(b) Show that in the Heisenberg picture, it is the *equal-time* commutation relation that yields the canonical value

$$[\hat{p}_H(t), \hat{x}_H(t')]_{t=t'} = \frac{\hbar}{i}$$

3.9 For clarity, the discussion of measurements has essentially been confined to one dimension. The simplest extension to three dimensions is to eigenstates of momentum with periodic boundary conditions. Here the eigenstates are the direct product $|\mathbf{k}\rangle = |k_x\rangle|k_y\rangle|k_z\rangle$, with $k_i = 2\pi n_i/L$ and $n_i = 0, \pm 1, \pm 2, \cdots$, and the statement of completeness factors as

$$\sum_{\mathbf{k}} |\mathbf{k}\rangle\langle\mathbf{k}| = \sum_{k_x}\sum_{k_y}\sum_{k_z} |k_x\rangle|k_y\rangle|k_z\rangle\langle k_x|\langle k_y|\langle k_z| = \hat{1}$$

With the central-force problem in three dimensions, the situation is a little more complicated. The maximal set of mutually commuting operators is taken as $(\hat{H}, \hat{\mathbf{L}}^2, \hat{L}_z)$, and the eigenstates are $|nlm\rangle = |nl\rangle|lm\rangle$ where[11]

$$\hat{\mathbf{L}}^2|lm\rangle = l(l+1)|lm\rangle \qquad ; \hat{L}_z|lm\rangle = m|lm\rangle \qquad ; \hat{H}|nl\rangle = E_{nl}|nl\rangle$$

These states satisfy the orthonormality relations $\langle lm|l'm'\rangle = \delta_{ll'}\delta_{mm'}$ and $\langle nl|n'l\rangle = \delta_{nn'}$. The completeness statement is

$$\sum_{nlm} |nlm\rangle\langle nlm| = \sum_{nlm} |nl\rangle|lm\rangle\langle nl|\langle lm| = \hat{1}$$

(a) Take the matrix element $\langle\mathbf{x}|\cdots|\mathbf{x}'\rangle$ of this last relation and reproduce the completeness statement in the coordinate representation

$$\sum_{nlm} R_{nl}(r)R_{nl}^{\star}(r')Y_{lm}(\theta, \phi)Y_{lm}^{\star}(\theta', \phi') = \frac{1}{r^2\sin\theta}\delta(r - r')\delta(\theta - \theta')\delta(\phi - \phi')$$

State carefully how the sums are to be carried out;

(b) Let $\hat{L}_{\pm} \equiv \hat{L}_x \pm i\hat{L}_y$. Show

$$\langle nlm|[\hat{\mathbf{L}}^2, \hat{L}_{\pm}]|n'l'm'\rangle = [l(l+1) - l'(l'+1)]\delta_{nn'}\langle lm|\hat{L}_{\pm}|l'm'\rangle = 0$$

Hence conclude that the matrices $\underline{L}_{\pm}$ are block-diagonal in this basis.

---

[11]The corresponding wave functions are $\langle\mathbf{x}|nlm\rangle = \langle r|nl\rangle\langle\theta\phi|lm\rangle = R_{nl}(r)Y_{lm}(\theta, \phi)$.

3.10 Show that both the trace and determinant of the hamiltonian matrix H_{mn} are invariant under a unitary transformation.

4.1 (a) Write the energy functional in the first of Eqs. (4.32) in spherical coordinates with a $\psi(r)$ as

$$\mathcal{E} = 4\pi \int_0^\infty r^2 dr \, \psi^\star \left[-\frac{\hbar^2}{2m} \frac{1}{r^2} \frac{d}{dr} r^2 \frac{d}{dr} + V(r) \right] \psi$$

Now partially integrate the first term with the wave function in Eq. (4.30) to verify the second of Eqs. (4.32)

$$\mathcal{E} = 4\pi \int_0^\infty r^2 dr \left[\frac{\hbar^2}{2m} \left| \frac{d\psi}{dr} \right|^2 + V(r) |\psi|^2 \right]$$

(b) Verify the numerical results in Fig. 4.3 and Eqs. (4.36).

4.2 Assume the potential is bounded from below. Show the energy functional satisfies $\mathcal{E} \geq V_{\min}$ where $V_{\min}$ is the minimum value of V.

4.3 Show through order H' that the wave functions $\psi_n^{(1)}$ in non-degenerate perturbation theory with an hermitian H' form an orthonormal system

$$\langle \psi_n^{(1)} | \psi_m^{(1)} \rangle = \delta_{nm} + O(H'^2) \qquad \text{; orthonormal}$$

4.4 (a) Show the second-order wave function in non-degenerate perturbation theory is given by

$$\psi_n^{(2)} = \phi_n + \sum_{m \neq n} \phi_m \frac{\langle \phi_m | H' | \phi_n \rangle}{E_n^0 - E_m^0} \left[1 - \frac{\langle \phi_n | H' | \phi_n \rangle}{E_n^0 - E_m^0} \right] +$$
$$\sum_{m \neq n} \sum_{p \neq n} \phi_m \frac{\langle \phi_m | H' | \phi_p \rangle \langle \phi_p | H' | \phi_n \rangle}{(E_n^0 - E_m^0)(E_n^0 - E_p^0)}$$

(b) Verify that this expression is well-defined if the system is non-degenerate [see Eq. (4.59)].

4.5 Prove that the second-order energy shift always lowers the energy of the ground state in non-degenerate perturbation theory.

4.6 Reproduce the λ^2 term in Eq. (4.77) using second-order non-degenerate perturbation theory.

4.7 Show that the Coulomb repulsion in Eqs. (4.92) and (4.95) has the following interpretation. It is the Coulomb potential provided by the second electron at a radial distance ρ_1 integrated over the charge density of the first electron in a shell of thickness $d\rho_1$.

4.8 As a model for the $(p\mu^- p)$ molecules ("mulecules") formed when μ^- are stopped in hydrogen, or for the $(pe^- p) \equiv H_2^+$ molecular ion, consider the problem of a particle of mass m at position $\mathbf{r}$ interacting through the Coulomb interaction with two heavy, fixed, point charges at positions $\mathbf{r}_1$ and $\mathbf{r}_2$. The hamiltonian for this problem is

$$H = \frac{\mathbf{p}^2}{2m} - \frac{e^2}{|\mathbf{r} - \mathbf{r}_1|} - \frac{e^2}{|\mathbf{r} - \mathbf{r}_2|} + \frac{e^2}{|\mathbf{r}_1 - \mathbf{r}_2|}$$

We shall compute the expectation value of this hamiltonian using a normalized linear combination of atomic wave functions with respect to each charge[12]

$$\psi(\mathbf{r}) = \mathcal{N} \left[\psi_{1s}(|\mathbf{r} - \mathbf{r}_1|) + \psi_{1s}(|\mathbf{r} - \mathbf{r}_2|) \right]$$

(a) Show the normalization constant is given by[13]

$$\frac{1}{\mathcal{N}^2} = 2 \left[1 + \left(1 + \Delta + \frac{1}{3}\Delta^2 \right) e^{-\Delta} \right] \qquad ; \; \Delta \equiv \frac{1}{a_0} |\mathbf{r}_2 - \mathbf{r}_1|$$

Here $\psi_{1s}(\mathbf{r}) = e^{-r/a_0}/(\pi a_0^3)^{1/2}$, $\varepsilon_{1s}^0 = -e^2/2a_0$, and $a_0 = \hbar^2/me^2$;

(b) Show the desired result can then be written

$$\text{B.E.} \equiv \frac{\langle \psi | H | \psi \rangle - \varepsilon_{1s}^0}{e^2/2a_0} = \frac{2}{\Delta} \left[\frac{(1 - 2\Delta^2/3)e^{-\Delta} + (1 + \Delta)e^{-2\Delta}}{1 + (1 + \Delta + \Delta^2/3)e^{-\Delta}} \right]$$

(c) Explain in what sense B.E. is an estimate of the binding energy of the molecule, and verify the plot of B.E.(Δ) in Fig. 11.4;

(d) How is the *exact* ground-state energy of this model problem related to what you have calculated?

(e) Discuss how you might improve this calculation, and this model;

---

[12]This problem is longer, but the results are well worth it. The calculation provides the simplest description of the *molecular bond*. It uses the "linear combination of atomic orbitals (LCAO)" method.

[13]*Hint*: Establish the first result below, and use the second from the tables

$$\int d\Omega_t \frac{e^{-\alpha|t - \mathbf{\Delta}|}}{|t - \mathbf{\Delta}|} = \frac{2\pi}{\alpha x_> x_<} \left[e^{-\alpha(x_> - x_<)} - e^{-\alpha(x_> + x_<)} \right]$$

$$\int^x x^n e^{-ax} \, dx = -\frac{e^{-ax}}{a^{n+1}} \left[(ax)^n + n(ax)^{n-1} + \cdots + n! \right]$$

(f) Compare with the experimental results for the H_2^+ ion (see Fig. 11.4)

$r_e = 1.06\,\text{Å}$; equilibrium internuclear separation

B.E. $= -2.65056$ eV

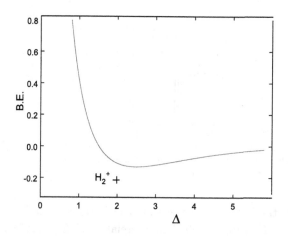

Fig. 11.4 The dimensionless quantity B.E.(Δ) in Prob. 4.8(c).

4.9 (a) Show the electromagnetic fields $(\mathbf{B}, \mathbf{E})$ are unchanged under the following gauge transformation of the potentials in Eqs. (4.130)

$$\mathbf{A} \to \mathbf{A} + \nabla\Lambda \qquad ; \ \Phi \to \Phi - \frac{1}{c}\frac{\partial\Lambda}{\partial t}$$

where $\Lambda(\mathbf{r}, t)$ is a scalar function of position and time;

(b) Show that the time-dependent Schrödinger equation is left *invariant* under the gauge transformation in part (a) accompanied by the following *local phase transformation* of the Schrödinger wave function

$$\Psi(\mathbf{r}, t) \to e^{ie\Lambda(\mathbf{r},t)/\hbar c}\Psi(\mathbf{r}, t)$$

4.10 Consider a particle of mass m in a one-dimensional box of size L with a potential $2mV/\hbar^2 = v\delta(x)$ located at the center (see Fig. 11.5).

(a) Calculate the energy functional $\mathcal{E}(\alpha)$ using the trial wave function

$$\psi(x) = \mathcal{N}\sin\alpha(x + L/2) \qquad ; \ -L/2 \le x \le 0$$
$$= \mathcal{N}\sin\alpha(L/2 - x) \qquad ; \ 0 \le x \le L/2$$

(b) Obtain the first-order perturbation theory result using $\alpha = \pi/L$;

(c) Minimize $\mathcal{E}(\alpha)$ with respect to α to find an improved variational estimate;

(d) Compare with the exact answer obtained by matching boundary conditions across the potential [see Eqs. (2.320)–(2.323)].

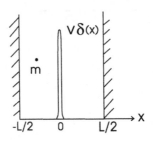

Fig. 11.5   Particle of mass m in a one-dimensional box of size L with an additional potential $2mV/\hbar^2 = v\delta(x)$ in the center.

4.11 Verify the behavior of the spherical harmonics under spatial reflection in Eq. (4.184) from their defining relation in Eq. (2.172).

4.12 Make a good numerical calculation and three-dimensional plot[14] of the charge density in the eigenstates $\chi_\pm(\mathbf{r})$ in Eqs. (4.200) (see Figs. 4.17–4.18.)

4.13 A particle moves in a two-dimensional circular box of radius a with perfectly rigid walls. Use the radial wave function $R(r) = [1 - (r/a)^\alpha]$, with α as a variational parameter, to calculate the ground-state energy. Compare with the exact answer obtained from the eigenvalue equation $J_0(ka) = 0$ (see Prob. 2.30).

4.14 Expand the wave function $\psi(x)$ in a truncated basis of known wave functions $\phi_n(x)$ satisfying the appropriate boundary conditions

$$\psi(x) = \sum_{n=1}^{N} a_n \phi_n(x)$$

and compute the energy functional

$$\mathcal{E}(\psi) = \frac{\langle \psi | H | \psi \rangle}{\langle \psi | \psi \rangle}$$

---

[14]Your choice as to just how to do this.

Treat the *coefficients* $(a_1, a_2, \cdots, a_N)$ as variational parameters, and show that the Euler-Lagrange equations for this variational problem are equivalent to diagonalizing the matrix H_{mn} in the truncated basis. Hence conclude that this analysis provides a variational basis for *matrix mechanics*.

4.15 The relativistic kinetic energy of an electron is given by

$$T = \left(\mathbf{p}^2 c^2 + m_0^2 c^4\right)^{1/2} - m_0 c^2 = \frac{\mathbf{p}^2}{2m_0} - \frac{\mathbf{p}^4}{8m_0^3 c^2} + \cdots$$

Treat the second term on the right as a perturbation, and calculate the first-order shift in the ground-state energy of hydrogen-like atoms (with charge Z on the nucleus). Compare with the unperturbed value.

4.16[15] A one-dimensional harmonic oscillator is perturbed by an extra potential energy γx^3. Calculate the change in each energy level to second order in the perturbation. Discuss.

4.17 Show the case where $E = E_n^0$ in Eq. (4.47) can be handled by adding another perturbation H'', diagonal in degenerate subspaces, and in the end taking the limit $H'' \to 0$.

5.1 Show that the asymptotic form of the scattered wave ψ_{scatt} in Eq. (5.16) satisfies the time-independent Schrödinger equation as $r \to \infty$[16]

$$(\nabla^2 + k^2)\psi_{\text{scatt}} = 0 \qquad ; r \to \infty$$

5.2 This problem provides an alternate derivation of the scattering Green's function. Define $\mathbf{r} \equiv \mathbf{x} - \mathbf{y}$, and introduce a new coordinate system whose origin is located at $\mathbf{y}$. Now consider the function $G_k^{(+)} \equiv e^{ikr}/4\pi r$.

(a) Show that for $r > 0$, this function satisfies

$$(\nabla^2 + k^2)\frac{e^{ikr}}{4\pi r} = 0 \qquad ; r > 0$$

(b) Take a small sphere of radius R about the origin, and use Gauss' law to show

$$\int_V d^3r \, \nabla^2 \frac{e^{ikr}}{4\pi r} = \int_A d\mathbf{A} \cdot \nabla \frac{e^{ikr}}{4\pi r}$$
$$= -1 \qquad ; R \to 0$$

---

[15]See Prob. 8.2 in [Schiff (1968)].
[16]*Hint*: Recall Eq. (2.200).

where $dA = e_r R^2 d\Omega$ is a little element of surface area, and the last equality holds in the limit $R \to 0$. Hence conclude that the differential Eq. (5.19) defining the Green's function with a Dirac delta-function source is satisfied;

(c) Show that the same derivation holds for $G_k^{(-)} \equiv e^{-ikr}/4\pi r$, and thus conclude that the proper choice of Green's function depends on the imposed boundary conditions.

5.3 (a) Carry out a calculation of $G_k^{(-)}(\mathbf{x}, \mathbf{y})$ in parallel with that for $G_k^{(+)}(\mathbf{x}, \mathbf{y})$ as detailed in the text, evaluate the residue of the pole at $t = -(k - i\eta)$, and hence derive Eq. (5.32);

(b) Show the contribution from the large semi-circle in Fig. 5.3 makes a vanishing contribution to the contour integral in Eq. (5.29) as $R \to \infty$.[17]

5.4 (a) Substitute the relation on the r.h.s. of Eq. (5.34) for $\psi_{\mathbf{k}}^{(+)}(\mathbf{y})$ in the integral, and derive the following *exact expression* for $\psi_{\mathbf{k}}^{(+)}(\mathbf{x})$

$$\psi_{\mathbf{k}}^{(+)}(\mathbf{x}) = e^{i\mathbf{k}\cdot\mathbf{x}} - \int d^3y \, G_k^{(+)}(\mathbf{x} - \mathbf{y})v(y)e^{i\mathbf{k}\cdot\mathbf{y}} +$$
$$\int d^3y \int d^3z \, G_k^{(+)}(\mathbf{x} - \mathbf{y})v(y)G_k^{(+)}(\mathbf{y} - \mathbf{z})v(z)\psi_{\mathbf{k}}^{(+)}(\mathbf{z})$$

(b) Repeat this process once more to obtain an exact second-order iteration, with a final term explicitly of $O(v^3)$.

5.5 (a) What is the wavenumber k of a nucleon with energy $E_{\text{inc}} = 20\,\text{MeV}$ incident on a heavy object? With $100\,\text{MeV}$? With $500\,\text{MeV}$?

(b) What is the maximum momentum transfer q in elastic scattering in each case?

(c) If one were to measure the radius R of the potential through the location of the first minimum in Fig. 5.9, what is the smallest value of R that could be measured in each case?

5.6 Calculate the scattering amplitude and differential cross section in Born approximation for the exponential potential, with either sign of V_0,

$$V(r) = V_0 \, e^{-\lambda r} \qquad ; \text{ exponential potential}$$

5.7 (a) Given a real potential $v(|\mathbf{x}|)$, and the asymptotic form of the wave function in the first of Eqs. (5.38), use the continuity equation and Gauss' law to derive the optical theorem

$$\frac{4\pi}{k}\text{Im} \, f_{\text{el}}(0) = \sigma_{\text{elastic}} \qquad ; \text{ optical theorem}$$

---

[17] See [Fetter and Walecka (2003a)].

where σ_{elastic} is the integrated elastic cross section;

(b) Generalize this result to complex v [see Eq. (5.247)]. Show[18]

$$\frac{4\pi}{k} \text{Im} f_{\text{el}}(k,0) = \sigma_{\text{total}} \qquad ; \text{optical theorem}$$

where σ_{total} is the *total* cross section, now including an absorptive part.

5.8 (a) Show that $G_k^{(\pm)}(\mathbf{x}-\mathbf{y})$ reproduce the Coulomb Green's function in the limit $k \to 0$;

(b) Take the limit $k \to 0$ in Eq. (5.79), and derive the generating function for the Legendre polynomials in Eq. (2.66).

5.9 Start from the expression for the scattering amplitude in Eq. (5.38). Insert the plane-wave expansion in Eq. (5.78), the partial-wave expansion of the scattering state in Eq. (5.96), and use the addition theorem to re-derive the partial-wave decomposition of the scattering amplitude in Eq. (5.105).[19]

5.10 (a) Locate a numerical program to calculate the differential cross section $d\sigma/d\Omega$ for scattering from a hard sphere, and reproduce the two curves shown in Fig. 11.6 for the cases $ka = 0.5$ and $ka = 10$;

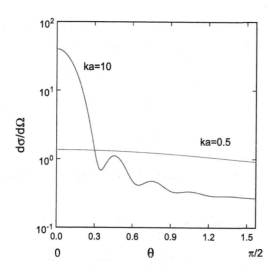

Fig. 11.6 Cross section $d\sigma/d\Omega$, measured in units of a^2, for scattering from a hard sphere of radius a. Two cases are plotted: $ka = 0.5$ snd $ka = 10$. Note that the plot only displays the region $0 \le \theta \le \pi/2$.

---

[18]Note Eq. (20.3) in [Schiff (1968)].

[19]Recall $\mathbf{x} = x\mathbf{e}_x$ and $\mathbf{k}' = k\mathbf{e}_x$.

(b) Extend the calculations in θ, and show that the integrated cross sections, in units of a^2, are 11.75 and 7.53 respectively;

(c) Extend the calculations as far as you can in ka.

5.11 (a) Show that at zero incident energy the differential cross section for scattering from a hard sphere of radius a is $d\sigma/d\Omega = a^2$, independent of angle;

(b) Show the total cross section is $\sigma = 4\pi a^2$;

(b) Compare with the numerical results in Prob. 5.10.

5.12 Show that as $k \to 0$, the phase shifts for scattering from a hard sphere of radius a behave as

$$\delta_l(k) \to -\frac{(2l+1)}{[(2l+1)!!]^2}(ka)^{2l+1} \qquad ; k \to 0$$

5.13 (a) Sketch the situation on the l.h.s. of Fig. 5.24 when the attractive potential gives rise to a positive scattering length;

(b) When will the scattering length a become infinite?

(c) Explain how it is possible to have an infinite zero-energy cross section in quantum mechanics.

5.14 The condition that the attractive square-well potential in Fig. 5.23 have just one bound state at zero energy was shown in Prob. 2.20 to be

$$\sqrt{v_0}R = \frac{\pi}{2} \qquad ; \text{zero-energy bound state}$$

Show from Eq. (5.162) that for this potential, the s-wave phase shift satisfies

$$k \cot \delta_0(k) = \frac{1}{2}Rk^2 + O(k^4)$$

Hence conclude that the scattering length and effective range in this case are given by

$$|a| = \infty \qquad ; r_0 = R$$

5.15 Derive the result for the effective range $r_0 = R$ for the potential in Prob. 5.14 from effective-range theory in Eq. (5.170).

5.16 Consider the hard-core–square-well potential in Fig. 11.7.

(a) Show that the condition for a zero-energy bound state is

$$\sqrt{v_0}\, b_w = \frac{\pi}{2} \qquad ; \text{zero-energy bound state}$$

(b) What is the appropriate modification of Eq. (5.162) in this case?

(c) Show from your result in (b) that the scattering length and effective range for this potential are given by

$$|a| = \infty \qquad ; \; r_0 = 2b + b_w$$

(d) Derive the result for the effective range in part (c) from effective-range theory in Eq. (5.170).

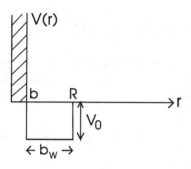

Fig. 11.7   Hard-core–square-well potential. Here $R = b + b_w$.

5.17 (a) In the WKB approximation, valid for large l, the effective potential is first modified to $\tilde{v}_{\text{eff}}(r)$ in the second of Eqs. (5.202). Show the WKB phase shift $\delta_l^{\text{WKB}}(k)$ is then obtained by adding a term of $O(1/k)$ to $\phi(r)$ in the phase of the wave functions in Eq. (5.195).

$$u_l^{(+)}(r) = \bar{a} \left\{ e^{ik[\phi(r)-\phi(r_0)+\pi/4k]} - e^{-ik[\phi(r)-\phi(r_0)+\pi/4k]} \right\}$$

(b) Show this wave function no longer vanishes at the classical turning point.

5.18 (a) Expand the exponential $e^{i\chi(b,k)}$ for $k \to \infty$, and show that Eqs. (5.239) then reproduce the Born approximation for the scattering amplitude when the momentum transfer lies in the transverse plane

$$f_{\text{BA}}(\mathbf{q}) = -\frac{2\mu}{4\pi\hbar^2} \int d^3y \, e^{-i\mathbf{q}\cdot\mathbf{y}} \, V(y)$$

(b) Discuss the improvement achieved through the use of the full set of Eqs. (5.239).

5.19 (a) The modification of Eqs. (5.256) for a circular *aperture* is

$$e^{i\chi(b,k)} = 0 \qquad ; \, b > R \qquad ; \, \text{complete absorption}$$
$$e^{i\chi(b,k)} = 1 \qquad ; \, b < R \qquad ; \, \text{no phase shift}$$

Show that one obtains the *same* diffraction pattern as in Eq. (5.259) and Fig. 5.33 for a black disc.[20] This is an example of *Babinet's principle* in classical optics.[21]

(b) Can you extend this result to black discs and complementary apertures of *any* shape?

5.20 (a) Start from Eqs. (5.269)–(5.272), and prove the optical theorem in Eq. (5.273);

(b) If the target is a probability sink, then the reaction cross section defined in Eq. (5.269) is a non-negative quantity. Use the superposition axiom to then establish the unitarity inequality satisfied by the partial-wave amplitudes in Eq. (5.274).

5.21 Show that if calculated with incoming-wave boundary conditions, the analog of Eq. (5.101) is

$$G_l^{(-)}(x, y; k) = -ikj_l(ky)h_l^{(2)}(kx) \qquad ; \, x > y$$
$$= -ikj_l(kx)h_l^{(2)}(ky) \qquad ; \, x < y$$

5.22 (a) Start from the integral equation satisfied by the partial-wave radial wave function $\psi_l^{(+)}(r; k)$. Show that if one defines

$$\psi_l^{(+)}(r; k) \equiv e^{i\delta_l(k)} \, \psi_l^{(s)}(r; k)$$

then $\psi_l^{(s)}(r; k)$ is *real*;

(b) Consider the partial-wave radial wave function $\psi_l^{(-)}(r; k)$ defined through the Green's function $e^{-ik|\mathbf{x}-\mathbf{y}|}/4\pi|\mathbf{x} - \mathbf{y}|$ with incoming-wave boundary conditions (see Prob. 5.21). Show

$$\psi_l^{(-)}(r; k) \equiv e^{-i\delta_l(k)} \, \psi_l^{(s)}(r; k)$$

Hence conclude that the partial-wave scattering amplitude $S_l(k)$ is obtained from

$$\psi_l^{(+)}(r; k) = e^{2i\delta_l(k)} \, \psi_l^{(-)}(r; k) = S_l(k) \, \psi_l^{(-)}(r; k)$$

---

[20] *Hint:* Use $\int_{\text{plane}} d^{(2)}b \, e^{i\mathbf{b}\cdot\mathbf{q}} = \delta^{(2)}(\mathbf{q})$.

[21] See [Fetter and Walecka (2003a)].

5.23 In an attempt to obtain a better description of the surface of the target than provided by the black disc, one may try a "diffuse grey disc" where the eikonal phase is defined by

$$e^{i\chi(b)} - 1 \equiv -\exp\left(-\frac{b^2 \ln 2}{R^2}\right) \qquad ; \text{ grey disc}$$

(a) Show that the elastic scattering cross section for the grey disc is given by

$$\left(\frac{d\sigma}{dq^2}\right)_{\text{el}} = \frac{\pi R^4}{4(\ln 2)^2} \exp\left(-\frac{q^2 R^2}{2 \ln 2}\right)$$

(b) Show the integrated elastic and reaction cross sections are given by

$$\sigma_{\text{elastic}} = \frac{\pi R^2}{2 \ln 2} \qquad ; \sigma_{\text{reaction}} = \frac{3\pi R^2}{2 \ln 2}$$

(c) Compare with the results for a black disc;

(d) Compare both the grey and black disc results with the experimental cross section for p-p scattering at $E_{\text{CM}} = 53.2\,\text{GeV}$ shown in Fig. 5.34. Note $-t \equiv (\hbar c \mathbf{q})^2$.

5.24 (a) Compute the classical attenuation of a beam going through a medium of scatterers with elementary cross sections σ_t and density distribution $\rho(x)$. Justify the identification of the classical transmission with $|e^{i\chi}|^2$, and hence show that the high-energy phase and *optical potential* for a composite system take the form

$$i\chi_{\text{opt}}(b) = -\frac{1}{2}\sigma_t \int_{-\infty}^{\infty} dz\,\rho(\sqrt{b^2 + z^2})$$
$$v_{\text{opt}}(x) = -ik\sigma_t\rho(x) \qquad ; \text{ optical potential}$$

(b) Use the optical theorem, and restore the full forward scattering amplitude, to obtain the following extended expression for the optical potential[22]

$$v_{\text{opt}}(x) = -4\pi f_{\text{el}}(k,0)\rho(x) \qquad ; \text{ optical potential}$$

When is the result in (a) then applicable?[23]

6.1 (a) Make a good numerical calculation of the second integral in Eqs. (6.39) to evaluate and plot $K_0(\alpha)$;

---

[22]See, for example, [Walecka (2004)].
[23] *Hint*: Contemplate Eq. (5.274) and the second of Eqs. (5.270).

(b) Take the derivative of your result in (a) to evaluate and plot $K_1(\alpha)$;

(c) Compare your numerical results with the two asymptotic expressions in Eqs. (6.41).

6.2[24] A hydrogen atom in its ground state is placed between the plates of a condenser. A voltage pulse is applied to the condenser so as to produce a homogeneous electric field that has a time dependence

$$\mathcal{E}(t) = 0 \qquad\qquad ; t < 0$$
$$\mathcal{E}(t) = \mathcal{E}_0\, e^{-t/\tau} \qquad ; t > 0$$

(a) Find the first-order probability that the atom is in the 2s-state $|200\rangle$ after a long time;

(b) What is the corresponding probability that it is one of the 2p-states?

6.3 Use the half-width of the central peak in Fig. 6.8 as a measure $\Delta\omega$ of the spread in frequencies corresponding to the time interval $T \equiv \Delta t$. Hence derive the *effective* uncertainty relation

$$\Delta E \Delta t \sim h \qquad ; \text{effective uncertainty relation}$$

Note, however, that this result is obtained in a significantly different fashion than the uncertainty principle in Eq. (3.221).

6.4 (a) Show the cross section for exciting the H-atom from the 1s to the 2s state through inelastic charged-lepton $l^{\pm}$ scattering is given to leading order in $\alpha = e^2/\hbar c$ by

$$\left(\frac{d\sigma}{d\Omega}\right)_{2s\leftarrow 1s} = \frac{\alpha^2}{q^4}\left(\frac{m_l c}{\hbar}\right)^2 \frac{k_f}{k_i} \frac{128(qa_0)^4}{[(qa_0)^2 + 9/4]^6}$$

Here $a_0 = \hbar^2/m_e e^2$ is the electron Bohr radius.

(b) Show the cross section for the $1s \to 2p$ transition is

$$\left(\frac{d\sigma}{d\Omega}\right)_{2p\leftarrow 1s} = \frac{\alpha^2}{q^4}\left(\frac{m_l c}{\hbar}\right)^2 \frac{k_f}{k_i} \frac{288(qa_0)^2}{[(qa_0)^2 + 9/4]^6}$$

6.5 Make a good numerical calculation and plot of the absolute square of the Coulomb inelastic atomic form factor in Eq. (6.129) for the following transitions: $1s \to 3s, 1s \to 3p$, and $1s \to 3d$. Compare and discuss.

---

[24]See [Schiff (1968)], Prob. 8.1.

6.6 Show the total inelastic cross sections in Prob. 6.4, at high incident energy where $ka_0 \to \infty$, become[25]

$$\sigma_{2s \leftarrow 1s} = \frac{128\pi}{5k^2} \left(\frac{m_l}{m_e}\right)^2 \left(\frac{2}{3}\right)^{10} \qquad ; ka_0 \to \infty$$

$$\sigma_{2p \leftarrow 1s} = \frac{576\pi}{k^2} \left(\frac{m_l}{m_e}\right)^2 \left(\frac{2}{3}\right)^{12} \ln(ka_0)$$

Hint: Convert from $\int \sin\theta \, d\theta$ to $\int q \, dq$.

6.7 A development analogous to that in the text can be used to describe *direct nuclear reactions*. Assume free plane-wave states for the projectile nucleon and single-nucleon states $\psi_{nlm}(\mathbf{r})$ for the target nucleons, of the type shown in Figs. 2.11 and 11.1. Assume the projectile and target nucleons interact through a two-nucleon potential of the form $V(|\mathbf{r}_p - \mathbf{r}|)$. Use the Golden Rule to show the cross section for the target transition $|n00\rangle \to |n'lm\rangle$ is given in plane-wave Born approximation by

$$\left(\frac{d\sigma}{d\Omega}\right)_{n'lm \leftarrow n00} = \frac{k_f}{k_i} \left|\frac{2m_p}{4\pi\hbar^2} \tilde{V}(q)\right|^2 |F_{n'lm, n00}(\mathbf{q})|^2 \qquad ; \text{plane-wave B.A.}$$

$$F_{n'lm, n00}(\mathbf{q}) = \delta_{m0}(-i)^l(2l+1)^{1/2} \int_0^\infty r^2 dr \, R_{n'l}(r) j_l(qr) R_{n0}(r)$$

$$; \text{inelastic nuclear f.f.}$$

Here m_p is the nucleon mass, and $\tilde{V}(q)$ is the Fourier transform of the two-body potential in Eqs. (5.43).

6.8 An improvement over the result in Prob. 6.7 is to let the initial and final projectile wave functions be distorted by the same nuclear potential seen by the target nucleon. Use the analysis in Eqs. (5.70)–(5.75) to show that the proper recipe for obtaining *distorted-wave Born approximation* is to make the following replacement in the projectile matrix elements of the two-body potential

$$\langle \mathbf{k}_f | V | \mathbf{k}_i \rangle \to \langle \psi_{\mathbf{k}_f}^{(-)} | V | \psi_{\mathbf{k}_i}^{(+)} \rangle \qquad ; \text{distorted-wave B.A.}$$

Here $\psi_{\mathbf{k}}^{(\pm)}(\mathbf{r}_p)$ are the scattering states of the projectile in the nuclear potential with outgoing and incoming-wave boundary conditions.

6.9 An example with *two* particles in the continuum in the final state is given by charged lepton $l^\pm$ scattering from the H-atom, where the atom is

---

[25]Compare [Schiff (1968)], Prob. 9.14.

ionized. Assume the final atomic electron is in a free plane-wave state with wavenumber $\mathbf{t}$. Thus the initial and final wave functions and energies are

$$\phi_i = \psi_{1s}(\mathbf{r}_e)\frac{1}{\sqrt{\Omega}}e^{i\mathbf{k}_i\cdot\mathbf{r}_l} \qquad ; \; E_i = \varepsilon_{1s}^0 + \frac{\hbar^2 k_i^2}{2m_l}$$

$$\phi_f = \frac{1}{\sqrt{\Omega}}e^{i\mathbf{t}\cdot\mathbf{r}_e}\frac{1}{\sqrt{\Omega}}e^{i\mathbf{k}_f\cdot\mathbf{r}_l} \qquad ; \; E_f = \frac{\hbar^2 t^2}{2m_e} + \frac{\hbar^2 k_f^2}{2m_l}$$

(a) Show the cross section is given to leading order in H_1 by

$$d\sigma_{\mathbf{t}\leftarrow 1s} = \frac{2\pi}{\hbar}|\langle\phi_f|H_1|\phi_i\rangle|^2\delta(E_f - E_i)\frac{\Omega d^3 t}{(2\pi)^3}\frac{\Omega d^3 k_f}{(2\pi)^3}\frac{1}{I_{\text{inc}}}$$

(b) Show the quantization volume Ω cancels from this expression;

(c) Do the integral over dt, and show that[26]

$$d\sigma_{\mathbf{t}\leftarrow 1s} = \frac{4\alpha^2}{q^4}\left(\frac{m_l c}{\hbar}\right)\left(\frac{m_e c}{\hbar}\right)\left(\frac{t}{k_i}\right)|\tilde{\psi}_{1s}(\mathbf{t}+\mathbf{q})|^2\frac{d^3 k_f}{(2\pi)^3}d\Omega_t$$

where $\tilde{\psi}_{1s}(\mathbf{t}+\mathbf{q})$ is the Fourier transform of the ground-state wave function

$$\tilde{\psi}_{1s}(\mathbf{t}+\mathbf{q}) = \int d^3 r_e\,\psi_{1s}(\mathbf{r}_e)e^{-i(\mathbf{t}+\mathbf{q})\cdot\mathbf{r}_e} \qquad ; \; \mathbf{q} = \mathbf{k}_f - \mathbf{k}_i$$

6.10 A typical nuclear gamma transition is of the order of 1 MeV. The nuclear radius is $R \approx 1.2B^{1/3} \times 10^{-13}$ cm, where B is the baryon number.

(a) Compare the nuclear size with the wavelength of the radiation;

(b) Compare with the atomic case.

6.11 Start from Eqs. (6.130), and define the probability density and probability current by[27]

$$\rho \equiv \Psi^*\Psi$$

$$\mathbf{S} \equiv \frac{1}{2m}\left\{\Psi^*\left(\mathbf{p} - \frac{e}{c}\mathbf{A}\right)\Psi + \left[\left(\mathbf{p} - \frac{e}{c}\mathbf{A}\right)\Psi\right]^*\Psi\right\}$$

Work in the coordinate representation, and show this current is conserved

$$\nabla\cdot\mathbf{S} + \frac{\partial\rho}{\partial t} = 0$$

As before, these equations provide the interpretation for a particle in an external electromagnetic field.

---

[26]Now $E_f = E_i$ determines $t(k_f, k_i)$. Recall the caveats in the text if $l^\pm = e^\pm$.

[27]Note that $(\mathbf{p} - e\mathbf{A}/c)/m = \mathbf{v}$ is the particle's *kinetic velocity* (see appendix B).

6.12 Make a unitary transformation from the Schrödinger to the interaction picture

$$|\Psi_I(t)\rangle = e^{i\hat{H}_0 t/\hbar}|\Psi(t)\rangle \qquad ; \text{ interaction picture}$$
$$\hat{H}_I(t) = e^{i\hat{H}_0 t/\hbar}\hat{H}_1(t)e^{-i\hat{H}_0 t/\hbar}$$

(a) Show the Schrödinger equation in the interaction picture is

$$i\hbar\frac{\partial}{\partial t}|\Psi_I(t)\rangle = \hat{H}_I(t)|\Psi_I(t)\rangle \qquad ; \text{ S-eqn}$$

(b) Show the time-development operator $\hat{U}(t,t_0)$ in Eq. (6.189) satisfies

$$i\hbar\frac{\partial}{\partial t}\hat{U}(t,t_0) = \hat{H}_I(t)\hat{U}(t,t_0)$$
$$\hat{U}(t_0,t_0) = 1$$

(c) Hence show that $|\Psi_I(t)\rangle = \hat{U}(t,t_0)|\Psi_I(t_0)\rangle$ provides the solution to the Schrödinger equation in the interaction picture that reduces to $|\Psi_I(t_0)\rangle$ at $t = t_0$.

7.1 Prove that the expansion of the vector potential $\mathbf{A}(\mathbf{x},t)$ in Eq. (7.23) has enough flexibility to match the arbitrary set of initial conditions in Eqs. (7.26).[28]

7.2 Start from Eq. (7.57). Introduce the expansion of $\mathbf{A}(\mathbf{x},t)$ in Eq. (7.36), and derive the expression for the momentum in the normal modes of the free, classical radiation field in Eq. (7.60).[29]

7.3 (a) Establish the vector identity

$$\boldsymbol{\nabla}\cdot(\mathbf{a}\times\mathbf{b}) = \mathbf{b}\cdot(\boldsymbol{\nabla}\times\mathbf{a}) - \mathbf{a}\cdot(\boldsymbol{\nabla}\times\mathbf{b})$$

(b) Integrate the Poynting vector in Eq. (7.55) over the surface of a small volume in space. Use Gauss' law and Maxwell's equations to reproduce the energy density in the free radiation field in Eq. (7.27).

7.4 (a) Prove the relation in Eq. (7.66), which states that at equal times the commutator of the vector field operator at two distinct spatial points vanishes;

(b) Use this to conclude that starting from Eq. (7.73), one has

$$\dot{\hat{A}}(\mathbf{x}) = \frac{i}{\hbar}[\hat{H},\hat{A}(\mathbf{x})] = \frac{i}{\hbar}[\hat{H}_{\text{rad}},\hat{A}(\mathbf{x})]$$

---

[28] Compare Prob. 4.2 in Vol. I.

[29] Recall Eq. (7.34).

7.5 (a) Take the matrix element of the electromagnetic current operator $\hat{\mathbf{j}}(\mathbf{x})$ in Eq. (7.96) between the states $|\phi_m; n_1 \cdots, n_\infty\rangle$. Use the hermiticity of $\hat{\dot{\mathbf{x}}}_e$, and show the result is

$$\langle \phi_m; n_1 \cdots n_\infty | \hat{\mathbf{j}}(\mathbf{x}) | \phi_m; n_1 \cdots n_\infty \rangle = \frac{e}{2c} \left\{ \phi_m^\star(\mathbf{x}) \dot{\mathbf{x}}\, \phi_m(\mathbf{x}) + [\dot{\mathbf{x}}\, \phi_m(\mathbf{x})]^\star \phi_m(\mathbf{x}) \right\}$$

Define all quantities;

(b) Compare with the results in Prob. 6.11. Show the charge density and electromagnetic current are related to the probability density and probability current by

$$\varrho(\mathbf{x}) = e\rho(\mathbf{x}) \qquad ; \hat{\mathbf{j}}(\mathbf{x}) = \frac{e}{c}\hat{\mathbf{S}}(\mathbf{x})$$

7.6 Show the transverse delta-function in Eq. (7.65) projects from a vector field $\mathbf{v}(\mathbf{x})$ that part that is *transverse*[30]

$$\mathbf{v}_T(\mathbf{x}) \equiv \int d^3x'\, \underline{\boldsymbol{\delta}}^T(\mathbf{x} - \mathbf{x}') \cdot \mathbf{v}(\mathbf{x}')$$
$$\boldsymbol{\nabla} \cdot \mathbf{v}_T(\mathbf{x}) = 0$$

7.7 (a) Show that the classical field vanishes for a state with a definite number of photons;

(b) Show that there is an uncertainty relation between field strength and photon number [recall Eq. (3.220)];

(c) Show that with many photons in the mode $\{\mathbf{k}, s\}$, produced, for example, by a laser, one can replace the creation and destruction operators for that mode by c-numbers[31]

$$\left(\frac{2\pi\hbar c^2}{\omega_k} \right)^{1/2} \hat{a}_{\mathbf{k}s} \to a(\mathbf{k}, s) \qquad ; \left(\frac{2\pi\hbar c^2}{\omega_k} \right)^{1/2} \hat{a}_{\mathbf{k}s}^\dagger \to a^\star(\mathbf{k}, s)$$

Hence, recover the classical field for that mode.

7.8 The Heisenberg picture for the free radiation field is obtained from the Schrödinger picture by

$$|\Psi_H\rangle = e^{i\hat{H}_0 t/\hbar}|\Psi_S(t)\rangle \qquad ; \text{Heisenberg picture}$$
$$\hat{O}_H(t) = e^{i\hat{H}_0 t/\hbar}\hat{O}_S e^{-i\hat{H}_0 t/\hbar} \qquad \hat{H}_0 \equiv \hat{H}_{\text{rad}}$$

---

[30]The dyadic product is defined by $[\underline{\mathbf{a}} \cdot \mathbf{b}]_i \equiv a_{ij}b_j$.

[31]See the discussion of Bose condensation in chapter 11 of Vol. II.

Use Eqs. (7.49) to show the free field operator in the Heisenberg picture is

$$\hat{\mathbf{A}}_H(\mathbf{x}, t) \equiv \frac{1}{\sqrt{\Omega}} \sum_{\mathbf{k}} \sum_{s=1}^{2} \left(\frac{2\pi \hbar c^2}{\omega_k} \right)^{1/2} \mathbf{e}_{\mathbf{k}s} \left(\hat{a}_{\mathbf{k}s} \, e^{i\mathbf{k}\cdot\mathbf{x} - i\omega_k t} + \hat{a}_{\mathbf{k}s}^\dagger \, e^{-i\mathbf{k}\cdot\mathbf{x} + i\omega_k t} \right)$$

Compare with the expression in Eq. (7.36).

7.9 Go to the coordinate representation, and verify the step leading from the first to the second of Eqs. (7.138).

7.10 (a) Show that the $1s \to 2p$ transition in H-like atoms makes up a fraction of $2^{13}/3^9 = 0.416$ of the dipole sum rule;

(b) Take out a mean excitation energy $\bar{\varepsilon}$, and use closure to re-write the dipole sum rule as

$$\sum_{n} (\varepsilon_n - \varepsilon_i) |\langle \phi_n | \mathbf{x} | \phi_i \rangle|^2 = \bar{\varepsilon} \, \langle \phi_i | r^2 | \phi_i \rangle = \frac{3\hbar^2}{2m_0}$$

Evaluate this expression for H-like atoms, and show that $\bar{\varepsilon} = |\varepsilon_{1s}|$, which is the magnitude of the binding energy of the ground state.

7.11 A particle moves in an isotropic three-dimensional harmonic oscillator. Show the dipole sum rule is saturated by transitions which go up one oscillator spacing.[32]

7.12 Given the mean life τ as a time interval Δt, and using the half-width γ from Fig. 7.14 as a measure of the spread in the photon energy ΔE, deduce an uncertainty relation for photoemission[33]

$$\Delta E \Delta t = \hbar \qquad ; \text{ photoemission}$$

7.13 Suppose one truncates the integral over the photon angular frequencies to an *asymmetric* region of integration in Eq. (7.217).

(a) Show the term left over from the first integral in Eq. (7.219) would add a small imaginary part to γ, so that $\gamma \to \gamma_R + i\gamma_I$;

(b) Show that the imaginary part $i\gamma_I$ would give rise to a small real energy shift of the levels. (But so do the electromagnetic self-energies of the levels, arising from emission and absorption of photons, which have been neglected here.)

7.14 (a) Establish the representation of the Dirac delta-function in Eq. (7.228);

---

[32] Recall Prob. 2.27.
[33] *Hint*: Recall Eq. (7.222).

(b) Use the result in (a) to justify Eqs. (7.201) and (6.176) for small, but finite, γ.

8.1 Give an argument that Eq. (8.20) represents the most general state vector with angular momentum L in the C-M system formed from a (π^+, π^-) pair.

8.2 (a) Review Probs. 9.6 and 9.7 in Vol. I. Show that the following η_λ are two-component eigenstates of the helicity operator $\boldsymbol{\sigma} \cdot (\mathbf{k}/k)$ with eigenvalues $\lambda = \pm 1$

$$\eta_{+1} = \begin{pmatrix} e^{-i\phi_k/2} \cos\theta_k/2 \\ e^{i\phi_k/2} \sin\theta_k/2 \end{pmatrix} \quad ; \quad \eta_{-1} = \begin{pmatrix} -e^{-i\phi_k/2} \sin\theta_k/2 \\ e^{i\phi_k/2} \cos\theta_k/2 \end{pmatrix}$$

(b) Now derive the results in Table 8.1.[34]

8.3 (a) Verify Eqs. (8.65);

(b) Verify Eq. (8.156);

(c) Verify the statement made after Eq. (8.141).

8.4 (a) Use the orthonormality of the spinors $\bar{\mathcal{U}}(\mathbf{k}\lambda)\mathcal{U}(\mathbf{k}\lambda') = \delta_{\lambda\lambda'}$ to prove from Eq. (8.152) that the 2×2 matrix $\alpha_{\lambda\lambda'}$ is unitary

$$\sum_{\lambda''} \alpha_{\lambda\lambda''} \alpha^\star_{\lambda'\lambda''} = \delta_{\lambda\lambda'}$$

(b) Then show Eq. (8.155) follows from Eqs. (8.150)–(8.154).

8.5 The following interaction lagrangian density

$$\hat{\mathcal{L}}_I(x) = ig\hat{\bar{\psi}}\gamma_5\boldsymbol{\tau}\hat{\psi} \cdot \hat{\boldsymbol{\phi}} + if\hat{\bar{\psi}}\gamma_5\gamma_\mu\boldsymbol{\tau}\hat{\psi} \cdot \frac{\partial\hat{\boldsymbol{\phi}}}{\partial x_\mu}$$

represents the most general Yukawa coupling between a nucleon and isovector pseudoscalar pion which is

- Lorentz invariant
- hermitian
- invariant under $\hat{P}$
- invariant under isospin rotations

Show that this $\hat{\mathcal{L}}_I(x)$ is *separately* invariant under $\hat{C}$ and $\hat{T}$.[35]

---

[34] Note that the overall phase of the last eigenfunction is a matter of convention.

[35] The charged field is written in terms of the hermitian fields as $\sqrt{2}\,\hat{\phi}^\star = \hat{\phi}_1 + i\hat{\phi}_2$.

8.6 Start from the assumed transformation properties of the interaction hamiltonian density in Eqs. (8.131), and derive the implied symmetry properties of the scattering operator in Eqs. (8.133).

8.7 Consider the reaction $a + b \rightarrow c + d$ for spinless particles, with S-matrix element $\langle \mathbf{k}_c \, \mathbf{k}_d | \hat{S} | \mathbf{k}_a \, \mathbf{k}_b \rangle$.

(a) Show the time-reversed states are given by

$$\hat{T} | \mathbf{k}_a \, \mathbf{k}_b \rangle = \eta_{t_a} \eta_{t_b} | -\mathbf{k}_a, -\mathbf{k}_b \rangle \qquad ; \quad \hat{T} | \mathbf{k}_c \, \mathbf{k}_d \rangle = \eta_{t_c} \eta_{t_d} | -\mathbf{k}_c, -\mathbf{k}_d \rangle$$

(b) Show that time-reversal symmetry of the scattering operator in the last of Eqs. (8.133) implies the following condition on the S-matrix elements

$$\langle \mathbf{k}_c \, \mathbf{k}_d | \hat{S} | \mathbf{k}_a \, \mathbf{k}_b \rangle = \eta_{t_a}^\star \eta_{t_b}^\star \eta_{t_c} \eta_{t_d} \langle -\mathbf{k}_a, -\mathbf{k}_b | \hat{S} | -\mathbf{k}_c, -\mathbf{k}_d \rangle$$

8.8 Show the two results on the right in Eqs. (8.42) follow from those on the left, even though $\hat{T}$ is anti-unitary.

8.9 (a) Construct the interaction hamiltonian density $\mathcal{H}_W(x)$ from the interaction lagrangian density $\mathcal{L}_W(x)$ in Eq. (8.164), and show that it is a scalar under the combined operations TCP as in Eq. (8.175);[36]

(b) Repeat for $\mathcal{H}_S(x)$. The free lagrangian density for the charged scalar field is $\mathcal{L}_0/c^2 = -(\partial \phi^\star / \partial x_\mu)(\partial \phi / \partial x_\mu) - (mc/\hbar)^2 \phi^\star \phi$. Note Eqs. (8.47).

9.1 Start from Eq. (9.7) and show

$$TU(0, -\infty)T^{-1} = U(+\infty, 0)^\dagger = U(0, +\infty)$$

9.2 (a) Explicitly demonstrate the result in Eqs. (9.22)–(9.23), used to prove Theorem Ia, up through $\nu = 4$;

(b) Demonstrate the corresponding result used to prove Theorem Ib.

9.3 (a) Use the fact that the Heisenberg states $|\underline{\Psi}^{(\pm)}\rangle$ are eigenstates of the four-momentum operator to show that all other states are orthogonal to the vacuum $|\underline{0}^{(\pm)}\rangle$, assumed to be non-degenerate.[37]

(b) Repeat the argument for the stable single-particle states $|\underline{p}^{(\pm)}\rangle$.

9.4 (a) Start from the expression for $U(t, t_0)$ for finite times in Eq. (9.6) and derive Eq. (9.47). Then use the iterated series in Eq. (9.7) and adiabatic damping in Eq. (9.8) to let $t_0 \rightarrow -\infty$, and derive Eq. (9.48);

---

[36]Recall we are henceforth suppressing the carets on the operators in abstract Hilbert space.

[37]As a nicety, you can give all particles some mass, even infinitesimal, in making this argument.

(b) Verify Eq. (9.52) for $|p_n^{(-)}\rangle$.

9.5 Consider elastic electron scattering from the nucleon (Fig. 11.8).

Fig. 11.8   Elastic electron scattering from the nucleon.

The S-matrix for this process is given by[38]

$$S_{fi} = \frac{(2\pi)^4}{\Omega^2}\delta^{(4)}(p_1 + k_1 - p_2 - k_2)\bar{u}(\mathbf{k}_2)\gamma_\mu u(\mathbf{k}_1)\frac{4\pi\alpha}{q^2}\left(\frac{M^2}{E_1 E_2}\right)^{1/2}(\mathcal{J}_\mu)_{fi}$$

$$(\mathcal{J}_\mu)_{fi} = \left(\frac{\Omega^2 E_1 E_2}{M^2}\right)^{1/2}\langle p_2|J_\mu(0)|p_1\rangle$$

Here $e_p J_\mu(0)$ is the electromagnetic current operator for the hadronic system. The matrix element is in the Heisenberg picture; it includes all Feynman diagrams contributing to the electromagnetic structure of the nucleon.

(a) Use Lorentz invariance, current conservation, and the Dirac equation to show

$$(\mathcal{J}_\mu)_{fi} = i\bar{\mathcal{U}}(\mathbf{p}_2\lambda_2)\left[F_1(q^2)\gamma_\mu - F_2(q^2)\sigma_{\mu\nu}q_\nu + iF_3(q^2)\gamma_5\sigma_{\mu\nu}q_\nu\right]\mathcal{U}(\mathbf{p}_1\lambda_1)$$

where $\bar{\mathcal{U}}\mathcal{U} = 1$;

(b) Use hermiticity of the current to prove the form factors F_i are real $(i = 1, 2, 3)$;

(c) Use invariance under P to prove $F_3 = 0$;

(d) Use invariance under T to *separately* prove $F_3 = 0$;

(e) Make a non-relativistic reduction of the current matrix element, and give a physical interpretation of the term proportional to F_3 as an electric dipole moment.

---

[38]Compare Eq. (8.51) in Vol. II. This problem is longer, but the results form an essential part of nuclear physics.

10.1 The Gell-Mann matrices $\underline{\lambda}^a$ for $a = 1, 2, \cdots, 8$ are, in order,

$$\begin{pmatrix} & 1 & \\ 1 & & \\ & & \end{pmatrix} \begin{pmatrix} & -i & \\ i & & \\ & & \end{pmatrix} \begin{pmatrix} 1 & & \\ & -1 & \\ & & \end{pmatrix} \begin{pmatrix} & & 1 \\ & & \\ 1 & & \end{pmatrix} \begin{pmatrix} & & -i \\ & & \\ i & & \end{pmatrix}$$

$$\begin{pmatrix} & & \\ & & 1 \\ & 1 & \end{pmatrix} \begin{pmatrix} & & \\ & & -i \\ & i & \end{pmatrix} \begin{pmatrix} 1/\sqrt{3} & & \\ & 1/\sqrt{3} & \\ & & -2/\sqrt{3} \end{pmatrix}$$

The corresponding non-zero structure constants are (recall they are anti-symmetric in the indices)

$$f^{123} = 1$$

$$f^{147} = -f^{156} = f^{246} = f^{257} = f^{345} = -f^{367} = \frac{1}{2}$$

$$f^{458} = f^{678} = \frac{\sqrt{3}}{2}$$

Pick any subset of these relations and verify them by explicit multiplication of the matrices involved.

10.2 (a) Use the quark field $\underline{\psi}$ from Eqs. (10.1)–(10.2), and the matrices $\underline{\lambda}^a$ from Prob. 10.1, to write out in detail the Yukawa coupling of quarks to gluons in the second of Eqs. (10.23). Leave the Dirac matrix products intact;

(b) Repeat for the equations of motion in Eq. (10.27).

10.3 Start from the equations of motion for QCD in Eqs. (10.27), (10.29), (10.31), and show that each of the currents in Eqs. (10.32)–(10.34) is *conserved*.

10.4 A more explicit notation is to write the quark wave functions as the direct product of a three-component color wave function, a four-component flavor wave function, and a four-component Dirac spinor

$$[\eta^{\text{color}} \otimes \zeta^{\text{flavor}} \otimes u^{\text{Dirac}}]_{ilr} = \eta_i \zeta_l u_r \qquad ; \text{ quark wave functions}$$

Matrices are then the direct product of matrices in each space

$$[a \otimes b \otimes c]_{ilr,jms} = a_{ij} b_{lm} c_{rs} \qquad ; \text{ direct product}$$

(a) Write the Dirac current in Eq. (10.33) in terms of the matrix $[\lambda \otimes 1 \otimes \gamma_\mu]$;

(b) Write Eq. (10.16) in terms of the matrix $[1 \otimes M \otimes 1]$;

(c) Write Eq. (10.34) in terms of the matrix $[1 \otimes \Sigma \otimes \gamma_\mu]$.

Unit matrices and direct-product symbols are conventionally suppressed.[39]

10.5 (a) Show the canonical momenta conjugate to the gluon field A_j^a in QCD are

$$\Pi_j^a \equiv \Pi_{A_j^a} = i\mathcal{F}_{4j}^a \qquad \text{; canonical momenta}$$

$$; j = 1, 2, 3 \qquad ; a = 1, \cdots, 8$$

(b) Use Eq. (10.31) to derive Gauss' law for the color fields

$$\boldsymbol{\nabla} \cdot \boldsymbol{\Pi}^a = -\frac{g}{2}\underline{\psi}^\dagger \underline{\lambda}^a \underline{\psi} + gf^{abc}\,\boldsymbol{\Pi}^b \cdot \mathbf{A}^c$$

$$\equiv -g(\rho_{\text{quark}}^a + \rho_{\text{gluon}}^a) \qquad \text{; Gauss' law}$$

10.6 (a) Show, with the aid of partial integration, that the hamiltonian density in QCD can be written

$$\mathcal{H}_{\text{QCD}} \doteq \frac{1}{2}\boldsymbol{\Pi}^a \cdot \boldsymbol{\Pi}^a + \frac{1}{4}\mathcal{F}_{ij}^a \mathcal{F}_{ij}^a + \underline{\psi}^\dagger \boldsymbol{\alpha} \cdot \left[\frac{1}{i}\boldsymbol{\nabla} - \frac{g}{2}\underline{\lambda}^a \mathbf{A}^a\right]\underline{\psi}$$

(b) Prove that any vector field $\boldsymbol{\Pi}$ can be separated into $\boldsymbol{\Pi}_T + \boldsymbol{\Pi}_L$ where[40]

$$\boldsymbol{\nabla} \cdot \boldsymbol{\Pi}_T = 0 \qquad ; \boldsymbol{\nabla} \times \boldsymbol{\Pi}_L = 0 \qquad ; \int_{\text{box}} d^3x\,\boldsymbol{\Pi}_L \cdot \boldsymbol{\Pi}_T = 0$$

(c) Hence, show the hamiltonian density can be written

$$\mathcal{H}_{\text{QCD}} \doteq \underline{\psi}^\dagger \boldsymbol{\alpha} \cdot \left[\frac{1}{i}\boldsymbol{\nabla} - \frac{g}{2}\underline{\lambda}^a \mathbf{A}^a\right]\underline{\psi} + \frac{1}{2}\boldsymbol{\Pi}_T^a \cdot \boldsymbol{\Pi}_T^a + \frac{1}{4}\mathcal{F}_{ij}^a \mathcal{F}_{ij}^a$$

$$+ \frac{g^2}{8\pi}\int d^3x \int d^3x'\,\rho^a(\mathbf{x})\frac{1}{|\mathbf{x} - \mathbf{x}'|}\rho^a(\mathbf{x}') \qquad \text{; hamiltonian density}$$

Here $\boldsymbol{\nabla} \cdot \boldsymbol{\Pi}^a = \boldsymbol{\nabla} \cdot \boldsymbol{\Pi}_L^a$ satisfies the constraint equation in Prob. 10.5(b), whose solution in terms of the color charge $\rho^a = \rho_{\text{quark}}^a + \rho_{\text{gluon}}^a$ at any instant in time is

$$\boldsymbol{\Pi}_L^a(\mathbf{x}) = \frac{g}{4\pi}\boldsymbol{\nabla}\int d^3x'\,\frac{1}{|\mathbf{x} - \mathbf{x}'|}\rho^a(\mathbf{x}')$$

---

[39]The order of display of the matrices $[a \otimes b \otimes c]$ in the text is a matter of convenience.

[40]See [Fetter and Walecka (2003a)]. Recall that in chapter 10, $\hbar = c = 1$.

Since ρ^a_{gluon} depends on $\mathbf{\Pi}_L$, this is an integral equation (or power series) for $\mathbf{\Pi}^a_L$.[41]

10.7 The quantization of quantum electrodynamics (QED) starting from lagrangian field theory is discussed in appendix C of volume II. In the Coulomb gauge, in H-L units ($\varepsilon_0 = 1$), the quantization condition on the E-M field is

$$[A_i(\mathbf{x}, t), \Pi_j(\mathbf{x}', t')]_{t=t'} = i\delta^T_{ij}(\mathbf{x} - \mathbf{x}')$$

$$\Pi_j \doteq \Pi^T_j = \frac{\partial A_j}{\partial t}$$

A representation of the fields satisfying these relations is then provided by Eqs. (C.67), with the quantization condition in Eq. (7.38).[42]

Now use the results of Prob. 10.6, and the analogy to QED, to discuss the quantization of QCD in the Coulomb gauge where $\nabla \cdot \mathbf{A}^a = 0$.

10.8 Consider a field theory with a Dirac fermion field ψ, a scalar boson field ϕ, and a Yukawa coupling between them

$$\mathcal{L} = \mathcal{L}_0 + \mathcal{L}_1 = -\bar{\psi}\left[\gamma_\mu\frac{\partial}{\partial x_\mu} + (M - g\phi)\right]\psi - \frac{1}{2}\left[\left(\frac{\partial\phi}{\partial x_\mu}\right)^2 + m^2\phi^2\right]$$

$$\mathcal{L}_1 = g\bar{\psi}\psi\phi$$

The generating functional for this theory is[43]

$$\tilde{W}(\bar{\zeta}, \zeta, J) = \exp\left\{i\int d^4x\left[g\left(-\frac{1}{i}\frac{\delta}{\delta\zeta(x)}\right)\left(\frac{1}{i}\frac{\delta}{\delta\bar{\zeta}(x)}\right)\left(\frac{1}{i}\frac{\delta}{\delta J(x)}\right)\right]\right\} \times$$
$$\tilde{W}_0(\bar{\zeta}, \zeta, J) / [\cdots]_{\zeta=\bar{\zeta}=J=0}$$

where

$$\tilde{W}_0(\bar{\zeta}, \zeta, J) = \exp\left\{-i\int\int d^4x d^4y\,\bar{\zeta}(x)S_F(x - y)\zeta(y)\right\} \times$$
$$\exp\left\{\frac{i}{2}\int\int d^4x d^4y\,J(x)\Delta_F(x - y)J(y)\right\}$$

---

[41] Note that, ultimately, one has an expression $\mathcal{H}_{\text{QCD}} = \mathcal{H}_{\text{QCD}}(\mathbf{A}^a, \mathbf{\Pi}^a_T)$.

[42] Compare Eqs. (7.65). See also the hamiltonian density in Eq. (C.68).

[43] The goal of this problem is to give the reader some proficiency in obtaining Feynman rules from the generating functional (compare appendix C in [Serot and Walecka (1986)]). Here $(\bar{\zeta}, \zeta)$ form Grassmann algebras, and the variational derivatives in $\tilde{W}$ are to be interpreted as "left derivatives".

(a) Expand the first exponential in $\tilde{W}$, and evaluate the numerator in $\tilde{W}$ to order g^2;

(b) Use this result to evaluate the denominator in $\tilde{W}$ to order g^2. Show to this order that the denominator in $\tilde{W}$ serves to precisely cancel the disconnected diagrams;

(c) Hence derive an expression to order g^2 for the fermion propagator

$$ iS'_F(x-y) = \left[\frac{1}{i}\frac{\delta}{\delta\bar{\zeta}(x)}\tilde{W}(\bar{\zeta},\zeta,J)\frac{1}{i}\frac{\delta}{\delta\zeta(y)}\right]_{\zeta=\bar{\zeta}=J=0} $$

(d) Interpret your result in terms of a set of coordinate-space Feynman rules for this theory.

A.1 (a) Consider the transformation to relative and center-of-mass coordinates in one dimension

$$ x = x_1 - x_2 \qquad ; \; X = \frac{m_1 x_1 + m_2 x_2}{m_1 + m_2} $$

Show that the magnitude of the jacobian for this transformation is unity. Hence conclude that

$$ dx\, dX = \left|\frac{\partial(x,X)}{\partial(x_1,x_2)}\right| dx_1 dx_2 = dx_1 dx_2 $$

(b) Now establish Eq. (A.9).

A.2 Show that the commutation relations in Eqs. (A.12) follow from those in Eqs. (A.11).

A.3 (a) Show $\mathbf{r}_1 = \mathbf{R} + \mu\mathbf{r}/m_1$, and $\mathbf{r}_2 = \mathbf{R} - \mu\mathbf{r}/m_2$;

(b) Define

$$ \psi(\mathbf{r}_1,\mathbf{r}_2) = \psi(\mathbf{R}+\mu\mathbf{r}/m_1,\, \mathbf{R}-\mu\mathbf{r}/m_2) \equiv \phi(\mathbf{R},\mathbf{r}) $$

Show by explicit calculation of the partial derivatives that

$$ \left(\frac{\hbar^2}{2m_1}\nabla_1^2 + \frac{\hbar^2}{2m_2}\nabla_2^2\right)\psi(\mathbf{r}_1,\mathbf{r}_2) = \left(\frac{\hbar^2}{2M}\nabla_R^2 + \frac{\hbar^2}{2\mu}\nabla_r^2\right)\phi(\mathbf{R},\mathbf{r}) $$

(c) As an example, take the free-particle solutions

$$ \psi(\mathbf{r}_1,\mathbf{r}_2) = \frac{1}{\sqrt{V}}e^{i\mathbf{k}_1\cdot\mathbf{r}_1}\frac{1}{\sqrt{V}}e^{i\mathbf{k}_2\cdot\mathbf{r}_2} $$

Determine $\phi(\mathbf{R},\mathbf{r})$, and verify the last relation in part (b).

B.1 Verify the vector identity Eq. (B.12).

Appendix A

The Two-Body Problem

In this appendix, we detail the transformation of the two-body problem to center-of-mass and relative coordinates.

A.1 Classical Mechanics

Consider two particles with mass and position $(m_1, \mathbf{r}_1)$ and $(m_2, \mathbf{r}_2)$ respectively, interacting through a potential $V(\mathbf{r}_1 - \mathbf{r}_2)$, as illustrated in Fig. A.1.

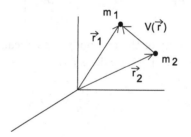

Fig. A.1 Two-body configuration with one particle of mass m_1 at $\mathbf{r}_1$ and second particle with mass m_2 at $\mathbf{r}_2$. The particles interact through a potential $V(\mathbf{r})$, where $\mathbf{r} = \mathbf{r}_1 - \mathbf{r}_2$ is the relative coordinate.

The energy of this system is

$$E = T + V = \tfrac{1}{2}m_1\dot{\mathbf{r}}_1^2 + \tfrac{1}{2}m_2\dot{\mathbf{r}}_2^2 + V(\mathbf{r}_1 - \mathbf{r}_2) \qquad (A.1)$$

The total mass and *reduced mass* are defined by

$$
\begin{aligned}
M &\equiv m_1 + m_2 && \text{; total mass} \\
\frac{1}{\mu} &\equiv \frac{1}{m_1} + \frac{1}{m_2} && \text{; reduced mass}
\end{aligned}
\qquad (A.2)
$$

447

Introduce the total momentum and relative coordinate

$$\mathbf{P} \equiv \mathbf{p}_1 + \mathbf{p}_2 \qquad ; \text{ total momentum}$$

$$\mathbf{r} \equiv \mathbf{r}_1 - \mathbf{r}_2 \qquad ; \text{ relative coordinate} \qquad (A.3)$$

The *relative momentum* is then defined by[1]

$$\mathbf{p} \equiv \mu\dot{\mathbf{r}} = \frac{m_1 m_2}{m_1 + m_2}(\dot{\mathbf{r}}_1 - \dot{\mathbf{r}}_2)$$

$$= \frac{m_2\mathbf{p}_1 - m_1\mathbf{p}_2}{m_1 + m_2} \qquad ; \text{ relative momentum} \qquad (A.4)$$

The center-of-mass (C-M) coordinate is defined by

$$\mathbf{R} \equiv \frac{m_1\mathbf{r}_1 + m_2\mathbf{r}_2}{m_1 + m_2} \qquad ; \text{ center-of-mass} \qquad (A.5)$$

It is then an algebraic identity that the kinetic energy of the system can be written as

$$\frac{\mathbf{p}_1^2}{2m_1} + \frac{\mathbf{p}_2^2}{2m_2} = \frac{\mathbf{p}^2}{2\mu} + \frac{\mathbf{P}^2}{2M}$$

$$= \frac{(m_1 + m_2)}{2m_1 m_2}\left[\frac{(m_2\mathbf{p}_1 - m_1\mathbf{p}_2)^2}{(m_1 + m_2)^2}\right] + \frac{(\mathbf{p}_1 + \mathbf{p}_2)^2}{2(m_1 + m_2)} \qquad (A.6)$$

Hence the classical energy of the two-body system can be written in terms of center-of-mass and relative quantities as

$$E = \frac{\mathbf{P}^2}{2M} + \left[\frac{\mathbf{p}^2}{2\mu} + V(\mathbf{r})\right] \qquad (A.7)$$

A.2   Quantum Mechanics

Quantum mechanics is obtained through the following steps:

(1) The classical hamiltonian is obtained from Eqs. (A.1) and (A.7) as

$$H = \frac{\mathbf{p}_1^2}{2m_1} + \frac{\mathbf{p}_2^2}{2m_2} + V(\mathbf{r}_1 - \mathbf{r}_2)$$

$$= \frac{\mathbf{P}^2}{2M} + \left[\frac{\mathbf{p}^2}{2\mu} + V(\mathbf{r})\right] \qquad (A.8)$$

---

[1]As usual here, $\mathbf{p}_1 = m_1\dot{\mathbf{r}}_1$ and $\mathbf{p}_2 = m_2\dot{\mathbf{r}}_2$.

(2) The volume element is preserved under the transformation to center-of-mass and relative coordinates (see Prob. A.1)

$$d^3r_1 d^3r_2 = d^3R \, d^3r \qquad ; \text{ unit jacobian} \qquad (A.9)$$

(3) Write the joint probability as[2]

$$|\psi(\mathbf{r}_1, \mathbf{r}_2)|^2 d^3r_1 d^3r_2 = |\phi(\mathbf{R}, \mathbf{r})|^2 d^3R \, d^3r \qquad ; \text{ probability} \qquad (A.10)$$

(4) Impose the *canonical commutation relations*

$$[\hat{p}_{\alpha i}, \hat{r}_{\beta j}] = \frac{\hbar}{i} \delta_{\alpha\beta} \delta_{ij} \qquad ; \alpha, \beta = 1, 2 \text{ (particles)}$$
$$; i, j = x, y, z \text{ (components)} \qquad (A.11)$$

It is a matter of algebra to show that these commutation relations imply the following (see Prob. A.2)

$$[\hat{P}_i, \hat{R}_j] = \frac{\hbar}{i} \delta_{ij} \qquad ; [\hat{P}_i, \hat{r}_j] = [\hat{p}_i, \hat{R}_j] = 0$$
$$[\hat{p}_i, \hat{r}_j] = \frac{\hbar}{i} \delta_{ij} \qquad (A.12)$$

(5) A representation of the commutation relations in Eqs. (A.12) is obtained as

$$\mathbf{P} = \frac{\hbar}{i} \boldsymbol{\nabla}_R \qquad ; \mathbf{p} = \frac{\hbar}{i} \boldsymbol{\nabla}_r \qquad ; \text{ coordinate rep} \qquad (A.13)$$

The hamiltonian and time-independent two-body Schrödinger equation are then written in the coordinate representation as

$$H = -\frac{\hbar^2}{2M} \nabla_R^2 + \left[-\frac{\hbar^2}{2\mu} \nabla_r^2 + V(\mathbf{r}) \right]$$
$$H\phi(\mathbf{R}, \mathbf{r}) = E\phi(\mathbf{R}, \mathbf{r}) \qquad (A.14)$$

(6) Separate variables, and look for a solution to these equations of the form

$$\phi(\mathbf{R}, \mathbf{r}) = \frac{1}{\sqrt{\Omega}} e^{i\mathbf{K}\cdot\mathbf{R}} \psi(\mathbf{r}) \qquad (A.15)$$

Here we work in a big cubical box of volume Ω and impose periodic boundary conditions on the center-of-mass coordinate, so that

$$\frac{1}{\Omega} \int_{\text{Box}} d^3R \, e^{i(\mathbf{K}-\mathbf{K}')\cdot\mathbf{R}} = \delta_{\mathbf{K},\mathbf{K}'} \qquad ; \text{ p.b.c.} \qquad (A.16)$$

---

[2]Here $\psi(\mathbf{r}_1, \mathbf{r}_2) = \psi(\mathbf{R} + \mu\mathbf{r}/m_1, \mathbf{R} - \mu\mathbf{r}/m_2) \equiv \phi(\mathbf{R}, \mathbf{r})$; see Prob. A.3.

Insertion of Eq. (A.15) into Eqs. (A.14) gives

$$\left[-\frac{\hbar^2}{2\mu}\nabla_r^2 + V(\mathbf{r})\right]\psi(\mathbf{r}) = \varepsilon\psi(\mathbf{r})$$

$$\varepsilon \equiv E - \frac{\hbar^2 K^2}{2M} \qquad (A.17)$$

This is just the one-body problem we have been studying! Only now (see Fig. A.2)

- The coordinate $\mathbf{r} = \mathbf{r}_1 - \mathbf{r}_2$ is the *relative coordinate*;
- The mass is the *reduced mass* $\mu = m_1 m_2/(m_1 + m_2)$;
- The energy $\varepsilon = E - \hbar^2 K^2/2M$ is the energy in the center-of-mass system.

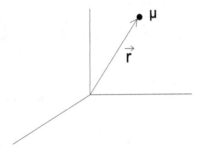

Fig. A.2 Equivalent one-body problem in the center-of-mass system with relative coordinate $\mathbf{r} = \mathbf{r}_1 - \mathbf{r}_2$ and reduced mass $\mu = m_1 m_2/(m_1 + m_2)$.

As an example, a solution to the time-independent Schrödinger Eq. (A.17) for a hydrogen-like atom with a lepton of mass m, and a nucleus of charge Z and mass M, gives a C-M eigenvalue spectrum

$$\varepsilon_n = -\frac{1}{2}\mu c^2 \frac{Z^2\alpha^2}{n^2} = -\frac{1}{2}mc^2\frac{Z^2\alpha^2}{n^2}\left(\frac{M}{M+m}\right) \quad ; \text{ hydrogen-like atom}$$

$$(A.18)$$

The final factor serves as a correction term for the nuclear motion in the two-body system. For a heavy nucleus with large M, this factor is unity.

Appendix B

Charged Particle in External Electromagnetic Field

In this appendix we examine in some detail the problem of a charged particle moving in an external, specified electromagnetic field. We start with the classical theory.

B.1 Classical Theory

Newton's second law and the Lorentz force equation combine to give the classical equation of motion[1]

$$m\frac{d^2\mathbf{x}}{dt^2} = e\boldsymbol{\mathcal{E}} + \frac{e}{c}\mathbf{v}\times\mathbf{B} \qquad ; \text{ Lorentz force}$$

$$\qquad\qquad\qquad\qquad ; \text{ Newton's second law} \qquad (\text{B.1})$$

The hamiltonian for a non-relativistic particle in a potential $V(\mathbf{x})$ in the absence of an electromagnetic field is

$$H = \frac{\mathbf{p}^2}{2m} + V(\mathbf{x}) \qquad ; \text{ no E-M field} \qquad (\text{B.2})$$

The electric and magnetic fields $(\boldsymbol{\mathcal{E}}, \mathbf{B})$ can be expressed in terms of the electromagnetic potentials $(\mathbf{A}, \Phi)$ through the relations

$$\mathbf{B} = \boldsymbol{\nabla}\times\mathbf{A}$$

$$\boldsymbol{\mathcal{E}} = -\boldsymbol{\nabla}\Phi - \frac{1}{c}\frac{\partial\mathbf{A}}{\partial t} \qquad\qquad (\text{B.3})$$

---

[1] Again, this is in c.g.s. units. Also, repeated Latin indices are here summed from 1 to 3; for example, $a_i b_i = \mathbf{a}\cdot\mathbf{b}$.

The electromagnetic fields are incorporated into the hamiltonian in Eq. (B.2) through the gauge-invariant replacements[2]

$$\mathbf{p} \to \mathbf{p} - \frac{e}{c}\mathbf{A} \qquad ; \text{ gauge-invariant replacements}$$
$$H \to H + e\Phi \qquad\qquad\qquad\qquad\qquad (\text{B.4})$$

This gives the classical hamiltonian

$$H = \frac{1}{2m}\left[\mathbf{p} - \frac{e}{c}\mathbf{A}(\mathbf{x}, t)\right]^2 + e\Phi(\mathbf{x}, t) + V(\mathbf{x}) \qquad (\text{B.5})$$

The goal of the first part of this appendix is to show that starting from this hamiltonian, Hamilton's equations give the correct classical equation of motion.

With a hamiltonian of the form $H(\mathbf{p}, \mathbf{x}; t)$, where $\mathbf{p}$ is the canonical momentum, Hamilton's equations read

$$\frac{dp_i}{dt} = -\frac{\partial H}{\partial x_i} \qquad ; \text{ Hamilton's eqns}$$
$$\frac{dx_i}{dt} = \frac{\partial H}{\partial p_i} \qquad ; i = 1, 2, 3 \qquad (\text{B.6})$$

These equations can be analyzed as follows:

(1) The second of Eqs. (B.6) expresses the particle's kinetic velocity v_i as

$$\frac{dx_i}{dt} = \frac{\partial H}{\partial p_i} = \frac{1}{m}\left(p - \frac{e}{c}A\right)_i \equiv v_i \qquad ; \text{ kinetic velocity} \quad (\text{B.7})$$

Note that this is *not* just p_i/m;

(2) The first of Eqs. (B.6) then gives

$$\frac{dp_i}{dt} = -\frac{\partial H}{\partial x_i} = \frac{e}{mc}\left(p - \frac{e}{c}A\right)_j \frac{\partial A_j}{\partial x_i} - e\frac{\partial \Phi}{\partial x_i} - \frac{\partial V}{\partial x_i}$$
$$= \frac{e}{c}v_j\frac{\partial A_j}{\partial x_i} - e\frac{\partial \Phi}{\partial x_i} - \frac{\partial V}{\partial x_i} \qquad (\text{B.8})$$

(3) Now take the total time derivative of the result in Eq. (B.7)

$$m\frac{d^2 x_i}{dt^2} = \frac{dp_i}{dt} - \frac{e}{c}\frac{dA_i}{dt} \qquad (\text{B.9})$$

The total time derivative of the vector potential $\mathbf{A}(\mathbf{x}, t)$, evaluated at the position of the particle, is obtained by writing out the total

---

[2]See Eqs. (6.140). In the classical theory, gauge invariance is established through the invariance of the mechanics under *canonical transformations*.

differential of $\mathbf{A}(\mathbf{x}, t)$ and then dividing by dt

$$\frac{d}{dt} A_i(\mathbf{x}, t) = \frac{\partial A_i}{\partial t} + \frac{\partial A_i}{\partial x_j} \frac{dx_j}{dt}$$

$$= \frac{\partial A_i}{\partial t} + \mathbf{v} \cdot \nabla A_i \qquad (B.10)$$

(4) Now substitute Eqs. (B.8) and (B.10) into Eq. (B.9)

$$m \frac{d^2 x_i}{dt^2} = \left[\frac{e}{c} \mathbf{v} \cdot \frac{\partial \mathbf{A}}{\partial x_i} - e \frac{\partial \Phi}{\partial x_i} - \frac{\partial V}{\partial x_i} \right] - \frac{e}{c} \left[\frac{\partial A_i}{\partial t} + \mathbf{v} \cdot \nabla A_i \right]$$

$$= e\mathcal{E}_i - \frac{\partial V}{\partial x_i} + \frac{e}{c} \left[\mathbf{v} \cdot \frac{\partial \mathbf{A}}{\partial x_i} - \mathbf{v} \cdot \nabla A_i \right] \qquad (B.11)$$

The second line identifies the electric field $\mathcal{E}$. The vectors in the last term can be manipulated to give

$$\mathbf{v} \cdot \frac{\partial \mathbf{A}}{\partial x_i} - \mathbf{v} \cdot \nabla A_i = [\mathbf{v} \times (\nabla \times \mathbf{A})]_i = [\mathbf{v} \times \mathbf{B}]_i \qquad (B.12)$$

Hence Eq. (B.11) becomes

$$m \frac{d^2 \mathbf{x}}{dt^2} = -\nabla V + e\mathcal{E} + \frac{e}{c} \mathbf{v} \times \mathbf{B} \qquad (B.13)$$

This is Newton's second law with the Lorentz force (and an additional potential V), whose derivation was the stated goal.

B.2 Quantum Theory

The time-dependent Schrödinger equation for the quantum theory reads

$$i\hbar \frac{\partial \Psi}{\partial t} = H\Psi \qquad ; \text{ S-eqn} \qquad (B.14)$$

where the hamiltonian is given by

$$H = \frac{1}{2m} \left[\mathbf{p} - \frac{e}{c} \mathbf{A}(\mathbf{x}, t) \right]^2 + e\Phi(\mathbf{x}, t) + V(\mathbf{x})$$

$$\equiv \frac{1}{2m} \left(\mathbf{p} - \frac{e}{c} \mathbf{A} \right) \cdot \left(\mathbf{p} - \frac{e}{c} \mathbf{A} \right) + e\Phi + V \qquad (B.15)$$

Canonical quantization in the coordinate representation gives the canonical momentum as

$$\mathbf{p} = \frac{\hbar}{i} \nabla \qquad ; \text{ coordinate rep} \qquad (B.16)$$

The potentials $(\mathbf{A}, \Phi, V)$ are all *real*, and H is a linear hermitian operator. The total probability is again conserved

$$\frac{d}{dt} \int d^3x \, |\Psi|^2 = \int d^3x \left[\Psi^\star \frac{\partial \Psi}{\partial t} + \frac{\partial \Psi^\star}{\partial t} \Psi \right]$$

$$= \frac{1}{i\hbar} \int d^3x \, [\Psi^\star H \Psi - (H\Psi)^\star \Psi] = 0 \qquad (B.17)$$

The equations of motion in the Schrödinger picture are now given by *Ehrenfest's theorem* in Eqs. (3.171). We proceed to investigate the implications of Ehrenfest's theorem for the set of operators $\{\mathbf{p}, \mathbf{x}\}$.[3]

(1) *For x_i*: Compute the commutator

$$\frac{i}{\hbar}[H, x_i] = \frac{i}{2m\hbar} \left[\left(p - \frac{e}{c}A \right)_j \left(p - \frac{e}{c}A \right)_j, x_i \right] \qquad (B.18)$$

The commutator on the r.h.s. can be written identically as

$$[\, , \,] = \left(p - \frac{e}{c}A \right)_j \left[\left(p - \frac{e}{c}A \right)_j, x_i \right] + \left[\left(p - \frac{e}{c}A \right)_j, x_i \right] \left(p - \frac{e}{c}A \right)_j$$

$$= \left(p - \frac{e}{c}A \right)_j [p_j, x_i] + [p_j, x_i] \left(p - \frac{e}{c}A \right)_j$$

$$= \frac{2\hbar}{i} \delta_{ij} \left(p - \frac{e}{c}A \right)_j \qquad (B.19)$$

Thus from Eqs. (3.171)

$$\frac{d}{dt} \langle x_i \rangle = \frac{1}{m} \left\langle \left(p - \frac{e}{c}A \right)_i \right\rangle = \langle v_i \rangle = \left\langle \frac{\partial H}{\partial p_i} \right\rangle \qquad (B.20)$$

Here the last two equalities follows from Eqs. (B.7).[4]

(2) *For p_i*: Compute the commutator

$$\frac{i}{\hbar}[H, p_i] = -e \frac{\partial \Phi}{\partial x_i} - \frac{\partial V}{\partial x_i} + \frac{i}{2m\hbar} \left[\left(p - \frac{e}{c}A \right)_j \left(p - \frac{e}{c}A \right)_j, p_i \right] \qquad (B.21)$$

As above, the commutator on the r.h.s. can be written

$$[\, , \,] = \left(p - \frac{e}{c}A \right)_j \left[\left(p - \frac{e}{c}A \right)_j, p_i \right] + \left[\left(p - \frac{e}{c}A \right)_j, p_i \right] \left(p - \frac{e}{c}A \right)_j$$

$$= \left(p - \frac{e}{c}A \right)_j \left[\left(-\frac{e}{c}A \right)_j, p_i \right] + \left[\left(-\frac{e}{c}A \right)_j, p_i \right] \left(p - \frac{e}{c}A \right)_j \qquad (B.22)$$

---

[3]Note that now $[\mathbf{p}, \mathbf{A}] \neq 0$. Recall that we work here in the coordinate representation.

[4]This relation defines the kinetic velocity operator v_i in the Schrödinger picture. The conserved probability current is constructed from v_i in Prob. 6.11.

This in turn, becomes

$$[\ ,\ ] = \frac{e}{c}\frac{\hbar}{i}\left\{\left(p - \frac{e}{c}A\right)_j\left(\frac{\partial A_j}{\partial x_i}\right) + \left(\frac{\partial A_j}{\partial x_i}\right)\left(p - \frac{e}{c}A\right)_j\right\} \quad \text{(B.23)}$$

Hence, from Eqs. (B.20),

$$[\ ,\ ] = \frac{2me\hbar}{ic}\frac{1}{2}\left(v_j\frac{\partial A_j}{\partial x_i} + \frac{\partial A_j}{\partial x_i}v_j\right) = \frac{2me\hbar}{ic}\left[\mathbf{v}\cdot\frac{\partial}{\partial x_i}\mathbf{A}\right]_{\text{sym}} \quad \text{(B.24)}$$

Here the symmetric product of two operators is defined by

$$[AB]_{\text{sym}} \equiv \frac{1}{2}(AB + BA) \quad \text{(B.25)}$$

Thus from Eqs. (B.21) and (B.24)

$$\frac{i}{\hbar}[H,\,p_i] = -e\frac{\partial\Phi}{\partial x_i} - \frac{\partial V}{\partial x_i} + \frac{e}{c}\left[\mathbf{v}\cdot\frac{\partial}{\partial x_i}\mathbf{A}\right]_{\text{sym}} \quad \text{(B.26)}$$

Ehrenfest's theorem in Eqs. (3.171) then gives

$$\frac{d}{dt}\langle p_i\rangle = \left\langle -e\frac{\partial\Phi}{\partial x_i} - \frac{\partial V}{\partial x_i} + \frac{e}{c}\left[\mathbf{v}\cdot\frac{\partial}{\partial x_i}\mathbf{A}\right]_{\text{sym}}\right\rangle = \left\langle -\frac{\partial H}{\partial x_i}\right\rangle \quad \text{(B.27)}$$

The last equality follows from the symmetrized version of Eq. (B.8).
(3) Now, as before, take the time derivative of Eq. (B.20)

$$m\frac{d^2}{dt^2}\langle x_i\rangle = \frac{d}{dt}\langle p_i\rangle - \frac{e}{c}\frac{d}{dt}\langle A_i\rangle \quad \text{(B.28)}$$

Equation (B.27) can be used for the first term on the r.h.s. For the second, since $\mathbf{A}(\mathbf{x}, t)$ has an *explicit* time dependence, one must use from Eqs. (3.171)

$$\frac{d}{dt}\langle A_i\rangle = \left\langle\frac{\partial A_i}{\partial t}\right\rangle + \left\langle\frac{i}{\hbar}[H,\,A_i]\right\rangle \qquad ; \text{note!} \quad \text{(B.29)}$$

Compute the required commutator

$$\frac{i}{\hbar}[H,\,A_i] = \frac{i}{2m\hbar}\left[\left(p - \frac{e}{c}A\right)_j\left(p - \frac{e}{c}A\right)_j,\,A_i\right] \quad \text{(B.30)}$$

Again, the commutator on the r.h.s. of Eqs. (B.30) is

$$[\,,\,] = \left(p - \frac{e}{c}A\right)_j \left[\left(p - \frac{e}{c}A\right)_j, A_i\right] + \left[\left(p - \frac{e}{c}A\right)_j, A_i\right]\left(p - \frac{e}{c}A\right)_j$$

$$= \frac{\hbar}{i}\left\{\left(p - \frac{e}{c}A\right)_j\left(\frac{\partial A_i}{\partial x_j}\right) + \left(\frac{\partial A_i}{\partial x_j}\right)\left(p - \frac{e}{c}A\right)_j\right\}$$

$$= \frac{2m\hbar}{i}\left[v_j\frac{\partial A_i}{\partial x_j}\right]_{\mathrm{sym}} \tag{B.31}$$

Hence Eq. (B.29) becomes

$$\frac{d}{dt}\langle A_i\rangle = \left\langle \frac{\partial A_i}{\partial t} + [\mathbf{v}\cdot\boldsymbol{\nabla}A_i]_{\mathrm{sym}} \right\rangle \tag{B.32}$$

(4) A combination of terms then gives Eq. (B.28) as

$$m\frac{d^2}{dt^2}\langle x_i\rangle = \left\langle -\frac{\partial V}{\partial x_i} - e\frac{\partial \Phi}{\partial x_i} - \frac{e}{c}\frac{\partial A_i}{\partial t} + \frac{e}{c}\left[\mathbf{v}\cdot\frac{\partial}{\partial x_i}\mathbf{A} - \mathbf{v}\cdot\boldsymbol{\nabla}A_i\right]_{\mathrm{sym}}\right\rangle \tag{B.33}$$

This is now in exactly the same form as the classical result in Eq. (B.11), and a similar development leads to a result analogous to Eq. (B.13)

$$m\frac{d^2}{dt^2}\langle\mathbf{x}\rangle = \left\langle -\boldsymbol{\nabla}V + e\boldsymbol{\mathcal{E}} + \frac{e}{c}[\mathbf{v}\times\mathbf{B}]_{\mathrm{sym}}\right\rangle \tag{B.34}$$

Several comments:

- In this expression, $\langle\mathbf{x}\rangle$ is the expectation value

$$\langle\mathbf{x}\rangle \equiv \int d^3x\,\Psi^\star(\mathbf{x},t)\,\mathbf{x}\,\Psi(\mathbf{x},t) \tag{B.35}$$

and similarly for the other terms;

- Equation (B.34) is an *exact result* that describes the motion of any quantum system in external electromagnetic fields $(\boldsymbol{\mathcal{E}}, \mathbf{B})$ and potential V;
- With a *well-localized wave packet*, the quantum result in Eq. (B.34) reproduces the classical result in Eq. (B.13);
- It is canonical quantization, together with Ehrenfest's theorem, that produces the same formal structure in Eqs. (B.20) and (B.27) as that obtained from the classical Hamilton's equations;
- Ultimately, this is the real justification for canonical quantization in quantum theory!

Bibliography

Abers, E. S., and Lee, B. W., (1973). *Phys. Rep.* **9**, 1

Bender, C. M., and Wu, T. T., (1969). *Phys. Rev.* **184**, 1231

Bethe, H. A., and Placzek, G., (1937). *Phys. Rev.* **51**, 450

Bethe, H. A., (1949). *Phys. Rev.* **76**, 38

Bjorken, J. D., and Drell, S. D., (1965). *Relativistic Quantum Fields*, McGraw-Hill, New York, NY

Bloch, F., (1928). *Zeit. für Phys.* **52**, 555

Blatt, J. M., and Weisskopf, V. F., (1952). *Theoretical Nuclear Physics*, John Wiley and Sons, New York, NY

Brack, M., and Bhaduri, R. K., (1997). *Semiclassical Physics*, Addison-Wesley, Reading, MA

Brillouin, L., (1926). *Compt. Rend. de l'Acad. des Sci.* **183**, 24

Cohen-Tannoudji, C., Diu, B., and Laloe, F., (2006). *Quantum Mechanics, 2 vols.*, Wiley-Interscience, New York, NY

Christenson, J. H., Cronin, J. W., Fitch, V. L., and Turlay, R., (1964). *Phys. Rev. Lett.* **13**, 138

Czyz, W., and Maximon, L. C., (1969). *Ann. Phys.* **52**, 59

Dirac, P. A. M., (1947). *The Principles of Quantum Mechanics, 3rd ed.*, Oxford University Press, New York, NY

Donnelly, T. W., Dubach, J., and Walecka, J. D., (1974). *Nucl. Phys.* **A232**, 355

Edmonds, A. R., (1974). *Angular Momentum in Quantum Mechanics*, 3rd printing, Princeton University Press, Princeton, NJ

Faddeev, L. D., and Popov, V. N., (1967). *Phys. Lett.* **25B**, 29

Feshbach, H., (1991). *Theoretical Nuclear Physics Vol. II, Nuclear Reactions*, John Wiley and Sons, New York, NY

Fetter, A. L., and Walecka, J. D., (2003). *Quantum Theory of Many-Particle Systems*, Dover Publications, Mineola, NY; originally published by McGraw-Hill, New York, NY (1971)

Fetter, A. L., and Walecka, J. D., (2003a). *Theoretical Mechanics of Particles and Continua*, Dover Publications, Mineola, NY; originally published by McGraw-Hill, New York, NY (1980)

Feynman, R. P., (1963). *Acta Phys. Polon.* **24**, 697

Feynman, R. P., and Hibbs, A. R., (1965). *Quantum Mechanics and Path Integrals*, McGraw-Hill, New York, NY

Fitch, V. L., and Rainwater, J., (1953). *Phys. Rev.* **92**, 789

Gasiorowicz, S., (2003). *Quantum Physics, 3rd ed.*, John Wiley and Sons, New York, NY

Glauber, R., (1959). *Lectures in Theoretical Physics, Vol. I*, Interscience, New York, NY

Goldberger, M. L., and Watson, K. M., (2004). *Collision Theory*, Dover Publications, Mineola, NY

Gottfried, K., (1966). *Quantum Mechanics, Vol. I*, W. A. Benjamin, New York, NY

Gottfried, K., and Yan, T.-M., (2004). *Quantum Mechanics: Fundamentals, 2nd ed.*, Springer, New York, NY

Griffiths, D. J., (2004). *Introduction to Quantum Mechanics, 2nd ed.*, Benjamin Cummings, San Francisco, CA

Gross, D. J., and Wilczek, F., (1973). *Phys. Rev. Lett.* **30**, 1343; *Phys. Rev.* **D8**, 3633

Hofstadter, R., (1956). *Rev. Mod. Phys.* **28**, 214

Jackson, J. D., (1998). *Classical Electrodynamics, 3rd ed.*, John Wiley and Sons, New York, NY

Jacob, M., and Wick, G. C., (1959). *Ann. of Phys.* **7**, 404

Koutschan, C., and Zeilberger, D., (2010). *http://astrophysics.fic.uni.lodz.pl/ 100yrs/pdf/04/076.pdf (viewed May 15, 2010)*

Kramers, H. A., (1926). *Zeit. für Phys.* **39**, 828

Landau, L. D., and Lifshitz, E. M., (1981). *Quantum Mechanics, 3rd ed.*, Butterworth-Heinemann, New York, NY

Lee, T. D., and Yang, C. N., (1956). *Phys. Rev.* **104**, 254

Leith, D. W. G. S., (1974). *Proceedings of the Summer Institute on Particle Physics, SLAC Report No. 179, ed. M. C. Zipf, Vol. 1*, p. 101, Stanford Linear Accelerator Center, Stanford, CA

Lüders, G., (1954). *Math. Fisik. Medd. Kgl. Danske Aked. Ved.* **28**, 5

Lüders, G., (1957). *Ann. of Phys.* **2**, 1

Merzbacher, E., (1998). *Quantum Mechanics, 3rd ed.*, John Wiley and Sons, New York, NY

Messiah, A., (1999). *Quantum Mechanics, 2 vols.*, Dover Publications, Mineola, NY

Morse, P. M., and Feshbach, F., (1953). *Methods of Theoretical Physics, 2 vols.*, McGraw-Hill, New York, NY

Newton, R., (1982). *Scattering Theory of Waves and Particles, 2nd ed.*, Springer-Verlag, New York, NY

Ohanian, H. C., (1995). *Modern Physics, 2nd ed.*, Prentice-Hall, Upper Saddle River, NJ

Particle Data Group, (2012). *http://pdg.lbl.gov/*

Pekeris, C. L., (1959). *Phys. Rev.* **115**, 1216

Politzer, H. D., (1973). *Phys. Rev. Lett.* **30**, 1346; (1974) *Phys. Rep.* **14**, 129

Sakurai, J. J., and Napalitano, J. J., (2010). *Modern Quantum Mechanics, 2nd ed.*, Addison-Wesley, Reading, MA

Schiff, L. I., (1968). *Quantum Mechanics, 3rd ed.*, McGraw-Hill, New York, NY

Schwinger, J. S., (1947). *Phys. Rev.* **72**, 742

Schwinger, J. S., (1951). *Phys. Rev.* **82**, 914

Serot, B. D., and Walecka, J. D., (1986). *Advances in Nuclear Physics* **16**, eds. J. W. Negele and E. Vogt, Plenum Press, New York, NY

Shankar, R., (1994). *Principles of Quantum Mechanics, 2nd ed.*, Springer, New York, NY

Walecka, J. D., (2001). *Electron Scattering for Nuclear and Nucleon Structure*, Cambridge University Press, Cambridge, UK

Walecka, J. D., (2004). *Theoretical Nuclear and Subnuclear Physics, 2nd ed.*, World Scientific Publishing Company, Singapore; originally published by Oxford University Press, New York, NY (1995)

Walecka, J. D., (2007). *Introduction to General Relativity*, World Scientific Publishing Company, Singapore

Walecka, J. D., (2008). *Introduction to Modern Physics: Theoretical Foundations*, World Scientific Publishing Company, Singapore

Walecka, J. D., (2010). *Advanced Modern Physics: Theoretical Foundations*, World Scientific Publishing Company, Singapore

Wentzel, G., (1926). *Zeit. für Phys.* **38**, 518

Whittaker, E. T., and Watson, G. N., (1969). *A Course of Modern Analysis, 4th ed.*, Cambridge University Press, London, UK

Wigner, E. P., and Weisskopf, V. F., (1930). *Zeit. für Phys.* **63**, 54

Wikipedia, (2011). http://en.wikipedia.org/wiki/(topic)

Wu, C. S., Ambler, E., Hayward, R. W., Hoppes, D. D., and Hudson, R. P., (1957). *Phys. Rev.* **105**, 1413

Yang, C. N., and Mills, R. L., (1954). *Phys. Rev.* **96**, 191

Index